£20
102

Free and moving boundary problems

Free and moving boundary problems

John Crank

Professor Emeritus
Brunel University

CLARENDON PRESS · OXFORD

Oxford University Press, Walton Street, Oxford OX2 6DP

Oxford New York Toronto
Delhi Bombay Calcutta Madras Karachi
Petaling Jaya Singapore Hong Kong Tokyo
Nairobi Dar es Salaam Cape Town
Melbourne Auckland
and associated companies in
Beirut Berlin Ibadan Nicosia

Oxford is a trademark of Oxford University Press

Published in the United States
by Oxford University Press, New York

© John Crank, 1984

First published 1984
First issued in paperback 1987

British Library Cataloguing in Publication Data
Crank, J.
Free and moving boundary problems
1. Differential equations, Nonlinear—
Numerical solutions 2. Boundary value
problems—Numerical solutions
I. Title
515.3'55 QA371
ISBN 0–19–853357–8
ISBN 0–19–853370–5 (pbk)

Library of Congress Cataloging in Publication Data
Crank, John.
Free and moving boundary problems.
Bibliography: p.
Includes index.
1. Boundary value problems. I. Title.
TA347.B69C73 1987 515.3'5 86–28484
ISBN 0–19–853357–8
ISBN 0–19–853370–5 (pbk)

Set and printed in
Northern Ireland by
The Universities Press (Belfast) Ltd.

Preface

THIS book aims to present a broad but reasonably detailed account of the mathematical solution of free and moving boundary problems. The moving boundaries occur mostly in heat-flow problems with phase changes and in certain diffusion processes. A free boundary does not move but its position has to be determined as part of the solution of a steady-state problem. Seepage through porous media is perhaps the most common but by no means the only source of problems of this kind.

The broad spectrum of active research workers includes three groups: engineers and others with practical problems; numerical analysts producing suitable numerical algorithms; and pure mathematicians who decide that certain problems and their solutions exist, are properly posed, and may even be unique. They also examine the convergence and stability properties of numerical schemes. A few people fit into all three groups. It is hoped that this book will help to alleviate the usual difficulties of communication between the various interested parties.

Both free and moving boundaries have been popular subjects for research in recent years, leading to an almost bewildering collection of new mathematical methods. This seems to be an opportune time to attempt a systematic presentation of them. Authors have tended to test their methods by solving a small number of model problems and so studies of melting ice, simple dams, oxygen diffusion with sorption, and electrochemical machining are referred to frequently in this book. The earliest mathematical work concentrated on one-dimensional problems. Modern computer developments, however, make it possible to handle free and moving boundaries in two and three space dimensions and to model practical systems more realistically. Such methods feature largely in this volume, though it makes no pretence to be a compendium of industrial problems.

Parallel studies of the mathematical properties of the differential equations and their solutions, and of the numerical algorithms, have been prolific. It has not been possible here to do more than indicate some of the established results and to include key references in the extensive bibliography. A separate volume is needed to do justice to this aspect of the subject.

It is a pleasure to acknowledge the generous help I have received from the Leverhulme Trust through the award of an Emeritus Fellowship. I am grateful to Mrs Lesley Barker who typed the text with great accuracy and intelligent skill from a manuscript which in parts came near to being illegible. I have been glad to be able to use the Brunel University Library and to have expert guidance from Assistant Librarian, Mr M. J. Dorling. Finally, I should like to thank all who have sent me reprints and unpublished copies of their work, and the staff of the Clarendon Press for their kind help and consideration.

Brunel University   J. C.
1982

Contents

1. MOVING-BOUNDARY PROBLEMS: FORMULATION   1

   1.1.   Introduction   1

   1.2.   A simple example: melting ice   2
      1.2.1.   Single phase   2
      1.2.2.   Two phases   4
      1.2.3.   The Stefan condition   4

   1.3.   Generalizations of the classical Stefan problem   5
      1.3.1.   Non-linear heat parameters   5
      1.3.2.   Linearized forms   8
      1.3.3.   Non-dimensional forms   9
      1.3.4.   Density change and convection   10
      1.3.5.   Ablation and the inverse Stefan problem   11
      1.3.6.   Multi-phase problems   12
      1.3.7.   Simultaneous diffusion and heat flow: alloys   14
      1.3.8.   Muskat's problem   16
      1.3.9.   Two and three space dimensions   16
      1.3.10. Implicit boundary conditions: oxygen diffusion problem   19
      1.3.11. Concentrated thermal capacities   23

2. FREE-BOUNDARY PROBLEMS: FORMULATION   30

   2.1.   Introduction   30

   2.2.   Simple rectangular dam   31
      2.2.1. Classical formulation: physical derivation   31
      2.2.2. Stream function   34
      2.2.3. Formulation on fixed domain: Baiocchi transformation   34
      2.2.4. Variational inequality formulation   38
      2.2.5. In terms of stream functions   40

   2.3.   Other two-dimensional dams   41
      2.3.1. Rectangular dam with toe drain   41
      2.3.2. Rectangular dam with sheetpile   44

2.3.3. Rectangular dam with toe drain and sheetpile   46
2.3.4. Dam with inlet face slanted   47
2.3.5. Dam with inlet face slanted and a toe drain   49
2.3.6. Dam with sloping base   50
2.3.7. Arbitrary-shaped dam   52
2.3.8. Arbitrary-shaped dam with toe drain; split-domain method   54

2.4. Three-dimensional dams   57

2.5. Coastal aquifer   60

2.6. Canals and ditches   63

2.7. Axisymmetric flow   66

2.8. Heterogeneous porous media   70
2.8.1. Stratified dams   71

2.9. Evaporation or infiltration   74
2.10. Unsaturated flow: capillary fringe   75
2.11. A generalized dam problem: a new formulation   79
2.12. Degenerate free-boundary problems   81
2.12.1. Hele-Shaw flow analogy   81
2.12.2. Hele-Shaw injection   82
2.12.3. Electrochemical machining   85
2.12.4. Evolution dam problem   90

2.13. Lubrication cavitation in a journal bearing   93
2.14. A more general look at the Baiocchi transformation   96
2.15. Connections between certain free-boundary problems   100

3. ANALYTICAL SOLUTIONS   101

3.1. Exact similarity solutions   101

3.2. Neumann's solution; generalizations; volume changes   101

3.3. Similarity solutions in cylindrical and spherical coordinates   110
3.4. Similarity solutions and moving heat sources   111
3.5. Integral-equation formulations   114
3.5.1. Integral transforms   114
3.5.2. Green's functions   117
3.5.3. Embedding technique   122
3.5.4. Heat-balance method: Goodman   128
3.5.5. Heat-balance method with spatial subdivision   133
3.5.6. Multi-dimensional problems   137

3.6. Approximate solutions   139
3.6.1. Steady-state approximation and improvements   139
3.6.2. Spacial subdivision   142

3.6.3. The inverse Stefan problem   144
3.6.4. Fictitious diffusivity approximations   147
3.6.5. Perturbation   methods;   asymptotic   solutions;
      stability   150

4. FRONT-TRACKING METHODS   163

  4.1. Numerical techniques   163
  4.2. Fixed finite-difference grid   163
  4.3. Modified grids   168
        4.3.1. Variable time step   168
        4.3.2. Variable space grid   170
        4.3.3. Finite-elements: adaptive meshes   173
  4.4. Method of lines   181

5. FRONT-FIXING METHODS   187

  5.1. One-dimensional problems   187
  5.2. Body-fitted curvilinear coordinates   192
  5.3. Examples of use of body-fitted coordinates   195
  5.4. Isotherm migration method   199
        5.4.1. IMM in one space dimension   199
        5.4.2. IMM in multi-dimensional problems   201

6. FIXED-DOMAIN METHODS   217

  6.1. Introduction   217
  6.2. Enthalpy method   217
        6.2.1. Weak solutions   220
        6.2.2. An explicit finite-difference scheme   223
        6.2.3. Alternative forms of the enthalpy function   226
        6.2.4. Other numerical schemes and multi-dimensional   228
              problems
        6.2.5. Accurate determination of phase-change boundary   235
        6.2.6. Body heating   239
        6.2.7. Other conservation forms and weak solutions   242
        6.2.8. Conditions not amenable to enthalpy formulation   252
  6.3. Truncation method: alternating phase   253
  6.4. Variational inequalities   257
        6.4.1. Introductory example   260
        6.4.2. Variational form   263
        6.4.3. Minimization formulation   265
        6.4.4. Review of equivalent forms   267
        6.4.5. One-phase Stefan problem   269

x   *Contents*

6.4.6. Two-phase Stefan problem   275
6.4.7. Mathematical results   281

7. ANALYTICAL SOLUTION OF SEEPAGE
   PROBLEMS   283

7.1. Introduction   283
7.2. Approximate methods   283
     7.2.1. Dupuit's approximation   283
     7.2.2. Boussinesq's equations   285
     7.2.3. Other approximate methods   287
7.3. Hodograph method: conformal transformations   288
     7.3.1. Hodograph plane   288
     7.3.2. Some useful transformations   293
     7.3.3. Trapezoidal drainage ditch   297
7.4. Polubarinova-Kochina's solution for the simple dam   300

8. NUMERICAL SOLUTION OF FREE-BOUNDARY
   PROBLEMS   313

8.1. Numerical solutions   313
8.2. Trial free-boundary methods   313
     8.2.1. Computation of approximate trial solutions   314
     8.2.2. Moving the free boundary   319
     8.2.3. Garabedian's modified boundary conditions   329
8.3. Boundary singularities   332
8.4. Methods using variable interchange   336
     8.4.1. Adaptive variational formulation   337
     8.4.2. Transformation of equation and flow region   343
8.5. Variational inequality and complementarity methods   355
     8.5.1. Finite-difference and finite-element forms of the
            Cryer algorithm   359
     8.5.2. Axisymmetric problems   372
8.6. A general algorithm for partially unsaturated flow   376
8.7. Integral equation methods   384
     8.7.1. Two-dimensional annular electrochemical machin-
            ing problem   385

REFERENCES   394

AUTHOR INDEX   409

SUBJECT INDEX   414

1. Moving-boundary problems: formulation

1.1. Introduction

PROBLEMS in which the solution of a differential equation has to satisfy certain conditions on the boundary of a prescribed domain are referred to as boundary-value problems. In many important cases, however, the boundary of the domain is not known in advance but has to be determined as part of the solution. The term 'free-boundary problem' is commonly used when the boundary is stationary and a steady-state problem exists. Moving boundaries, on the other hand, are associated with time-dependent problems and the position of the boundary has to be determined as a function of time and space. The terms 'free' and 'moving' are used in this separate way in this book. Some authors, however, prefer to include both types of problem under the single term 'free-boundary problem' and a very few authors have used 'moving boundary' in a composite sense.

In all cases, two conditions are needed on the free or moving boundary, one to determine the boundary itself and the other to complete the definition of the solution of the differential equation. Suitable conditions on the fixed boundaries and, where appropriate, an initial condition are also prescribed as usual.

In this book, a free-boundary problem requires the solution of an elliptic partial differential equation. For a moving-boundary problem the equation is of parabolic type. The practical applications are mainly but not exclusively concerned with fluid flow in porous media and with diffusion and heat flow incorporating phase transformations or chemical reactions. Many other important free and moving boundary problems arising, for example, as shock waves in gas dynamics, as cracks in solid mechanics, or as optimal stopping problems in decision theory have perforce had to be excluded, in order to keep the book within reasonable bounds.

Moving-boundary problems are often called Stefan problems, with reference to the early work of J. Stefan who, around 1890, was interested in the melting of the polar ice cap. Successive authors have referred, somewhat indiscriminately, to Stefan's publications. Stefan (1889*a*) derived the large latent-heat approximation given in eqn (3.158). In his paper (1889*b*) he extended this solution to include a time-dependent surface temperature. Stefan (1889*b*) also quoted the error-function type of solution discussed in §3.2 for the single-phase problem, and derived the associated subsidiary equations (3.14) and (3.15), together with a further second-order approximation of (3.14). In the same paper, Stefan also pointed out that if a semi-infinite liquid solidifies in such a way that the solidification boundary proceeds at a constant rate, the temperature, *u*, can be expressed in the form

$$u = A(e^{at-mx} - 1)/a,$$

where A, a, and m are constants, and the velocity x/t of solidification is a/m. The full solution is quoted in eqn (3.183). Finally, Stefan (1889*b*) developed a general solution in the form of two Taylor series. The whole of this paper published in 1889 is reproduced in Stefan (1891).

Surveys of the early literature with numerous references dating from the time of Stefan have been written by Bankoff (1964), Muehlbauer and Sunderland (1965), and Boley (1972). Rubinstein's classic book (1971) gives a systematic presentation of the mathematical developments in Stefan problems up to that time. More recent surveys are given by Furzeland (1977*a*), Hoffman (1977), Fox (1979), and Crank (1981), all with useful bibliographies. Reports on several conferences (Ockendon and Hodgkins 1975; Wilson, Solomon, and Boggs 1978; Furzeland 1979*a*; Magenes 1980; Albrecht, Collatz, and Hoffman 1980; Fasano and Premicerio 1983) contain up-to-date accounts of mathematical developments and of wide-ranging applications to problems in physical and biological sciences, engineering, metallurgy, soil mechanics, decision and control theory, etc. which are of practical importance in sundry industries.

1.2. A simple example: melting ice

1.2.1. Single phase

A simple version of a Stefan problem is the melting of a semi-infinite sheet of ice, initially at the melting temperature, taken to be zero, the surface of which is raised at time $T = 0$ to a temperature above zero, at which it is subsequently maintained. A boundary surface or interface on which melting occurs, moves from the surface into the sheet and separates a region of water from one of ice at zero temperature as in Fig. 1.1(a). The path of the melting interface, $S(T)$, in the X–T plane is shown in Fig. 1.1(b), where $S(T)$ denotes the thickness of the water phase at time T and X is the space coordinate measured from the outer surface of the sheet, $X = 0$.

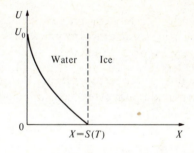

FIG. 1.1(a). Simple Stefan problem

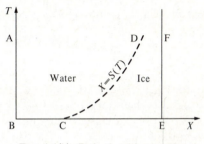

FIG. 1.1(b). Path of melting interface

If $U(X, T)$ denotes the temperature distribution in the water at time T, the problem is to find the pair of unknowns $U(X, T)$ and $S(T)$ by solving the heat-flow equation

$$c\rho \frac{\partial U}{\partial T} = K \frac{\partial^2 U}{\partial X^2}, \quad 0 < X < S(T), \quad T > 0, \tag{1.1}$$

subject to a fixed boundary condition

$$U = U_0, \quad X = 0, \quad T > 0, \tag{1.2}$$

where U_0 is the constant surface temperature, and initial conditions

$$U = 0, \quad X > 0, \quad T = 0, \tag{1.3}$$

$$S(0) = 0. \tag{1.4}$$

In eqn (1.1) the specific heat c, density ρ, and heat conductivity K are assumed to be constants in this simple example. Two further conditions are needed on the moving interface, one to provide the second boundary condition necessary for the solution of the second-order equation (1.1) and the other to determine the position of the interface itself. In this

problem they are

$$U = 0, \qquad\qquad\qquad \left.\begin{array}{l} \\ \\ \end{array}\right\} X = S(T), \qquad T > 0, \qquad (1.5)$$
$$-K\, \partial U/\partial X = L\rho\, dS/dT \qquad \qquad\qquad (1.6)$$

where L is the latent heat required to melt ice. Equation (1.6) is known as the 'Stefan condition' and it expresses the heat balance on the interface as explained in §1.2.3.

1.2.2. Two phases

If the ice is initially at a temperature below the melting temperature, not necessarily uniform, heat flow occurs in both the water and ice phases. This two-phase problem is to find the triple $\{U_1(X, T), U_2(X, T), S(T)\}$, where U_1 and U_2 denote temperatures in the water and ice phases respectively. A typical example is a finite sheet of ice occupying the space $0 \leqslant S(T) \leqslant X \leqslant \ell$, where

$$c_i\rho_i \frac{\partial U_i}{\partial t} = K_i \frac{\partial^2 U_i}{\partial X^2}, \qquad i = 1, 2 \qquad (1.7)$$

and, with reference to Fig. 1.1(b), AB is $X = 0$, EF is $X = \ell$, CD is $X = S(T)$ and $i = 1$ refers to the water phase $0 < X < S(T)$, and $i = 2$ to the ice phase $S(T) < X < \ell$. It is assumed that the water and ice phases together always occupy the space $0 \leqslant X \leqslant \ell$. The Stefan conditions are

$$U_1 = U_2 = 0, \qquad\qquad\qquad\left.\begin{array}{l} \\ \\ \end{array}\right\} X = S(T) \qquad (1.8)$$
$$K_2 \frac{\partial U_2}{\partial X} - K_1 \frac{\partial U_1}{\partial X} = L\rho \frac{dS}{dT}, \qquad\qquad (1.9)$$

and for the time being volume changes on melting are assumed negligible, i.e. ice and water have the same density ρ, so that $\rho_1 = \rho_2 = \rho$ (see §1.3.4).

1.2.3. The Stefan condition

The Stefan condition (1.9) is easily derived by referring to Fig. 1.2 which shows the boundary moving a distance δX in time δT. In order to

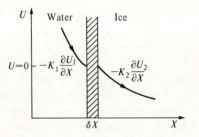

FIG. 1.2. Stefan condition

melt the ice contained per unit area perpendicular to X in the shaded region an amount of heat $L\rho\,\delta X$ is required. An amount of heat $-K_1\delta T\,\partial U_1/\partial X$ enters the shaded element from the water phase and $-K_2\delta T\,\partial U_2/\partial X$ escapes into the ice. Assuming there are no heat sources on the interface the heat balance of the shaded element requires that

$$- K_1\,\partial U_1/\partial X + K_2\,\partial U_2/\partial X = L\rho\,\mathrm{d}S/\mathrm{d}T, \tag{1.10}$$

which is (1.9).

1.3. Generalizations of the classical Stefan problem

1.3.1. Non-linear heat parameters

Practically important generalized forms of the two-phase problem posed in §1.2.2 incorporate any or all of the following non-linear features:

(i) The heat parameters in (1.7) K_i, c_i, ρ_i may all be functions of U, X, T, e.g. $K_i(U_i, X, T)$.

(ii) On the moving interface the temperature U_m may be space and time dependent, i.e.

$$X = S(T), \; U_i = U_m(S, T), \qquad i = 1, 2. \tag{1.11}$$

(iii) There may be a heat source $q(U_m, S, T)$ (or sink) on the moving boundary, possibly coupled with variable thermal properties there, e.g. (1.9) becomes

$$K_2(U_m, S, T)\frac{\partial U_2}{\partial X} - K_1(U_m, S, T)\frac{\partial U_1}{\partial X} = L(U_m, S, T)\rho\frac{\mathrm{d}S}{\mathrm{d}T} - q(U_m, S, T).$$
$$\tag{1.12}$$

(iv) Body heating terms $h(U, X, T)$ may be required on the right-hand side of (1.1) in a single-phase problem or in each of the phases in (1.7).

The effect of such a heat production term may be that there will no longer be a sharp boundary between the liquid and solid phases. What has been called a 'mushy' region may develop such that the sharp melting boundary is replaced by a finite region throughout which the material is at its phase-change temperature. Atthey (1974) discusses the spot welding of two sheets of metal as an example of a mushy region. The sheets are gripped between two electrodes and the electric current passing between them produces joule heating of the metal. This practical situation is described in detail in §6.2.6 and used as an example to illustrate a convenient way of formulating and solving problems containing mushy regions.

Rubinstein (1982a) expressed concern about the precise physical significance of the term mushy region, which is considered to be at a uniform temperature but has phase changes occurring within it, on the grounds that it is incompatible with the concept of a classical solution. In the Stefan boundary condition (1.9), for example, a uniform temperature means that the right-hand side is zero and hence there is no movement of the solid–liquid interface and no melting or freezing can occur. If the mushy region is considered to consist of small solid and liquid regions separated by many interfaces, the solid particles must be slightly below the melting temperature and the liquid slightly above it, so that local heat changes occur and new interfaces can form and existing ones disappear. Elliott and Ockendon (1982) visualize blobs of pure liquid and slightly superheated solid in an unstable equilibrium, with an 'averaged' temperature being zero over a region containing many blobs but forming only a small part of the whole of the mushy region. They draw a comparison with the forming of dendrites in the solidification of alloys. Mathematically, as in §6.2.6, the mushy region has been interpreted to be a region where the local enthalpy takes values in the range between those of the pure solid and liquid and the temperature is uniformly constant and equal to the phase-change temperature.

Rubinstein (1982a) proposed some more sophisticated mathematical models of solid–liquid zones in single component media and binary alloys. Solomon *et al.* (1982) described a semi-empirical model of a mushy region with an exact similarity solution. They considered a semi-infinite liquid in the region $x \geqslant 0$ initially at its solidification temperature T_m and from time $t = 0$ the temperature on the surface $x = 0$ is maintained constant at $T_s < T_m$. Subsequently, three zones are distinguished: (1) liquid at temperature T_m in $x \geqslant Y(t)$; (2) solid at temperature $T(x, t) < T_m$ in $0 \leqslant x < X(t) \leqslant Y(t)$; (3) the mushy region at temperature T_m in $X(t) \leqslant x \leqslant Y(t)$. Guided by microscopic studies of Thomas and Westwater (1963), Solomon *et al.* make the following assumptions; (a) the material in the mushy region contains a fixed fraction θL of the total latent heat $L(0 < \theta < 1)$; (b) the width of the mushy region is inversely proportional to the temperature gradient in the solid region at $x = X(t)$, i.e.

$$\{Y(t) - X(t)\}T_x(X(t), t) = \gamma = \text{constant.}$$

The constants θ, γ are characteristics of the material and their assumption of θ constant implies a uniform packing of crystals throughout the mushy region. Energy conservation then leads to

$$KT_x(X(t), t) = \rho L \left\{ \theta \frac{dX}{dt} + (1 - \theta) \frac{dY}{dt} \right\}.$$

Taking other conditions to be

$$T_t(x, t) = \alpha T_{xx}(x, t), \qquad 0 < x < X(t)$$
$$T(x, t) = T_m, \qquad x \geqslant X(t),$$
$$T(0, t) = T_s, \qquad t > 0, \qquad T(x, 0) = T_m, \qquad 0 < x < \infty,$$

Solomon *et al.* (1982) obtain the solution

$$T(x, t) = T_s + \Delta \, \text{erf}(x/2\sqrt{\alpha t})/\text{erf } \lambda, \qquad 0 < x < X(t)$$
$$X(t) = 2\lambda\sqrt{\alpha t}, \qquad Y(t) = 2\mu\sqrt{\alpha t},$$

with

$$\mu = \lambda + \{\gamma\sqrt{\pi}e^{\lambda^2} \, \text{erf } \lambda\}/2\Delta T$$

and λ the unique root of the equation

$$\frac{St}{\sqrt{\pi}} = \lambda e^{\lambda^2} \, \text{erf } \lambda + \frac{\gamma(1-\theta)}{\Delta T} \frac{\sqrt{\pi}}{2} (e^{\lambda^2} \, \text{erf } \lambda)^2.$$

Here $\Delta T = T_m - T_s$ and $St = c\Delta T/L$ is the Stefan number which is seen to be related to λ for values of the dimensionless parameter $\gamma(1-\theta)/\Delta T = \omega$ say. Solomon *et al.* plot St as a function of λ for various values of ω when $\theta = 0$ and also show the dimensionless relative mushy region width

$$\frac{\Delta T}{\gamma} \frac{\mu - \lambda}{\lambda} = \frac{\sqrt{\pi} \, e^{\lambda^2}}{2} \frac{\text{erf } \lambda}{\lambda}$$

as a function of λ.

Primicerio (1981*b*) reported some preliminary thoughts on the theoretical investigation of mushy regions in one-dimensional problems and on the possible existence of both classical and weak solutions with or without the presence of distributed heat sources.

(v) The initial condition may be space dependent, i.e. $U = U_0(X)$ at $T = 0$, and on any of the fixed boundaries there may be time dependent or non-linear conditions, e.g.

$$U = f(T) \qquad \text{or} \qquad \partial U/\partial X = \phi(U, T), \qquad X = 0. \qquad (1.13)$$

(vi) Rubinstein (1979) in discussing the connection between multidimensional Stefan problems and the theory of partial differential equations of the first order, draws attention to the interesting possibility that the Stefan boundary condition might include a second-order derivative, e.g. the condition (1.9) above could become

$$L\rho \frac{dS}{dT} = K_2 \frac{\partial U_2}{\partial X} - K_1 \frac{\partial U_1}{\partial X} - \frac{\partial S}{\partial Y} \left(\frac{\partial U_2}{\partial Y} - \frac{\partial U_1}{\partial Y} \right) + b^2 \frac{\partial^2 S}{\partial Y^2}$$

on $Y = S(X, T)$ and b^2 is a small parameter. He quotes a practical example relating to the transfer of fluid across a deformable, semi-permeable membrane of large curvature. Details of the formulation and numerical solution of such biological problems were discussed by Rubinstein (1974) and Geiman and Rubinstein (1974) and surveyed by Rubinstein (1980b).

(vii) Fasano (1980) assembled several of the above generalizations and extensions of Stefan boundary conditions and examined theoretical properties for one-dimensional problems. He gave an extensive list of references several of which are sources of further reading.

(viii) Results on the homogenization of Stefan problems were obtained by Damlamian (1980). He dealt with a mixture of two or more media which have different properties and are laid out in a periodic pattern. He established new constitutive laws for the homogeneous problem approached in the limit as the size of the periodic mesh goes to zero.

1.3.2. Linearized forms

Little systematic work is available on the cases listed above in which the parameters are functions of temperature. Only equations and boundary conditions linearized in the sense that the parameters are space and time dependent have been more widely studied. They provide a good description of many practical systems; they may also be useful as approximations to temperature-dependent systems often as successive stages of iterative treatments.

The 'linear' two-phase generalized Stefan problem considered, for example, by Meyer (1976) is given in the region $b_1 \leqslant X \leqslant b_2$ by

$$\frac{\partial}{\partial X}\left(K_1 \frac{\partial U_1}{\partial X}\right) - c_1 \frac{\partial U_1}{\partial T} = h_1, \qquad b_1 < X < S(T), \qquad T > 0, \quad (1.14)$$

$$\frac{\partial}{\partial X}\left(K_2 \frac{\partial U_2}{\partial X}\right) - c_2 \frac{\partial U_2}{\partial T} = h_2, \qquad S(T) < X < b_2, \qquad T > 0, \quad (1.15)$$

$$U_1 = \beta_1(T) \frac{\partial U_1}{\partial X} + \alpha_1(T), \qquad X = b_1, \qquad T > 0, \qquad (1.16)$$

$$U_2 = \beta_2(T) \frac{\partial U_2}{\partial X} + \alpha_2(T), \qquad X = b_2, \qquad T > 0, \qquad (1.17)$$

where K_1, c_1, h_1, K_2, c_2, h_2 may be functions of X and T, e.g. $K_1 = K_1(X, T)$, etc. Also

$$\left. \begin{aligned} U_1 &= \mu_1(S(T), T); \qquad U_2 = \mu_2(S(T), T), \\ K_1 \frac{\partial U_1}{\partial X} &- K_2 \frac{\partial U_2}{\partial X} + L(S(T), T) \frac{dS}{dT} = \mu_3(S(T), T), \end{aligned} \right\} \begin{aligned} & X = S(T), \quad T > 0, \\ & \end{aligned}$$

$$(1.18)$$
$$(1.19)$$

together with initial conditions of any suitable kind.

1.3.3 Non-dimensional forms

So far, the variables and the heat parameters have denoted quantities expressed in physical units, e.g. X in centimetres, T in seconds. Many authors present Stefan problems in non-dimensional variables.

Thus, introduction of the new variables

$$x = \frac{X}{\ell}, \quad t = \frac{K}{c\rho} \frac{T}{\ell^2}, \quad u = \frac{U}{U_0}, \quad s = \frac{S}{\ell}, \tag{1.20}$$

where ℓ is some standard length, reduces equations (1.1–6) to the following:

$$\frac{\partial u}{\partial t} = \frac{\partial^2 u}{\partial x^2}, \qquad 0 < x < s(t), \qquad t > 0, \tag{1.21}$$

$$u = 1, \qquad x = 0, \qquad t > 0, \tag{1.22}$$

$$u = 0, \qquad x > 0, \qquad t = 0, \tag{1.23}$$

$$s(0) = 0, \tag{1.24}$$

$$\left. \begin{array}{l} u = 0, \\[2mm] -\dfrac{\partial u}{\partial x} = \lambda \dfrac{ds}{dt}, \end{array} \right\} x = s(t), \qquad t > 0, \tag{1.25}$$
$$\tag{1.26}$$

where $\lambda = L/(cU_0)$ is a dimensionless 'latent heat' and $1/\lambda$ is the Stefan number. The choice of $\lambda = 1$ implies $U_0 = L/c$, i.e. the boundary condition (1.2) on $X = 0$ becomes $U = L/c$.

The two-phase system specified by equations (1.7–9) can be similarly expressed in terms of the variables

$$x = \frac{X}{\ell}, \qquad t = \frac{K_0}{c_0\rho} \frac{T}{\ell^2}, \qquad u_i = \frac{U_i}{U_0}, \qquad s = \frac{S}{\ell}, \tag{1.27}$$

where U_0, K_0, c_0 are any standard values of the respective variables. Then the two equations (1.7) become

$$\frac{\partial u_i}{\partial t} = \kappa_i \frac{\partial^2 u_i}{\partial x^2}, \qquad i = 1, 2, \tag{1.28}$$

where $\kappa_i = (K_i/K_0)(c_0/c_i)$, and (1.8, 9) become

$$\left. \begin{array}{l} u_1 = u_2 = 0, \\[2mm] \gamma_2 \dfrac{\partial u_2}{\partial x} - \gamma_1 \dfrac{\partial u_1}{\partial x} = \dfrac{ds}{dt}, \end{array} \right\} x = s(t), \tag{1.29}$$
$$\tag{1.30}$$

where $\gamma_i = K_i/(\lambda K_0)$. When the heat parameters are not constants, e.g. $K_i = K_i(U_i, X, T)$ as in §1.3.1, the corresponding non-dimensional parameters are also functions of the variables, e.g.

$$\kappa_i = \kappa_i(u_i, x, t), \quad \gamma_i = \gamma_i(u(s(t)), s(t), t). \tag{1.31}$$

1.3.4. Density change and convection

The physical derivation of the Stefan condition (1.10) in §1.2.3 is a particular case of conservation of energy in which both phases are assumed to be incompressible and at rest. More generally, across any phase-change boundary energy, momentum, and mass must be conserved. For the case of two incompressible phases, one solid and one liquid, of different densities ρ_1 and ρ_2 respectively, the mass condition can only be satisfied if the liquid phase moves with a velocity v given by

$$\rho_1 \frac{ds}{dt} = \rho_2 \left(\frac{ds}{dt} - v\right), \qquad x = s(t). \tag{1.32}$$

In the liquid region a convective term must be added to the heat-flow equation, so that, for example, (1.21) becomes

$$\frac{\partial u}{\partial t} = \frac{\partial^2 u}{\partial x^2} - v \frac{\partial u}{\partial x}. \tag{1.33}$$

In a one-dimensional problem the continuity and momentum equations for an inviscid liquid require that $\partial v/\partial x = 0$. The flow problem and the convective term are much more complicated in three dimensions. Equation (1.33) represents heat flow in one dimension accompanied by convection whatever the latter is caused by, and v need not necessarily be a consequence of a density change as in (1.32).

Although density changes occur frequently in practical systems, most mathematical papers have neglected to take them into account. A few authors have discussed one-dimensional problems. Chambré (1956) assumed the solid phase to be stationary and to have an infinite heat conductivity so that no equations were formulated in it. Similarity solutions were obtained in the liquid phase. Horvay (1962) was critical of Chambré's physical assumption of an incompressible fluid in a constant pressure field and produced alternative similarity solutions. Later Horvay (1965) formulated equations for temperature, velocity, and pressure in the liquid phase and included the influence of pressure on the freezing temperature. Tao (1979) used a new variable $y(x, t) = (x + \varepsilon S(t))/(1 + \varepsilon)$, where $S(t)$ denotes the position of the phase-change boundary in the two-phase system, and $\varepsilon = (\rho_2 - \rho_1)/\rho_2$, so that $y(S(t), t) = S(t)$, which simplifies the Stefan condition at the moving boundary. Dankwerts (1950), reported by Crank (1975a), introduced coordinates at rest in the two phases and obtained similarity solutions for a number of one-dimensional problems in infinite media. Wilson (1982) also used coordinates at rest to obtain similarity solutions of a multi-phase problem with phases of different constant densities. Some details of Wilson's method and some other similarity solutions are given in §3.2.

1.3.5. *Ablation and the inverse Stefan problem*

The term ablation refers to the removal of any part of the surface of a body by melting or evaporation. In the single-phase, melting-ice problem described in §1.2.1 and Fig. 1.1(a), the water formed on melting stays in position with its outer surface at $x = 0$ and occupies the region previously occupied by ice. In the case of an ablating sheet of ice the melt is continuously and immediately removed from the surface and the receding outer surface of the ice forms the moving boundary. Ablation occurs if a solid is melted by a hot, well-stirred fluid through which the temperature is everywhere uniform. Andrews and Atthey (1975a,b) describe the formation of a hole in a metal by evaporated ablation produced by a high-power laser beam (see §3.6.5(i)).

In general, in one dimension, the heat input, $q(t)$, to the solid is prescribed on the unknown ablating boundary $x = s(t)$. This is a one-phase Stefan problem. For the ablation of a slab originally occupying the space $0 \le x \le \ell$ and with heat, $q(t) > 0$, falling on the surface, initially at $x = \ell$, the equations have the form

$$\frac{\partial u}{\partial t} = \frac{\partial^2 u}{\partial x^2}, \qquad 0 < x < s(t), \tag{1.34}$$

$$u = g(t), \qquad \frac{\partial u}{\partial x} = \lambda \frac{ds}{dt} + q(t), \qquad x = s(t), \tag{1.35}$$

$$u = \phi(x) < 0, \qquad 0 < x < \ell, \qquad t = 0, \tag{1.36}$$

$$u = f(t) < 0, \qquad x = 0, \qquad t > 0, \tag{1.37}$$

$$s(0) = \ell. \tag{1.38}$$

An inverse Stefan problem is one in which the motion of the melting interface is known and some other boundary condition has to be determined, e.g. to solve the system (1.34–8) but with $s(t)$ a known function of t and $f(t)$ to be determined. Physically, the boundary condition on $x = 0$ has to be determined such that the melting interface moves in a prescribed way. An alternative inverse problem is to determine the heat source $q(t)$ on the melting interface given both $s(t)$ and $f(t)$. A solution of an inverse problem is, of course, the solution of some Stefan problem and may provide an indirect method of solving the ablation problem posed by equations (1.34–8).

In a wider context, an inverse problem can arise in a two-phase situation. For example, a direct Stefan probem is to solve (1.7), (1.8), and (1.9) and in particular to find $S(T)$ for prescribed conditions on the fixed boundaries such as

$$\partial U_1 / \partial X = -f_1(T), \qquad X = 0, \qquad \partial U_2 / \partial x = -f_2(T), \qquad X = \ell,$$

and appropriate initial conditions, where $f_1(T)$, $f_2(T)$ are given functions. An inverse problem would be to find $f_1(T)$, $f_2(T)$ such that a prescribed motion $S(T)$ is produced. The automatic, real-time control problem posed by this inverse situation is of practical importance, e.g., in steel casting, and is elaborated by Hoffman and Sprekels (1982) who present numerical solutions. A wider survey of the application of control techniques to parabolic systems is given by Hoffmann and Niezgódka (1983).

1.3.6. Multi-phase problems

The two-phase problem formulated in §1.2.2 above can be easily extended to more than two phases and moving interfaces. More equations of type (1.7) and more Stefan conditions (1.9) together with appropriate conditions on fixed boundaries and at $t = 0$ determine $U_1, U_2, \ldots, U_{n+1}$ and $S_1, S_2, \ldots, S_n$. The essential feature of a multi-phase problem is that each domain and the solution of the corresponding parabolic equation is connected to every neighbouring domain and solution through a set of relations expressing the physics or chemistry of the problem being considered, e.g. the Stefan condition. A simple example is provided by a collection of ice cubes in a glass of water. There is one heat-flow equation in the water phase and one in each ice cube. In the most general case the 'cubes' have different heat parameters, and equations and solutions in each one are linked with the water-domain equation and solution by Stefan conditions on each cube–water interface. Cannon (1978) relates the analysis of multi-phase problems to a problem in which two substances in solution diffuse and react quickly and completely on a moving boundary.

The ablation of the alloy walls of a space vehicle leads to a three-phase problem with solid, liquid, and vapour phases and two moving boundaries (Koh *et al.* 1969).

Bonnerot and Jamet (1981) discuss a simple, one-dimensional problem involving three phases, solid, liquid, and vapour, which can appear and disappear. They consider a solid material which initially occupies the region $0 < X < a$, where X is the space coordinate in the direction perpendicular to the wall. The wall is at a known temperature initially and then it is heated from the right (Fig. 1.3(a) by a given heat flow on $X = a$. The surface $X = 0$ is thermally insulated.

When the temperature on the right side of the wall, $X = a$, reaches the melting temperature a liquid phase appears. Assuming no density change on melting, the solid occupies the space $0 < X < a_1(T)$ and the liquid is in $a_1(T) < X < a$, where $X = a_1(T)$ is the melting interface (Fig. 1.3b). This interface moves to the left and temperature increases at any fixed point.

If the melting interface $X = a_1(T)$ reaches the left side of the wall $X = 0$ at $T = T_f$ before the temperature on $X = a$ reaches the vaporization temperature, the solid phase disappears and the wall collapses.

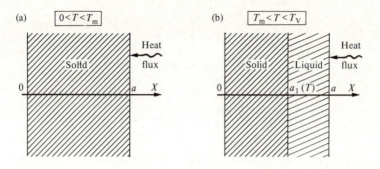

Fig. 1.3(a),(b). The wall before and after the appearance of the liquid phase

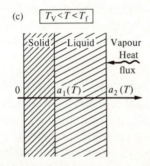

Fig. 1.3(c). The wall after the appearance of the vapour

On the other hand, if the vaporization temperature is reached in the liquid at $X = a$ before the wall collapses, a vapour phase appears which is separated from the liquid by a second moving interface $X = a_2(T)$. If the vapour is removed as soon as it appears only two liquid and solid phases remain, the solid occupying the region $0 < X < a_1(T)$ and the liquid $a_1(T) < X < a_2(T) < a$ as in Fig. 1.3(c). Ultimately, the interface $X = a_1(T)$ reaches $X = 0$ and the wall collapses. The relevant system of equations, using the nomenclature of §1.2.2 and denoting the solid and liquid phases by $i = 1$, $i = 2$ respectively, is:

$$c_i \rho_i \frac{\partial U_i}{\partial T} = K_i \frac{\partial^2 U_i}{\partial X^2}, \qquad i = 1, 2, \tag{1.39}$$

$$U(X, 0) = U_0(X), \qquad 0 < X < a, \tag{1.40}$$

$$\frac{\partial U}{\partial X} = 0, \qquad X = 0, \qquad T > 0, \tag{1.41}$$

$$K_1 \frac{\partial U}{\partial X} = F(T), \qquad X = a, \qquad 0 < T < T_m, \tag{1.42}$$

$$K_2 \frac{\partial U}{\partial X} = F(T), \qquad X = a, \qquad T_m < T < T_v, \tag{1.43}$$

$$\left. \begin{array}{l} U = U_m, \\[2mm] K_1 \dfrac{\partial U_1}{\partial X} - K_2 \dfrac{\partial U_2}{\partial X} = L_m \rho \dfrac{da_1}{dT}, \end{array} \right\} \quad x = a_1(T), \tag{1.44}$$

$$\left. \begin{array}{l} U = U_v, \\[2mm] K_2 \dfrac{\partial U_2}{\partial X} - F(T) = L_v \rho \dfrac{da_2}{dT}, \end{array} \right\} \quad x = a_2(T), \tag{1.45}$$

$$U(a, T_m) = U_m, \; U(a, T_v) = U_v, \; a_1(T_f) = 0. \tag{1.46}$$

Here L_m, L_v are latent heats of melting and vaporization respectively. Numerical methods used by Bonnerot and Jamet are discussed in §4.3.3. Tayler (1975) discusses the degeneration of phases in a slightly more general way.

1.3.7. Simultaneous diffusion and heat flow: alloys

Alloy solidification problems differ from classical Stefan problems in that the melting temperature is not known in advance; it depends on the composition of the alloy. Typically, an alloy is considered to comprise a pure substance containing a small concentration of one or more secondary substances, conveniently referred to as 'impurities'. The solidification of an alloy calls for a simultaneous study of the processes of heat flow and the diffusion of impurities. An introduction to the essential thermodynamics can conveniently be based on the eutectic diagram for one impurity shown in Fig. 1.4.

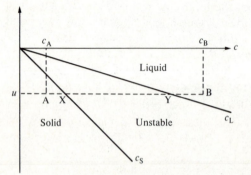

FIG. 1.4. Eutectic diagram

If at any point in the alloy the concentration of impurity is c_A and the temperature is u represented by the point A in Fig. 1.4 this point is in the solid region. If the concentration is increased to a value c_B at the same temperature, the point B is reached in the liquid region. Equilibrium between the impurity and the pure alloy is not possible in the range $c_S < c < c_L$. This is a region of instability but liquid and solid phases can coexist in equilibrium at the same temperature u with respective concentrations $c_S(u)$ and $c_L(u)$, e.g. at points X, Y in Fig. 1.4. Thus, the melting temperature will depend on the concentration of the impurity and at a phase-change boundary there will be a discontinuous change in the concentration related to the melting temperature through the eutectic diagram, i.e. $c_L(u) - c_S(u)$ for a temperature u.

A simple two-phase alloy solidification process can now be expressed in non-dimensional terms as follows, where u_1, c_1, κ_1 refer to the solid phase and u_2, etc. to the liquid phase.

$$\frac{\partial u_1}{\partial t} = \kappa_1 \frac{\partial^2 u_1}{\partial x^2}, \qquad \frac{\partial c_1}{\partial t} = D_1 \frac{\partial^2 c_1}{\partial x^2}, \qquad 0 < x < s(t), \qquad (1.47)$$

$$\frac{\partial u_2}{\partial t} = \kappa_2 \frac{\partial^2 u_2}{\partial x^2}, \qquad \frac{\partial c_2}{\partial t} = D_2 \frac{\partial^2 c_2}{\partial x^2}, \qquad s(t) < x < 1, \qquad (1.48)$$

$$\left. \begin{aligned} \gamma_2 \frac{\partial u_2}{\partial x} - \gamma_1 \frac{\partial u_1}{\partial x} &= \frac{ds}{dt}, \\ \beta_2 \frac{\partial c_2}{\partial x} - \beta_1 \frac{\partial c_1}{\partial x} &= (c_1 - c_2) \frac{ds}{dt}, \\ u_1 = u_2 = g, \qquad c_1 = c_S(g), \qquad c_2 &= c_L(g), \end{aligned} \right\} \quad x = s(t), \qquad (1.49)$$

together with appropriate conditions on $x = 0, 1$ and $t = 0$. Tayler (1975) discusses some special limiting cases of the general formulation. Fix (1978) outlines a simplified model of alloy solidification due to Langer and Turski (1977) to illustrate ideas of non-linear instabilities associated with Mullins and Sekerka (1963, 1964) on the melting front and related to the growth of dendrites in multi-dimensional problems. The simplifications in the Langer model are: (a) heat conductivity is everywhere infinitely large; (b) the diffusion coefficient for the impurity is constant; and (c) the concentration jump $c_L(U_m) - c_S(U_m)$ is constant, independent of the melting temperature. Fix then develops an alternative weak variational formulation of the complete alloy solidification problem. Ockendon (1978) also derived a weak formulation in one space dimension. An enthalpy formulation of alloy solidification and some solutions are presented in §6.2.7(ii).

Other examples of simultaneous diffusion and heat flow have been examined by Mikhailov (1975, 1976) and in flame studies by Crowley (1980, 1981) and Buckmaster (1979) (see §6.2.7(iii)).

1.3.8. Muskat's problem

A problem that only involves fluid flow but which is otherwise akin to the alloy solidification problem in that the concentration on the moving boundary is not explicitly specified is usually referred to as the Muskat problem (1934, 1937). The theory is that when two immiscible fluids in a porous medium are in contact with each other, the movement of the interface formed and of the fluids constitutes a moving boundary problem. A typical system of equations given by Cannon (1978) to determine the pressures $p_i(x, t)$, $i = 1, 2$, in the two media 1 and 2 is:

$$\frac{\partial \phi_i}{\partial t} = \frac{\partial}{\partial x}\left\{ a_i(x, p_i)\frac{\partial p_i}{\partial x}\right\}, \qquad i = 1, 2, \tag{1.50a}$$

where $i = 1$ refers to $0 < x < s(t)$, $i = 2$ to $s(t) < x < 1$, and $0 < t \leqslant T$ in both cases;

$$\left.\begin{aligned} p_1(s(t), t) &= p_2(s(t), t)\\ a_1\,\partial p_1/\partial x &= a_2\,\partial p_2/\partial x,\\ \phi\dot{s}(t) &= -a_1\,\partial p_1/\partial x \end{aligned}\right\} x = s(t), \tag{1.50b}$$

where a_i are given functions of x, p_i, and $\phi\{s(t), p(s(t), t)\}$ is known, together with specified initial values of p_i, $i = 1, 2$, at $t = 0$, $p_1(0, t)$ and $p_2(1, t)$ given functions of time, and $s(0) = s_0$, $0 < s_0 < 1$. Muskat's general problem is not pursued further in this book but a particular case of the motion of two immiscible fluids is considered in §2.12.1 in relation to the flow in a Hele-Shaw cell.

1.3.9. Two and three space dimensions

As an example of the extension of the various one-dimensional problems formulated above in §1.2 and §1.3, the two-phase moving-boundary problem with heat generation in each phase and on the moving boundary can be expressed as follows, with $\mathbf{x}$ the space coordinate vector:

$$\rho c \frac{\partial u}{\partial t} = \nabla(K\nabla u) + Q, \qquad \mathbf{x} \in D_i, \qquad i = 1, 2, \qquad 0 < t < T, \tag{1.51}$$

$$\frac{\partial u_i}{\partial n} - hu_i = g_i(\mathbf{x}, t), \qquad \mathbf{x} \in \partial D_i, \qquad 0 < t < T, \tag{1.52}$$

$$u_1(\mathbf{x}, t) = u_2(\mathbf{x}, t) = u_{\mathrm{m}}, \tag{1.53}$$

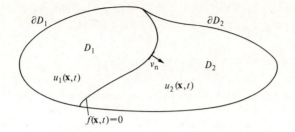

FIG. 1.5. Two-phase region with a moving boundary $f(\mathbf{x}, t) = 0$.

and

$$\left[K\frac{\partial u}{\partial n}\right]_1^2 = K_2\frac{\partial u_2}{\partial n} - K_1\frac{\partial u_1}{\partial n} = -\rho L v_n + q \tag{1.54}$$

on the moving boundary $f(\mathbf{x}, t) = 0$, $0 < t < T$,

$$\left.\begin{array}{c} u(\mathbf{x}, 0) = u_0(\mathbf{x}), \\ f(\mathbf{x}, 0) = f_0(\mathbf{x}), \end{array}\right\} t = 0. \tag{1.55}$$

With reference to Fig. 1.5, $D = D_1 \cup D_2$, $\partial D = \partial D_1 \cup \partial D_2$ denote the interior and boundary of the relevant domain. The operator ∇ encompasses different coordinate systems, e.g. rectangular, cartesian, cylindrical and spherical polar, and generalized curvilinear. In heat conduction nomenclature u_i denotes the temperature in each phase, $i = 1, 2$, u_m the phase-change temperature, L the latent heat, n the outward normal to the moving boundary $f(\mathbf{x}, t) = 0$, and v_n the velocity of this boundary along the normal. The thermal properties c, ρ, K and the heat generation terms Q, q may be functions of u, $\mathbf{x}$, and t and may be different in the two phases.

Patel (1968) expressed the condition (1.54) on the moving boundary in the more usable form

$$\left\{1 + \left(\frac{\partial s}{\partial x}\right)^2 + \left(\frac{\partial s}{\partial y}\right)^2\right\}\left|K\frac{\partial u}{\partial z}\right|_1^2 = -\rho L\frac{\partial s}{\partial t}, \; z = s(x, y, t) \tag{1.56}$$

for the case $q = 0$. This relationship connects explicitly $\partial u/\partial z$ with derivatives of the moving-boundary surface $f(\mathbf{x}, t) = 0$ written as $z = s(x, y, t)$. Similar relations hold for the other two components $\partial u/\partial x$ and $\partial u/\partial y$. Patel's derivation of (1.56) for $q = 0$ is as follows, denoting temperatures by u_i, $i = 1, 2$, and considering the three-dimensional case in which the moving boundary is given by $f(x, y, z, t) = 0$. Thus we have

$$u_i(x, y, z, t) = u_m, \quad i = 1, 2, \tag{1.56a}$$

$$K_1 \frac{\partial u_1}{\partial n} - K_2 \frac{\partial u_2}{\partial n} = \rho L v_n, \tag{1.56b}$$

$$\text{on} \quad f(x, y, z, t) = 0, \tag{1.56c}$$

where n is the outward normal and v_n the normal velocity. Then

$$\mathbf{n} = \nabla f / |\nabla f| = \nabla u_i / |\nabla u_i|, \tag{1.56d}$$

$$\frac{\partial u_i}{\partial n} = \nabla u_i \cdot \mathbf{n} = (\nabla u_i \cdot \nabla f) / |\nabla f| = |\nabla u_i|, \tag{1.56e}$$

$$v_n = \mathbf{v} \cdot \mathbf{n} = (\mathbf{v} \cdot \nabla f) / |\nabla f| = (\mathbf{v} \cdot \nabla u_i) / |\nabla u_i|. \tag{1.56f}$$

Remembering from (1.56c) that

$$df = \frac{\partial f}{\partial x} dx + \frac{\partial f}{\partial y} dy + \frac{\partial f}{\partial z} dz + \frac{\partial f}{\partial t} dt = 0,$$

and hence that

$$\dot{x} \frac{\partial f}{\partial x} + \dot{y} \frac{\partial f}{\partial y} + \dot{z} \frac{\partial f}{\partial z} + \frac{\partial f}{\partial t} = 0,$$

we see from (1.56f) that

$$|\nabla f| \, v_n = (\mathbf{i}\dot{x} + \mathbf{j}\dot{y} + \mathbf{k}\dot{z}) \left(\mathbf{i} \frac{\partial f}{\partial x} + \mathbf{j} \frac{\partial f}{\partial y} + \mathbf{k} \frac{\partial f}{\partial z} \right) = -\frac{\partial f}{\partial t},$$

and similarly for $|\nabla u_i| \, v_n$, so that

$$v_n = -(\partial f / \partial t) / |\nabla f| = -(\partial u_i / \partial t) / |\nabla u_i|. \tag{1.56g}$$

Using (1.56e,g), eqn (1.56b) becomes

$$K_1 \nabla u_1 \cdot \nabla f - K_2 \nabla u_2 \cdot \nabla f = -\rho L (\partial f / \partial t), \tag{1.56h}$$

and by differentiating (1.56a), (1.56c) we find

$$\frac{\partial u_i}{\partial x} = \frac{\partial f / \partial x}{\partial f / \partial z} \frac{\partial u_i}{\partial z}, \frac{\partial u_i}{\partial y} = \frac{\partial f / \partial y}{\partial f / \partial z} \frac{\partial u_i}{\partial z}, \tag{1.56i}$$

and hence by substituting (1.56i) into (1.56h)

$$\left[1 + \left(\frac{\partial f / \partial x}{\partial f / \partial z} \right)^2 + \left(\frac{\partial f / \partial y}{\partial f / \partial z} \right)^2 \right] \left[K_1 \frac{\partial u_1}{\partial z} - K_2 \frac{\partial u_2}{\partial z} \right] = -\rho L \frac{\partial f / \partial t}{\partial f / \partial z}, \tag{1.56j}$$

which is the same as (1.56) above for $f(x, y, z, t) = z - s(x, y, t) = 0$. For

$q \neq 0$ in (1.54) Patel derives the additional term

$$-\left\{\left(\frac{\partial f/\partial x}{\partial f/\partial z}\right)q_x + \left(\frac{\partial f/\partial y}{\partial f/\partial z}\right)q_y + q_z\right\}$$

on the right-hand side of (1.56j). Crank and Gupta (1975) derive the two-dimensional single-phase form of (1.56) by a simpler argument. By combining (1.56e) with (1.56f), equation (1.56b) can be written in the form quoted by Carslaw and Jaegar (1959, p. 284)

$$|K_1\nabla u_1| - |K_2\nabla u_2| = -\rho L\frac{\partial u_1/\partial t}{\nabla u_1} = -\rho L\frac{\partial u_2/\partial t}{\nabla u_2}. \qquad (1.56k)$$

1.3.10. Implicit boundary conditions: oxygen diffusion problem

A feature common to all the problems formulated above in §§1.2 and 1.3 is that the Stefan condition and its generalizations connect the position or velocity of the boundary with the dependent variable u or its derivatives, e.g. with the temperature gradient $\partial u/\partial x$ on the boundary in the one-dimensional case. Problems exist, however, in which such an explicit relationship does not occur and Sackett (1971) introduced the term 'implicit free-boundary problems' to describe them. They correspond to the special case $L = 0$ in (1.12).

A much-quoted example of an implicit condition occurs in the oxygen diffusion problem. Practical details are given by Crank and Gupta (1972a,b) and by Constable and Evans (1975, 1976). First, oxygen is allowed to diffuse into a medium which absorbs and immobilizes oxygen at a constant rate. The concentration of oxygen at the surface of the medium is maintained constant. A moving boundary marks the innermost limit of oxygen penetration. This first phase of the problem continues until a steady state is reached in which the oxygen does not penetrate any further into the medium. The surface of the medium is then sealed so that no further oxygen passes in or out. The medium continues to absorb the available oxygen already diffusing in it and, as a consequence, the boundary marking the depth of penetration in the steady state recedes towards the sealed surface. The major problem is that of tracing this movement of the boundary and of determining the distribution of oxygen as a function of time. A secondary problem in the application of numerical techniques is associated with the discontinuity in the derivative boundary condition which results from the abrupt sealing of the outer surface.

If $C(X, T)$ denotes the concentration of oxygen free to diffuse at a distance X from the outer surface of the medium at time T, D is a constant diffusion coefficient, and m, the rate of consumption of oxygen per unit volume of the medium, is also assumed constant for $C(X, T) > 0$, the steady state is defined by a solution of

$$D\frac{d^2C}{dX^2} - m = 0$$

which satisfies the conditions

$$C = \partial C/\partial X = 0, \qquad X \geqslant X_0,$$

where X_0 is the innermost extent of oxygen penetration, and on the outer surface

$$C = C_0 = \text{constant}, X = 0.$$

The required solution is readily seen to be given by

$$C = \frac{m}{2D}(X - X_0)^2, \qquad X_0 = \sqrt{\left(\frac{2DC_0}{m}\right)}.$$

After the surface $X = 0$ has been sealed, the position of the receding boundary is denoted by $X_0(T)$ and the problem to be solved becomes

$$\frac{\partial C}{\partial T} = D\frac{\partial^2 C}{\partial X^2} - m, \qquad 0 \leqslant X \leqslant X_0(T), \tag{1.57a}$$

$$\frac{\partial C}{\partial X} = 0, \qquad X = 0, \qquad T \geqslant 0, \tag{1.58a}$$

$$C = \frac{\partial C}{\partial X} = 0, \qquad X = X_0(T), \qquad T \geqslant 0, \tag{1.59a}$$

$$C = \frac{m}{2D}(X - X_0)^2, \qquad 0 \leqslant X \leqslant X_0, \qquad T = 0, \tag{1.60a}$$

where $T = 0$ is the moment when the surface is sealed. By making the changes of variables,

$$x = \frac{X}{X_0}, \qquad t = \frac{D}{X_0^2}T, \qquad c = \frac{D}{mX_0^2} = \frac{C}{2C_0},$$

and denoting by $s(t)$ the value of x corresponding to $X_0(T)$, the above system is reduced to the following non-dimensional form:

$$\frac{\partial c}{\partial t} = \frac{\partial^2 c}{\partial x^2} - 1, \qquad 0 \leqslant x \leqslant s(t), \tag{1.57}$$

$$\frac{\partial c}{\partial x} = 0, \qquad x = 0, \qquad t \geqslant 0, \tag{1.58}$$

$$c = \partial c/\partial x = 0, \qquad x = s(t), \qquad t \geqslant 0, \tag{1.59}$$

$$c = \tfrac{1}{2}(1 - x)^2, \qquad 0 \leqslant x \leqslant 1, \qquad t = 0, \tag{1.60}$$

where c is the concentration of oxygen free to diffuse. It is the absence of $\partial s/\partial t$ in (1.59) that renders this problem implicit.

Schatz (1969) uses the transformations

$$w = \partial u/\partial x \quad \text{or} \quad w = \partial u/\partial t \tag{1.61}$$

to express an implicit problem in an explicit form. Thus, in the ablation problem of §1.3.5 for $\lambda = 0$ but $q \neq 0$ in (1.35) the use of $w = \partial u/\partial x$ gives

$$\left.\begin{array}{l} \dfrac{\partial w}{\partial t} = \dfrac{\partial^2 w}{\partial x^2}, \quad 0 < x < s(t), \\[2ex] w = q(t), \quad \dfrac{\partial w}{\partial x} = -q\dot{s} + \dot{g}, \quad x = s(t), \\[2ex] \dfrac{\partial w}{\partial x} = \dot{f}(t), \quad x = 0, \quad t > 0, \\[2ex] w = \partial\phi/\partial x, \quad t = 0, \quad 0 < x < 1. \end{array}\right\} \tag{1.62}$$

If $q = 0$ the appropriate transformation is $w = \partial u/\partial t$ provided $\dot{g} \neq 0$ and u is differentiable at $t = 0$, $x = 0$ and 1, which requires $f(0) = \phi(0)$ and $g(0) = \phi(1)$, $q(0) = \phi'(1)$. Otherwise, the Schatz transformation introduces singularities. Tayler (1975) and Ockendon (1975) use the oxygen diffusion problem to illustrate these statements. Thus for $\lambda = 0$, $g = t$, $q = 0$, and $\partial u/\partial x = 0$ on $x = 0$, use of the transformation $w = \partial u/\partial t$ would be straightforward. However, the initial condition (1.60) used by Crank and Gupta (1972a) coupled with (1.58) means that c is not differentiable at $x = 0$, $t = 0$. The Schatz substitution, $w = \partial c/\partial t$, reduces the system (1.57–60) to

$$\left.\begin{array}{l} \partial^2 w/\partial x^2 = \partial w/\partial t, \quad 0 < x < s(t), \\[1ex] w = 0, \quad \partial w/\partial x = -ds/dt, \quad x = s(t), \\[1ex] w = 0, \quad s = 1, \quad t = 0, \\[1ex] \partial w/\partial x = \delta(t), \quad x = 0, \end{array}\right\} \tag{1.63}$$

where $\delta(t)$ expresses the discontinuous jump in $\partial c/\partial x$ from -1 to zero given by (1.60) and (1.58) at $x = 0$, $t = 0$.

Proofs that the transformations will convert an implicit problem into an explicit one and vice versa together with an examination of the necessary conditions are given by Schatz (1969).

Various methods of obtaining numerical solutions of equations (1.57–60) are described in §§3.5.2, 4.2, 4.3.1–3, 5.1, 6.2.7(i), 6.3, 6.4.1, and 6.4.2. Results are presented in Tables 3.1, 3.2, 4.1, 4.2, 6.3, and 6.6–8.

A two-dimensional version of the oxygen diffusion problem (Evans and Gourlay 1977) is referred to in §6.3.

A slightly more general version of the oxygen problem in cylindrically shaped sections of tissue is described by Galib, Bruch, and Sloss (1981). One application relates to oxygen diffusion after rubber tourniquets are removed from rabbits under anaesthesia (Rahmer *et al.* 1977). In non-dimensional terms the mathematical problem is defined by the equations in cylindrical coordinates (r, θ)

$$\frac{\partial c}{\partial t} = \frac{\partial^2 c}{\partial r^2} + \frac{1}{r}\frac{\partial c}{\partial r} + \frac{1}{r^2}\frac{\partial^2 c}{\partial \theta^2} + f(r, \theta, t, c),$$

where $c(r, \theta, t)$ is oxygen concentration in the tissue, $f(r, \theta, t, c)$ describes the rate of absorption of oxygen, and, with $\rho(\theta, t)$ denoting the moving boundary, r_1 the sealed capillary radius, and r_0 the outer tissue radius (Fig. 1.6),

$$c(\rho, \theta, t) = 0, \frac{\partial c}{\partial r}(\rho, \theta, t) = 0, \frac{\partial c}{\partial \theta}(\rho, \theta, t) = 0,$$

$$\frac{\partial c}{\partial r}(r_1, \theta, t) = 0, \qquad c(r_0, \theta, t) = 0.$$

Also $c(r, \theta, 0) = c_0(r, \theta)$ specifies the oxygen distribution in the tissue when the capillary surface is sealed.

Galib *et al.* (1981) use a semi-inverse method to obtain analytical solutions. First a mathematical form for the shape of the moving boundary is assumed. Then an expression for the concentration $c(r, \theta, t)$ which satisfies the conditions on the fixed boundaries and which is zero with zero normal derivative on the moving boundary assumed is taken e.g.

$$c(r, \theta, t) = \tfrac{1}{2}\{\rho(\theta, t) - r\}^2 - \frac{\{\rho(\theta, t) - r\}^3}{3\{\rho(\theta, t) - r_i\}}.$$

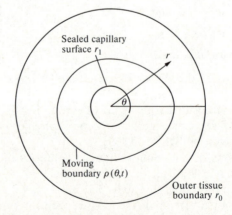

Sealed capillary
surface r_1

r

θ

Moving
boundary $\rho(\theta,t)$

Outer tissue
boundary r_0

FIG. 1.6. Cross-section of capillary, surrounding tissue, and moving boundary

The corresponding absorption function $f(r, \theta, t, c)$ is found by substituting this expression for c into the partial differential equation. Details are given by Galib *et al.* (1981). In order to model different experimental systems some expressions chosen for $\rho(\theta, t)$ are

$$\rho = 4/(t+1),$$

$$\rho = \frac{4}{(t+1)^{0.7}} + \frac{t-1}{(t+1)^2} + \frac{1}{(t+1)^6},$$

$$\rho = (3 + \cos \theta)/(t+1).$$

Finally Galib *et al.* (1981) used the truncation method of Berger *et al.* (1975) described in §6.3 to obtain numerical solutions for $c(r, \theta, t)$ and $\rho(\theta, t)$ for the functions $f(r, \theta, c)$ derived by their inverse method. The numerical results checked satisfactorily with the analytical expressions and in some cases with experimental data available.

1.3.11. Concentrated thermal capacities

If a thermally isotropic body A is in contact over part of its boundary S_{AB} with a second body B of much greater thermal conductivity, the temperature in B may be considered to be a function of time only and not of the space coordinates. This is the simplest example of what Tichonov (1950) described as a concentrated capacity B located on the boundary of A. More generally, the body B may be assumed to be anisotropic with a thermal conductivity infinitely large along lines orthogonal to S_{AB}. The the temperature in B depends on the coordinates of points on S_{AB} as well as on time. The boundary conditions on S_{AB} coupling the heat-flow fields in A and B express continuity of temperature and heat flux in the usual way. The solution of the heat-flow problem in A must satisfy conditions prescribed on the boundary of A including those on S_{AB}. If the latter are written in a form which includes the differential equation to be satisfied in B they contain derivatives of order equal to or possibly greater than the highest order of derivative appearing in the differential equation describing the heat flow in A. Tichonov (1950) based his definition of problems with concentrated capacities on the appearance of these higher-order derivatives in the boundary conditions.

Rubinstein (1971, 1979) drew attention to the practical importance of problems involving concentrated capacities in which phase changes occur. Thus, relatively thin, oil-saturated strata can be considered as concentrated capacities on the boundaries of impermeable rocks. Tertiary methods of oil production lead to problems with free boundaries moving inside a concentrated capacity as paraffin sediments in oil-saturated media are melted by the injection of hot liquid or vapour.

Fasano, Primicerio, and Rubinstein (1980) discussed the following model problem. A thin horizontal layer of a porous medium constitutes a concentrated capacity on the boundary of an impermeable medium. An incompressible liquid flows through the porous layer with a prescribed velocity and causes a phase change which absorbs latent heat at a given fixed temperature. The two phases are liquids with different specific heats. Fasano *et al.* (1980) extended a well-known scheme by Lauwerier (1955) for a similar problem which did not include a phase change. Lauwerier's basic assumptions were:

(1) horizontal heat transfer in the impermeable medium is negligible compared with vertical transfer;
(2) in the concentrated capacity temperature changes in the vertical direction are negligible;
(3) horizontal heat conduction can be neglected compared with heat-transfer by forced convection in the concentrated capacity.

Fasano *et al.* (1980) adopted the first two assumptions but omitted the third and took into account both conduction and forced convection.

The horizontal and vertical coordinates are denoted by x and z respectively, the thin porous layer occupies the region $-\infty < x < \infty$, $-2h < z < 0$, and the surrounding impermeable medium extends to infinity on both sides. The subscript 0 refers to the impermeable medium and subscripts 1 and 2 refer to the two liquid phases in the porous layer. The temperature fields are assumed symmetric with respect to the plane $z = -h$ and only the impermeable region $z > 0$ is considered. Lauwerier's first assumption gives

$$k_0 u_{0zz} = u_{0t}, \qquad -\infty < x < \infty, \qquad z > 0, \qquad t > 0, \qquad (1.64)$$

and in the porous layer we have

$$k_i u_{ixx} + k_i^* u_{izz} = c_i u_{it} + c_i^* v u_{ix}, \qquad i = 1, 2, \qquad (1.65)$$

$$u_{iz} = 0, \qquad z = -h, \qquad i = 1, 2, \qquad (1.66)$$

$$u_1 = u_0, \qquad k_1^* u_{1z} = k_0 u_{0z}, \qquad z = 0, \qquad -\infty < x < s(t), \qquad (1.67)$$

$$u_2 = u_0, \qquad k_2^* u_{2z} = k_0 u_{0z}, \qquad z = 0, \qquad s(t) < x < \infty, \qquad (1.68)$$

where k_i^*, c_i^* are respectively conductivities in the z-direction and specific heats per unit volume in the two liquid phases $i = 1, 2$, while v in (1.65) is the velocity of percolation in the x-direction, assumed constant. On integrating (1.65) with respect to z between $-h$ and 0 and using (1.66–8) we obtain

$$k_i u_{ixx} + (k_0/h) u_{0z}|_{z=0} - c_i^* v u_{ix} - c_i u_{it} = 0 \qquad (1.69)$$

after dropping the dependence of u_1, u_2 on z by Lauwerier's assumption (2). In non-dimensional variables, but retaining the nomenclature x, z, u_i, equations (1.64) and (1.68) together with necessary boundary conditions form the system

$$u_{0zz} - u_{0t} = 0, \qquad -\infty < x < \infty, \qquad z > 0, \qquad t > 0, \qquad (1.70)$$

$$u_{1xx} - b_1 u_{1x} - u_{1t} = -a_1 u_{0z}(x, 0, t), \, -\infty < x < s(t), \qquad t > 0, \quad (1.71)$$

$$u_{2xx} - b_2 u_{2x} - u_{2t} = -a_2 u_{0z}(x, 0, t), \qquad s(t) < x < \infty, \qquad t > 0,$$
$$(1.72)$$

where b_i, a_i ($i = 1, 2$) are constants and u_1, u_2 are functions of x, t only,

$$u_0(x, +0, t) = u_i(x, t), \qquad z = 0, \qquad i = 1, 2, \qquad (1.73)$$

$$u_0(x, z, 0) = f(x, z), \qquad -\infty < x < \infty, \qquad z > 0, \qquad t = 0 \qquad (1.74)$$

$$s(0) = 0, \qquad u_1(x, 0) = f_1(x), \qquad x < 0, \qquad u_2(x, 0) = f_2(x), \qquad x > 0,$$
$$(1.75)$$

$$u_1(s(t), t) = u_2(s(t), t) = 0, \qquad t > 0 \qquad (1.76a)$$

$$\dot{s}(t) = u_{2x}(s(t), t) - u_{1x}(s(t), t), \qquad t > 0. \qquad (1.76b)$$

The compatability conditions

$$f(x, 0) = \begin{cases} f_1(x), & x < 0, \\ f_2(x), & x > 0, \end{cases} f_1(0) = f_2(0) = 0, \qquad (1.77)$$

are also assumed.

Fasano *et al.* (1980) proceeded to analyse the above problem but only for the case in which $x = s(t)$ is a known, time-dependent, phase-change boundary, and $s(t)$ is assumed sufficiently smooth. They defined

$$u^*(x, t; s) = \begin{cases} u_1(x, t), & x \leq s(t) \\ u_2(x, t), & x > s(t), \end{cases} \qquad (1.78)$$

and assumed (1.76a) so that

$$u^*(s(t), t; s) = 0. \qquad (1.79)$$

In $-\infty < x < +\infty$, for the solution $u_0(x, z, t)$ of equation (1.70) together with (1.74) and (1.73) written as $u_0(x, 0, t) = u^*(x, t; s)$, and assuming $f(x, z)$ is continuously differentiable with respect to z, we can write

$$u_{0z}(x, z, t) = f(x, 0)(\pi t)^{-\frac{1}{2}} \exp(-z^2/4t) + \int_0^\infty f_z(x, \zeta) N(z, t, \zeta, 0) \, d\zeta$$

$$- \int_0^t u^*(x, \tau; s) G_{\zeta z}(z, t, 0, \tau) \, d\tau, \quad (1.80)$$

where N and G are the usual Neumann and Green functions in the quarter plane $z > 0$, $t > 0$. After integrating by parts the last integral in

(1.80) and putting

$$u^{**}(x, t; s) = \begin{cases} u_t^*(x, t; s), & x \neq s(t) \\ 0, & x = s(t), \end{cases} \tag{1.81}$$

Fasano *et al.* (1980) obtained

$$\int_0^t u^*(x, \tau, ; s)G_{\zeta z}(z, t; 0, \tau)\, d\tau = f(x, 0)N(z, t, 0, 0)$$

$$-\int_0^t u^{**}(x, \tau; s)N(z, t, 0, \tau)\, d\tau,$$

which combined with (1.71), (1.72) after letting $z \to 0$ yields

$$u_{ixx} - b_i u_{ix} - u_{it} = a_i F(x, t) - a_i \pi^{-\frac{1}{2}} \int_0^t u^{**}(x, \tau; s)(t - \tau)^{-\frac{1}{2}}\, d\tau \quad \text{in} \quad D_i,$$

$$i = 1, 2, \quad (1.82)$$

where $i = 1$ in $-\infty < x < s(t)$, $i = 2$ in $s(t) < x < \infty$, $t > 0$, and

$$F(x, t) = (\pi t)^{-\frac{1}{2}} \int_0^\infty f_z(x, \zeta)\exp(-\zeta^2/4t)\, d\zeta.$$

They prove theorems of existence and uniqueness of (1.82). In an appendix to their paper (1980) J. R. Ockendon suggested an asymptotic formulation of a slightly modified form of the above problem in which the free boundary is taken to be $x = S(z, t)$, $S \neq 0$.

Rubinstein, Geiman, and Shachaf (1980) obtained numerical solutions of the system of equations (1.70–6b). They introduced new coordinates moving with the free boundary given by

$$\xi = x - s(t), \qquad \eta = y, \qquad \tau = t,$$

and used an iterative scheme written with x, y, t now denoting the new ξ, η, τ in the form

$$u_{0zz}^n + u_{0x}^n \dot{s}^n = u_{0t}^n, \qquad n = 1, 2, \ldots$$

$$u_{ixx}^n - (\alpha - \dot{s}^n)u_{ix}^n + \beta u_{0z}^{n-1} = u_{it}^n, \qquad n = 1, 2$$

$$u_i^n(0, t) = 0, \qquad u_0^n(x, 0, t) = u_i^n(x, t), \qquad n = 0, 1, 2, \ldots$$

$$u_0^n(x, z, 0) = f_0(x, z), \qquad u_i^n(x, 0) = f_i(x), \qquad n = 0, 1, 2, \ldots \quad (1.83)$$

$$\dot{s}^n(t) \equiv u_{2x}^{n-1}(s^{n-1}(t), t) - u_{1x}^{n-1}(s^{n-1}(t), t), \qquad s^n(0) = 0, n = 1, 2, \ldots$$

$$u_0^0(x, y, t) = u_0(x, z, t - \tau); \; u_i^0(x, t) = u_{-i}(x, t - \tau).$$

The functions $u_0(x, z, t - \tau)$ etc. used as starting values for the iterative process come from a previously obtained non-iterative approximation in which first-order space derivatives are taken at time $t - \tau(\tau > 0)$; e.g. the

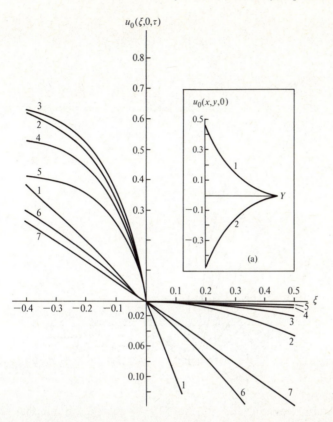

FIG. 1.7. Temperatures in concentrated capacity: $u_0(x, y, 0) = \tanh(-x)\mathrm{erfc}\{y/(2\sqrt{0.2})\}$; for curves 1–5, $\alpha = 12.5$, $\beta = 1.0$, (1) $\tau = 0$, (2) $\tau = 0.1$, (3) $\tau = 0.2$, (4) $\tau = 0.5$, (5) $\tau = 1.0$; for curves 6 and 7, $\alpha = 1.0$, $\beta = 0.1$, (6) $\tau = 0.5$, (7) $\tau = 1.0$; inset, (1) $x = -0.5$, (2) $x = 0.5$

term $u_{0x}(x, z, t)\dot{s}(t)$ is replaced by $u_{0x}(x, z, t - \tau)\dot{s}(t)$ in the equation on which the first of (1.83) is based.

The authors presented graphical solutions for a number of particular cases showing the motion of the free boundary and temperature behaviour in the impermeable medium. Figure 1.7 shows temperatures in the concentrated capacity at various times which vary monotonically. In contrast, Fig. 1.8 shows temperatures in the impermeable medium that develop spatial oscillations, reaching a minimum at $\xi = 0$ which is the free boundary. The authors attribute this behaviour to the combination of the effects of the relatively large speed of the free boundary, the loss of heat from the impermeable layer alternating with the uptake of heat from the concentrated capacity due to convective heat transfer, and the absorption of latent heat. The oscillations appear only for certain combinations of

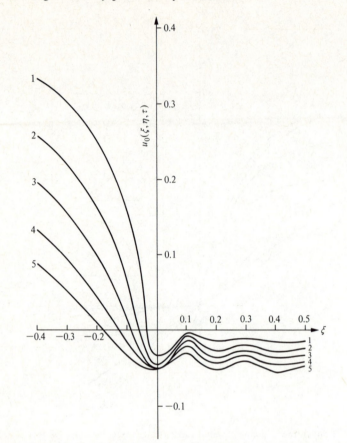

FIG.   1.8. Oscillating   temperatures   in   impermeable   medium: $u_0(x, y, 0) =$ tanh$(-x)$erfc$\{y/(2\sqrt{0.2})\}$; $\alpha = 12.5$, $\beta = 1.0$, $\tau = 1.0$; (1) $\eta = 0.1$, (2) $\eta = 0.2$, (3) $\eta = 0.3$, (4) $\eta = 0.4$, (5) $\eta = 0.5$

parameters. Thus in Fig. 1.8, $\alpha = 12.5$ and $\beta = 1.0$. For $\alpha = 1.0$, $\beta = 0.1$ there are no oscillations and temperatures vary monotonically. These are provisional results and further studies are needed for a complete understanding of the causes of the oscillations.

Rubinstein (1980a) analysed a similar axially symmetric model problem in which hot liquid is injected into the oil-saturated porous layer, the concentrated capacity, through a well of infinitely small radius along the axis $r = 0$. Allowing for symmetry about $z = -h$, the porous layer occupies the space $0 \leqslant r < \infty$, $-h < z < 0$ and the impermeable medium is in $0 \leqslant r < \infty$, $0 < z < \infty$. The melting interface where $u = u_m =$ constant is defined by $r = R(t)$, $-h < z < 0$, and divides the porous layer into two parts $0 \leqslant r < R(t)$, $R(t) < r < \infty$, indicated by $i = 1, 2$. The liquid densities on the two sides of the interface are the same but the specific heats have different constant values. The following non-dimensional system of equations specifies the problem:

$$u_{0zz} = u_{0t}, \qquad 0 < r < \infty, \qquad 0 < z < \infty, \qquad t > 0,$$

$$u_{irr} + (1 - 2v) r^{-1} u_{ir} + \beta u_{0z} \big|_{z=0} = u_{it}, \qquad t > 0, \qquad i = 1, 2$$

$u_1(o, t) = h(t)$, the temperature of the injected liquid, $t > 0$,

$$u_i(R(t), t) = 0, \qquad i = 1, 2, \qquad t > 0,$$

$$u_0(r, 0, t) = u_i(r, t), \qquad i = 1, 2,$$

$$u_i(r, 0) = f_i(r), \qquad i = 1, 2; \qquad u_0(r, z, 0) = f_0(r, z),$$

$$\dot{R}(t) = u_{2r}(R(t), t) - u_{1r}(R(t), t), \qquad t > 0; \qquad R(0) = R_0 > 0,$$

and $f_{0z}(0, 0) = 0$; $f_{1r}(0) = 0$; $f_0(r, 0) = f_i(r)$, $f_i(R_0) = 0$, $i = 1, 2$; $f_1(r) > 0$, $f_2(r) < 0$; $f_i(0) = 0$, $i = 1, 2$; $\beta = \text{constant} > 0$.

Rubinstein, Geiman, and Shachaf (1980) used the same type of procedure as for the plane case discussed above to obtain numerical solutions for the axisymmetric problem, except that iteration was found to be unnecessary for the parameters chosen. No special features were noted.

Rubinstein (1979) formulated a concentrated capacity problem which he considered to be a better description of the practical oil production process. No analysis or numerical solution was attempted.

2. Free-boundary problems: formulation

2.1. Introduction

THE problems introduced in this chapter are not time dependent as in Chapter 1. Typically, a free-boundary problem consists of a partial differential equation of elliptic type to be satisfied within a bounded domain together with the necessary boundary conditions. One section of the boundary, the free boundary, is unknown and must be determined as part of the solution. To make this possible, an additional condition has to be specified on the free boundary.

Flow through porous media is an important source of free-boundary problems, most frequently in relation to seepage phenomena that occur in nature. Examples are seepage through earth dams; seepage out of open channels such as rivers, canals, ponds, and irrigation systems, or into wells. Practical interest in free-boundary problems, however, is not confined to natural seepage but extends for example to topics in plasma physics, semiconductors, and electrochemical machining.

Free boundaries arise in porous flow when the porous medium is occupied by two fluids separated by a sharp interface, which is the free boundary. Water/air interfaces are the most common, but water/water vapour, oil/water, oil/gas, and salt water/fresh water interfaces are also important. The assumption that a sharp interface exists implies that the porous flow is saturated, i.e. any particular part of the porous medium is either saturated by one fluid or contains none of it, so that there are sharp interfaces, the free boundaries, between the regions occupied by the different fluids. For example, in a saturated water/air flow any one portion of the medium is either wet or dry. Free-boundary formulations of porous flows can therefore be regarded as approximations to partially saturated flows in which the fluid content of part of the medium may be between none and the maximum value corresponding to saturation. Sometimes, comparisons between the mathematical solution of a free-boundary model and the solutions for corresponding partially saturated flows can indicate whether the assumption of saturated flow is justified.

Comprehensive and directly relevant reviews of free-boundary prob-

lems, mainly in the context of saturated porous flow, and their solution
are given by Baiocchi *et al.* (1973*a*), Cryer (1976*a,b*), and Bruch (1980),
and each of these is a valuable source of other more detailed references.
Cryer prepared a series of surveys on various aspects of free-boundary
problems including Cryer and Fetter (1977), which dealt with the variational inequality solution of axisymmetric porous flow in well problems.
Cryer (1977) compiled a bibliography of free-boundary problems with
five appendices. Standard references for background information on
porous-flow problems are Muskat (1937), Polubarinova-Kochina (1962),
Harr (1962), and Bear (1972). Cryer (1976a) gives more specialist
references.

In this chapter several free-boundary problems are formulated
mathematically and their physical origins are introduced briefly. Discussion of methods of obtaining mathematical solutions is deferred to
Chapters 7 and 8 for the most part.

2.2. Simple rectangular dam

The simplest case of seepage through a dam separating two reservoirs
at different levels can be regarded as the stationary saturated flow of an
incompressible fluid through a homogeneous isotropic medium. The
simple rectangular dam which has parallel vertical walls and an impervious horizontal base has been much studied and adopted by many authors
as a model problem for assessing new methods of solving free-boundary
problems.

2.2.1. Classical formulation: physical derivation

The physical derivation of the mathematical equations will first be
given with reference to Fig. 2.1. The dam is regarded as being long

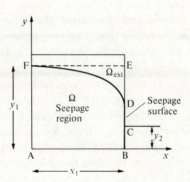

FIG. 2.1. Simple rectangular dam

enough for the flow to be considered as two-dimensional in the (x, y) plane. The flow is taken to be laminar and to be governed by Darcy's law (1856) expressed in the form

$$\mathbf{q}_1 = \frac{-K}{\rho g} \operatorname{grad} h = \frac{-k}{\mu} \operatorname{grad}\{p + \rho g y\} \tag{2.1}$$

where $\mathbf{q}_1$ is the velocity vector, p is the pressure in the fluid, h is the hydraulic head, ρ is density, μ is the viscosity of the fluid, the scalar constant K is called the hydraulic conductivity, and $k = \mu K/\rho g$ is the permeability of the medium. The vertical coordinate y is positive upwards. For the assumptions already made about the fluid k is constant and so the function

$$\phi_1 = (k/\mu)(p + \rho g y) \tag{2.2}$$

is a velocity potential and $\mathbf{q}_1 = -\operatorname{grad} \phi_1$. But the equation of continuity

$$\operatorname{div} \mathbf{q}_1 = 0 \tag{2.3}$$

must also be satisfied by the incompressible flow and hence from (2.1–3)

$$\operatorname{div}(\operatorname{grad} \phi_1) = \nabla^2 \phi_1 = 0, \tag{2.4}$$

i.e. ϕ_1 satisfies Laplace's equation in the seepage region Ω in Fig. 2.1. It is convenient to introduce a modified velocity potential

$$\phi(x, y) = \frac{\mu \phi_1(x, y)}{k \rho g} = \frac{p}{\rho g} + y \tag{2.5}$$

so that

$$\mathbf{q} = -\operatorname{grad} \phi \quad \text{and} \quad \nabla^2 \phi = 0, \tag{2.6}$$

where $\mathbf{q} = \mu \mathbf{q}_1/k \rho g$, a modified flow rate.

The seepage region is bounded by parts of the walls of the dam AF and BD and its base AB but also by the free surface FD whose shape and position are to be determined, including the location of the 'point of detachment' D on the wall BE. The part of the boundary CD, known as the 'seepage surface', must exist for physical reasons.

The conditions to be satisfied by $\phi(x, y)$ on the different parts of the boundary of the region Ω are derived as follows. Since there can be no flow across an impervious surface the normal derivative $\partial \phi/\partial n$ must be zero on any such surface, e.g. on the base AB, $\partial \phi/\partial y = 0$. Since the free surface FD is the interface between the water in flow region Ω and the air above, into which no water penetrates, the condition $\partial \phi/\partial n$ holds on FD also. The second condition on the free boundary is that pressure must be continuous across it. But outside the flow region and above the two reservoirs the pressure is constant and may be taken to be zero. Putting

$p = 0$ in (2.5) gives $\phi = y$ on FD. Similarly on the seepage surface CD $\phi = y$, since $p = 0$, because again the water there is in contact with air. The condition along a porous boundary in contact with a reservoir depends on the fact that the fluid velocity in a reservoir is negligible compared to that in the porous medium. A zero velocity means the velocity potential $\phi = y + p/(g\rho)$ is constant and since $p = 0$ at a reservoir surface we have $\phi = y_1$ or $\phi = y_2$ along AF or BC respectively (Fig. 2.1).

The model problem of the simple rectangular dam is therefore defined with reference to Fig. 2.1 by the following equation and conditions:

$$\nabla^2 \phi = 0 \quad \text{in } \Omega, \qquad 0 < x < x_1, \qquad 0 < y < f(x), \tag{2.7}$$

$$\phi = y_1 \quad \text{on AF}, \quad \phi = y_2 \quad \text{on BC}, \tag{2.8}$$

$$\phi = y, \qquad \partial \phi / \partial n = 0 \quad \text{on FD}, \tag{2.9}$$

$$\phi = y \quad \text{on CD}, \tag{2.10}$$

$$\partial \phi / \partial y = 0 \quad \text{on AB}, \tag{2.11}$$

$$\partial \phi / \partial x \leq 0 \quad \text{on CD}. \tag{2.12}$$

The condition (2.12) expresses the fact that some fluid leaves the porous medium on the seepage face CD, but none can enter the medium there. The complete solution includes the determination of the free boundary $y = f(x)$ such that

$$f(0) = y_1, \qquad f(x_1) > y_2, \qquad \frac{\mathrm{d}}{\mathrm{d}x} f(0) = 0, \qquad \frac{\mathrm{d}}{\mathrm{d}x} f(x_1) = \infty. \tag{2.13}$$

The final condition in (2.13) is discussed more fully in §8.3. In this example, orthogonal to the equipotential curves $\phi = \text{constant}$ there will be streamlines along which $\partial \phi / \partial n$ is zero. It follows that the impervious base and the free surface are both streamlines. Fluid particles move along a streamline always in the same direction from the upper to the lower reservoir which is the direction of x increasing in Fig. 2.1. Thus the fluid velocity is positive in the flow region and ϕ must decrease along any streamline. In particular, along the streamline free surface ϕ is a decreasing function of x, and since on FD $\phi = y = f(x)$, it follows that $f(x)$ is monotone decreasing.

Mathematical properties of the free boundary $y = f(x)$ have been examined for the simple dam and other problems by several authors including Friedman and Jensen (1977), Jensen (1977, 1980), and Boieri and Gastaldi (1980). Jensen (1980) gives references to a series of earlier papers on which his own is based. In many cases of steady flow it has been proved analytically that f is a continuous function strictly decreasing, with $f(0) = y_1$ and $f(x_1) = y_2$, and that f is concave. Different results have been proved in some evolution dam problems (see §2.9). Collatz

(1980) used monotonicity to establish upper and lower bounds for some free-boundary problems.

Finally, because the base and the free surface are streamlines they define a stream tube through any section of which the fluid flow is the same. In particular through any vertical section of the dam we have a constant flow rate given by

$$-\int_0^{f(x)} \frac{\partial}{\partial x} \phi(x, y) \, dy = \text{constant} = q \qquad (2.14)$$

and sometimes referred to as the 'discharge'.

2.2.2. Stream function

The components of the velocity $\mathbf{q}$ in (2.6) are given by

$$u = -\partial\phi/\partial x, \qquad v = -\partial\phi/\partial y. \qquad (2.15a)$$

There also exists a stream function ψ which satisfies Laplace's equation $\nabla^2\psi = 0$, and for which

$$u = -\partial\psi/\partial y, \qquad v = \partial\psi/\partial x. \qquad (2.15b)$$

The Cauchy–Riemann equations $\phi_x = \psi_y$ and $-\phi_y = \psi_x$ follow from (2.15a) and (2.15b) as usual.

The simple dam problem defined by equations (2.7–12) can also be expressed in terms of the two functions ϕ and ψ as

$$\left.\begin{aligned}
&\partial\phi/\partial x - \partial\psi/\partial y = 0, \qquad \partial\phi/\partial y + \partial\psi/\partial x = 0 \quad \text{in } \Omega, \\
&\phi = y_1 \quad \text{on AF}, \qquad \phi = y_2 \quad \text{on BC}, \\
&\phi = y, \qquad \psi = 0 \quad \text{on FD}, \\
&\phi = y \quad \text{on CD}, \\
&\psi = q \quad \text{on AB},
\end{aligned}\right\} \qquad (2.16)$$

where q is the flow rate through the dam defined in (2.14).

It is important to note that, in this example, q can be explicitly determined from the dimensions of the dam and the water levels. In fact, Charni (1951) showed that

$$q = (y_1^2 - y_2^2)/(2x_1) \qquad (2.17)$$

with reference to Fig. 2.1.

2.2.3. Formulation on fixed domain: Baiocchi transformation

In 1971 Baiocchi proposed a new way of formulating flow problems through porous media which has proved highly effective in both theoretical and numerical respects. It allows existence and uniqueness theorems to be proved rigorously; it also yields new numerical algorithms which are simple and efficient to use and which compete well with older schemes based on the classical formulations of equations (2.7–12) or (2.16). Baiocchi (1972) developed the theoretical side of his idea for the simple rectangular dam problem and the numerical application is due to Comincioli *et al.* (1971). Extensive systematic developments and applications to more general problems were subsequently pursued by Baiocchi and his colleagues at the Laboratorio di Analisi Numerica del Consiglio Nazionale della Richerche, Pavia, Italy. The early theoretical results were reported by Baiocchi, Comincioli, Magenes, and Pozzi (1973b). Baiocchi, Comincioli, Guerri, and Volpi (1973a) developed numerical methods and presented solutions for a number of problems of practical importance, some giving rise to mathematical and numerical difficulties which do not arise in the simple dam problem. These authors also surveyed the theoretical results of the companion paper by Baiocchi, Comincioli, Magenes, and Pozzi (1973b). A most helpful survey of more recent work by Bruch (1980) outlines the basic ideas of the Baiocchi method and its application to a wide range of important practical problems. Solutions are presented graphically and in tables and some are compared with results of earlier workers. Chapter 8 is concerned with numerical methods of solution including some based on Baiocchi's ideas.

The Baiocchi approach is introduced here with reference to the simple rectangular dam (Fig. 2.1). The unknown region Ω, in which the problem has been formulated by equations (2.7–12), is extended into a known region with fixed boundaries, in this case ABEF. The dependent variable is extended continuously across the free surface and thus defined over the whole of the known extended region.

The domain Ω in Fig. 2.1 is extended to the known region ABEF denoted by $D(0 < x < x_1, 0 < y < y_1)$ so that $D \equiv \Omega + \Omega_{ext} + \Gamma$, where Γ is the free boundary FD including the points F and D, and Ω_{ext} is defined by $0 < x < x_1$, $f(x) < y < y_1$. The dependent variable ϕ is extended continuously and defined on $\bar{D}$, i.e. the domain D plus its boundaries, by putting

$$\tilde{\phi}(x, y) = \phi(x, y) \quad \text{in } \bar{\Omega},$$
$$\tilde{\phi}(x, y) = y \quad \text{in } \bar{D} - \bar{\Omega}. \tag{2.18}$$

A weak solution of equations (2.7–12) is a triplet $\{f, \Omega, \phi\}$ such that: $f(x)$ is a continuous strictly decreasing function in $0 \le x \le x_1$ with $f(0) = y_1$, $f(x_1) \ge y_2$; Ω is the open domain already indicated; $\phi(x, y)$ is square summable together with its first derivatives in Ω, continuous in $\bar{\Omega}$, and in this sense satisfies (2.8), (2.10), and the first of (2.9); $\phi(x, y)$ satisfies

$$\iint_{\Omega} (\phi_x \bar{\phi}_x + \phi_y \bar{\phi}_y) \, dx \, dy = 0 \tag{2.19}$$

for all $\bar{\phi}(x, y)$ defined on $[0, x_1] \times [0, y_1]$ with continuous first derivatives and with $\bar{\phi} = 0$ on a neighbourhood of AF and BE. As usual, the weak solution (2.19) implies that $\nabla^2 \phi = 0$ on Ω and ϕ satisfies (2.11) and the second of (2.9).

The Baiocchi transformation is now used to define a new dependent variable w which is

$$w(x, y) = \int_y^{y_1} \{\bar{\phi}(x, \eta) - \eta\} \, d\eta. \tag{2.20}$$

First, certain properties of $w(x, t)$ are investigated. On the free surface Γ and in $\bar{D} - \bar{\Omega}$, $w = 0$, and so w is continuous on $\bar{D}$ because (2.20) implies

$$w(x, y) = \int_y^{f(x)} \{\bar{\phi}(x, \eta) - \eta\} \, d\eta \quad \text{in } \bar{\Omega}. \tag{2.20a}$$

Furthermore, substitution of $\bar{\phi} = y + p/\rho g$ in (2.20) gives

$$w(x, y) = \int_y^{f(x)} \frac{p(x, \eta)}{\rho g} \, d\eta,$$

and hence, since p and ρg are positive quantities in Ω, $w(x, y) > 0$ in Ω. Differentiation of (2.20a) with respect to x yields

$$w_x(x, y) = f_x(x)\{\bar{\phi}(x, f(x)) - f(x)\} + \int_y^{f(x)} \bar{\phi}_x(x, \eta) \, d\eta$$

$$= \int_y^{f(x)} \bar{\phi}_x(x, \eta) \, d\eta, \tag{2.21}$$

since, from the first of (2.9), $\bar{\phi}(x, f(x)) = f(x)$. Similarly

$$w_y = -\{\bar{\phi}(x, y) - y\}. \tag{2.22}$$

Thus

$$w_x = w_y = 0 \text{ on } y = f(x) \text{ and in } \bar{D} - \bar{\Omega},$$

and the first derivatives of w are continuous throughout D.

A convenient starting point for the derivation of the differential equation satisfied by w is the Green's theorem

$$\iint_D \nabla \bar{\phi} \cdot \nabla w \, dx \, dy = \int_{\delta D} \bar{\phi} \frac{\partial w}{\partial n} \, dS - \iint_D \bar{\phi} \nabla^2 w \, dx \, dy$$

$$= -\iint_D \bar{\phi} \nabla^2 w \, dx \, dy \tag{2.23}$$

by virtue of the boundary conditions for $\partial w/\partial n$ and $\bar{\phi}$. Introduction of a function $\Phi(x, y)$ which is given by

$$\Phi = \int_0^y \bar{\phi}(x, \eta)\, d\eta, \tag{2.24}$$

i.e.

$$\bar{\phi}(x, y) = \Phi_y(x, y), \tag{2.25}$$

leads to a second expression for $\iint \nabla \bar{\phi} \cdot \nabla w\, dx\, dy$ as follows:

$$\iint_D \nabla \bar{\phi} \cdot \nabla w\, dx\, dy = \iint_D \{\bar{\phi}_x w_x + \bar{\phi}_y w_y\}\, dx\, dy$$

$$= \iint_D \{w_x \Phi_{xy} + w_y \Phi_{yy}\}\, dx\, dy = \iint_\Omega \{w_y \Phi_{xx} + w_y \Phi_{yy}\}\, dx\, dy, \tag{2.26}$$

after integration by parts and noting that Φ vanishes together with all its derivatives on FA, AB, and BC and that w vanishes on FD. Now substitution from (2.22), (2.18), and (2.25) into (2.26) gives

$$\iint_D \nabla \bar{\phi} \cdot \nabla w\, dx\, dy = -\iint_\Omega [(\phi - y)\Phi_{xx} + (\phi - y)\Phi_{yy}]\, dx\, dy$$

$$= \iint_\Omega (\phi_x \Phi_x + \phi_y \Phi_y - \Phi_y)\, dx\, dy$$

$$= \iint_\Omega -\bar{\phi}\, dx\, dy = \iint_D -\bar{\phi}\chi_\Omega\, dx\, dy \tag{2.27}$$

by a further use of Green's theorem, inserting the boundary conditions and $\nabla^2 \phi = 0$ in Ω and where $\chi_\Omega = 1$ in Ω and zero outside Ω. Finally, (2.23) and (2.27) together yield

$$\left.\begin{aligned} \nabla^2 w &= 1 \quad \text{in } \Omega, & w &> 0, \\ \nabla^2 w &= 0 \quad \text{in } D - \Omega, & w &= 0. \end{aligned}\right\} \tag{2.28}$$

To complete the formulation of the simple dam problem we need to establish the conditions satisfied by $w(x, t)$ on the boundaries of D. Thus

on AF, $\bar{\phi}(0, y) = y_1$, $w(0, y) = \displaystyle\int_y^{y_1} (y_1 - \eta)\, d\eta = \tfrac{1}{2}(y_1 - y)^2$,

$$\tag{2.29}$$

on BC, $\bar{\phi}(x_1, y) = y_2$, $w(x_1, y) = \displaystyle\int_y^{y_2} (y_2 - \eta)\, d\eta = \tfrac{1}{2}(y_2 - y)^2$,

$$\tag{2.30}$$

on CD, $\tilde{\phi}(x_1, y) = y,$ $w(x_1, y) = 0,$ (2.31)

on DE, $\tilde{\phi}(x_1, y) = y,$ $w(x_1, y) = 0,$ (2.32)

on FE, $\tilde{\phi}(x, y_1) = y,$ $w(x, y_1) = 0.$ (2.33)

Since AB is a streamline

$$w_x(x, 0) = \int_0^{y_1} \tilde{\phi}_x(x, \eta) \, d\eta = -q,$$

where q is the flow rate through the dam and is independent of x. Using (2.29) and (2.30) with $y = 0$ shows that

$$w_x = -\tfrac{1}{2}(y_1^2 - y_2^2)/x_1,$$

and on integration with respect to x, remembering that at $x = 0$, $w(0, 0) = \tfrac{1}{2}y_1^2$, we have finally

on AB, $\tilde{\phi}_y(x, 0) = 0,$ $w(x, 0) = \tfrac{1}{2}y_1^2 - \tfrac{1}{2}(y_1^2 - y_2^2)x/x_1.$ (2.34)

The properties of w and the equations in (2.28) can be expressed by the following differential equation and inequalities in the domain D:

$$\left. \begin{aligned} & w(1 - \nabla^2 w) = 0, \\ & 1 - \nabla^2 w(x, y) \geqslant 0, \qquad w(x, y) \geqslant 0. \end{aligned} \right\} \tag{2.35}$$

A numerical algorithm for finding $w(x, y)$ satisfying (2.35) and the boundary conditions (2.29–33) with (2.34) is given by the relations (8.86) in Chapter 8. We remark only at this stage that once such a solution is found, the domain Ω can be identified as the region in which $w > 0$, i.e. $\Omega = \{(x, y) \mid (x, y) \text{ in } D; \, w(x, y) > 0\}$; also the free boundary $y = f(x)$ is given by

$$f(x) = \sup\{y \mid (x, y) \text{ in } \Omega\}, \qquad 0 < x < x_1,$$

with

$$f(0) = \lim_{x \to 0^+} f(x) \quad \text{and} \quad f(x_1) = \lim_{x \to x_1^-} f(x).$$

2.2.4. Variational inequality formulation

It is well established (Duvaut and Lions, 1976; Mosco, 1973) that if $w(x, y)$ satisfies (2.35) and the conditions (2.29–33) with (2.34) it also satisfies the variational inequality

$$a(w, v - w) \geqslant (v - w) \tag{2.36}$$

for all functions v which, together with their first derivatives, exist and are square summable on D and which agree with w on the boundary of D. The function w and its first derivatives are likewise square summable on

D. The nomenclature in (2.36) is that

$$a(w, v-u) = \iint_D \nabla w \cdot \nabla(v-u) \, dx \, dy$$

$$= \iint_D \{w_x(v_x - w_x) + w_y(v_y - w_y)\} \, dx \, dy \quad (2.37)$$

and

$$(v - w) = -\iint_D (v - w) \, dx \, dy. \quad (2.38)$$

A proof outlined by Bruch (1980) runs as follows. Introduce the function $\beta = v - w$, which is thus zero on the boundary of D; then

$$\iint_D \{\nabla\beta \cdot \nabla w\} \, dx \, dy + \iint_D \beta \, dx \, dy$$

$$= \iint_D \{\nabla\beta \cdot \nabla w\} \, dx \, dy + \iint_D \beta\chi_\Omega \, dx \, dy + \iint_{D-\Omega} v \, dx \, dy \quad (2.39)$$

since $w = 0$ in $\bar{D} - \bar{\Omega}$. Defining $\Phi(x, y) = \int_0^y \beta(x, \eta) \, d\eta$ and hence $\beta(x, y) = \Phi_y(x, y)$ and noting that Φ vanishes together with all its derivatives on FA, AB, and BC while w vanishes on FE, the argument leading to equation (2.26) above gives again

$$\iint_D \{\nabla\beta \cdot \nabla w\} \, dx \, dy = \iint_D (w_y \Phi_{xx} + w_y \Phi_{yy}) \, dx \, dy. \quad (2.40)$$

Substituting from (2.18) and (2.22) in (2.40) and repeating the derivation of (2.27) leads to

$$\iint_D \{\nabla\beta \cdot \nabla w\} \, dx \, dy = -\int_D \beta\chi_\Omega \, dx \, dy. \quad (2.41)$$

The proof of (2.36) follows immediately by using (2.41) in (2.39) and remembering that $v \geqslant 0$ in $\bar{D} - \bar{\Omega}$.

Reverting to the formulation of the problem, because of the symmetry in the inequality (2.36), the required solution $w(x, y)$ is also a solution of the minimization problem

$$J(w) \leqslant J(v), \quad (2.42)$$

where the functions $v(x, y)$ are as in (2.36) and

$$J(v) = \frac{1}{2} \iint_D (v_x^2 + v_y^2) \, dx \, dy + \iint v \, dx \, dy. \tag{2.43}$$

This follows by writing (2.36) in the form

$$-\iint_D (w_x^2 + w_y^2) \, dx \, dy - \iint w \, dx \, dy$$

$$\geq -\iint_D (w_x v_x + w_y v_y) \, dx \, dy - \iint_D v \, dx \, dy \tag{2.44}$$

using (2.37). The general inequality $(a-b)^2 \geq 0$, i.e. $2ab \leq a^2 + b^2$, gives $2w_x v_x \leq w_x^2 + v_x^2$ and also $2w_y v_y \leq w_y^2 + v_y^2$ and hence (2.44) implies (2.42).

Proofs of the existence and uniqueness of a solution of the inequality (2.36) and hence of the minimization problem (2.42) are given by Baiocchi *et al.* (1973*b*, 1976), Lions and Stampacchia (1967), and Glowinski (1976).

Baiocchi (1980*a*) extended the treatment of the simple rectangular dam to include a derivative condition $\partial \phi / \partial x = \lambda(\phi - y_1)$, $\lambda > 0$, on the inlet face AF (Fig. 2.1) instead of the condition (2.8), $\phi = y_1$. Apparently the influence of impurities in the reservoir when carried to the face AF is better expressed by the derivative condition. Baiocchi derives the variational inequality

$$\iint_D \nabla w \cdot \nabla (v - w) \, dx \, dy + \lambda \int_0^{y_1} [w(0, y)\{v(0, y) - w(0, y)\}] \, dy$$

$$\geq \iint_D (w - v) \, dx \, dy - \lambda \int_0^{y_1} [(y_1 y - y^2/2)\{v(0, y) - w(0, y)\}] \, dy,$$

where D is the fixed domain of the whole dam. Baiocchi (1980*a*) finds this problem to be well posed, that existence and uniqueness can be proved, and that numerical solution is straightforward.

Baiocchi's extensive list of references gives a good impression of the wide application to which his transformation and method have been put.

2.2.5. *In terms of stream functions*

An alternative definition of the Baiocchi transformation (2.20) incorporates the stream function ψ introduced in §2.2.2 and extended to the fixed domain D by setting

$$\left.\begin{array}{l} \tilde{\psi}(x, y) = \psi(x, y) \quad \text{in } \bar{\Omega}, \\ \tilde{\psi}(x, y) = 0 \quad \text{in } \bar{D} - \bar{\Omega} \end{array}\right\} \tag{2.45}$$

analogous to the extension of $\phi(x, y)$ in (2.18). The alternative form of (2.20) is

$$w(P) = \int_{\widehat{FP}} [-\tilde{\psi}\, dx + (y - \tilde{\phi})\, dy] \tag{2.46}$$

where, with reference to Fig. 2.1, P is a point in D and $\widehat{FP}$ is a smooth path joining F to P. The integration in (2.46) is independent of the path because $\tilde{\phi}_x - \tilde{\psi}_y = 0$ from the Cauchy–Riemann equation.

The first derivatives of (2.46), namely

$$w_x(P) = -\tilde{\psi}(x, y), \qquad w_y(P) = y - \tilde{\phi},$$

are seen to be continuous through $\bar{D}$ and equal to zero on the free surface and throughout $\bar{D} - \bar{\Omega}$.

The differential inequalities (2.35) together with the conditions (2.29–33) and (2.34) can be derived also from the form (2.46) of the Baiocchi transformation. In addition

$$\psi(x, y) = -w_x(x, y) \quad \text{in } \bar{\Omega} \tag{2.47}$$

could be deduced from the solution for $w(x, y)$.

2.3. Other two-dimensional dams

Free-boundary problems presented by porous flow through other types of dam, more complicated than the simple rectangular dam, will now be formulated. They provide examples of other boundary conditions which in some cases call for modifications in the methods of solution described later in Chapter 8.

2.3.1. Rectangular dam with toe drain

In Fig. 2.2 the portion AB of the base is impervious but BC is the toe drain through which fluid can escape. The free surface does not meet the vertical wall and so there is no separation point to be considered. Instead, the free surface meets the toe drain at a point C which is unknown *a priori*. Accordingly, the solution domain is extended to a point D along the drain, as well as up to $y = y_1$ in the y direction.

The problem can be formulated in terms of the modified potential ϕ and stream function ψ as for the simple dam. Also, the Baiocchi-transformation is still defined by (2.20) or (2.46) where now D denotes

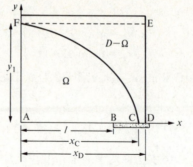

FIG. 2.2. Dam with toe drain

the extended domain $0 < x < x_D$, $0 < y < y_1$ and $\tilde{\phi}(x, y)$ is defined by

$$\left. \begin{aligned} \tilde{\phi}(x, y) &= \phi(x, y) \quad \text{in } \bar{\Omega} \\ &= y \quad \text{in } \bar{D} - \bar{\Omega}, \end{aligned} \right\} \tag{2.48}$$

and $\tilde{\psi}(x, y)$ by

$$\begin{aligned} \tilde{\psi}(x, y) &= \psi(x, y) \quad \text{in } \bar{\Omega} \\ &= 0 \quad \text{in } \bar{D} - \bar{\Omega}. \end{aligned} \tag{2.49}$$

The formulations in terms of ϕ, ψ, and w are given together for convenience. The problem is defined in the two ways as follows

$$\nabla^2 \phi = 0 \quad \text{or} \quad \phi_x - \psi_y = 0, \quad \phi_y + \psi_x = 0 \quad \text{in } \Omega, \tag{2.50a}$$

$$\phi = y_1, \quad w = \tfrac{1}{2}(y - y_1)^2 \quad \text{on AF}, \tag{2.50b}$$

$$\phi = 0 \quad \text{on BC}, \quad w_y = 0 \quad \text{on BC}, \tag{2.50c}$$

$$\phi_y = 0, \quad \psi = q, \quad w = \tfrac{1}{2}y_1^2 - qx \quad \text{on AB}, \tag{2.50d}$$

$$\phi = y, \partial\phi/\partial n = 0 \quad \text{or} \quad \psi = 0 \quad \text{on FC}, \tag{2.50e}$$

$$w = 0 \quad \text{on } \bar{D} - \bar{\Omega} \text{ including FC}. \tag{2.50f}$$

The condition (2.50c), $\phi = 0$ on BC, follows from (2.5) because it is usual to put $p = 0$ on the drain and $y = 0$ there; $w_y = 0$ immediately from (2.22). Furthermore, $w_x(x, 0) = \int_0^{y_1} \phi_x(x, \eta) \, d\eta = -q$, where q is the constant flow rate through all vertical sections of the dam in Ω, and so the condition for w on AB (2.50d) follows by integrating w_x with respect to x and ensuring that $w(0, 0) = \tfrac{1}{2}y_1^2$ to agree with (2.50b) at $y = 0$.

The variational inequality (2.36) is to be satisfied by $w(x, y)$ for this problem. Proofs of this and the uniqueness of the solution are similar to those for the simple dam. Polubarinova-Kochina (1962) and Harr (1962) approximate the flow rate q for this problem by

$$q = \frac{y_1^2}{\ell + (\ell^2 + \frac{1}{3}y_1^2)^{\frac{1}{2}}}, \qquad (2.50\text{g})$$

where $\ell = \text{AB}$ in Fig. 2.2, and this can be inserted as the known value of q into the boundary condition (2.50d). We note that this implies the value $w(\ell, 0) = q^2/6$ at the point B, where the drain starts. Thus (2.50d) gives $w(\ell, 0) = \frac{1}{2}y_1^2 - q\ell$ at B while (2.50g) yields $y_1^2 = \frac{1}{3}q^2 + 2q\ell$, so that $w(\ell, 0)$ becomes simply $q^2/6$.

Bruch and Caffrey (1979) used both the finite-difference and finite-element algorithms (8.90) and (8.91) to solve a typical problem. They compared their results with the analytical solution (Harr 1962) for two mesh sizes (see Fig. 8.17(a)).

Before proceeding, it is convenient to comment on the approximate expression (2.50g) for the flow rate q. It emerges as a special case in a very lengthy treatment of the much more general problem of a trapezoidal dam with a trapezoidal toe drain due to Numerov (1942) and reproduced by Harr (1962, pp. 210–26). Sloss and Bruch (1978) give a more direct derivation of the value $w(\ell, 0) = \frac{1}{6}q^2$ based on Kozeny's conformal transformation (1931). Its use in the present problem is strictly an approximation because Kozeny's solution is for a dam with a parabolic inlet face and a horizontal toe drain. The free surface is also found to be parabolic but the flow pattern near the toe drain is assumed to be insensitive to the geometry of the inlet face.

Kozeny's transformation from the $z = x + iy$ plane to the $w = \phi + i\psi$ plane, suitably adapted for the origin at A in the (x, y) plane (Fig. 2.2) and with $\psi = q$ along AB is $z - \ell = \alpha[\phi + i(\psi - q)]^2$ so that

$$x - \ell = \alpha[\phi^2 - (\psi - q)^2], \qquad y = 2\alpha\phi(\psi - q). \qquad (2.50\text{h})$$

The condition (2.50e), i.e. $\phi = y$, $\psi = 0$ on the free surface gives immediately $\alpha = -1/(2q)$ from the second of (2.50h). Since $\phi = 0$ on BC (2.50c) the first of (2.50h) gives

$$\psi = -[2q(x - \ell)]^{1/2} + q \quad \text{on BC}, \qquad (2.50\text{i})$$

where the negative sign is needed since $\psi = 0$ at C on the free surface. Also, combining (2.50e) with (2.50h) on FC gives $x - \ell = -(y^2 - q^2)/(2q)$, and so

$$x_C = \frac{1}{2}q + \ell. \qquad (2.50\text{j})$$

Finally, the line integral (2.46) taken along the path FCB gives, by using (2.50i), (2.50j) and $\psi = 0$, $\phi = y$ on FC,

$$w_B = w(\ell, 0) = -\int_{x_C}^{\ell} \psi \, dx = \frac{1}{6}q^2.$$

Substitution of $w = \frac{1}{6}q^2$, $x = \ell$ in (2.50d) gives $y_1^2 = \frac{1}{3}q^2 + 2q\ell$ which also follows directly by rewriting (2.50g) and so the flow rate q is confirmed by the present derivation.

2.3.2. Rectangular dam with a sheetpile

The dam shown in Fig. 2.3 has an impervious sheetpile along the part GE' of the inlet face. This problem introduces a significant new feature. The point F where the free boundary meets the inlet face is not known *a priori*. This means that the flow rate q through the dam is not known, as in the two previous examples, but is an additional unknown to be determined as part of the solution. The formulation is as follows where Ω is again the flow region ABCDFGA and D is the extended domain ABEE'A:

$$\nabla^2 \phi = 0 \quad \text{in } \Omega \tag{2.51a}$$

$$\phi_x = 0, \qquad w_x = 0 \quad \text{on GF apart from the points G, F,} \tag{2.51b}$$

$$\phi_y = 0, \qquad w = \tfrac{1}{2}y_2^2 + q(x_1 - x) \quad \text{on AB,} \tag{2.51c}$$

$$\phi = y_1, \qquad w = \tfrac{1}{2}y_2^2 + qx_1 + \tfrac{1}{2}(y_1 - y)^2 - \tfrac{1}{2}y_1^2 \quad \text{on AG,} \tag{2.51d}$$

$$\phi = y_2, \qquad w = \tfrac{1}{2}(y_2 - y)^2 \quad \text{on BC,} \tag{2.51e}$$

$$\phi = y, \qquad w = 0 \quad \text{on CD,} \tag{2.51f}$$

$$\phi = y, \qquad \partial\phi/\partial n = 0 \quad \text{on FD,} \tag{2.51g}$$

$$w = 0 \quad \text{on } \bar{D} - \bar{\Omega} \text{ including FD.} \tag{2.51h}$$

Since AB is a streamline, the condition $w_x(x, 0) = -q = $ the constant discharge rate still holds but q is not known. The condition (2.51c) follows as in §2.3.1 remembering that at B, $x = x_1$, $y = 0$, $w = \frac{1}{2}y_2^2$ from

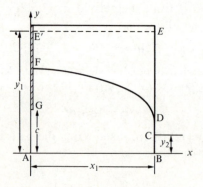

FIG. 2.3. Dam with sheetpile

(2.51e). On AG we have $w_y = -(\phi - y)$ and hence

$$[w]_0^y = -\int_0^y (\phi - \eta) \, d\eta = [\tfrac{1}{2}(y_1 - y)^2]_0^y.$$

But $w(0, 0) = \tfrac{1}{2}y_2^2 + qx_1$ from (2.51c) and so (2.51d) follows.

The variational inequality (2.36) is to be satisfied by $w(x, y)$ in the present problem also. However, Baiocchi *et al.* (1973*b*) obtained the mathematical result that, provided q is chosen such that

$$q \geq \{y_1^2 - y_2^2 - (y_1 - AG)^2\}/2x_1, \tag{2.52}$$

there is a unique solution w_q and there is a unique value of q in the range defined in (2.52) for which w_q and its first derivatives are continuous on the boundary of the domain D. Baiocchi *et al.* (1973*a*) gave a discretized interpretation of these mathematical statements which is convenient for use in a numerical algorithm. A suitable discretized form of the definition of q, where

$$-q = \int_0^{f(x)} \phi_x(x, \eta) \, d\eta, \qquad 0 < x < x_1, \tag{2.53}$$

is

$$-q = \sum_{j=1}^{N_2} \frac{\phi_{i,j} - \phi_{i-1,j}}{\Delta x} \Delta y, \qquad i = 1, 2, \ldots, N_1, \tag{2.54}$$

where $N_1 \Delta x = x_1$, $N_2 \Delta y = y_1$. The corresponding discretized form of (2.22) is

$$\phi_{i,j} = j \, \Delta y + (w_{i,j-1} - w_{i,j})/\Delta y, \qquad i = 0, 1, \ldots, N_1, \qquad j = 1, 2, \ldots, N_2. \tag{2.55}$$

Substitution of (2.55) into (2.54) with $i = 1$ and summing over j leads to

$$-q \, \Delta x = w_{1,0} - w_{1,G} - w_{0,0} + w_{0,G} \tag{2.56}$$

because all intermediate terms cancel leaving only the w values for $x = 0$, Δx at $y = 0$ and at $y = AG$ denoted by $w_{0,G}$, $w_{1,G}$. The summation terminates at the lower edge of the sheetpile because $\phi_x = 0$ on GF (see 2.51b). Using (2.51c) gives $(w_{1,0} - w_{0,0}) = -q \, \Delta x$ and so finally (2.56) yields $w_{0,G} - w_{1,G} = 0$, which is a compatibility condition to ensure that both the Dirichlet condition along AG and the Neumann condition along GF are satisfied. The mathematical statement quoted by Baiocchi *et al.* (1973*a*) can now be interpreted as implying that q must be chosen so that a numerical solution satisfies the condition

$$f_h(q) = w_{1,G}(q) - w_{0,G}(q) = 0 \tag{2.57}$$

where $f_h(q)$ is a discrete analogy to the continuous function of q in the theoretical analysis. For a fixed q within the range specified in (2.52) the Cryer algorithm will converge to the solution of the variational inequality, for example, but (2.57) will not necessarily be satisfied. Baiocchi *et al.* (1973*a*) inserted an outer iteration on q into the Cryer algorithms and used the modified procedure detailed in §8.5.1(iii). They obtained several sets of numerical results (see Tables 8.13–16).

2.3.3. *Rectangular dam with toe drain and sheetpile*

This problem is a combination of the two previous ones but its formulation contains the essential feature of an unknown flow rate q as in §2.3.2. Indeed, Bruch (1980) considers the two problems of the sheetpile with and without a toe drain together. The essential difference between them is that when there is a toe drain, as in Fig. 2.4, (2.51c) is replaced by

$$\phi_y = 0, \qquad w = \tfrac{1}{6}q^2 + q(x_b - x), \qquad (2.58c)$$

and (2.51d) by

$$\phi = y_1, \qquad w = \tfrac{1}{6}q^2 + qx_b + \tfrac{1}{2}(y - y_1)^2 - \tfrac{1}{2}y_1^2. \qquad (2.58d)$$

The derivation of (2.58c) and (2.58d) is as for (2.51c) and (2.51d) except that w is taken to have the value $\tfrac{1}{6}q^2$ at B, since the flow pattern near the drain is similar to that in §2.3.1. Bruch (1980) points out that this assumption could be avoided by using a compatability condition at B between the boundary conditions on AB and BC. The variational inequality for the present problem is still (2.36) and the rest of the discussion in §2.3.2 applies. The same numerical algorithm, incorporating (2.57), was used as described in §8.5.1(iii) by Bruch and Caffrey (1979) who give graphical results (see Fig. 8.17(b)).

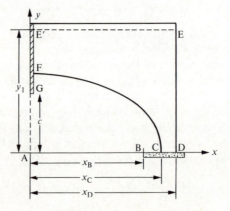

FIG. 2.4. Dam with toe drain and sheetpile

2.3.4. Dam with inlet face slanted

The different geometry of this dam, depicted in Fig. 2.5, gives rise to new features in the formulation. Although $\phi(x, y)$ still has the constant value y_1 along AF, the Dirichlet condition (2.29) for w on AF can no longer be obtained by the integration leading to $w(0, y)$. Instead we have a Neumann condition

$$\phi = y_1, \qquad w_y = y - y_1 \quad \text{on AF}, \qquad (2.59a)$$

in which $w(x, y)$ is redefined to be

$$w(x, y) = -q(x - x_1) + \tfrac{1}{2}y_2^2 - \int_0^y (\tilde{\phi} - \eta)\, d\eta. \qquad (2.59b)$$

Here the flow rate q is again unknown and is contained in the boundary condition on AB which is

$$\tilde{\phi}_y(x, 0) = 0, \qquad w(x, 0) = -q(x - x_1) + \tfrac{1}{2}y_2^2, \qquad 0 \le x < x_1, \qquad (2.59c)$$

Also

$$\phi = y_2, \qquad w = \tfrac{1}{2}(y_2 - y)^2 \quad \text{on BC}. \qquad (2.59d)$$

Other definitions in the extended domain D, taken to be the trapezium AFEB, and conditions on its boundary except AF are as for the simple dam §2.2.3, but the variational inequality is more complicated. For test functions $v(x, y)$ which agree with $w(x, y)$ on the boundary of D except AF and otherwise have the properties of §2.2.4, the variational inequality is

$$a(w_q, v - w_q) \ge (v - w_q), \qquad (2.60)$$

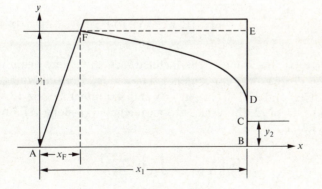

FIG. 2.5. Dam with front face slanted

where now, with the suffix q temporarily dropped for convenience,

$$a(w, v-w) = \iint\limits_{D} \{w_x[v-w]_x + w_y[v-w]_y$$

$$+ (y_1/x_f)w_x[v-w]_y - (y_1/x_f)w_y[v-w]_x\}\, dx\, dy \quad (2.60a)$$

and

$$(v-w) = -\iint\limits_{D}[v-w]\, dx\, dy + \frac{y_1^2 + x_f^2}{x_f y_1} \int_A^F (y-y_1)(v-w)\, dy.$$

$$(2.60b)$$

A proof is outlined below but first we remark that Baiocchi *et al.* (1973*a*) quoted mathematical results for this problem similar to those in §2.3.2 including restrictions on q. There is a unique value of q for $q \geqslant -y_2^2/(2x_1)$ for which (2.60) has a unique solution w_q which, together with its first derivatives, is continuous on the whole boundary of D. The compatability condition at F corresponding to (2.57) and obtained in the same way is

$$f_h(q) = w_{F,N_2-1} - \tfrac{1}{2}(\Delta y)^2, \quad (2.61)$$

which is to be incorporated in the numerical algorithm discussed in §8.5.1. Baiocchi *et al.* (1973*b*) and Bruch (1980) base a proof of the inequality (2.60) on the form of Green's theorem

$$\iint\limits_{D} \{\beta\nabla^2 w - \nabla \cdot [g\beta w_y - g\beta w_x] + \nabla\beta \cdot \nabla w\}\, dx\, dy$$

$$= \int_{\partial D} \{\beta[w_x - gw_y]\, dy - \beta[w_y + gw_x]\, dx\}, \quad (2.62)$$

where $g = y_1/x_F$, the slope of the inclined face and where again the suffix q has been dropped for convenience, i.e. $w = w_q$. For the boundary lines AB, BC, CE, and EF, $\beta = v - w = 0$ and so, remembering (2.59*a*), the right-hand side of (2.62) reduces to an integral I_{AF} along AF only, given by

$$I_{AF} = \frac{y_1^2 + x_F^2}{x_F y_1} \int_A^F \beta(y-y_1)\, dy \quad (2.63)$$

since $dx = (x_F/y_1)\, dy$ along AF. The second integral over the domain D

on the left-hand side of (2.62) can be expanded as follows:

$$-\iint_D \nabla \cdot (g\beta w_y - g\beta w_x)\, dx\, dy = -\iint_D \{\beta[g_x w_y - g_y w_x]$$

$$+ g[\beta_x w_y - \beta_y w_x] + g\beta[w_{xy} - w_{yx}]\}\, dx\, dy$$

$$= -\iint_D \{g[\beta_x w_y - \beta_y w_x]\}\, dx\, dy \quad (2.64)$$

since $g = y_1/x_F$ and the first derivatives of w are continuous. Finally we recall that

$$\iint_D \beta \nabla^2 w\, dx\, dy \leqslant \iint_D (v - w)\, dx\, dy, \qquad (2.65)$$

and combine (2.63), (2.64), and (2.65) to establish (2.60). We note that (2.60) is associated with a non-symmetric bilinear form and so this problem cannot be formulated as a minimization problem. Accordingly, Baiocchi *et al.* (1973*a*) approximated the complementarity formulation as in §8.5.1(iii) and this is convenient for use of (2.61). The matrix in the successive overrelaxation method (SOR) with projection is no longer symmetric but the method is still convergent. Miellou (1971) and Baiocchi *et al.* (1973*a*) present some numerical results and a demonstration of convergence for one particular dam (see Table 8.17).

2.3.5. Dam with inlet face slanted and a toe drain

Sloss and Bruch (1978) obtained numerical results for this problem but Bruch (1980) quoted an alternative form of variational inequality in which, with reference to (2.60),

$$a(w, v - w) = \iint_D \{\nabla w \cdot \nabla(v - w) - g[(v - w)_x w_y - (v - w)_y w_x]$$

$$+ (v - w)[-g_x w_y + g_y w_x]\}\, dx\, dy, \quad (2.66a)$$

$$(v - w) = -\iint_D (v - w)\, dx\, dy + \int_A^F (v - w)(y - y_1)\left(\frac{y_1^2 + x_F^2}{x_F y_1}\right) dy,$$

$$(2.66b)$$

where now $g = y/x$. A uniqueness proof was given by Bruch (1980). Sloss and Bruch (1978) showed results agreeing well with Polubarinova-Kochina's (1962) analytical solution for one dam and with results of Jeppson (1968*c*) and of Shaug and Bruch (1976), who used an inverse formulation, for a second dam (see Table 8.18 for iteration details).

2.3.6. Dam with sloping base

The two problems depicted in Figs. 2.6(a),(b) are outlined by Baiocchi *et al.* (1973*a*) with references to Baiocchi *et al.* (1973*b*) for details. Both problems are essentially the same for numerical treatment as for the dam with the slanting face of §2.3.4. Again the flow rate q has to be determined by a compatability condition. The conditions are $w = 0$ on FE and EC (Figs. 2.6a,b); $w = \frac{1}{2}(y_1 - y)^2$ on AF; $w = \frac{1}{2}(y_2 - y)^2$ on BC; $w_x = -q$ on AB.

For Fig. 2.6(a) the variational inequality is

$$\iint_D [w_x(v-w)_x + w_y(v-w)_y - (a/y_3)\{w_x(v-w)_y - w_y(v-w)_x\}]\,dx\,dy$$

$$\geq \iint_D -(v-w)\,dx\,dy - q\{1-(a/y_3)\}\int_A^B (v-w)\,dx, \quad (2.67)$$

where the domain D is here the trapezium ABEF.

The corresponding inequality for Fig. 2.6(b) is the same as (2.67) except that the line-integral term along AB becomes

$$-q\left(\frac{y_3}{a}+\frac{a}{y_3}\right)\int_A^B (v-w)\,dx. \quad (2.67a)$$

In both (2.67) and (2.67a) w signifies w_q, the solution corresponding to the value of q which satisfies the compatability condition

$$f_h(q) = w_{1,y_1} - \tfrac{1}{2}(y_1 - y_3)^2 + q\,\Delta x, \quad (2.68)$$

where w_{1,y_1} is the value of w at the mesh point $x = \Delta x$, $y = y_1$, where AF

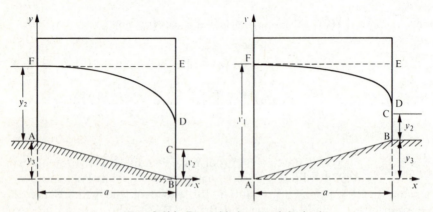

FIG. 2.6(a). Dam with downward sloping base
FIG. 2.6(b). Dam with upward sloping base

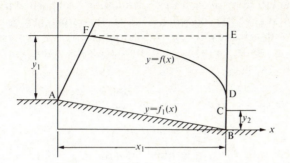

FIG. 2.6(c). Dam with sloping base and inlet face

and BE are lines on a mesh with increments Δx, Δy where $\Delta y/\Delta x = y_3/a$. Baiocchi *et al.* (1973*a*) give results for one example of each dam (see Tables 8.19, 8.20).

The problem of the dam with a sloping base and a sloping inlet face (Fig. 2.6(c)) contains a jump discontinuity in a fixed-boundary condition which precludes a variational formulation. Thus Baiocchi's transformation for the problem depicted in Fig. 2.6(c) leads to the equations

$$\nabla^2 w = \chi_\Omega \quad \text{in } D \tag{2.69a}$$

$$w = 0 \quad \text{in } D - \Omega, \qquad w > 0 \quad \text{in } \Omega, \tag{2.69b}$$

$$w_x = -q \quad \text{on } AB \tag{2.69c}$$

$$w_y = y - y_1 \quad \text{on } AF, \tag{2.69d}$$

$$w = \tfrac{1}{2}(y_2 - y)^2 \quad \text{on } BC, \qquad w = 0 \quad \text{on } CE \text{ and } FE. \tag{2.69e}$$

The jump discontinuity occurs in the direction of the oblique derivative as shown by (2.69c) and (2.69d) on AB, AF excluding the end points. In a manner analogous to the use of an enthalpy function in Stefan problems, e.g. equations (6.6), (2.69a), and (2.69b) can be combined into the non-linear equation

$$\nabla^2 w \in H(w) \quad \text{in } D, \tag{2.69f}$$

where

$$H(w) = 0, \ w < 0, \ 0 \le H(w) \le 1, \ w = 0, \ H(w) = 1, \ w > 0.$$

A discretized form of (2.69f) can be written as $\{f\} - [A]\{w\} \in H(\{w\})$ where $\{f\}$ is the vector containing the known terms from the boundary conditions and $[A]$ is an irreducibly diagonally dominant matrix with components $a_{ij} \le 0$, $i \ne j$ and $a_{ii} > 0$. Thus

$$f_i - \sum_{j=1}^{n} a_{ij} w_j \in H(w_i), \qquad i = 1, 2, \ldots, n,$$

where n is the number of points in the discretization of D.

Comincioli (1975) proposed this formulation, established mathematical properties, and gave the following SOR scheme for a numerical solution. Starting with an arbitrary vector $\{w_{q^{(0)}}^{(0)}\}$, where $q^{(0)}$ is a first estimate of q, a sequence of vectors $\{w_{q^{(k)}}^{(k)}\}$ is generated and subsequently $q^{(r)}$ and $\{w_{q^{(r)}}^{(k)}\}$ from the equations

$$\omega f_i + (1-\omega)a_{ii}w_{q^{(r)},i}^{(k)} - \omega \sum_{j=1}^{i-1} a_{ij}w_{q^{(r)},j}^{(k+1)} - \omega \sum_{j=i+1}^{n} a_{ij}w_{q^{(r)},j}^{(k)}$$
$$- a_{ii}w_{q^{(r)},i}^{(k+1)} \in \omega H(w_{q^{(r)},i}^{(k+1)}), \qquad i=1,2,\ldots,n. \quad (2.69\text{g})$$

An outer iteration for $q^{(r)}$ is built into the algorithm as described in §8.5.1. Thus a non-linear equation is to be solved for $w_{q^{(r)},i}^{(k+1)}$ for each i and r, but the solution can be given explicitly because of the definition of H. The problem has become one of finding $w_{q^{(r)},i}^{(k+1)}$ which minimizes the function

$$F(t) = \tfrac{1}{2}a_{ii}t^2 + \omega t^+ - \left[\omega f_i + (1-\omega)a_{ii}w_{q^{(r)},i}^{(k)} - \omega \sum_{j=1}^{i-1} a_{ij}w_{q^{(r)},j}^{(k+1)}\right.$$
$$\left. - \omega \sum_{j=i+1}^{n} a_{ij}w_{q^{(r)},j}^{(k)}\right]t, \quad (2.69\text{h})$$

where $t^+ = \tfrac{1}{2}(t+|t|)$. This formulation could be used in the other problems of this chapter.

2.3.7. Arbitrary-shaped dam

Baiocchi (1975a) formulated the problem of porous flow through a dam with an arbitrary shape and an impermeable horizontal base separating two reservoirs using the quasi-variational inequalities introduced by Bensoussan and Lions (1973). The cross-section of the dam (Fig. 2.7) has the form $D(x, y)$ with $0 < x < x_1$, $0 < y < Y(x)$, where Y is a smooth

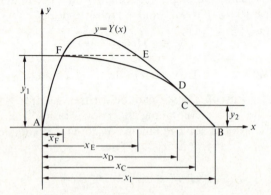

FIG. 2.7. Arbitrary-shaped dam

concave function of x in $0 < x < x_1$ and, with reference to Fig. 2.7, $Y(x_F) = y_1$, $Y(x_C) = y_2$ are the heights of the two reservoirs respectively. The flow region Ω is defined by ABDF, an extended variable $\tilde{\phi}$ is defined as $\tilde{\phi} = \phi$ in $\bar{\Omega}$ and $\tilde{\phi} = y$ in $\bar{D} - \bar{\Omega}$, and the Baiocchi transformation defines $z(x, y)$ by

$$z(x, y) = \int_0^y (\eta - \tilde{\phi}) \, d\eta \quad \text{in } \bar{D},$$

As before $\nabla^2 z = 1$ in Ω and $\nabla^2 z = 0$ in $\bar{D} - \bar{\Omega}$, so that $0 \le \nabla^2 z \le 1$ in D. The conditions for $z(x, y)$ are

$$z(x, y) \ge z(x, Y(x)), \qquad 0 \le x \le x_1, \qquad 0 \le y \le Y(x), \qquad (2.70a)$$

$$z = 0 \quad \text{on AB}, \qquad (2.70b)$$

$$z_y = y - y_2 \quad \text{on BC}, \qquad z_y = y - y_1 \quad \text{on AF}, \qquad (2.70c)$$

$$z_y = 0 \quad \text{on FEC}, \qquad (2.70d)$$

and $z(x, Y(x))$ is concave as a function of x in $x_F \le x \le x_C$. Also

$$q = z_x(x, Y(x)), \qquad x_F \le x \le x_C. \qquad (2.70e)$$

The variational statement obtained by Baiocchi (1975*a*) for the functions $z(x, y)$, using test functions $v(x, y)$ which are square summable on D together with their first derivatives, and with $v(x, y) = 0$ on AB, is written as

$$a(z, v - z) + j(z, v) \ge j(z, z) + (v - z) \qquad (2.71)$$

where

$$a(z, v - z) = \iint_D \{(z_x - Y_x z_y)(v - z)_x + (z_y + Y_x z_x)(v - z)_y$$

$$- Y_{xx} z_y (v - z)\} \, dx \, dy, \qquad (2.71a)$$

$$(v - z) = - \iint_{D_1 \cup D_2} (v - z) \, dx \, dy + \int_0^{x_F} (1 + Y_x^2)(Y(x) - y_1)$$

$$\times \{v(x, Y(x)) - z(x, Y(x))\} \, dx$$

$$+ \int_{x_C}^{x_1} (1 + Y_x^2)\{Y(x) - y_2\}\{v(x, Y(x)) - z(x, Y(x))\} \, dx, \qquad (2.71b)$$

$$j(z, v) = \iint_{D_3} [v(x, y) - z(x, Y(x))]^+ \, dx \, dy, \qquad (2.71c)$$

$$j(z, z) = \iint_{D_3} [z(x, y) - z(x, Y(x))]^+ \, dx \, dy, \qquad (2.71d)$$

where the domains D_1, D_2, D_3 are the parts of D between $0 < x \leqslant x_F$, $x_C \leqslant x < x_1$, $x_F < x < x_C$ respectively and in (2.71c) and (2.71d), $[t]^+ = t$ for $t \geqslant 0$, $[t]^+ = 0$ for $t \leqslant 0$. The term 'quasi-variational inequality' is used in this formulation because the function $z(x, Y(x))$ depends itself on the unknown function z.

Baiocchi (1975a) gave a derivation of (2.71) which was reproduced by Bruch (1980). Baiocchi (1975a, 1975b), Baiocchi, Comincioli, and Maione (1975), and Comincioli (1975) established the existence of maximal and minimal solutions for a range of problems, and Comincioli (1975) and Baiocchi (1975a) gave numerical algorithms to obtain approximations to them. Uniqueness of the solutions of their quasi-variational inequalities is conjectured because the numerical experiments showed the maximal and minimal solutions to be the same (Baiocchi 1975c; Gilardi 1976; Baiocchi, Brezzi, and Comincioli 1976).

The problem of the dam with slanting inlet face §2.3.4 is the special case $x_C = x_1 = x_\ell$ (Fig. 2.7) of (2.71) and was solved in this way by Baiocchi (1975a) and Comincioli (1975) who gave algorithms and numerical results for the maximal and minimal solutions. Bruch (1980) indicated how the quasi formulation can be reduced to the formulation given above in §2.3.4 when $Y_x(x) = y_1/x_F$. Baiocchi (1975a) showed how the variational inequality (2.36) for the simple dam problem is regained from the general quasi theory when D is a rectangle.

The dam with vertical inlet face and horizontal base but slanting exit face also presents a quasi-variational problem. In Baiocchi *et al.* (1976) and Baiocchi and Magenes (1974, 1975) the problem is expressed as a family of variational inequalities depending on two parameters q and the horizontal distance to the intersection of the free surface with the seepage face. Baiocchi *et al.* (1975) solved a quasi-variational inequality for the dam with both faces slanted on a horizontal base.

In §§2.11 and 8.6 generalized formulations are given of problems in arbitrary shaped dams by Brezis *et al.* and independently by Alt who constructed a general purpose algorithm.

Various aspects of steady flow in a fairly general region, leading to a truncation algorithm (§6.3), were studied by Rogers (1980, pp. 333–82).

2.3.8. *Arbitrary-shaped dam with toe drain: split-domain method*

In order to avoid the complexities of quasi-variational inequalities Bruch *et al.* (1982) proposed to use the Baiocchi transformation and formulation only in the part of the solution domain containing the free surface, and in the remaining part of the domain to apply classical methods for solving fixed boundary-value problems. The two regions thus defined have an overlap which permits an alternating iterative numerical scheme to be adopted. The approach was first used by Remar *et al.*

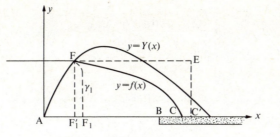

FIG. 2.8. Split-domain method

(1982) (see §8.5.2) and was applied to the arbitrary-shaped dam with toe drain by Bruch *et al.* (1982).

With reference to Fig. 2.8 the flow domain Ω is the region AFCBA defined by $0<x\leqslant x_F$, $0<y\leqslant Y(x)$; $x_F<x<x_C$, $0<y<f(x)$, where $y = f(x)$ is the free surface and $f(x_F) = Y_F$,

$$f(x_C) = 0, \; df(x_F)/dx = -1/Y_x(x_F), \; df/dx = -\infty \quad \text{at} \quad x = x_C.$$

The solution region Ω is extended to the known region D, $0<x\leqslant x_F$, $0<y<Y(x)$; $x_F<x<x_C$, $0<y<y_F$, i.e. AFEC'. The problem could be formulated in terms of the velocity potential ϕ and stream function ψ or in terms of the Baiocchi variable $w(x, y)$ with equations similar to those for the rectangular dam with toe drain in §2.3.1.

Instead, Bruch *et al.* (1982) split the domain D into two regions, D_ϕ and D_W, with a common overlap where D_ϕ is defined to be the region $0<x\leqslant x_F$, $0<y<Y(x)$; $x_F<x<g(y)$, $0<y<y_F$, where $x = g(y)$ is the equation of the chosen curve γ between F and F_1 in Fig. 2.8. This leaves D_W which is the rectangular region FEC'F'$_1$, $x_F<x<x_C$, $0<y<y_F$ and finally $\Omega_{f.s.}$ is the region $x_F<x<x_C$, $0<y<f(x)$. The fixed boundary-value problem in D_ϕ is

$$\nabla^2\phi = 0 \quad \text{in } D_\phi \tag{2.72}$$

$$\phi = y_F \quad \text{on AF}, \qquad \phi = y - w_y \quad \text{on } \gamma \tag{2.72a}$$

$$\phi_y = 0 \quad \text{on AF}_1, \qquad 0<x<x_{F_1}. \tag{2.72b}$$

In D_W the problem in terms of the Baiocchi variable $w(x, y)$ given by (2.20) or (2.46) is

$$\nabla^2 w = 1 \quad \text{in } \Omega_{f.s.}; \qquad \nabla^2 w = 0 \quad \text{in } D - \bar{\Omega}_{f.s.} \tag{2.73a}$$

$$w = \tfrac{1}{6}q^2 + q(x_B - x) \quad \text{on } F'_1 B, \tag{2.73b}$$

$$w_y = 0 \quad \text{on BC}, \qquad x_B<x<x_C, \tag{2.73c}$$

$$w = \int_y^{y_F} (\phi - \eta) \, d\eta \quad \text{on FF}_1', \tag{2.73d}$$

$$w = 0 \quad \text{in } \bar{D}_W - \bar{\Omega}_{\text{f.s.}} \tag{2.73e}$$

$$w \geqslant 0 \quad \text{in } D_W, \qquad w > 0 \quad \text{in } \Omega_{\text{f.s.}} \tag{2.73f}$$

Thus in D_W we have the differential inequalities

$$w(x, y) \geqslant 0, \qquad 1 - \nabla^2 w(x, y) \geqslant 0, \qquad \text{and} \qquad w(1 - \nabla^2 w) = 0. \tag{2.73g}$$

The associated variational inequality is

$$\iint_{D_W} \nabla(v - w) \cdot \nabla w \, dx \, dy \geqslant - \iint_{D_W} (v - w) \, dx \, dy, \tag{2.74}$$

where v agrees with w on $F_1'B$ and FF_1', $v = 0$ on FE and EC′, and $v \geqslant$ a.e. on D_W. Bruch and Sloss (1981) used (2.74) to show the uniqueness and existence of the solution $w(x, y)$.

This is another problem in which the flow rate q is unknown and the necessary compatability condition similar to that used by Sloss and Bruch (1978) is

$$f_h(q) = w_q(x_F, y_F - \Delta y) - \tfrac{1}{2}(\Delta y)^2. \tag{2.75}$$

The essential feature of the split domain is that information is interchanged between the two domains along the line FF_1' and the curve γ. The solution of (2.73) or (2.74) in D_W provides values of w to be used in the derivative boundary condition (2.72a) on γ in $\bar{D}_\phi$; and correspondingly the solution of the equations (2.72) yields values of ϕ to be used in the boundary condition (2.73d) in D_W.

Bruch, Sloss, and Remar (1982) applied their split-field approach to the particular example in which $Y_F = 30$ ft, $x_F = 30$ ft, $x_B = 60$ ft. They took $q^{(0)} = 15 \, \text{ft}^3/\text{sec}$ per foot depth normal to the plane of flow and $q^{(1)} = 16 \, \text{ft}^3/\text{sec}$ per foot depth. An alternating method was used in which an SOR scheme was applied to a triangular finite-element mesh with linear shape functions in the D_ϕ region coupled with a finite-difference SOR scheme with projection in the D_W region. The relaxation parameter was 1.85. For each value of q, discretized solutions of (2.72–72b) for ϕ and (2.73a–f) for w are subjected to the compatability condition (2.75), using the secant method (see equation (8.92)). The alternating method between solutions in D_W and D_ϕ was similar to that described by Remar *et al.* (1982). Bruch *et al.* (1982) gave computational and iteration information for two grid sizes $\Delta x = \Delta y = 2.5$ ft and $\Delta x = \Delta y = 1.0$ ft. In a comparison with earlier results obtained by solving for w throughout the whole domain (Sloss and Bruch 1978), the calculated position of the free boundary was found to be the same for the two methods with $\Delta x = \Delta y = 2.5$ ft. With $\Delta x = \Delta y = 1$ ft the point of intersection C of the free surface with the drain was the only point of difference and the split-field value was more accurate and its computing time was shorter.

2.4. Three-dimensional dams

The formulation of problems in three-dimensional dams introduces no essentially new features that do not occur in the two-dimensional problems. The differences are matters of complexity rather than of principle. When the Baiocchi transformation and method are to be used the *a priori* unknown solution region is extended across the unknown free surface into a known domain which is three dimensional. The transformation becomes

$$w(x, y, z) = \int_z^{g(x,y)} [\phi(x, y, \eta) - \eta] \, d\eta, \tag{2.76}$$

where $z = g(x, y)$ is the free surface and ϕ is the modified velocity potential (2.5). Stampacchia (1974) formulated a variational inequality and gave existence and uniqueness proofs. Gilardi (1977) expressed a three-dimensional flow through porous media as a quasi-variational inequality and established an existence theorem. Bruch (1980) mentioned other theoretical papers by Caffarelli (1976) and by Carbone and Valli (1976, 1977), who considered a three-dimensional non-homogeneous dam with variable cross-section.

Stampacchia's (1974) formulation, reproduced by Caffrey and Bruch (1979) and outlined below for the dam shown in Fig. 2.9, is seen to be closely similar to the two-dimensional rectangular dam problem formulated in §§2.2.1 and 2.2.3. The width of the dam between the impervious sides is denoted by a, $x = \Phi_1(y)$ is the inlet face, and $x = \Phi_2(y)$ the outlet

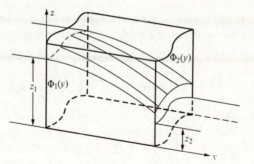

FIG. 2.9. Three-dimensional flow

face. The wetted parts of the inlet and outlet faces of the dam are denoted by F_1, F_2, and F_2^+ respectively, where

$$F_1(x, y, z) \text{ is defined by } x = \Phi_1(y), \quad 0 \leq y \leq a, \qquad 0 \leq z \leq z_1,$$
(2.77a)

$$F_2(x, y, z) \text{ is defined by } x = \Phi_2(y), \qquad 0 \leq y \leq a, \qquad 0 \leq z \leq z_2,$$
(2.77b)

and

$$F_2^+ \text{ by } x = \Phi_2(y), \qquad 0 \leq y \leq a, \qquad z_2 \leq z \leq g(\Phi_2(y), y). \quad (2.77c)$$

The wetted sides of the dam are denoted by S_1 and S_2 where

$$S_1(x, y, z) \text{ is defined by } \Phi_1(0) \leq x \leq \Phi_2(0), \qquad y = 0, \qquad 0 \leq z \leq g(x, 0),$$
(2.78a)

and

$$S_2(x, y, z) \text{ by } \Phi_1(a) \leq x \leq \Phi_2(a), \qquad y = a, \qquad 0 \leq z \leq g(x, a).$$
(2.78b)

The flow region Ω is defined by $\Phi_1(y) < x < \Phi_2(y)$, $0 < y < a$, $0 < z < g(x, y)$, and $g(\Phi_1(y), y) = z_1$, $g(\Phi_2(y), y) = z_2$ as in Fig. 2.9. The Baiocchi variable $w(x, y, z)$ is defined by (2.76) in Ω and is zero in $D - \Omega$, where D is the extended region of the dam up to the height of the inlet reservoir and is the domain $\Phi_1(y) < x < \Phi_2(y)$, $0 < y < a$, $0 < z < z_1$.

The problem is defined by the equations

$$\left. \begin{aligned}
&\nabla^2 \phi = 0 \quad \text{in } \Omega \\
&\phi = z_1 \quad \text{on } F_1, \qquad \phi = z_2 \text{ on } F_2, \qquad \phi = z \text{ on } F_2^+, \\
&\phi_z = 0 \quad \text{on the base B}, \qquad \Phi_1(y) < x < \Phi_2(y), \qquad 0 < y < a, \qquad z = 0, \\
&\phi_y = 0 \quad \text{on } S_1, S_2, \\
&\phi = z, \qquad \partial \phi / \partial n = 0 \quad \text{on the surface } z = g(x, y).
\end{aligned} \right\}$$
(2.79)

The formulation in terms of $w(x, y, z)$ follows closely the reasoning of §2.2.3 for the two-dimensional problem. Slightly more detail is given by Caffrey and Bruch (1979). Thus starting from (2.76) we derive

$$w_x = \int_z^{g(x,y)} \frac{\partial \phi}{\partial x}(x, y, \eta) \, d\eta, \qquad w_y = \int_z^{g(x,y)} \frac{\partial \phi}{\partial y}(x, y, \eta) \, d\eta,$$

$$w_z = z - \phi(x, y, z). \tag{2.80}$$

It follows immediately that in $D - \bar{\Omega}$, $w_x = w_y = w_z = 0$, and since $w = 0$ by definition in $D - \bar{\Omega}$, so w and its first derivatives are continuous across the free surface.

The second derivatives of w are

$$
\left.
\begin{aligned}
w_{xx} &= \int_z^{g(x,y)} \frac{\partial^2}{\partial x^2} [\phi(x, y, \eta)] \, d\eta + \frac{\partial}{\partial x} [\phi(x, y, g(x, y))] \frac{\partial g}{\partial x}, \\[6pt]
w_{zz} &= -\frac{\partial}{\partial z} [\phi(x, y, z)] + 1, \\[6pt]
w_{yy} &= \int_z^{g(x,y)} \left\{ -\frac{\partial^2}{\partial \eta^2} [\phi(x, y, \eta)] - \frac{\partial^2}{\partial x^2} [\phi(x, y, \eta)] \right\} d\eta \\[6pt]
&\quad + \frac{\partial}{\partial y} [\phi(x, y, g(x, y))] \frac{\partial g}{\partial y},
\end{aligned}
\right\}
\tag{2.81}
$$

where the equation $\nabla^2 \phi = 0$ has been used to obtain w_{yy}. Furthermore,

$$
\int_z^{g(x,y)} \frac{\partial^2}{\partial \eta^2} [\phi(x, y, \eta)] \, d\eta = \frac{\partial}{\partial z} [\phi(x, y, g)] - \frac{\partial}{\partial z} [\phi(x, y, z)],
$$

so that finally, by expanding the condition $\partial \phi / \partial n = 0$ on $z = g(x, y)$ in terms of its components $\partial \phi / \partial x$, $\partial g / \partial x$, etc., we have

$$
\nabla^2 w = 1 \quad \text{in } \Omega, \qquad \nabla^2 w = 0 \quad \text{in } D - \bar{\Omega}. \tag{2.82}
$$

The boundary conditions for w on the sides of Ω follow readily by combining appropriate items in (2.79) and (2.80) and are listed in (2.85) below. The condition for w on the base B of the dam, defined in (2.79), is less obvious. We know that in Ω, ϕ satisfies the weak form of $\nabla^2 \phi = 0$, so that

$$
\iiint \left\{ \frac{\partial \phi}{\partial x} \frac{\partial v}{\partial x} + \frac{\partial \phi}{\partial y} \frac{\partial v}{\partial y} + \frac{\partial \phi}{\partial z} \frac{\partial v}{\partial z} \right\} dx \, dy \, dz = 0 \tag{2.83}
$$

for all functions $v(x, y, z)$ with continuous derivatives in Ω and which are zero near the faces F_1, F_2, and F_2^+. Confining attention to the base B of Ω by choosing $v(x, y, z) = v_0(x, y)$, where v_0 has continuous derivatives in $\bar{B}$ and vanishes on the curves $x = \Phi_1(y)$, $x = \Phi_2(y)$, $0 \leq y \leq a$, (2.83) reduces to

$$
\iint_B \left\{ \frac{\partial v_0}{\partial x} \int_0^{g(x,y)} \frac{\partial \phi}{\partial x} \, d\eta + \frac{\partial v_0}{\partial y} \int_0^{g(x,y)} \frac{\partial \phi}{\partial y} \, d\eta \right\} dx \, dy = 0,
$$

which after using (2.80) becomes

$$
\iint_B \left\{ \frac{\partial v_0}{\partial x} \frac{\partial w_0}{\partial x} + \frac{\partial v_0}{\partial y} \frac{\partial w_0}{\partial y} \right\} dx \, dy = 0, \quad \text{i.e.} \ \ \frac{\partial^2 w_0}{\partial x^2} + \frac{\partial^2 w_0}{\partial y^2} = 0, \tag{2.84}
$$

where $w_0(x, y) = w(x, y, 0)$ in B.

The complete formulation in terms of w derived by Stampacchia (1974) is therefore

$$\nabla^2 w = 1 \quad \text{in } \Omega, \qquad \nabla^2 w = 0 \quad \text{in } D - \bar{\Omega}, \tag{2.85a}$$

$$w \geqslant 0 \quad \text{in } D, \qquad w = 0 \quad \text{in } D - \Omega, \tag{2.85b}$$

$$w = \tfrac{1}{2}(z_1 - z)^2 \quad \text{on } F_1, \qquad w = \tfrac{1}{2}(z_2 - z)^2 \quad \text{on } F_2, \tag{2.85c}$$

$$w = 0 \quad \text{on } F_2^+, \qquad w_y = 0 \quad \text{on } S_1, S_2, \tag{2.85d}$$

$$w_{xx} + w_{yy} = 0 \quad \text{on } B. \tag{2.85e}$$

The boundary conditions for the solution of (2.85e) are $w_y = 0$ on $(x, 0, 0)$ and on $(x, a, 0)$ together with $w(\Phi_1(y), y, 0) = \tfrac{1}{2}z_1^2$ and $w(\Phi_2(y), y, 0) = \tfrac{1}{2}z_2^2$. Conditions (2.85a), (2.85b) can be expressed as the inequalities

$$w \geqslant 0, \qquad 1 - \nabla^2 w \geqslant 0, \qquad w(1 - \nabla^2 w) = 0 \quad \text{in } D. \tag{2.85f}$$

Stampacchia (1974) established the equivalent variational inequality for w which takes the usual form

$$\iiint_D \nabla w \cdot \nabla(v - w) \, dx \, dy \, dz \geqslant - \iiint_D (v - w) \, dx \, dy \, dz \tag{2.86}$$

for all functions $v(x, y, z)$ satisfying the usual conditions for such test functions and where v vanishes near the boundaries of D apart from S_1 and S_2. The proof follows closely that of (2.36) in §2.2.4. Stampacchia established some of the mathematical properties of w.

In Chapter 8, §8.5.1(ii) contains two numerical algorithms and a graphical solution for two different three-dimensional dam problems (see Figs. 8.15, 8.16(a) and (b)). Caffrey and Bruch (1980) treat the same dam with a toe drain.

2.5. Coastal aquifer

The extraction of fresh water by means of a well from a layer of porous rock or soil jutting out into the sea is of practical importance. The situation in such an aquifer, shown in Fig. 2.10(a), is that the upper part of the porous medium contains fresh water and sea water has penetrated the lower part. The region from which fresh water can be extracted without contamination by salt water is thus bounded below by the upper surface of the entrapped sea water. The simplest problem, therefore, is to determine the extent of the sea water in the absence of a well.

The mathematical treatment of this problem presented by Baiocchi *et al.* (1973*a*) is closely similar to that for the simple rectangular dam in §2.2. The known quantities are $y_1 = AF$, the height above sea level of the inlet fresh water, and the densities ρ_f and ρ_s of the fresh and salt water,

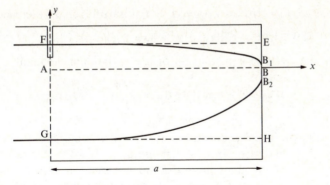

F IG . 2.10(a). Coastal aquifer

which can be different. Baiocchi *et al.* assumed that the distance to the sea is great enough for the streamlines to the left of GF to be considered horizontal, and so the velocity potential is constant on GF. The free surfaces FB_1 and GB_2 are unknown *a priori*, including the points B_1, B_2, and G. Taking Ω to denote the flow region to be studied defined by $0 < x < a$, $f_2(x) < y < f_1(x)$, where $y = f_1(x)$, $y = f_2(x)$ are the free surfaces FB_1, GB_2 respectively, we note first that f_1, f_2 are continuous functions in $0 \le x \le a$, that $f_1(0) = y_1$, f_1 is strictly decreasing, and $f_1(a) \ge 0$, whereas f_2 is strictly increasing, $f_2(a) \le 0$, and $f_2(0) = y_1/\rho$, where $\rho = 1 - \rho_s/\rho_f$. The extended domain $\bar{D}$ is the rectangle FEHG. The extension $\tilde{\phi}$ of the velocity potential $\phi(x, y)$ into $\bar{D}$ is defined by

$$\tilde{\phi} = y, \qquad f_1(x) \le y \le y_1, \qquad 0 \le x \le a, \qquad (2.87)$$

$$\tilde{\phi} = \rho y, \qquad (1/\rho)y_1 \le y \le f_2(x), \qquad 0 \le x \le a, \qquad (2.88)$$

and the Baiocchi transformation is $w(x, y) = \int_y^{y_1} (\phi - \eta) \, d\eta$.

The problem can be formulated in terms of $\phi(x, y)$ by the equations

$$\nabla^2 \phi = 0 \quad \text{in } \Omega, \qquad (2.89a)$$

$$\phi = f_1(x) \quad \text{on } FB_1, \qquad \phi = \rho f_2(x) \quad \text{on } GB_2, \qquad (2.89b)$$

$$\phi = y \quad \text{on } BB_1, \qquad \phi = \rho y \quad \text{on } B_2B, \qquad \phi = y_1 \quad \text{on } GF, \qquad (2.89c)$$

$$\partial \phi / \partial n = 0 \quad \text{on } FB_1 \text{ and } GB_2, \qquad (2.89d)$$

and the conditions for $w(x, y)$ are

$$w = \tfrac{1}{2}(y_1 - y)^2 \quad \text{on GF}, \qquad (2.90a)$$

$$w = \tfrac{1}{2}(1 - \rho)y_2^2 \quad \text{on HB}, \qquad (2.90b)$$

$$w = \tfrac{1}{2}(1 - \rho)y_1^2/\rho^2 + \tfrac{1}{2}(x - a)(1 - \rho)y_1^2/(\rho a) \quad \text{on GH}, \qquad (2.90c)$$

$$w = 0 \quad \text{elsewhere on the boundary of } D.$$

Baiocchi *et al.* (1973*b*) also proved that w satisfies in $\bar{D}$ the inequality

$$0 \leqslant w(x, y) \leqslant \tfrac{1}{2}(1-\rho)y^2 + \tfrac{1}{2}(x-a)(1-\rho)y_1^2/(\rho a). \qquad (2.91)$$

It follows that $w(x, y)$ is a solution of the variational inequality

$$\iint_D [\nabla w \cdot \nabla(v-w)]\,\mathrm{d}x\,\mathrm{d}y \geqslant -\iint_D (v-w)\,\mathrm{d}x\,\mathrm{d}y \qquad (2.92)$$

for all v with the usual properties and for which v agrees with w on the boundary of D and also lies in the range in (2.91). The converse is true and there is an equivalent minimization formulation closely similar to (2.42), (2.43) for the simple dam. The only difference between the two problems is the convex space in which $w(x, y)$ is to be found. The discharge q of the aquifer across any vertical section is given by

$$-q = \int_{f_2(x)}^{f_1(x)} \phi_x(x, \eta)\,\mathrm{d}\eta = w_x \quad \text{along GH}$$

$$= \tfrac{1}{2}(1-\rho)y_1^2/(\rho a) \qquad (2.93)$$

from (2.90*c*), since $\phi_x = 0$ outside $f_2(x) \leqslant y \leqslant f_1(x)$. Baiocchi *et al.* (1973*a*) used a numerical algorithm similar to the one for the simple dam and they gave tabulated values of $f_2(x)$, the lower free surface in Fig. 2.10(a), for an aquifer corresponding to a practical situation (Table 8.8).

Comincioli (1980) discussed the practically more important problem in which the fresh water is being extracted by a well, as in Fig. 2.10(b). He posed two problems: (1) given the height H_1 of the bottom of the well above sea level, find the fresh-water flow region Ω, the discharge q of the well, and the pressure distribution $p(x, y)$; (2) given the discharge q, find

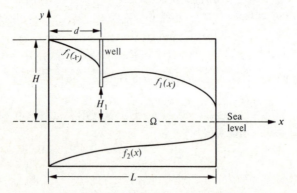

FIG. 2.10(b). Coastal aquifer and well

H_1, Ω, p. He finds that Problem (2) can be formulated as a variational inequality and that the numerical solution presents no difficulties. Problem (1) on the other hand leads to a quasi-variational inequality and the iterative method of solution has to incorporate a compatability condition, in order to find q, closely similar to condition (2.57) in §2.3.2, and which ensures continuity of pressure p.

The discharge of the well q is given by $q = q(x < d) - q(x > d)$, where $q(x)$ takes constant values in the intervals $0 \le x \le d$ and $d \le x \le L$ (Fig. 2.10b) and is defined as in (2.93) by the integral

$$-q(x) = \int_{f_2(x)}^{f_1(x)} (p + \rho_f y)_x \, \mathrm{d}y = \int_{f_2(x)}^{f_1(x)} \phi_x \, \mathrm{d}y.$$

More details of the mathematical and numerical aspects of Problems (1) and (2) are given by Comincioli (1980) and by Baiocchi, Comincioli, and Maione (1977), who also present some numerical results. The approach is essentially that used by Comincioli (1975) and described above to derive (2.69h).

Ground water pollution by the hydrodynamic dispersion of solutes in aquifers is studied by Bear (1980).

2.6. Canals and ditches

Another set of free boundary problems in porous flow relates to seepage from canals or ditches through permeable soil into a drain at a finite depth or into a water table. Particular cases of seepage from trapezoidal and rectangular channels into drains were studied by Bruch *et al.* (1978) who presented numerical algorithms and results based on their variational-inequality formulation. Bruch and Sloss (1978) gave similar information for a single triangular channel and Bruch (1979*a*) considered an array of parallel identical triangular channels. Bruch (1979*b*) obtained similar results for a single channel over a shallow water table.

The treatment of all these examples can be illustrated by considering a channel with an arbitrary-shaped bottom, part of which, near the axis of symmetry, is horizontal as in Fig. 2.11. Only half the system is shown because of symmetry about the y-axis. Torelli (1974) gave theoretical results and Bruch (1980) established a variational inequality for the Baiocchi variable, $w(x, t)$.

The formulation is closely analogous to that of the dam problems considered in previous sections. The flow region Ω is ABCF specified by $0 < x \le x_F$, $0 < y < Y(x)$, together with $x_F < x < x_C$, $0 < y < f(x)$, where $y = Y(x)$ is the equation of the channel bottom and $y = f(x)$ is the free

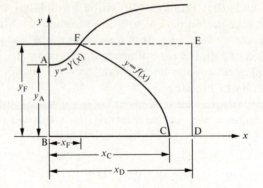

FIG. 2.11. Seepage from open channel

surface. The conditions for ϕ and ψ are

$$\phi_x = \psi_y, \qquad \phi_y = -\psi_x \quad \text{in } \Omega \tag{2.94a}$$

$$\phi = 0 \quad \text{on BC,} \tag{2.94b}$$

$$\phi = y_F = Y(x_F) \quad \text{on AF,} \tag{2.94c}$$

$$\phi = y, \quad \psi = 0 \quad \text{on FC,} \tag{2.94d}$$

$$\psi = \tfrac{1}{2}q \quad \text{on AB.} \tag{2.94e}$$

The reason for the condition (2.94b) on the drain BC is the same as for (2.50c) on a toe drain.

Bruch (1980) suggested the region ABDE is a convenient one to take as the extended domain D for numerical work and extensions of the variables ϕ and ψ are defined as usual by $\bar{\phi} = \phi$ in $\bar{\Omega}$, $\bar{\phi} = y$ in $\bar{D} - \bar{\Omega}$; $\bar{\psi} = \psi$ in $\bar{\Omega}$, $\bar{\psi} = 0$ in $\bar{D} - \bar{\Omega}$. The formulation in terms of $w(x, t)$ defined as usual is

$$\nabla^2 w = 1 \quad \text{in } \Omega, \qquad \nabla^2 w = 0 \quad \text{in } D - \Omega, \tag{2.95a}$$

$$w > 0 \quad \text{in } \Omega, \qquad w = 0 \quad \text{in } \bar{D} - \bar{\Omega}, \tag{2.95b}$$

$$w_x = -\bar{\psi}, \qquad w_y = y - \bar{\phi} \quad \text{in } D, \tag{2.95c}$$

$$w_y = 0 \quad \text{on BC,} \qquad w_y = y - y_F \quad \text{on AF,} \qquad y = Y(x), \tag{2.95d}$$

$$w_x = -\tfrac{1}{2}q \quad \text{on AB,} \tag{2.95e}$$

since BC is only half the complete drain. Because q is unknown *a priori*, a compatability condition is needed. Bruch and Sloss (1978) derived the condition

$$f_h = w_{N,M-1} - \tfrac{1}{2}(\Delta y)^2 = 0 \tag{2.95f}$$

for a triangular channel, which applies also to the present problem;

$w_{N,M-1}$ denotes $w(x, y)$ at the point $N \Delta x$, $(M-1) \Delta y$ on a mesh for which $x_F = N \Delta x$, $y_F = M \Delta y$. They discretized the expression for the discharge

$$\tfrac{1}{2}q = \int_{0,y_F}^{x_F,y_F} \phi_y \, dx - \int_{x_F,y_A}^{x_F,y_F} \phi_x \, dy,$$

and, following the derivation of (2.56) in §2.3.2, incorporated the discretized form of (2.95e), $(w_{0,j} - w_{-1,j})/\Delta x = -\tfrac{1}{2}q$, where $j \Delta y = y_A$, to produce (2.95f).

With $v(x, y)$ having the usual general properties and with $v \geq 0$ in D, $v = 0$ on FE and ED, and defining for convenience $\beta(x, y) = (y/Y) \, dY/dx$, $\beta = 0$ on AB, BD, where the second derivatives are continuous in $\bar{D}$, the variational inequality formulation is

$$\iint_D \nabla w \cdot \nabla(v - w) \, dx \, dy \geq - \iint_D (v - w) \, dx \, dy$$

$$- \int_A^F (1 + Y_x^2)(y - y_F)(v - w) \, dx - \tfrac{1}{2}q \int_A^B (v - w) \, dy. \quad (2.95g)$$

Computational details and some numerical results for channels of particular rectilinear shapes are to be found in the references given at the beginning of this section. These papers contain also many references to other relevant papers. The SOR algorithm with projection given in equations (8.90), (8.91) in Chapter 8 is coupled with use of the secant method (8.97) to handle compatability conditions of type (2.95f).

Where the drain BC is replaced by a water table (Bruch 1979b), the base becomes impermeable and Fig. 2.12 indicates that the free surface FE must become asymptotically parallel to the water table. Bruch made the assumption that at $x = L$ the free surface is not quite horizontal and that, in order to carry away the seepage fluid and make a steady-state flow feasible, the base has a downward slope beyond $x = L$. Furthermore

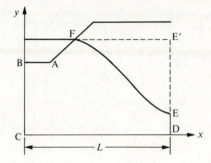

FIG. 2.12. Channel over shallow water table

DE was taken to be an equipotential line. Again, the discharge q is not known and a compatability condition is needed. Bruch (1979b) gives details of a Baiocchi-type formulation with a numerical scheme and compares results with other published data. Bruch and Sloss (1981) applied their split-field method (§2.3.8) to seepage from open channels. There is the added advantage that a compatability condition is not needed because the condition (2.95e) is replaced by $\phi_x = 0$ on AB and q can be found from the solution. Seepage from partially lined canals is studied by Bruch and Mirnateghi (1982).

2.7. Axisymmetric flow

Free boundary problems in which the porous flow is radial are provided by seepage into wells and out of ponds. Wells have been extensively studied and standard general references are Polubarinova-Kochina (1962), Harr (1962), Hantush (1964). Cryer (1976a), and Cryer and Fetter (1977) give an extensive list of references to analytical and numerical solutions. Jeppson (1968a,c), Neuman and Witherspoon (1970), and Remar *et al.* (1982) discuss axisymmetric seepage from a pond by numerical methods.

When an axisymmetric well is sunk in a porous layer or aquifer, the water in the aquifer flows towards the well from which it is removed by pumping. After a time the flow becomes steady and the water level in the well is maintained at a constant height h_W. The assumption of saturated-unsaturated flow is made here as in the dam problem. A free boundary, the position of which is to be determined, separates the upper, dry part of the aquifer from the wet, saturated lower part. Figure 2.13 shows half the symmetric situation and illustrates the nomenclature.

In order to preserve a steady state, the water pumped from the cell must be balanced by a flow into a surrounding region of radius R in Fig. 2.13. It is usual to assume that outside this region the water is effectively

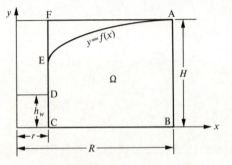

FIG. 2.13. Fully penetrating well

stationary, and that the velocity potential for $x \geq R$ is $\phi = H = $ constant. The well is seen to be the axially symmetric equivalent of a rectangular dam, and there is a seepage face DE at the well. Thus, at the well face, $x = r$, it is assumed that $\phi = h_w$ if $y \leq h_w$ and $\phi = y$ if $y \geq h_w$. In practice, wells are often lined by a porous material and the resulting reduction in flow can be allowed for empirically by putting $\phi = y + \phi_\ell$ on the seepage face, where ϕ_ℓ is a positive constant. The radial form of Laplace's equation has to be satisfied by $\phi(x, y)$, the velocity potential, but otherwise the mathematical formulation is essentially the same as for the simple dam in §2.2 and equations (2.7–12). The formulation of the well problem as posed by Cryer and Fetter (1977), allowing for a variable soil permeability, is

$$\operatorname{div}(k \operatorname{grad} \phi) = \frac{\partial}{\partial x}\left(k \frac{\partial \phi}{\partial x}\right) + \frac{\partial}{\partial y}\left(k \frac{\partial \phi}{\partial y}\right) = 0 \quad \text{in } \Omega, \qquad (2.96a)$$

$$\phi = H \quad \text{on AB}, \qquad (2.96b)$$

$$\phi = h_w \quad \text{on CD}, \qquad \phi = y \quad \text{on DE}, \qquad (2.96c)$$

$$\partial \phi / \partial y = 0 \quad \text{on BC}, \qquad (2.96d)$$

$$\phi = y, \qquad \partial \phi / \partial n = 0 \quad \text{on EA}. \qquad (2.96e)$$

The condition (2.96d) implies that the well extends to the bottom of the aquifer and is called a 'fully penetrating well'. A partially penetrating well is mentioned later. As usual, Ω is the unknown flow region, $0 < y < f(x)$, $r < x < R$. The form of eqn (2.96a) implies that $k = xK$, where K is the soil permeability. Cryer and Fetter (1977) postulate that k is of the form

$$k(x, y) = \exp\{\gamma(x) + g(y)\} \qquad (2.97)$$

where $\gamma(x)$ and $g(\gamma)$ are continuously differentiable and $g'(y) \geq 0$. For the particular case of constant permeability, $K = 1$ say, then $k(x, y) = x = \exp(\ln x)$ and

$$\gamma(x) = \ln x, \qquad g(y) = 0. \qquad (2.98)$$

A weak solution of the well problem is the function $\phi(x, y)$ which satisfies

$$\iint_\Omega k \nabla \phi \cdot \nabla v \, \mathrm{d}x \, \mathrm{d}y = 0 \qquad (2.99)$$

for all functions $v(x, y)$ which satisfy the usual general conditions and which agree with ϕ on AB, CD, DE. The solution ϕ of (2.99) satisfies (2.96b), (2.96c), and $\phi = y$ on EA.

Cryer and Fetter (1977) following Benci (1973, 1974) express the weak problem as a variational inequality. The appropriate Baiocchi variable defined on the rectangular domain D given by $r < x < R$, $0 < y < H$ is

$$w(x, y) = \int_y^{f(x)} \exp(g(\eta))[\phi(x, t) - \eta] \, d\eta \quad \text{in } \Omega,$$

$$= 0 \quad \text{in } D - \Omega. \tag{2.100}$$

By arguments similar to those used in §§2.2.3 and 2.2.4 Cryer and Fetter (1977) give the detailed derivation of a differential equation for $w(x, y)$ and then of the corresponding variational inequality. The formulation of the well problem finally takes the form

$$L_W(x, y) = (\exp[\gamma(x) - g(y)]w_x)_x + (\exp[\gamma(x) - g(y)]w_y)_y$$

$$= \exp[\gamma(x)] \quad \text{in } \Omega, \tag{2.101a}$$

$$L_W = 0 \quad \text{in } D - \bar{\Omega}, \tag{2.101b}$$

$$w_x = w_y = \partial w/\partial n = 0 \quad \text{on EA}, \tag{2.101c}$$

$$w(x, H) = 0 \quad \text{on FA}, \tag{2.101d}$$

$$w(R, y) = \int_y^H \exp(g(t))(H - \eta) \, d\eta \quad \text{on AB}, \tag{2.101e}$$

$$w(r, y) = \int_y^{h_W} \exp(g(t))(h_W - \eta) \, d\eta \quad \text{on CD}, \tag{2.101f}$$

$$w(r, y) = 0 \quad \text{on DE}, \tag{2.101g}$$

$$w(x, 0) = w(r, 0) + \{w(R, 0) - w(r, 0)\}F(x)/F(R) \quad \text{on CB}, \tag{2.101h}$$

where

$$F(x) = \int_r^x \exp[-\gamma(\eta)] \, d\eta, \tag{2.101i}$$

$$w \geq 0 \quad \text{in } \Omega \text{ provided} \quad g'(y) \geq 0.$$

Introducing functions $v(x, y) \geq 0$ which together with their first derivatives are square summable on D and which agree with $w(x, y)$ on $\bar{D}$ there follows the variational inequality statement

$$a(w, v - w) + j(v - w) \geq 0, \tag{2.102}$$

where

$$a(u, v) = \iint_D \exp[\gamma(x) - g(y)] \, \text{grad } u \, \text{grad } v \, dx \, dy \tag{2.102a}$$

and

$$j(v) = \iint_D \exp[\gamma(x)]v \, dx \, dy. \tag{2.102b}$$

In this problem the boundary conditions for $\phi(x, y)$ and $w(x, y)$ do not involve the flow rate q to the well and so no additional compatibility condition is needed. This is because q can be expressed in terms of the dimensions and water levels, e.g. for constant permeability,

$$q = \frac{\pi(H^2 - h_w^2)}{\ln(R/r)}.$$

(2.103)

Because a in (2.102) is symmetric there is a minimization statement of the well problem, i.e.

$$J(w) \leqslant J(v),$$

(2.104)

where $J(v) = a(v, v) + 2j(v)$.

Cryer and Fetter (1977) prove existence and uniqueness and then approximate the problem using piecewise-linear finite elements. Numerical results are quoted for a particular example (see §8.5.2 and Tables 8.21a,b,c).

Less attention appears to have been paid to partially penetrating wells. Boreli (1955), Taylor and Brown (1967), and Cooley (1971) obtained numerical solutions but presented them graphically. Also Cryer (1976a)

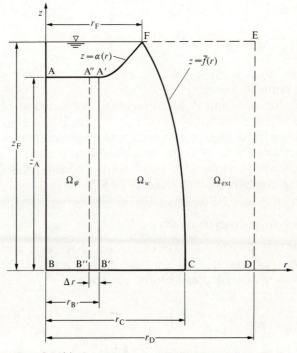

FIG. 2.14(a). Seepage from pond by split-domain method

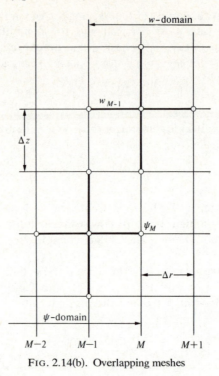

FIG. 2.14(b). Overlapping meshes

referred only to a paper by Youngs (1971) for the three-dimensional problem of multiple wells.

The split-domain method of formulating the arbitrary-shaped dam problem in §2.3.8 was used by Remar *et al.* (1982) to solve the axisymmetric problem of seepage from a pond with a bottom of arbitrary shape. With reference to Fig. 2.14(a), the variable ψ is sought in Ω_ψ, i.e. ABB′A′ and the Baiocchi variable w in the extended domain A″B″DEF. More details of the computation are given in §8.5.2.

2.8. Heterogeneous porous media

So far in this chapter the porous medium has been assumed to be homogeneous and isotropic. If this is not so, a more general form of the equation of flow is needed and can be written as

$$\frac{\partial}{\partial x}\left[k_{11}(x,y)\frac{\partial\phi}{\partial x}+k_{12}(x,y)\frac{\partial\phi}{\partial y}\right]+\frac{\partial}{\partial y}\left[k_{21}(x,y)\frac{\partial\phi}{\partial x}+k_{22}(x,y)\frac{\partial\phi}{\partial y}\right]=0,$$

where for an isotropic medium $k_{11}=k_{22}=1$, $k_{12}=k_{21}=0$, and for a homogeneous medium $k(x, y)$ is a constant. An example of one kind of inhomogeneity has been discussed in §2.7, where $k(x, y)$ is of the form $k_1(x)k_2(y)$. Before the work of Cryer and Fetter (1977), Benci (1973, 1974) had solved the rectangular dam problem for this type of $k(x, y)$ function. Other treatments are by Baiocchi (1976) and Baiocchi and Friedman (1977).

Other cases of practical importance and for which solutions have been obtained refer to stratified porous media. The permeabilities have different constant values in different strata and so $k(x, y)$ is a discontinuous function. Problems in horizontally or vertically stratified dams have received attention by various authors including Baiocchi *et al.* (1973*a,b*), Kikuchi (1977), Caffarelli and Friedman (1978*a,b*), Harr (1962), Jeppson (1968*b*, 1969). Mauersberger (1965), and Neuman and Witherspoon (1970). Outmans (1964) obtained flow rates for stratified dams.

2.8.1. *Stratified dams*

Vertically and horizontally stratified dams (Figs. 2.15(a), (b)) were formulated as variational problems by Baiocchi *et al.* (1973*a*) and some numerical solutions presented. The two formulations have much in common and so they are here described together. Figures 2.15(a) and (b) are labelled correspondingly. The permeabilities k_1 and k_2 refer to the domains denoted by D_1 and D_2 in each case but these domains are different. In Fig. 2.15(a), D_1 is the domain $0 < x < d, 0 < y < y_1$, and D_2 is $d < x < a, 0 < y < y_1$; in Fig. 2.15(b) D_1 is the domain $0 < x < a, 0 < y < d$, and D_2 is $0 < x < a, d < y < y_1$. In each case Ω denotes the flow region $0 < x < a, 0 < y < f(x)$, where $y = f(x)$ is the free boundary. The equations

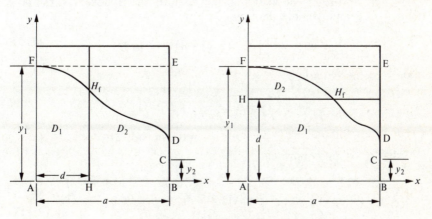

FIG. 2.15(a). Vertically stratified dam
FIG. 2.15(b). Horizontally stratified dam

and conditions common to the two dams are

$$\frac{\partial}{\partial x}\left(k\frac{\partial \phi}{\partial x}\right)+\frac{\partial}{\partial y}\left(k\frac{\partial \phi}{\partial y}\right)=0 \quad \text{in } \Omega, \tag{2.105}$$

where

$$k = k_1 \quad \text{in } D_1 \quad \text{and} \quad k = k_2 \quad \text{in } D_2, \tag{2.106}$$

$$\phi = y_1 \quad \text{on AF}, \quad \phi = y_2 \quad \text{on BC}, \tag{2.107}$$

$$\phi = y \quad \text{on CD and FD}, \tag{2.108}$$

$$\partial \phi/\partial n = 0 \quad \text{on AB and FD}. \tag{2.109}$$

The appropriate weak solution is

$$\iint_\Omega k\left(\frac{\partial \phi}{\partial x}\frac{\partial v}{\partial x}+\frac{\partial \phi}{\partial y}\frac{\partial v}{\partial y}\right)dx\,dy = 0, \tag{2.110}$$

where $v(x, y)$ has the usual properties in $\bar{D}$ and vanishes on AF and BE. If we denote by Ω_1 the part of Ω in D_1 and by ϕ_1 the variable $\phi(x, y)$ in Ω_1, with similar significance for Ω_2 and ϕ_2, then (2.110) implies $\nabla^2\phi_1 = 0$ in Ω_1, $\nabla^2\phi_2 = 0$ in Ω_2, and also the continuity conditions on HH_f, which are $k_1\,\partial\phi_1/\partial x = k_2\,\partial\phi_2/\partial x$ in Fig. 2.15(a), and $k_1\,\partial\phi_1/\partial y = k_2\,\partial\phi_2/\partial y$ in Fig. 2.15(b). The usual extension of $\phi(x, y)$ is defined for both dams by

$$\tilde{\phi}(x, y) = \phi(x, y) \quad \text{in } \bar{\Omega}$$
$$= y \quad \text{in } \bar{D}-\bar{\Omega},$$

but the details of the rest of the formulation are different for the two cases.

For the vertical stratification (Fig. 2.15(a)) the Baiocchi variable is defined as for the simple dam by

$$w(x, y) = \int_y^{y_1} [\tilde{\phi}(x, y)-\eta]\,d\eta \quad \text{in } \bar{D}. \tag{2.111}$$

It satisfies the equation

$$\frac{\partial}{\partial x}\left(k\frac{\partial w}{\partial x}\right)+\frac{\partial}{\partial y}\left(k\frac{\partial w}{\partial y}\right)=k \quad \text{in } \Omega$$

$$=0 \quad \text{outside } \Omega,$$

and the following conditions

$$\left.\begin{aligned}
&w = \tfrac{1}{2}(y_1-y)^2 \quad \text{on AF}, \qquad w = \tfrac{1}{2}(y_2-y)^2 \quad \text{on BC},\\
&w = \tfrac{1}{2}y_1^2 - qx/k_1 \quad \text{on AH},\\
&w = \tfrac{1}{2}y_2^2 + q(a-x)/k_2 \quad \text{on HB},\\
&w = 0 \quad \text{elsewhere on the boundary of } D.
\end{aligned}\right\} \tag{2.112}$$

It can be proved that the function $k \int_0^{f(x)} \phi_x(x, \eta) \, d\eta$ is a constant, denoted by $-q$, where q is the flow rate and

$$q = \frac{k_1 k_2 (y_1^2 - y_2^2)}{2[dk_2 + (a - d)k_1]}. \tag{2.113}$$

The appropriate variational inequality is

$$\iint_D k \left[\frac{\partial w}{\partial x} \left(\frac{\partial v}{\partial x} - \frac{\partial w}{\partial x} \right) + \frac{\partial w}{\partial y} \left(\frac{\partial v}{\partial y} - \frac{\partial w}{\partial y} \right) \right] dx \, dy + \iint_D k(v - w) \, dx \, dy \geq 0. \tag{2.114}$$

For $v(x, y)$ satisfying the usual conditions in these problems (2.114) has a unique solution with a continuous first derivative on $\bar{D}$. The equivalent minimum problem is to find w such that $J(w) \leq J(v)$, where

$$J(v) = \frac{1}{2} \iint_D k \left[\left(\frac{\partial v}{\partial x} \right)^2 + \left(\frac{\partial v}{\partial y} \right)^2 \right] dx \, dy + \iint_D kv \, dx \, dy.$$

For the horizontal stratification (Fig. 2.15(b)), a necessary modified definition of $w(x, y)$ is

$$w(x, y) = \int_y^{y_1} k(\eta)[\bar{\phi}(x, y) - \eta] \, d\eta. \tag{2.115}$$

It satisfies the equation

$$k \left[\frac{\partial}{\partial x} \left(\frac{1}{k} \frac{\partial w}{\partial x} \right) + \frac{\partial}{\partial y} \left(\frac{1}{k} \frac{\partial w}{\partial y} \right) \right] = k \quad \text{in } \Omega$$

$$= 0 \quad \text{outside } \Omega,$$

and the following conditions

$$
\left.
\begin{aligned}
w &= \int_y^{y_1} k(\eta)(y_1 - \eta) \, d\eta \quad \text{on AF,} \\
w &= \int_y^{y_2} k(\eta)(y_2 - \eta) \, d\eta \quad \text{on BC,} \\
w &= \int_0^{y_1} k(\eta)(y_1 - \eta) \, d\eta - qx \quad \text{on AB,} \\
w &= 0 \quad \text{elsewhere on the boundary of } D
\end{aligned}
\right\} \tag{2.116}
$$

We recall that k is the function which takes the values k_1 in D_1 and k_2 in D_2. So far in formulating the problem for horizontal stratification it has been supposed that $k_2 > k_1 > 0$. This condition will be mentioned later in this section. Again, the function $\int_0^{f(x)} k(\eta)\phi_x(x, \eta)\,d\eta$ is constant through-out the dam and equal to $-q$, where

$$q = \frac{1}{a}\left\{(y_1 - y_2)\int_0^{y_2} k(\eta)\,d\eta + \int_{y_2}^{y_1} k(\eta)(y_1 - \eta)\,d\eta\right\}. \quad (2.117)$$

In this case the variational inequality is

$$\iint_D \frac{1}{k}\left[\frac{\partial w}{\partial x}\left(\frac{\partial v}{\partial x} - \frac{\partial w}{\partial x}\right) + \frac{\partial w}{\partial y}\left(\frac{\partial v}{\partial y} - \frac{\partial w}{\partial y}\right)\right] dx\,dy + \iint_D (v - w)\,dx\,dy \geq 0, \quad (2.118)$$

with the usual restrictions on $v(x, y)$. There is a unique solution for $w(x, y)$, continuous on $\bar{D}$.

The equivalent minimum problem this time is $J(w) \leq J(v)$, where

$$J(v) = \frac{1}{2}\iint_D \frac{1}{k}\left[\left(\frac{\partial v}{\partial x}\right)^2 + \left(\frac{\partial v}{\partial y}\right)^2\right] dx\,dy + \iint_D v\,dx\,dy. \quad (2.119)$$

For the case $k_1 > k_2$ theoretical results appeared not to be complete (Baiocchi *et al.* 1973*b*) but Baiocchi *et al.* (1973*a*) concluded from their numerical results that the method can be applied even though $k_1 > k_2$. Some results are extracted in Tables 8.9–12.

2.9. Evaporation or infiltration

Porous flow through a simple rectangular dam between two reservoirs when evaporation of water from the free surface or precipitation on to it occurs, has been studied by a few authors, chiefly by Pozzi (1974*a,b*). Friedman (1976) and Jensen (1977) examined the shape of the free boundary which is non-monotone in general.

Pozzi took evaporation into account by replacing the usual condition $\partial\phi/\partial n = 0$ or $\psi = 0$ on the free surface, $y = f(x)$, by the condition

$$\psi = cx \qquad \text{or} \qquad \partial\phi/\partial n = c\{1 + (\partial f/\partial x)^2\}^{-\frac{1}{2}},$$

where c is the evaporation rate. He used the Baiocchi formulation by introducing the extended variables $\tilde{\phi}$ and $\tilde{\psi}$ defined by

$$\tilde{\phi}(x, y) = \phi(x, y), \qquad \tilde{\psi}(x, y) = \psi(x, y) \quad \text{in } \bar{\Omega},$$
$$\tilde{\phi}(x, y) = y, \qquad \tilde{\psi}(x, y) = cx \quad \text{in } D - \bar{\Omega}.$$

The Baiocchi variable is defined by (2.20) in §2.2.3 or by the modified form of (2.46)

$$w(x, y) = \int_{FP} [c\xi - \tilde{\psi}(\xi, \eta)]\,d\xi + \int_{FB} [\eta - \tilde{\phi}(\xi, \eta)]\,d\eta,$$

which is still independent of the path of integration. The differential relationships for $w(x, y)$ are

$$\nabla^2 w = 1 + c \quad \text{in } \Omega, \qquad \nabla^2 w = 0 \quad \text{in } \bar{D} - \Omega,$$

$$w > 0 \quad \text{in } \Omega, \qquad w = 0 \quad \text{in } \bar{D} - \Omega.$$

The corresponding variational inequality for $w(x, t)$ is

$$\iint_D \nabla w \cdot \nabla(v - w) \, dx \, dy - (1 + c) \iint_D (v - w) \, dx \, dy,$$

and the variables ϕ and ψ can be regained from the solution by using

$$\phi(x, y) = y - w_y(x, y), \qquad \psi(x, y) = cx - w_x(x, y) \quad \text{in } \bar{\Omega}.$$

Boundary conditions for several values of the evaporation rate c are given by Pozzi (1974a). He investigated mathematical properties and obtained some numerical results using the Cryer algorithm.

Pozzi (1974b) observed that the problem of the dam when there is infiltration or accretion ($c < 0$), e.g. due to rain, may not be well posed as usually formulated and this may be true if the rate of evaporation is too large. With an impervious base, Pozzi found that if

$$c > \left(\frac{y_1 + y_2}{b - a}\right)^2 \equiv c_3$$

with reference to Fig. 2.1, the free boundary has three line segments: from (a, y_1) to $(\alpha, 0)$, $(\alpha, 0)$ to $(\beta, 0)$, and $(\beta, 0)$ to (b, y_2), where α and β can be calculated explicitly. If $c < c_3$, the free boundary does not meet the x-axis but is non-monotone in general. Friedman (1976) considered the more general case which permits a porous base to the dam through which water can move at a rate $\ell(x)$; if $\ell(x) > 0$ water moves downwards and upwards if $\ell(x) < 0$, corresponding to a boundary condition $(\partial \phi / \partial y)_{y=0} = \lambda \ell(x)$, $a < x < b$, λ non-negative. Friedman's main result was that if $c \geqslant 0$, $\lambda \ell(x) \leqslant 1$, $\lambda \ell(x) + c \geqslant 0$, and if the free boundary does not meet the axis, there exists a number α, $a < \alpha \leqslant b$, such that $f(x)$ is strictly monotone decreasing if $a \leqslant x \leqslant \alpha$ and $f(x)$ is strictly monotone increasing if $\alpha \leqslant x \leqslant b$; if $\alpha < b$, $f(b) = y_2$. Jensen (1977) generalized the work of Friedman and Pozzi and determined regions where $f'(x) \geqslant 0$ and $f'(x) \leqslant 0$; he estimated the number of sign changes of $f'(x)$.

2.10. Unsaturated flow: capillary fringe

So far in this chapter it has been assumed that the flow is saturated, i.e. the porous medium is either wet or dry. In some practical situations the porous flow is unsaturated and some allowance for the presence of air in the flow region must be made. In order to avoid the general two-phase flow problem which is complicated and may not be a free-boundary problem (see §§2.11 and 8.6) the idea of a 'capillary fringe' was introduced (see Bear 1972). It is assumed that the flow is unsaturated only in a narrow zone, the capillary fringe, above the water level or table. The term is used because the fringe zone is attributed to capillary forces which draw up the water in the porous medium as happens in a capillary tube. Three regions are distinguished, namely the saturated region, the capillary region, and the dry region. It is assumed that the governing equations are the same in the capillary and saturated regions. The unknown free boundary is the interface between the dry and capillary regions. The pressure is still taken to be zero on the boundary between the saturated and capillary regions. In the fringe the pressure is negative, $-p_C$, where p_C is the known capillary pressure. Thus the expression (2.5) in §2.2.1 defining ϕ becomes

$$\phi = y - p_C/(\rho g) = y - y_C, \tag{2.120}$$

where y_C is the capillary rise.

Comincioli and Guerri (1976) formulated and solved the problem of the rectangular dam with a capillary fringe included as in Fig. 2.16. Their list of references gives a valuable guide to other mathematical and numerical papers on free boundary problems in porous flow, most of them using the Baiocchi method. The extended region $\bar{D}$ has the rectangular boundary ABE′F′, and they extended the definition of ϕ such that

$$\tilde{\phi} = \phi(x, y) \quad \text{in } \bar{\Omega}, \qquad \tilde{\phi} = y - y_C \quad \text{in } \bar{D} - \bar{\Omega}. \tag{2.121}$$

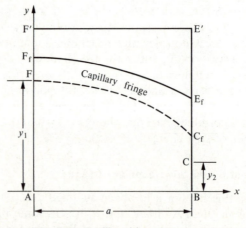

FIG. 2.16. Capillary fringe

The Baiocchi variable is defined by

$$w(x, t) = \int_{y}^{y_1 + y_C} [\tilde{\phi}(x, \eta) - \eta + y_C] \, d\eta. \tag{2.122}$$

Comincioli and Guerri set out the boundary conditions satisfied by ϕ and the derivatives of $w(x, y)$ leading after integration to the following formulation in terms of $w(x, y)$:

$$w(x, 0) = -qx + d \quad \text{where} \quad d = w(0, 0), \tag{2.123a}$$

$$w(0, y) = \tfrac{1}{2}(y - y_1 - y_C)^2 + d - \tfrac{1}{2}(y_1 + y_C)^2, \quad 0 \leqslant y \leqslant y_1, \tag{2.123b}$$

$$w(a, y) = \tfrac{1}{2}(y - y_2 - y_C)^2 - qa + d - \tfrac{1}{2}(y_2 + y_C)^2, \quad 0 \leqslant y \leqslant y_2, \tag{2.123c}$$

$$w(a, y) = -y_C y + \tfrac{1}{2} y_C^2 - qa + d - \tfrac{1}{2}(y_2 + y_C)^2 + y_C y_2, \quad y_2 \leqslant y \leqslant s, \tag{2.123d}$$

$$\nabla^2 w = 1 \quad \text{in } \Omega, \qquad \nabla^2 w = 0 \quad \text{in } D - \Omega, \tag{2.123e}$$

$$w > 0 \quad \text{in } \Omega, \qquad w = 0 \quad \text{in } \bar{D} - \bar{\Omega}. \tag{2.123f}$$

In (2.123a) both the flow rate q and d are unknown as also is s, the ordinate of the point C_f in (2.123d). These unknown parameters necessitate an extension to previous methods of solution. They reflect the fact that the points F_f, E_f, and C_f are not known.

If q, d, and s are supposed known, the variational inequality

$$\iint_D \nabla w \cdot \nabla(v - w) \, dx \, dy + \iint_D (v^+ - w^+) \, dx \, dy \geqslant 0,$$

$$v^+ = \tfrac{1}{2}(v + |v|), \quad (2.124)$$

and the minimum statement

$$J(w) \leqslant J(v), \qquad J(v) = \frac{1}{2} \iint_D |\text{grad } v|^2 \, dx \, dy + \iint_D v^+ \, dx \, dy \tag{2.125}$$

can be obtained and provide a weak solution of the problem

$$\nabla^2 w = 1, \quad w > 0, \quad \nabla^2 w = 0, \quad w < 0, \quad 0 \leqslant \nabla^2 w \leqslant 1, \quad w = 0,$$

$$w = g \quad \text{on } \Gamma_{D,s}, \qquad w_x = 0 \Gamma_{N,s}, \tag{2.126}$$

where g takes the boundary values of w specified in (2.123a–d) and for any fixed s, $\Gamma_{N,s}$ signifies the boundary lines FF' and C_fE', $\Gamma_{D,s} = \partial D - \Gamma_{N,s}$.

In the absence of a capillary fringe, $y_C = 0$, and we revert to the simple dam problem for which (2.124), (2.125), or (2.126) can be solved by the Cryer algorithm (8.86). If only q is unknown as, for example, in the dam with a sheetpile (§2.3.2), an extra compatibility condition is introduced to obtain q. For the problem with a capillary fringe two compatibility conditions were derived by Comincioli and Guerri (1976) and used to find q and d for an assumed s. They employed a numerical shooting method to find s.

The compatibility conditions expressing the continuity of w_x at $F(0, y_1)$ and $C_f(a, s)$ are derived by the arguments used to obtain (2.57) in §2.3.2. A finite-difference solution is obtained on a grid with mesh sizes $\Delta x = h_1$, $\Delta y = h_2$ such that

$$N_a = a/h_1, \qquad N_1 = y_1/h_2, \qquad N_2 = y_2/h_2, \qquad N_C = (y_1 + y_C)/h_2,$$

and if $w(i, j)$ denotes the value of $w(x, y)$ at the grid point (ih_1, jh_2), the compatibility conditions are

$$w(0, N_1) = w(1, N_1), \qquad w(N_a, j(s)) = w(N_a - 1, j(s)). \qquad (2.127)$$

Comincioli (1974) and Comincioli and Guerri (1976) used the algorithm described in §8.5.1(iii) to introduce these compatibility conditions into their basic SOR algorithm for an assumed value of s which they then improved by a shooting method based on the physical boundary conditions.

Thus the equations and conditions (2.126) together with the compatibility conditions (2.127) are first solved using s_0, an estimated value of s, to give a solution $w_{s_0}(i, j)$. This solution is examined on the boundary CE' (Fig. 2.16). We take $s_{11} = j(s_1)h_2$, with j the smallest integer such that $j(s_{11}) \geqslant N_2$ and $\partial w_{s_0}(N_a, j)/\partial x > 0$ and apply a heuristic argument based on the two boundary conditions

$$w_x < 0 \quad \text{on } CC_f, \qquad w_y > -y_C \quad \text{on } C_f E'$$

in Fig. 2.16. Thus if

$$\frac{\partial}{\partial x}\{w_{s_0}(N_a, j)\} < 0, \qquad N_2 \leqslant j \leqslant j(s_{11})$$

and

$$\frac{\partial}{\partial y}\{w_{s_0}(N_a, j)\} > -y_C, \qquad N_C > j > j(s_{11}),$$

we accept s_{11} as the new estimate of s. If, however,

$$\frac{\partial}{\partial y}\{w_{s_0}(N_a, j)\} < -y_C, \qquad N_C \geqslant j > j(s_{11}),$$

we take $s_{12} = j(s_{12})h_2$, with j now the greatest integer for which $j(s_{12}) \leqslant N_C$ and $\partial\{w_{s_0}(N_a, j)\}/\partial y < -y_C$. Some interpolated value between s_{11} and s_{12} is then adopted as s_1, the value of s to be used instead of s_0 in a second iteration of the SOR scheme with the compatibility conditions.

The process continues till the conditions on CC_f and C_fE' are simultaneously satisfied by a solution $w(i, j)$.

2.11. A generalized dam problem: a new formulation

Most of the problems in this chapter, so far, have been formulated in terms of the Baiocchi variable. In some cases quasi-variational inequalities have been involved and have led to mathematical and computational complications. In efforts to circumvent these difficulties, new approaches, which are closely similar to each other, were developed independently by Alt (1977, 1979, 1980a,b) and Brezis, Kinderlehrer, and Stampacchia (1978). The essence of the paper by Brezis *et al.* (1978) is given in this chapter. Alt (1980) constructed a general-purpose numerical algorithm to solve his new formulation and this, together with some of his numerical results, are presented in §8.6.

The dam problem, generalized as in Fig. 2.17, was formulated by Brezis *et al.* (1978) as follows. The symbol S denotes the boundary of the dam and Γ that of the flow region Ω. Thus S is composed of S_1, the impervious base, S_2, the parts of S in contact with air, and S_3, those parts occupied by water. Thus S_3 coincides with Γ_3, while Γ_1 is the wet part of the base, S_1, Γ_2 is the free boundary, and Γ_4 the seepage region. The pressure $p(x, y)$ satisfies the relationships

$$\nabla^2 p = 0 \quad \text{in } \Omega, \tag{2.128}$$

$$\frac{\partial}{\partial \nu}(p + y) = 0 \quad \text{on } \Gamma_1, \tag{2.129}$$

$$p = \frac{\partial}{\partial y}(p + y) = 0 \quad \text{on } \Gamma_2, \tag{2.130}$$

$$p = \phi \quad \text{on } \Gamma_3, \qquad \phi(x, y) = H(x) - y, \tag{2.131}$$

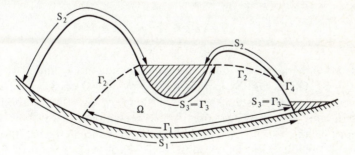

FIG. 2.17. General dam

where $H(x)$ is the height of the water in a reservoir,

$$p = 0, \frac{\partial}{\partial \nu}(p+y) \leq 0 \quad \text{on } \Gamma_4, \tag{2.132}$$

because no water enters the dam over the seepage face. In the rectangular dam this condition is imposed implicitly, since $p = 0$ on Γ_4 and $p > 0$ in Ω. By the maximum principle, $p > 0$ on Ω, since $\phi \geq 0$ on Γ_3 and $\partial p/\partial \nu > 0$ on Γ_1, since ν is always the outward normal.

We introduce a test function $\zeta(x, y)$ which is continuous over the whole of the dam and its boundaries, i.e. over $\bar{D}$, and which satisfies $\zeta = 0$ on Γ_3, $\zeta \geq 0$ on Γ_4. Then

$$\iint_\Omega \nabla p \cdot \nabla \zeta \, dx \, dy + \iint_\Omega \frac{\partial \zeta}{\partial y} \, dx \, dy$$

$$= \int_{\Gamma_1 + \Gamma_2 + \Gamma_4} (\partial p/\partial \nu + \nu \cdot y)\zeta \, ds = \int_{\Gamma_4} \frac{\partial}{\partial \nu}(p+y)\zeta \, ds \leq 0. \tag{2.133}$$

Brezis *et al.* (1978) extended $p(x, y)$ such that $p = 0$ in $D - \Omega$ and introduced the Heaviside function, H, so that

$$\iint_D \nabla p \cdot \nabla \zeta \, dx \, dy + \iint_D H(p)\zeta_y \, dx \, dy$$

$$= \iint_\Omega \nabla p \cdot \nabla \zeta \, dx \, dy + \iint_\Omega \zeta_y \, dx \, dy \leq 0. \tag{2.134}$$

Since Γ_4 is unknown, they put $\zeta = 0$ on $\Gamma_3 = S_3$ and $\zeta \geq 0$ on S_2. The problem now becomes to find $p(x, y)$ belonging to the space of functions which together with their first derivatives are square summable on D, $p \geq 0$ on D, and to find a second function g satisfying $g = 1$, $p > 0$, and $0 \leq g \leq 1$, $p = 0$; such that

$$\iint_D \nabla p \cdot \nabla \zeta \, dx \, dy + \iint_D g\zeta_y \, dx \, dy \leq 0 \tag{2.135}$$

for all ζ belonging to the same space of functions as p, and $\zeta \geq 0$ on S_2, $\zeta = 0$ on S_3. On S_2 and S_3, $p = \phi$, and we put $\phi = 0$ on S_2.

Brezis *et al.* (1978) established the existence of a unique solution to the problem and demonstrated that for a rectangular dam their unique solution yields a Baiocchi variable $w(x, y)$ which satisfies the usual differential inequalities, i.e. conditions (2.35) above.

Visintin (1979) proposed an analogous formulation based on a suitable extension of the pressure function and Quarteroni and Visintin (1980) studied the numerical solution of the problem.

2.12. Degenerate free-boundary problems

Generally, a stationary, free-boundary problem of the type considered so far in this chapter is associated with an elliptic partial differential equation, whereas in a moving boundary problem the equation is parabolic. There are, however, some important problems in which the boundary is moving but the equation is elliptic, i.e. they are degenerate problems. Examples are provided by the Hele-Shaw flow associated with the injection of fluid into a narrow channel and a problem in electrochemical machining. Their formulation in terms of a variational inequality reveals a new feature which can be particularly advantageous in numerical work.

2.12.1. Hele-Shaw flow analogy

In 1897, Hele-Shaw devised a cell (see Lamb 1932) consisting essentially of two closely spaced glass plates between which he studied the steady-state flow of viscous, incompressible fluids. Recognizing that at small Reynolds numbers the Navier–Stokes equations of motion for his apparatus take the form of Darcy's law, Hele-Shaw made use of the mathematical analogy to study two-dimensional problems in porous flow, particularly ground-water flow around complicated structures. The Hele-Shaw analogy has been used by other authors to study various free-boundary porous-flow problems (Saffman and Taylor 1958; Taylor and Saffman 1959; Taylor 1961; Richardson 1972). Saffman and Taylor (1958) were particularly interested in the stability of the interface between one fluid and another which is driving it forwards, when penetrating tongues can form.

In order to examine the mathematical analogy more carefully the situation shown in Fig. 2.18 is considered. The Hele-Shaw flow occurs

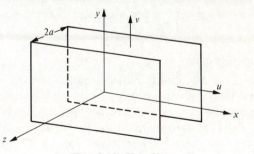

FIG. 2.18. Hele-Shaw cell

between two vertical plates, separated by a small distance $2a$ so that the flow can be considered two-dimensional. With reference to axes indicated in Fig. 2.18, where y is vertically upwards, u and v denote mean velocity components between the plates. It follows easily from the general equations of flow that (see Saffman and Taylor 1958)

$$u = \frac{-a^2}{12\mu}\frac{\partial p}{\partial x}, \qquad v = \frac{-a^2}{12\mu}\left(\frac{\partial p}{\partial y} + \rho g\right), \tag{2.136}$$

corresponding to a velocity potential given by

$$\phi = \frac{a^2}{12\mu}(p + \rho g y). \tag{2.137}$$

Comparison with (2.2) shows the direct analogy with flow in a porous medium of permeability $a^2/12$, remembering that u and v are mean velocities in the Hele-Shaw flow.

It remains to examine the conditions on an interface between two fluids, for example when fluid 1 is being pushed out completely by fluid 2. The mean velocity u_1 in fluid 1, ahead of the interface, is

$$u_1 = \frac{-a^2}{12\mu_1}\,\mathrm{grad}(p + \rho_1 g y) = \mathrm{grad}\,\phi_1,$$

and in fluid 2, behind the interface, the mean velocity is

$$u_2 = \frac{-a^2}{12\mu_2}\,\mathrm{grad}(p + \rho_2 g y) = \mathrm{grad}\,\phi_2.$$

If a is small, the width of the projection of the meniscus on to the plates is small, and the expressions for u can be assumed to hold right up to the interface, which can be regarded as a sharp line. It follows that the components of u_1 and u_2 normal to the interface are continuous across it. Furthermore, if surface tension effects are negligible, the pressure p is constant across the interface. Thus the motion of the two fluids reproduces that of the interface between two fluids of viscosities μ_1 and μ_2 in a porous medium with permeability $a^2/12$.

2.12.2. Hele-Shaw injection

Richardson (1972) considered the injection of fluid into an infinite cell by a point source placed in the centre of an initially fluid blob which subsequently expands. His practical interest was in injection moulding. Elliott and Janovsky (1979) formulated the problem of a finite injection source in a horizontal finite cell in terms of an elliptic variational inequality. The time-dependent nature of the problem appears only in

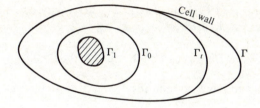

FIG. 2.19. Injection into horizontal cell

the condition on the moving boundary, which is the surface of the expanding fluid blob.

With reference to Fig. 2.19, the fluid initially occupies the domain D_0 lying between the two curves Γ_0 and Γ_I and more fluid is subsequently injected at a constant rate $Q > 0$ through the surface Γ_I. The region between Γ_I and the outer wall of the cell Γ is denoted by D. The moving boundary is Γ_t, for time $t > 0$; at $t = 0$, $\Gamma_t \equiv \Gamma_0$; D_t denotes the domain between Γ_I and Γ_t.

Applying the general discussion of §2.12.1 to the horizontal cell we can write

$$v = -\nabla p$$

where $v = (v_x, v_y)$ is the fluid velocity and p is a modified pressure defined by $a^2/(12\mu)P$, where $P(x, y)$ is the true pressure. In the incompressible fluid

$$\nabla^2 p = 0. \tag{2.138}$$

The velocity of the moving boundary, Γ_t, is taken to be v_n in the direction of the outward pointing normal n, and the pressure on Γ_t is assumed to be always zero. When the fluid reaches the cell wall the flow is assumed tangential and the normal derivative of pressure, p_n, vanishes. It is convenient to define Γ_t, $t > 0$, by

$$\Gamma_t = s(x, t) = t - \ell(x) = 0, \tag{2.139}$$

with x denoting the space coordinate vector (x_1, x_2), such that at time t the fluid occupies the part of D in which x satisfies $\ell(x) < t$: the function $\ell(x)$ is unknown *a priori*. The classical formulation of this problem is

$$\nabla^2 p = 0 \quad \text{in} \quad D_t, \tag{2.140a}$$

$$p = 0, p_n = -v_n \quad \text{or} \quad \nabla p \cdot \nabla s = \frac{\partial s}{\partial t} \quad \text{or} \quad \nabla p \cdot \nabla \ell = -1 \quad \text{on} \quad \Gamma_t, \tag{2.140b}$$

$$p_n = Q \quad \text{on} \quad \Gamma_I, \qquad p_n = 0 \quad \text{on} \quad \Gamma, \tag{2.140c}$$

$$\ell(x) = 0 \quad \text{in} \quad D_0. \tag{2.140d}$$

Following Duvaut (1973) and Elliott (1980) a new dependent variable $u(x, t)$ is introduced defined by

$$u(x, t) = \int_0^t p(x, \tau) \, d\tau \quad \text{in } D, \qquad t > 0. \qquad (2.141)$$

On the fixed and moving boundaries

$$u_n = 0 \quad \text{on } \Gamma, \qquad u_n = Qt \quad \text{on } \Gamma_1, \qquad u = u_n = 0 \quad \text{on } \Gamma_t. \quad (2.142)$$

Assuming p and $\ell(x)$ to be smooth, it follows from (2.141) by using (2.140b), (2.140a) that

$$\nabla u = \int_{\ell(x)}^t \nabla p(x, \tau) \, d\tau \quad \text{in } D_t, \qquad \ell(x) < t < T, \qquad (2.143)$$

and

$$\begin{rcases} \nabla^2 u = 1 \quad \text{in } D_t, \qquad \ell(x) < t < T, \\ \nabla^2 u = 0 \quad \text{in } D_0, \qquad 0 < t < T, \end{rcases} \qquad (2.144)$$

where $t = T$ is the final time at which the cell is completely filled. Thus in the domain D, $u(x, t)$ satisfies the inequalities

$$\begin{aligned} (-\nabla^2 u - f) &\geq 0, \qquad u \geq 0, \\ (-\nabla^2 u - f)u &= 0, \qquad \text{on } D, \end{aligned} \qquad (2.145)$$

since $u(x, t) = 0$ in $D - D_t$ and where $f(x) = 0$ in D_0 and $f(x) = -1$ elsewhere.

The elliptic variational inequality corresponding to (2.142) and (2.145) is

$$\iint_D \nabla u \cdot \nabla(v - u) \, dx_1 \, dx_2 \geq \iint_D f(v - u) \, dx_1 \, dx_2 + \int_{\Gamma_1} Qt(v - u) \, ds, \quad (2.146)$$

where the functions u and v together with their first derivatives are square integrable in D. There is the equivalent minimization statement

$$J(u) \leq J(v), \qquad (2.147)$$

where

$$J(v) = \frac{1}{2} \iint_D |\nabla v|^2 \, dx_1 \, dx_2 - \iint_D fv \, dx_1 \, dx_2 - \int_{\Gamma_1} Qtv \, ds.$$

The most interesting feature of (2.146) and (2.147) is that time t enters only as a parameter in Qt and thus the free boundary Γ_t may be found at any required time t^* by solving one elliptic free boundary problem with $Qt = Qt^*$, without the need to determine the solution at times prior to t^*.

Mathematical properties of the solutions and their existence, uniqueness, and regularity were examined by Elliott and Janovsky (1979, 1981, 1983). They expressed (2.146), (2.147) as a discrete linear complementarity problem arising from a finite-element discretization and obtained solutions by using the SOR with projection algorithm described in §8.5.1(i). They included an error analysis and presented solutions graphically for two problems: (i) for an elliptic source inside a square container; and (ii) for the injection of fluid into a mould lying between two semicircles.

2.12.3. Electrochemical machining

As an alternative to mechanical machining, a piece of metal can sometimes be shaped by using it as an anode in an electrolytic cell with an appropriately shaped cathode. This is a moving boundary problem because the anode surface changes with time. A detailed account of the physical and practical details of the process and its industrial uses are given by McGeough (1974). Essentially, the anode is moved towards the cathode, or vice versa, at a constant velocity and the products of the erosion of the anode are swept away by the electrolyte which is pumped through the space between the electrodes. The cations combine with the electrolyte to form gases and so the cathode shape does not change.

Several authors have studied a quasi-steady mathematical model in which the electrodes are assumed to be equipotential surfaces, and the electrolyte to be homogeneous and isotropic with constant conductivity. Hougaard (1977) gave references to pioneer papers and formulated two-dimensional problems on the complex-potential plane. The transformed boundary conditions depend on the cathode shape and may be so complicated that classical analytical or semi-analytical methods of solution have to be replaced by purely numerical methods. Hougård presented methods and solutions based on integral equations for a number of cathode shapes. Two basic geometrical arrangements have been investigated. The two-dimensional annular problem of shaping a cylindrical anode by placing it inside a long cylindrical cathode has been treated by Christiansen and Rasmussen (1976), who formulated the problem as an integral equation of the first kind (see §8.7.1). Hansen and Holm (1980) used an integral equation of the second kind. Meyer (1978e) applied his method of lines (§4.4) to the Poisson equation with electrochemical machining as an example. An enthalpy-type formulation by Crowley (1979) is described in §6.2.7(i).

Elliott (1980) formulated a variational inequality for this problem and drew attention to the close analogy with the Hele-Shaw injection problem discussed above in §2.12.2. This analogy can be anticipated from the similarity between the generalized Fig. 2.20 for electrochemical machining and Fig. 2.19. The anode is the shrinking region $A(t)$ with moving boundary Γ_t and Γ_0 denotes the initial anode surface at $t = 0$. The region inside the cathode surface C is denoted by D and the region occupied by

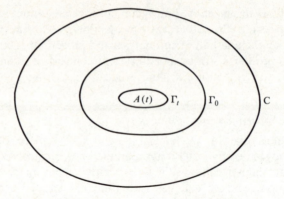

FIG. 2.20. Annular electrochemical machining

the electrolyte by D_t, so that D includes $A(t)$ and D_t. If $\phi(x, y)$ is the electric potential field, the mathematical problem is to find $\phi(x, t)$ and $\ell(x)$ such that

$$\nabla^2 \phi = 0 \quad \text{in } D_t, \tag{2.148a}$$

$$S(t, x) \equiv t - \ell(x) = 0 \quad \text{on } \Gamma_t, \tag{2.148b}$$

$$\ell(x) = 0 \quad \text{in } D_0, \tag{2.148c}$$

$$\phi = 0 \quad \text{on } C, \qquad \phi = V(t) \quad \text{on } \Gamma_t \tag{2.148d}$$

$$\left.\begin{array}{c} \nabla\phi \cdot \nabla S = -L \, \partial S/\partial t \\ \nabla\phi \cdot \nabla\ell = L. \end{array}\right\} \tag{2.148e}$$

Here $V(t)$ is the potential difference between the electrodes and $L > 0$ is a known constant of the system. This formulation is based on Faraday's and Ohm's laws (McGeough and Rasmussen 1974).

Elliott (1980) formulated a variational inequality for this problem following the arguments used in §2.12.2. The maximum principle for Laplace's equation (2.148a) states that

$$0 \leq \phi(x, t) \leq V(t).$$

For each t, $\phi(x, t)$ is extended to all of D by $\phi(x, t) = V(t)$ in $A(t)$ and so $V(t) - \phi(x, t) \geq 0$ in D. The Baiocchi variable $u(x, t)$ in this case is given by

$$u(x, t) = \int_0^t \{V(\tau) - \phi(x, \tau)\} \, d\tau \quad \text{in } D, \qquad t > 0. \tag{2.149}$$

Corresponding to (2.145) we can deduce that

$$\begin{array}{cc} -\nabla^2 u - f \geq 0, & u \geq 0, \\ (-\nabla^2 u - f)u = 0 & \text{on } D, \end{array} \tag{2.150}$$

where $f(x) \equiv -L$ in A_0 and $f(x) \equiv 0$ in D_0. Finally, using the conditions (2.148d), (2.148e) we have

$$u(x, t) = \int_0^t V(\tau) \, d\tau \equiv W(t) \quad \text{on C}, \qquad u = u_n = 0 \quad \text{on } \Gamma_t.$$

$$(2.151)$$

Thus, as in the Hele-Shaw problem, the anode surface at any specified time can be obtained by solving a single elliptic free boundary problem and t enters only through the boundary condition, $u = W(t)$.

The elliptic variational inequality follows easily since, if $u(x, t)$ is a classical solution of (2.150) and (2.151),

$$\iint_D \nabla u \cdot \nabla(v-u) \, dx_1 \, dx_2 = \iint_{D_t} \nabla u \cdot \nabla(v-u) \, dx_1 \, dx_2$$

$$= \iint_{D_t} -\nabla^2 u(v-u) \, dx_1 \, dx_2 + \int_{C+\Gamma_t} u_n(v-u) \, ds$$

$$= \iint_{D_t} f(v-u) \, dx_1 \, dx_2 \geqslant \iint_D f(v-u) \, dx_1 \, dx_2,$$

i.e. for each $t \geqslant 0$, u is the solution of

$$\iint_D \nabla u \cdot \nabla(v-u) \, dx_1 \, dx_2 \geqslant \iint_D f(v-u) \, dx_1 \, dx_2, \qquad (2.152)$$

where u and v are, together with their first derivatives, square integrable in D, $v \geqslant 0$ in D, and $v = W(t)$ on C. The corresponding minimization statement is

$$J(u) \leqslant J(v), \qquad (2.153)$$

where

$$J(v) = \frac{1}{2} \iint_D |\nabla v|^2 \, dx_1 \, dx_2 - \iint_D fv \, dx_1 \, dx_2.$$

Elliott (1980) proved some regularity results for $u(t)$ and gave some numerical results obtained by a finite-element approximation to (2.152) with error estimates. His results for one problem are compared with those of other authors in Table 8.25.

The second geometric arrangement considered calls for the solution of a two-dimensional planar problem between rectangular electrodes, theoretically of infinite area. The quasi-steady-state model for this case

formulated by McGeough and Rasmussen (1974) was extended by Forsyth and Rasmussen (1979) to include cathodes which contain insulated
parts. These latter authors also referred to several earlier papers
using different methods of solution, including perturbation techniques for
smoothing problems in which variations on the electrode surfaces are
small compared with the electrolyte-filled gap. Their results for a typical
partially insulated cathode are shown in Fig. 2.21.

In rectangular coordinates (x, y), $y = h(x)$ denotes the cathode surface
and $y = g(x, t)$ the anode, which moves towards the cathode with a

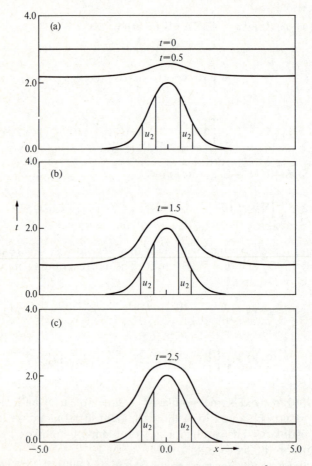

FIG. 2.21. Time evolution of the ECM system with $h(x) = 2 \exp(-x^2)$, $g(x, 0) = 3.0$, $\alpha = 2$.
The insulated section of the cathode is labelled u_2

constant velocity α. The potential ϕ is given by

$$\nabla^2\phi = \frac{\partial^2\phi}{\partial x^2} + \frac{\partial^2\phi}{\partial y^2} = 0,$$

$$\phi = 0, \qquad y = h(x), \qquad \phi = 1, \qquad y = g(x, t). \tag{2.154}$$

If there are insulated parts of the cathode, U_2, and conducting parts, U_1, the conditions in (2.154) become

$$\phi = 0 \quad \text{in } U_1, \qquad \partial\phi/\partial n = 0 \quad \text{in } U_2, \qquad y = h(x). \tag{2.155}$$

The boundary condition on the moving anode is

$$\frac{\partial g}{\partial t} = \frac{\partial\phi}{\partial y}\bigg)_g - \frac{\partial g}{\partial x}\frac{\partial\phi}{\partial x}\bigg)_g - \alpha = \nabla g \cdot \nabla\phi - \alpha, \tag{2.155a}$$

where $\partial\phi/\partial y)_g$ denotes the value of the derivative at $y = g(x, t)$, etc. Forsyth and Rasmussen (1979) confined attention to functions $h(x)$, $g(x, t)$ which approach straight lines of zero slope as $x \to \pm\infty$.

In principle, the solution proceeds in small time steps and at each step the classical, fixed boundary-value problem defined by $\nabla^2\phi = 0$ and the boundary conditions appropriate to the time is solved with the anode treated as a fixed surface. The condition (2.155a) is then used to determine a new anode surface which becomes the fixed boundary in the next time step. Forsyth and Rasmussen (1979) used a variational method to solve Laplace's equation in a transformed domain with straight boundaries by using the body-fitted curvilinear coordinates discussed in relation to Stefan problems in §5.2. The new feature is that the coordinate transformation is applied directly to the variational integral. These authors presented numerical and graphical solutions for different cathode shapes.

Later Hansen (1983) was critical of the simple mathematical model studied in these earlier papers because no account was taken of the flow of electrolyte between the electrodes. He generalized his integral equation method (Hansen 1980) (see §8.7.1) in order to deal with practically realistic geometric shapes, and obtained numerical results which could be compared with experiments. He concluded that at the upstream end of the electrode gap the quasi-steady state model, with no allowance for electrolyte flow, predicts the anode profile reasonably well. In the middle and downstream regions, however, the predictions are poor because, Hansen thought, the bubbles of gas, evolved at the cathode and swept along in the electrolyte, form an insulating screen which reduces the local current density. More details of the integral equation method in this context are given in §8.7.

2.12.4. Evolution dam problem

When a dam separates two reservoirs in which the water levels depend on time the problem is a degenerate one. The velocity potential is harmonic in the flow region but the free surface varies with time and is not a streamline. The condition $\partial\phi/\partial n = 0$ on the free surface is replaced by $\phi_t - \nabla\phi \cdot \nabla p = 0$. Unsteady flow of an incompressible fluid in a rectangular dam (Fig. 2.22) was studied theoretically by Friedman and Torelli (1977) and Torelli (1975, 1977a,b); he also allowed for fluid to move across the base AB, i.e. $\phi_y(x, 0, t) = -\ell(x, t)$. Torelli (1977$a$) used Baiocchi's approach and formulated the problem in an extended fixed domain in space and time using the variable $w(x, y, t)$, where

$$w(x, y, t) = \int_0^t [\tilde{\phi}(x, t+y-\tau, \tau) - (t+y-\tau)]\,\mathrm{d}\tau$$

in the domain $0 < x < x_1$, $0 < y < y_3$, $0 < t < T$, where x_1 is the width of the dam, y_3 its height, and T some final value of time $t > 0$. He formulated differential and variational inequalities and gave some theoretical results. Friedman and Torelli (1977, 1978) established the regularity and physical properties of Torelli's solution and proposed a finite-difference scheme.

Transient problems have been studied from the numerical point of view by several authors including Taylor and Luthin (1969), Cooley (1971), Neuman and Witherspoon (1971), Verma and Brutsaert (1971), Todsen (1971), and Neuman (1975). In a transient dam problem, the conservation condition on the moving interface analogous to the Stefan condition can be written

$$\ell_x k_x \frac{\partial\phi}{\partial x} + \ell_y k_y \frac{\partial\phi}{\partial y} + \ell_z k_z \frac{\partial\phi}{\partial z} = -v_n\beta$$

$$= -\beta\left(\ell_x \frac{\mathrm{d}x}{\mathrm{d}t} + \ell_y \frac{\mathrm{d}y}{\mathrm{d}t} + \ell_z \frac{\mathrm{d}z}{\mathrm{d}t}\right),$$

(2.156a)

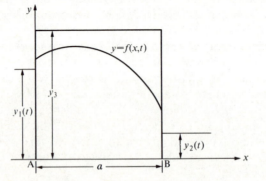

FIG. 2.22. Unsteady flow through porous dam

where ℓ_x, ℓ_y, ℓ_z are the direction cosines of the outward normal to the interface on which $p = 0$, β is a porosity constant for the medium, and k_x, k_y, k_z are directional permeabilities. France *et al.* (1971) and Taylor *et al.* (1973) incorporated this boundary condition into finite-element schemes for both transient and steady-state problems (see §8.2.2).

Meyer (1980) used Patel's (1968) form of boundary condition discussed in §1.3.9. Thus in two space dimensions, for example, and substituting $l_x/l_y = (\partial p/\partial x)/(\partial p/\partial y)$, (2.156a) becomes

$$k_x \frac{\partial p}{\partial x}\frac{\partial \phi}{\partial x} + k_y \frac{\partial p}{\partial y}\frac{\partial \phi}{\partial y} = -\beta\left(\frac{\partial p}{\partial x}\frac{\partial x}{\partial t} + \frac{\partial p}{\partial y}\frac{\partial y}{\partial t}\right) = \beta\frac{\partial p}{\partial t} \qquad (2.156b)$$

on the surface $p = 0$ where also we have $\partial y/\partial t = -(\partial p/\partial t)(\partial y/\partial p)$. Thus (2.156b) gives

$$\frac{\partial f}{\partial t} = -\frac{1}{\beta}\left[-k_x \frac{\partial y}{\partial x}\frac{\partial p}{\partial x} + k_y \frac{\partial \phi}{\partial y}\right] = \frac{1}{\beta}[k_x(1-\phi_y)(\partial y/\partial x)^2 - k_y\phi_y], \quad (2.156c)$$

remembering that $\phi = p + y$ and with the free boundary denoted by $y = f(x, t)$. Meyer applied his method of lines (§4.3.4) to a transient dam problem by using (2.156c).

Yet another form of moving-boundary condition was used by Comincioli and Torelli (1979). Thus by writing

$$\frac{\partial p}{\partial t} = \left(i\frac{\partial p}{\partial x} + j\frac{\partial p}{\partial y}\right) \cdot \left(i\frac{\partial x}{\partial t} + j\frac{\partial y}{\partial t}\right)$$

$$= -\left(i\frac{\partial p}{\partial x} + j\frac{\partial p}{\partial y}\right) \cdot \left(i\frac{\partial \phi}{\partial x} + j\frac{\partial \phi}{\partial y}\right), \qquad (2.156d)$$

we obtain $p_t = p_x^2 + p_y^2 + p_y$ on using $\phi = p + y$.

Comincioli and Torelli (1979) considered the various numerical algorithms to be of a heuristic nature and not sufficiently rigorous from a mathematical point of view. Accordingly, they developed a new numerical algorithm based on a finite-element discretization of a variational equality.

Formulated in terms of pressure and with reference to Fig. 2.22 the equation to be solved is

$$\nu p_t - \nabla^2 p = 0 \quad \text{in } \Omega(t),$$

where $\Omega(t)$ is the flow region $0 < x < a$, $0 < t < T$, $0 < y < f(x, t)$, $y = f(x, t)$ is the free boundary, and a is the width of the dam. The constant ν is the retentivity coefficient ($\nu \geqslant 0$). On the free boundary $p = 0$, and from (2.156d), $p_x^2 + p_y^2 + p_y = p_t$, while on the faces of the dam, $f(0, t) \geqslant y_1(t)$, $f(a, t) \geqslant y_2(t)$, together with

$$p(0, y, t) = (y_1(t) - y), \qquad p(a, y, t) = (y_2(t) - y), \qquad p \geqslant 0.$$

On the bottom boundary, $p_y(x, 0, t) = -1$, and finally the initial conditions are $f(x, 0) = f_0(x)$, and, for $\nu > 0$, $p(x, y, 0) = p_0(x, y)$. The problem is to find a pair (f, p) satisfying the above relationships where $y_1(t)$, $y_2(t)$ $f_0(x)$, $p_0(x, y)$ are given functions.

The Baiocchi-type variable

$$w(x, y, t) = \int_0^t \tilde{p}(x, y+t-\tau, \tau) \, d\tau$$

is introduced, where $\tilde{p} = p$ in Ω and $\tilde{p} = 0$ in $D - \Omega$, where D is the extended domain $0 < x < a$, $0 < y < y_3$. The existence, uniqueness, and regularity of the solution of an appropriate variational inequality for w follow from the theoretical papers referred to above. Then *a priori* estimates and a convergence theorem are established for a finite-element approximate problem of implicit type but an explicit approximation could be introduced equally well. The final algorithm involves the iterative solution of a non-linear set of equations for w_{ij} when the mesh point (i, j) lies in the domain D and a simpler set of equations for points (i, j) on the bottom boundary. The first case is equivalent to a minimum problem in one variable and can be solved directly. Values finally adopted at the new time level incorporate a relaxation parameter of 1.7, though this is not known to be optimum. Solutions are shown graphically for four different situations involving rapid raising or lowering of reservoir heights from an initial steady-state position of the free boundary. A fifth problem starts from a non-equilibrium free-boundary position specified as

$$f_0(x) = 1 + 0.3 \sin 2\pi x$$

with $y_1(t) = y_2(t) = 0$, $t \geq 0$.

The extension of the Baiocchi approach to time-dependent problems in dams of general shape presents difficult quasi-variational problems. Gilardi (1979, 1980) therefore adapted the formulation of Alt (1977, 1979, 1980a,b) given in §8.6 and of Brezis *et al.* (1978) outlined in §2.11. He arrived at the following formulation of the problem: find a pair of functions (p, χ) satisfying

$$p \geq 0, \qquad 0 \leq \chi \leq 1, \qquad p(1 - \chi) = 0 \quad \text{in } \Omega,$$

$$p = G \quad \text{on } \Sigma_0, \qquad G \text{ bounded and } \geq 0 \quad \text{on } \bar{\Omega},$$

$$\chi(0) = \chi_0 \quad \text{in } D,$$

$$\int_\Omega [\nabla p \cdot \nabla v + \chi(v_y - v_t)] \, dx \leq 0,$$

for any smooth v such that $v(0) = v(T) = 0$, $v \geq 0$ on Σ_0.

This variational inequality reduces to that of Brezis *et al.* (1978) in §2.11 when $v_t = 0$ and where v and χ take the places of ζ and g. In the time-dependent problem the space–time domain Ω stands on the area domain D of the dam and extends over $0 < t < T$ in time, ∂D the boundary of D comprises two parts Γ_0 and Γ_1, and Σ_0, Σ_1 are the corresponding parts of the boundary surface of Ω. The function χ_0 specifies the initial wet or partially saturated region. The condition $p \geqslant 0$ indicates saturated flow in regions where $p > 0$ but elsewhere unsaturated flow is allowed. An existence theorem for this evolution problem is proved in Gilardi (1979).

Visintin (1980*a,b*) extended his formulation of the stationary problem (Visintin 1979) to the time-dependent problem with reference to the Fig. 2.17 by introducing a new system of coordinates (ξ, η, τ) corresponding to a rotation of $\pi/4$ in the (y, t) plane, i.e.

$$\begin{pmatrix} \xi \\ \eta \\ \tau \end{pmatrix} = \zeta \begin{pmatrix} x \\ y \\ t \end{pmatrix} = \begin{pmatrix} x \\ (y + t)/\sqrt{2} \\ (-y + t)/\sqrt{2} \end{pmatrix}.$$

The final variational inequality involved two unknown functions z^+ and z^- satisfying appropriate boundary conditions and Visintin proved an existence theorem. He claimed that possible generalizations include three-dimensional flow, a vertical sheetpile, and a capillary fringe.

The non-stationary flow of a compressible fluid through a simple rectangular porous dam, with fluid again moving through the base at a rate $\ell(x, t)$, was studied by Torelli (1977*b*). He used the Baiocchi transformation

$$w(x, y, t) = \int_0^t \tilde{p}(x, t + y - \tau, \tau) \, d\tau,$$

where $\tilde{p}$ is the extended form of the pressure over a fixed space–time domain, to obtain theoretical results based on a variational inequality.

Rasmussen and Salhani (1981) described three different numerical methods for more general unsteady, two-dimensional porous flow with a free surface. They used a Rayleigh–Ritz expansion, a Kantorovitch expansion, and a coordinate transformation.

2.13. Lubrication cavitation in a journal bearing

This phenomenon which has been of practical and theoretical interest to engineers for almost a century presents a classical elliptic free-boundary problem (Pinkus and Sternlich 1961). It is of historic interest to mathematicians in that Christopherson (1941) introduced for the first time the idea that the differential equation must be solved subject to an inequality constraint that the solution be non-negative. In fact, he used essentially what is now known as the Cryer algorithm, with relaxation instead of SOR, and in spirit, though not explicitly, formulated a linear complementarity problem. SOR was first used by Gnanados and Osborne (1964). Cryer (1971b) gave a mathematical analysis of Christopherson's approach.

The journal and its bearing are shown in Figs. 2.23(a) and (b). Between the rotating cylinder, the journal, and the bearing surface is a thin film of

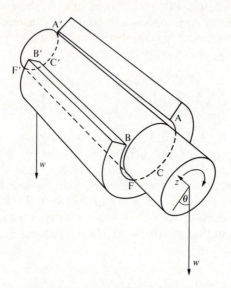

FIG. 2.23(a). Partial journal bearing

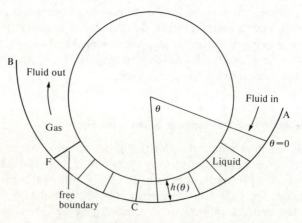

FIG. 2.23(b). End view of bearing

lubricating fluid which is fed in along AA' (Fig. 2.23(a)) and flows out through BB' and the ends ACB and A'C'B'. If the width of the fluid film decreases between A and C and then increases to B, the pressure can be expected to increase between A and C and to decrease between C and B (Fig. 2.23(b)). Along AA', BB', ACB, and A'C'B' the lubricant is in contact with the atmosphere and the boundary condition there is $p = 0$, if pressure is normalized to zero atmospheric pressure. The lubricant occurs in both liquid and vapour phases and the free boundary is the interface FF' between liquid and vapour. It is assumed that vaporization occurs at zero pressure so that the inequality $p \geqslant 0$ is satisfied everywhere.

Where $p > 0$ the liquid phase exists and the flow is determined by Reynolds's equation in two dimensions if, for a thin film, the pressure is assumed not to vary across the gap. Then the equation is to be solved in the region AA'F'F (Fig. 2.24), where the x-axis lies along the length of the cylindrical journal, θ is the angular distance along the film, and $h(\theta)$ is the film thickness, a known function. The liquid region is denoted by Ω_+, the gaseous region by Ω_0, and the whole rectangle AA'B'B by Ω.

The problem is defined by the following equations and Fig. 2.24:

$$Lp = \nabla \cdot (h^3 \nabla p) - dh/d\theta = 0 \quad \text{in } \Omega_+, \tag{2.157a}$$

$$p = 0 \quad \text{in } \Omega_0, \qquad p = 0 \quad \text{on the boundary of } \Omega, \tag{2.157b}$$

$$p = \partial p/\partial n = 0 \quad \text{on FF'}, \qquad \theta = s(x), \tag{2.157c}$$

where $\theta \equiv s(x)$ is the free surface to be determined together with $p(\theta, x)$. The cavitation conditions (2.157c) ensure non-negative pressures. The conditions $dh/d\theta < 0$, $\theta_A < \theta < \theta_C$ and $dh/d\theta > 0$, $\theta_C < \theta < \theta_B$ ensure that the free boundary lies above $\theta = \theta_C$ (Elliott 1976).

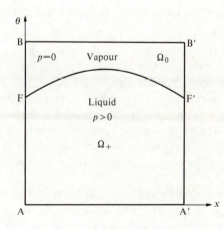

FIG. 2.24. Journal bearing: solution domain

We note that this problem is already in variational form in the original variable $p(\theta, x)$ because of the conditions (2.157c). In fact,

$$-\nabla \cdot (h^3 \nabla p) + dh/d\theta \geqslant 0, \qquad p \geqslant 0 \quad \text{in } \Omega,$$
$$p\{-\nabla \cdot (h^3 \nabla p) + dh/d\theta\} = 0,$$

(2.158)

and finally

$$\iint_{\Omega} h^3 \nabla p \cdot \nabla(v-p) \, dx \, d\theta \geqslant - \iint_{\Omega} (dh/d\theta)(v-p) \, dx \, d\theta \qquad (2.159)$$

for all $v \geqslant 0$ in Ω, square integrable together with its first derivatives, and $v = 0$ on the boundary of Ω.

Figure 2.24 refers to a bearing of finite length, $-L \leqslant x \leqslant L$ say. If L is very large so that the pressure variation in the axial direction is small the problem is treated as one dimensional with $p = p(\theta)$ and describes an infinite journal bearing. Cryer proved the existence of a unique solution for this case and developed the theory of his SOR algorithm to find numerical solutions of a discretized formulation.

Meyer (1978d) used his method of lines (see §4.4) to solve the two-dimensional problem and showed that his solution converged to that of the variational inequality (2.159). He also stated that the method of lines could be applied with little change to more realistic technical problems.

Some unsolved free-boundary problems from the theory of lubrication, including the conditions on a cavitation free boundary, are discussed by Capriz and Cimatti (1980).

2.14. A more general look at the Baiocchi transformation

Several examples of the use of the Baiocchi transformation and method have been described in this chapter. The aim in this section is to present a more general statement of free boundary problems in terms of variational inequalities. The individual physical situations emerge as special cases. In particular, two questions are examined systematically: first what determines the need for a Baiocchi transformation in some problems but not in others; second, in what direction, i.e. with reference to what independent variable, should the integration in the Baiocchi transformation be performed? What follows is based on an elucidation by Baiocchi himself (1980b).

A typical free-boundary problem can be stated as follows, without any reference to a particular physical situation for the moment. We denote the Laplace operator for n independent variables by $\nabla^2 u = \sum_{i=1}^{n} \partial^2 u/\partial x_i^2$ in a fixed open domain D. Given f and ψ, two real, smooth functions on $\bar{D}$, i.e. the domain D together with its boundary ∂D, we wish to find a pair $\{\Omega, u\}$, where Ω is an open sub-domain of D with boundary $\partial \Omega$ also included in $\bar{D}$, u is a function defined in $\bar{\Omega}$, and both Ω and u are to be smooth. The conditions to be satisfied are:

$$-\nabla^2 u = f \quad \text{in } \Omega, \tag{2.160}$$

$$u = \psi \qquad \text{on} \qquad \partial \Omega \cap D, \tag{2.161}$$

$$\text{grad } u = \text{grad } \psi \qquad \text{on} \qquad \partial \Omega \cap D, \tag{2.162}$$

$$u = 0 \qquad \text{on} \qquad \partial \Omega \cap \partial D. \tag{2.163}$$

The 'fixed' part of the boundary of Ω is also part of ∂D and written as $\partial \Omega \cap \partial D$ in (2.163) and other types of this condition are admissible, e.g. a Neumann derivative condition. The 'free boundary' is denoted by $\partial \Omega \cap D$ and (2.161), (2.162) are the two conditions needed because this boundary is unknown.

A more general free-boundary problem arises if the Laplace operator is replaced by an operator L so that (2.160) becomes

$$Lu = \sum_{i,j=1}^{n} -(a_{ij}u_{x_i})_{x_i} + \sum_{i=1}^{n} b_i u_{x_i} + cu = f \quad \text{in } \Omega, \tag{2.164}$$

and instead of (2.162) we take

$$\sum_{i,j=1}^{n} a_{ij}(u-\psi)_{x_i} + \sum_{j=1}^{n} g_j(u-\psi)_{x_j} = 0 \quad \text{on } \partial \Omega \cap D, \tag{2.165}$$

for a prescribed $g \equiv (g_j(x))$. Here $u_{x_i} \equiv \partial u/\partial x_i$ etc., x denotes the vector $x = (x_1, x_2, \ldots, x_i, \ldots, x_n)$, and $g \equiv (g_1, g_2, \ldots, g_i, \ldots, g_n)$. For the Laplace operator $a_{ij} = 1$, $i = j$; $a_{ij} = 0$, $i \neq j$; and with $g \equiv 0$, (2.165) is equivalent to (2.162). By (2.161), $u - \psi = 0$ on the free boundary and therefore $\text{grad}(u - \psi)$ is parallel to ν_Ω, the unit outward normal to $\partial \Omega$. Thus, (2.165) can be written in the form

$$[a \cdot \text{grad}(u-\psi) - g] \cdot \nu_\Omega = 0 \quad \text{on } \partial \Omega \cap D, \tag{2.166}$$

where $a \cdot \text{grad}(u - \psi)$ is the vector with components $\sum_{j=1}^{n} a_{ij}(u-\psi)_{x_j}$ with $i = 1, \ldots, n$.

If now we denote by K a set of admissible functions v, which are square summable together with their first derivatives, and $v = 0$ on ∂D, $v \geq \psi$ in D, we can write

$$K = \{v \in H_0'(D) : v \geq \psi\},$$

where $H_0'(D)$ is the usual Sobolev space of functions which are zero on ∂D. Then the free boundary problem becomes to find $u \subset K$ such that

$$a(u, u-v) \leq (u-v) \qquad \text{for all} \qquad v \in K \tag{2.167}$$

where

$$a(u, v) = \iint\limits_{D} \operatorname{grad} u \cdot \operatorname{grad} v \, dx \, dy, \qquad (u - v) = \iint f(u - v) \, dx \, dy.$$

The proof of (2.167) is as for (2.36) outlined in §2.2.4.

Several physical problems have been formulated in this way in the preceeding sections. We rely on the original theorem by Stampacchia (1964) for existence and uniqueness results on u. In order to show that with u a solution of (2.167) and Ω defined to be the sub-domain of D for which $u > \psi$, the pair $\{\Omega, u\}$ solves the problem of equations (2.160–3), we need to investigate further the smoothness of u. We appeal to the theoretical work of Lewy and Stampacchia (1969) for the important result that the solution of (2.167) implies u is in $C^1(D)$, i.e. has continuous first derivatives in D. We require also that ψ is in $C^1(D)$.

If we take u to be the restriction of u to $\bar{\Omega}$, then (2.161) holds by the definition of Ω; (2.163) is true because $u \in H_0^1(D)$; (2.162) holds because $u - \psi$ is a C^1 function which is non-negative, and so $\operatorname{grad}(u - \psi) = 0$ where $u - \psi = 0$, i.e. on the free boundary. Finally, we refer, for example, to Baiocchi *et al.* (1973a) for the proof that the solution pair (Ω, u) of (2.167) provides a weak solution of the original problem (2.160–3).

We noted that the conditions (2.161) and (2.162), which lead to the continuous first derivatives of u on the free boundary that are essential for a variational formulation, correspond to $g \equiv 0$ in the generalized free boundary condition (2.166). This point was noted by Levy and Stampacchia (1969). The implication is that problems in which g is not zero cannot be solved directly through a variational inequality. In general we must expect limited continuity of u across the free boundary (Brezis, 1972 a,b). Thus, if we define $L^2(\Omega)$ to be the space of square integrable functions in Ω the solution of (2.167) belongs to $H^2(\Omega)$ provided f belongs to $L^2(\Omega)$.

We now briefly categorize some of the free boundary problems considered earlier in this chapter and some of the moving boundary problems formulated in Chapter 1 as special cases of the general problem defined by (2.164) and (2.166). Especially, we see whether they are immediately amenable to variational solution and if not what form the Baiocchi variable should take.

(i) *Journal bearing.* The equations (2.157a–c) describing the cavitation problem in a journal bearing (§2.13) are in the form of (2.164) and (2.166) with $\psi = 0$ and $g = 0$ and it was directly expressible as the variational inequality (2.159).

(ii) *Oxygen diffusion.* Similarly, the diffusion of oxygen in absorbing tissue, formulated in equations (1.57–60) corresponds to $L = \partial/\partial t - \partial^2/\partial x^2$,

$f = -1$, $\psi = 0$, and from (1.58), $g = 0$, and is formulated directly in the variational form of (6.141).

(iii) *Simple rectangular dam.* With suitable change of nomenclature, e.g. ϕ replaced by u, y by x_1, and x by x_2, equations (2.7) and (2.9) in §2.2 are like (2.160) with $f = 0$, $\psi = x$, and the second condition in (2.9) on the free boundary can be put in the form of (2.166) with $g_1 = -1$, $g_i = 0$ for $i \neq 1$. The Baiocchi variable (2.20) therefore had to be defined in order to introduce the continuity necessary for the variational form (2.36). The direction of integration in (2.20) is x_1, e.g. g_1, and this was found necessary and useful in other steady dam problems formulated in §2.3. Difficulties which still arose in some cases were due to the conditions on the fixed boundaries.

(iv) *Evolution dam problem.* The dam is denoted by D_0 in two or three space dimensions ($m = 2$ or 3) for the non-steady dam problem of §2.12.4, and D denotes the space–time domain on the base D_0 and extending in time t over the range $0 \leqslant t \leqslant T$. If the coordinate x_n in (2.164), for example, is written $x_n = t$ and $n = m + 1$, the evolution dam problem still has the form of (2.160) and (2.161) with $f = 0$, $\psi = x_1$, the vertical component, but ∇^2 involves the space variables only. The only t derivative appears in the free boundary condition $u_t = \text{grad } u \cdot \text{grad}(u - x_1)$ and grad involves only space variables. This condition is like (2.166) with $g_1 = -1$, $g_i = 0$ for $1 < i < n$, and $g_n = -1$. Torelli (1977*a*) used the variable $t + y - \tau$ for the integration in the Baiocchi variable (see §2.12.4).

(v) *One-phase Stefan problem.* By retaining the nomenclature of the previous problem (iv) but with D_0 denoting the ice–water region and D its extension to the space–time domain for $0 \leqslant t \leqslant T$, the melting ice problem described in one dimension, for example, in §1.2.1 is (2.161), (2.164), and (2.166) with $f = \psi = 0$, $L = \partial/\partial t - \nabla^2$, and $\Omega(x, t)$ is the water region. In §1.3.9 the usual condition (1.56b) on the melting boundary is expressed in the form (1.56k). For a one-phase problem in the present nomenclature this becomes $(\text{grad } u)^2 = \mathscr{L}u_t$, with $\mathscr{L}$ a heat parameter, which has the form (2.166) with $g_i = 0$ for $i < n$, $g_n = -\mathscr{L}$. This indicates immediately the need for Duvaut (1973) transformation (6.159) where the integration is in the t-direction.

To summarize, the direction of integration in the examples (i)–(v) was always in the direction of g, with no integration needed if $g = 0$. Baiocchi (1976) established this as a general result for free-boundary problems; a similar investigation for Stefan-like, moving boundary problems was carried out by Gastaldi (1979). In terms of p, a continuous extension of u defined by

$$p = u - \psi \quad \text{in } \bar{\Omega}, \qquad p = 0 \quad \text{in } \bar{D} - \bar{\Omega}, \tag{2.168}$$

the general problem defined by (2.164), (2.161), (2.166) was expressed by Stampacchia (1964) as

$$Lp = \chi_\Omega(f - L\psi) - g \cdot \text{grad } \chi_\Omega, \tag{2.169}$$

where χ_Ω is the characteristic function of Ω in D, i.e. $\chi_\Omega = 1$ in Ω and $\chi_\Omega = 0$ in $D - \Omega$. The relation (2.169) follows from (2.164), (2.166), (2.168) and confirms that a Baiocchi integration along the direction of g can remove the difficult term grad χ_Ω.

Elliott and Ockendon (1982) comment on the difficulties of generalizing the above derivation of variational inequalities. They also illustrate the possible existence of more than one conservation formulation for a particular problem, e.g. the rectangular dam problem. Only conservation laws with appropriate discontinuities in the dependent variables can be diagnosed as amenable to weak and variational formulations. In general there are no firm rules for guidance.

2.15. Connections between certain free-boundary problems

Rogers (1980) examined systematically the connections between the single-phase Stefan problem, the problems of oxygen diffusion and electrochemical machining, and the time-dependent dam problem which have been formulated earlier in this chapter and in Chapter 1. Rogers proposed truncation algorithms for the numerical solution of these problems which have close affinity with linear complementarity algorithms as noted, for example, in §6.4.4. Rogers's findings on the connection between the evolution dam problem and that of electrochemical machining reflect the use of Hele-Shaw flow as an analogy for two-dimensional porous flow on the one hand (see §2.12.1) and its similarity of mathematical formulation to that of electrochemical machining on the other (see §2.12.3).

3. Analytical solutions

3.1. Exact similarity solutions

VERY few analytical solutions are available in closed form. They are mainly for the one-dimensional cases of an infinite or semi-infinite region with simple initial and boundary conditions and constant thermal properties. These exact solutions usually take the form of functions of the single variable $x/t^{\frac{1}{2}}$ and are known as similarity solutions. Corresponding solutions which are functions of $r/t^{\frac{1}{2}}$ only are available in cylindrical and spherical coordinates. Some similarity solutions have been obtained for combined heat and mass transfer in a semi-infinite region. A good collection of similarity solutions and references is to be found in Carslaw and Jaeger (1959).

3.2. Neumann's solution; generalizations; volume changes

A similarity solution which predates the work of Stefan himself was apparently presented in lectures given by Franz Neumann in the 1860s (Riemann-Weber 1912). It refers to a semi-infinite region, $x > 0$, initially occupied by a liquid at a constant temperature greater than the melting point and with the surface, $x = 0$, subsequently maintained at a constant temperature below that of melting. Neumann's solution for this problem is reproduced by Carslaw and Jaeger (1959, Chapter IX) together with Stefan's results (1891) for the special case in which the liquid is initially at its melting point. In fact, this latter result had been obtained previously by Lamé and Clapeyron (1831). The corresponding problem of melting a solid in the region $x > 0$, initially at a temperature below the melting point, is presented here to illustrate Neumann's method. It is convenient to refer to ice and water purely for descriptive reasons and to adopt the nomenclature of §1.2.2, except that x, t, k etc. will be used for convenience even when physical variables with appropriate dimensions are signified.

The problem is formulated by the system

$$\frac{\partial u_1}{\partial t} = k_1 \frac{\partial^2 u_1}{\partial x^2}, \qquad 0 < x < s(t), \tag{3.1}$$

$$\frac{\partial u_2}{\partial t} = k_2 \frac{\partial^2 u_2}{\partial x^2}, \qquad x > s(t), \tag{3.2}$$

$$-K_1 \frac{\partial u_1}{\partial x} + K_2 \frac{\partial u_2}{\partial x} = L\rho \frac{ds}{dt}, \qquad x = s(t), \tag{3.3}$$

$$u_1 = U_1, \qquad x = 0, \qquad t \geqslant 0, \tag{3.4}$$

$$u_2 = -U_2, \qquad x \to \infty, \qquad t \geqslant 0, \tag{3.5}$$

$$u_1 = u_2 = 0, \qquad x = s(t), \qquad t \geqslant 0, \tag{3.6}$$

where $k_i = K_i/\rho c_i$, $i = 1, 2$. No initial conditions are specified at this stage of Neumann's solution. They emerge later. Equations (3.1) and (3.4) are satisfied by

$$u_1 = U_1 + A \operatorname{erf} \frac{x}{2(k_1 t)^{\frac{1}{2}}}, \tag{3.7}$$

and (3.2) and (3.5) by

$$u_2 = -U_2 + B \operatorname{erfc} \frac{x}{2(k_2 t)^{\frac{1}{2}}}, \tag{3.8}$$

where A and B are constants to be determined. Condition (3.6) requires

$$A \operatorname{erf} \frac{s}{2(k_1 t)^{\frac{1}{2}}} = -U_1, \qquad B \operatorname{erfc} \frac{s}{2(k_2 t)^{\frac{1}{2}}} = U_2. \tag{3.9}$$

The two relationships (3.9) can only be satisfied for all t if

$$s = \alpha t^{\frac{1}{2}}, \tag{3.10}$$

where α is a constant. By differentiating (3.7) and (3.8) and using (3.3) and (3.9) the constant α is seen to be given by the root of

$$\frac{U_1 K_1 e^{-\alpha^2/4k_1}}{(\pi k_1)^{\frac{1}{2}} \operatorname{erf}(\alpha/2k_1^{\frac{1}{2}})} - \frac{U_2 K_2 e^{-\alpha^2/4k_2}}{(\pi k_2)^{\frac{1}{2}} \operatorname{erfc}(\alpha/2k_2^{\frac{1}{2}})} = \frac{L\rho\alpha}{2}, \tag{3.11}$$

where ρ is the density of both ice and water implying that no volume change occurs on melting.

Table 3.7 gives values of $\lambda = \alpha/(2k_1^{\frac{1}{2}})$ computed by Churchill and Evans (1971) for a wide selection of parameters. Some roots for water and ice and different initial and surface temperatures are given by Carslaw and Jaeger (1959, p. 286). Once α has been found from (3.11), u_1 and u_2 can be written down from (3.7), (3,8), and (3.9). They are

$$u_1 = U_1 - \frac{U_1}{\text{erf}(\alpha/2k_1^{\frac{1}{2}})} \, \text{erf} \, \frac{x}{2(k_1 t)^{\frac{1}{2}}}, \qquad (3.12)$$

$$u_2 = -U_2 + \frac{U_2}{\text{erfc}(\alpha/2k_2^{\frac{1}{2}})} \, \text{erfc} \, \frac{x}{2(k_2 t)^{\frac{1}{2}}}. \qquad (3.13)$$

It is now possible to see what initial conditions are satisfied by (3.10) and (3.13). At $t = 0$ they give $s = 0$ and $u_2 = -U_2$, i.e. the whole region $x > 0$ is solid at uniform temperature $-U_2$. The special case in which the solid is initially at its melting temperature, so that $U_2 = 0$, is the single-phase problem introduced in §1.2.1 and (3.11) reduces to

$$\lambda e^{\lambda^2} \, \text{erf} \, \lambda = U_1 c_1/(L\pi^{\frac{1}{2}}), \qquad (3.14)$$

where $\lambda = \alpha/(2k_1^{\frac{1}{2}})$. Carslaw and Jaeger (1959, p. 287) show a graph of the left-hand side of (3.14) as a function of λ from which values of λ can be read for any given value of $U_1 c_1/(L\pi^{\frac{1}{2}})$. They also point out that for small values of λ, and hence of the right-hand side of (3.14), use of the first term in the series expansion for erf λ gives approximately

$$\lambda^2 = U_1 c_1/2L. \qquad (3.15)$$

The solutions (3.12), (3.13) and the equation (3.11) are quoted by Carslaw and Jaeger (1959, p. 288). Their solution of the corresponding solidification of a liquid in $x > 0$ initially at a uniform temperature above freezing is deducible by suitable changes of nomenclature.

Carslaw and Jaeger (1959, Chapter XI) also present Neumann-type solutions for other physically important problems including the region $x > 0$ initially liquid and $x < 0$ solid, the case in which melting occurs over a temperature range, and three-phase problems. They solve the commonly neglected problem in which motion of the liquid results from a change of volume on solidification, due to a difference between the densities of solid and liquid. As an example they consider the freezing of a semi-infinite liquid in the region $x > 0$ when the density of the solid, ρ_1, exceeds that of the liquid, ρ_2, and both phases are incompressible. Thus the problem is defined by equation (3.1) in the solid phase but in the liquid phase (3.2) is replaced by

$$k_2 \frac{\partial^2 u_2}{\partial x^2} + \frac{(\rho_1 - \rho_2)}{\rho_2} \frac{ds}{dt} \frac{\partial u_2}{\partial x} - \frac{\partial u_2}{\partial t} = 0, \qquad x > s(t),$$

(see equations (1.32) and (1.33)). On the solidification boundary, $x = s(t)$, they take $u_1 = u_2 = U_M$, together with (3.3). For their solidification problem, instead of (3.4), (3.5), they have $u_1 = 0$, $x = 0$, $u_2 \to U$ as $x \to \infty$. The similarity solution based on $s = 2\lambda(k_1 t)^{\frac{1}{2}}$ leads to the equation for λ

$$\frac{e^{-\lambda^2}}{\text{erf }\lambda} - \frac{(U - U_M)K_2 k_1^{\frac{1}{2}} e^{-\lambda^2 \rho_1{}^2 k_1/\rho_2{}^2 k_2}}{U_M K_1 k_2^{\frac{1}{2}} \text{ erfc}(\lambda\rho_1 k_1^{\frac{1}{2}}/\rho_2 k_2^{\frac{1}{2}})} = \frac{\lambda L \pi^{\frac{1}{2}}}{c_1 U_M},$$

which reduces to (3.11) if $\rho_1 = \rho_2$, remembering that $\alpha = 2\lambda k_1^{\frac{1}{2}}$.

Rubenstein (1971) in considering the crystallization of a supercooled melt in contact with its solid phase (see Chambré 1956) also took density changes into account but did not impose the condition of incompressibility in the liquid phase. Thus the condition (1.32) still holds on the phase-change boundary with $v = v(s(t), t)$, but an equation of motion

$$\frac{\partial v}{\partial t} + v\frac{\partial v}{\partial x} = \nu\frac{\partial^2 v}{\partial x^2}, \qquad s(t) < x < \infty, \qquad t > 0$$

is to be satisfied by $v(x, t)$. Rubinstein takes the solid phase to occupy the space $-\infty < x < s(t)$ for $t > 0$ and the liquid phase $s(t) < x < \infty$, coupled with initial and boundary conditions

$$v = 0, \qquad u_1 = U_1, \qquad u_2 = U_2, \qquad t = 0,$$
$$v \to 0, \qquad u_1 \to U_1, \qquad x \to \infty, \qquad u_2 \to U_2, \qquad x \to -\infty.$$

The heat flow equations are

$$\frac{\partial u_2}{\partial t} = k_2\frac{\partial^2 u_2}{\partial x^2}, \qquad -\infty < x < s(t), \qquad t > 0,$$

$$\frac{\partial u_1}{\partial t} + v\frac{\partial u_1}{\partial x} = k_1\frac{\partial^2 u_1}{\partial x^2}, \qquad s(t) < x < \infty, \qquad t > 0,$$

and conditions (3.3) and $u_1 = u_2 = U_M$ still hold on the phase-change boundary. Rubinstein derives a similarity solution based on

$$s(t) \equiv 2\beta(\nu t)^{\frac{1}{2}}, \quad v = 2(\nu/\pi t)^{\frac{1}{2}}f(\eta), \quad u_i = u_i(\eta), \quad i = 1, 2$$
$$\eta = x/\{2(\nu t)^{\frac{1}{2}}\},$$

where β, $f(\eta)$, and $u_i(\eta)$ are to be determined from

$$\ddot{f}(\eta) - (4/\pi^{\frac{1}{2}})\dot{f}(\eta)f(\eta) + 2\{f(\eta) + \eta\dot{f}(\eta)\} = 0,$$
$$\ddot{u}_1(\eta) + \{2\sigma_1\eta - (4/\pi^{\frac{1}{2}})\sigma_1 f(\eta)\}\dot{u}_1(\eta) = 0,$$
$$\ddot{u}_2(\eta) + 2\sigma_2\eta\dot{u}_2(\eta) = 0,$$
$$u_i(\beta) = u_M, \qquad u_i(\eta) \to U_i, \qquad |\eta| \to \infty, \qquad i = 1, 2,$$
$$f(\beta) = -\tfrac{1}{2}\pi^{\frac{1}{2}}\varepsilon\beta, \qquad \text{where} \qquad \varepsilon = (\rho_2 - \rho_1)/\rho_1,$$
$$\dot{f}(\eta) \to f(\eta) \to 0, \eta \to \infty; -\mu\beta = \{\dot{u}_1(\eta) - \gamma\dot{u}_2(\eta)\}, \eta = \beta,$$

with

$$\sigma_1 = \nu/k_1, \qquad \sigma_2 = \nu/k_2, \qquad \gamma = K_2/K_1, \qquad \mu = 2\nu\lambda\rho_2/K_1.$$

It follows easily that

$$f(\eta) = -\frac{\varepsilon\beta \exp\{-(\eta^2 - \beta^2)\}}{2/\pi^{\frac{1}{2}} + \varepsilon\beta(\text{erf}\ \eta - \text{erf}\ \beta)\exp(\beta^2)},$$

$$u_1(\eta) = U_1 + (u_M - U_1)R(\eta, \beta),$$

where

$$R(\eta, \beta) = \int_\eta^\infty r(t)\,\mathrm{d}t \Big/ \int_\beta^\infty r(t)\,\mathrm{d}t$$

with

$$r(t) = e^{-\sigma_1 t^2}\{2/\pi^{\frac{1}{2}} + \varepsilon\beta(\text{erf}\ t - \text{erf}\ \beta)e^{\beta^2}\}^{-2\sigma_1},$$

and finally

$$u_2(\eta) = U_2 + (u_M - U_2)\frac{1 + \text{erf}\{\eta\sqrt{\sigma_2}\}}{1 + \text{erf}\{\beta\sqrt{\sigma_2}\}}.$$

These relationships yield an equation for β. Rubinstein pursues the simplest case of $\sigma_1 = 1$, which he considers relevant for a gas, and obtains

$$u_1 = U_1 + (u_M - U_1)\frac{(2/\sqrt{\pi})\text{erfc}\ \eta}{\{2/\sqrt{\pi} + \varepsilon\beta(\text{erf}\ \eta - \text{erf}\ \beta)\exp(\beta^2)\}\text{erfc}\ \beta}$$

and

$$\{\mu - \varepsilon(u_M - U_1)\}\beta = \frac{2}{\sqrt{\pi}}\left\{(u_M - U_1)\frac{\exp(-\beta^2)}{\text{erfc}\ \beta} + \frac{\gamma\sigma_2^{\frac{1}{2}}(u_M - U_2)\exp(-\sigma_2\beta^2)}{1 + \text{erf}(\beta\sqrt{\sigma_2})}\right\}.$$

Considerable simplification occurs if $U_2 = u_M$, i.e. the solid phase is initially at the crystallization temperature throughout.

In a problem with several phases of different densities, Wilson (1982) denoted by S_i, $i = 1, 2, \ldots, n-1$, the free boundaries separating phases of densities ρ_i, $i = 1, 2, \ldots, n$, together with $S_0 = 0$ and S_n is the right-hand boundary, not necessarily at infinity. If the whole slab is of mass M, conservation of mass is expressed by

$$S_n(t) = M/\rho_n + \sum_{j=1}^{n-1} S_j(t)(\rho_{j+1} - \rho_j)/\rho_n$$

as long as $S_{n-1}(t)$ does not exceed $S_n(t)$. When $S_{n-1}(t)$ determined by the free-boundary problem becomes equal to $S_n(t)$, the nth phase has disappeared and a new free-boundary problem with $(n-1)$ phases is to be considered. Similarly, a growing number of phases could be considered.

The first local coordinate $\xi_1(x, t)$ is simply $\xi_1(x, t) = x$, $0 \le x \le s_1(t)$, and the position of the first moving boundary is denoted by $\Gamma_1(t)$ in the local system so that $\Gamma_1(t) = \xi_1(S_1(t), t) = S_1(t)$. Introducing $\mu_i = \rho_i/\rho_{i+1}$, $i = 1, 2, \ldots, n-1$, we further define $\xi_2(x, t) = x - (1 - \mu_1)\Gamma_1(t)$ and $\Gamma_2(t) = \xi_2(S_2(t), t) = S_2(t) - (1 - \mu_1)\Gamma_1(t)$ for $S_1(t) \leqslant x \leqslant S_2(t)$. In general for $S_{i-1}(t) \leqslant x \leqslant S_i(t)$,

$$\xi_i(x, t) = x - \sum_{j=1}^{i-1} (1 - \mu_j)\Gamma_j(t),$$

$$\Gamma_i(t) = S_i(t) - \sum_{j=1}^{i-1} (1 - \mu_j)\Gamma_j(t).$$

For each i, ξ_i is at rest in the ith phase because of the conservation of mass at the various interfaces.

The position of the $(i-1)$th interface in the coordinate system at rest with respect to the ith phase is given by

$$\xi_i(S_{i-1}(t), t) = S_{i-1}(t) - \sum_{j=1}^{i-1} (1 - \mu_j)\Gamma_j(t)$$

$$= \Gamma_{i-1}(t) + \sum_{j=1}^{i-2} (1 - \mu_j)\Gamma_j(t) - \sum_{j=1}^{i-1} (1 - \mu_j)\Gamma_j(t) = \mu_{i-1}\Gamma_{i-1}(t).$$

This can be made to include $i = 1$ by taking $\Gamma_0 = 0$. In order to retrieve x and S_i after computation in the local coordinates we need

$$x = \xi_1(x, t) + \sum_{j=1}^{i-1} (1 - \mu_j)\Gamma_j(t), \qquad \mu_{i-1}\Gamma_{i-1} < \xi_i < \Gamma_i,$$

$$S_i(t) = \Gamma_i(t) + \sum_{j=1}^{i-1} (1 - \mu_j)\Gamma_j(t),$$

for $i = 1, 2, \ldots, n$ and with $\Gamma_0 = 0$. It follows that if each Γ_i is proportional to t^γ for some power γ, then so is S_i. The linear system for the S_i in terms of the Γ_i may be inverted to give

$$\Gamma_i = \rho_i^{-1} \sum_{j=1}^{i} \rho_j(S_j - S_{j-1}), \qquad i = 1, 2, \ldots, n-1.$$

A problem considered by Weiner (1955) and for which Wilson (1978) established the existence and uniqueness of similarity solutions can be used as an example. The one-dimensional heat flow equation is to be solved in consecutive regions of the half-space $x > 0$ separated by phase-change boundaries whose motions are to be determined. The constant densities of the separate phases can differ from each other. As phase changes occur the neighbouring regions are assumed to remain in contact but energies associated with acceleration, friction, etc. are neglected. The dependent variable, u, may undergo a finite discontinuous jump at an interface between specified, constant, limiting values on the two sides and is also constant at the fixed boundary, $x = 0$. Initially all free boundaries coincide at $x = 0$, and $u = U_n = $ constant for $x > 0$, for $t = 0$ and $u \to U_n$ as $x \to \infty$, $t \geq 0$. Here $S_0 = 0$ and $S_n = \infty$.

By using the nomenclature defined above and assuming the usual Stefan condition to hold on each interface, with L_i denoting latent heat per unit mass of the ith phase for the change from the ith to the $(i+1)$th phase, the problem can be stated as follows:

$$u_{it} = k_i u_{i\xi\xi}, \qquad u_{i-1}\Gamma_{i-1}(t) < \xi < \Gamma_i(t), \qquad t > 0, \qquad i = 1, \ldots, n,$$

$$u_i(t, \mu_{i-1}\Gamma_{i-1}^+(t)) = U_{i-1}, \qquad t > 0, \qquad i = 1, \ldots, n,$$

$$u_i(t, \Gamma_i^-(t)) = U_i + \Delta_i, \qquad t > 0, \qquad i = 1, \ldots, n-1,$$

$$u_n(t, \xi) \to U_n, \qquad \xi \to \infty, \qquad t > 0,$$

$$\rho_i L_i \dot{\Gamma}_i = -K_i u_{i\xi}(t, \Gamma_i^-) + K_{i+1} u_{i+1,\xi}(t, \mu_i \Gamma_i^+), \qquad i = 1, \ldots, n-1,$$

$$\Gamma_0(t) = 0, \qquad \Gamma_n(t) = +\infty, \qquad t \geq 0, \qquad \Gamma_i(0) = 0, \qquad i = 1, \ldots, n-1,$$

$$u_n(0, \xi) = U_n, \qquad \xi > 0, \qquad t = 0.$$

Here K_i denotes heat conductivity and k_i heat diffusivity. Weiner's (1955) similarity solution is expressed by Wilson (1982) as

$$u_i(t, \xi) = (U_{i-1} + \alpha_i)[\mathrm{erfc}\{\xi/\sqrt{(4k_i t)}\}/\mathrm{erfc}\{a_{i-1}\mu_{i-1}\sqrt{(k_{i-1}/k_i)}\}] - \alpha_i,$$

$$t > 0, \qquad \mu_{i-1}\Gamma_{i-1}(t) < \xi < \Gamma_i(t), \qquad i = 1, \ldots, n-1,$$

$$\Gamma_i(t) = a_1\sqrt{(4k_i t)}, \qquad i = 1, \ldots, n-1,$$

and for $\xi > \Gamma_{n-1}(t)$, $t > 0$

$$U_n(t, \xi) = \begin{cases} U_n & \text{if} \quad k_n = 0, \\[2mm] U_n + \dfrac{(U_{n-1} - U_n)\mathrm{erfc}\{\xi/\sqrt{(4k_n t)}\}}{\mathrm{erfc}\{a_{n-1}\mu_{n-1}\sqrt{(k_{n-1}/k_n)}\}} & \text{if} \quad k_n > 0. \end{cases}$$

In these equations, for $i = 1, \ldots, n-1$,

$$\alpha_1 = -U_{i-1} + (U_{i-1} - U_i - \Delta_i)\mathrm{erfc}\{a_{i-1}\mu_{i-1}\sqrt{(k_{i-1}/k_i)}\}/\Phi_i,$$

$$\Phi_i = \mathrm{erf}(a_i) - \mathrm{erf}\{a_{i-1}\mu_{i-1}\sqrt{(k_{i-1}/k_i)}\},$$

and the a_i $(i = 1, \ldots, n-1)$ are the unique solution of the non-linear equations

$$-\rho_i L_i k_i a_i \sqrt{\pi} = -K_i(U_{i-1} - U_i - \Delta_i)\exp(-a_i^2)/\Phi_1\sqrt{k_i}$$

$$+ K_{i+1}(U_i - U_{i+1} - \Delta_{i+1})\exp(-a_i^2 \mu_i^2 k_i/k_{i+1})/(\Phi_{i+1}\sqrt{k_{i+1}}),$$

$$i = 1, \ldots, n-2,$$

and

$$\rho_{n-1}L_{n-1}k_{n-1}a_{n-1}\sqrt{\pi} + K_{n-1}(U_{n-2} - U_{n-1} - \Delta_{n-1})\exp(-a_{n-1}^2)/(\Phi_{n-1}\sqrt{k_{n-1}})$$

$$= \begin{cases} 0 & \text{if} \quad k_n = 0 \\ K_n(U_{n-1} - U_n)\exp(-a_{n-1}^2\mu_{n-1}^2 k_{n-1}/k_n)/(\Phi_n\sqrt{k_n}) & \text{if} \quad k_n \neq 0. \end{cases}$$

In this last expression consistency with the previous definition is achieved by taking $\text{erf}(a_n) \equiv 1$ and $\Phi_n = 1 - \text{erf}\{a_{n-1}\mu_{n-1}\sqrt{(k_{n-1}/k_n)}\}$. In the definition of Φ_1 the dummy parameters a_0, k_0 are taken to be zero.

Wilson (1978) established the existence and uniqueness of the similarity solution subject to the following conditions: $k_i > 0$, $K_i > 0$, $i = 1, \ldots, n-1$, and either $k_n > 0$, $K_n > 0$, or $k_n = K_n = 0$; and either $U_{i-1} > U_i + \Delta_i$, $L_i \geqslant 0$, $i = 1, \ldots, n-1$ and $U_{n-1} \geqslant U_n$; or $U_{i-1} < U_i + \Delta_i$, $L_i \leqslant 0$, $i = 1, \ldots, n-1$ and $U_{n-1} \leqslant U_n$; provided that if $U_{n-1} = U_n$ then $L_{n-1} \neq 0$; and if $k_n = K_n = 0$ then $U_{n-1} = U_n$

Reverting to the original frame of reference we see that

$$S_i(t) = 2\left\{a_i\sqrt{k_i} + \sum_{j=1}^{i-1}(1-\mu_j)a_j\sqrt{k_j}\right\}\sqrt{t}, \qquad i = 2, \ldots, n,$$

and $S_1(t) = \Gamma_1(t) = 2a_1\sqrt{(k_1t)}$.

Although a similarity solution has been obtained for this example, Wilson (1982) suggests the use of the same local coordinates in combination with finite elements or finite differences in more complicated problems.

A problem relating to the manufacture of glass in which volume changes are substantial enough to be of practical importance is reported by Gelder and Guy (1975).

Carslaw and Jaeger (1959, p. 287) derive a similarity solution for the solidification of a supercooled liquid initially in $x > 0$, i.e. the region $x > 0$ contains material at temperature $V < T_1$ which is in the liquid state even though its temperature V is below the solidification temperature T_1. Solidification starts at the plane $x = 0$ and moves to the right; no heat is removed from the solidified material and so it will have the constant temperature T_1 throughout. In the supercooled liquid there will be a temperature profile of the form

$$v = V + A\,\text{erfc}\,\frac{x}{2(k_2t)^{\frac{1}{2}}}$$

and the usual Stefan conditions expressing heat conservation on the boundary, $s(t) = 2\lambda(k_2t)^{\frac{1}{2}}$, between the solid and liquid phases, leads to the relations $V + A\,\text{erfc}\,\lambda = T_1$ and $Ae^{-\lambda^2} = \lambda L\pi^{\frac{1}{2}}/c_2$. It follows that λ is the root of $\lambda e^{\lambda^2}\,\text{erfc}\,\lambda = (T_1 - V)c_2/L\pi^{\frac{1}{2}}$ and Carslaw and Jaeger give a graph of $\lambda e^{\lambda^2}\,\text{erfc}\,\lambda$.

The analytical properties of more general Stefan problems, which include superheated solids and supercooled liquids, have been examined by Sherman (1970). In his paper reference is made to several earlier papers of a similar nature.

The following equations describe the melting of a superheated solid:

$$\frac{\partial u}{\partial t} = \frac{\partial^2 u}{\partial x^2}, \qquad 0 < x < s(t), \qquad t > 0,$$

$$\frac{\partial u}{\partial x} = 0, \qquad x = 0, \quad t > 0,$$

$$u = \phi(x) \geqslant 0, \qquad \phi(a) = 0, \qquad 0 \leqslant x \leqslant s(0) \equiv a, \qquad t = 0,$$

$$u = 0, \qquad \frac{ds}{dt} = \frac{\partial u}{\partial x}, \qquad x = s(t), \qquad t > 0.$$

The solid initially occupies the region $0 \leqslant x \leqslant a$, and melting starts at $x = a$, assuming $\phi'(a) > 0$, due to heat arriving from the solid. The boundary condition stipulates that ds/dt is negative which corresponds to the physical situation that the melting front proceeds in the direction of x decreasing.

The same system of equations can describe the freezing of a liquid which has a 'negative' latent heat. In order to freeze such a liquid a positive amount of heat must be added. Thus if the region $0 \leqslant x \leqslant a$ were occupied initially by a liquid with 'negative' latent heat and temperature $u = \phi(x) \geqslant 0$, freezing would start at $x = a$ and progress in the direction of x decreasing according to the condition at $x = a$, $-\partial u/\partial x =$ heat arriving $=$ rate of freezing $= -ds/dt$, all in non-dimensional terms. Mathematically the problem of freezing with negative latent heat is the same as the melting of a superheated liquid.

The solidification or melting of a supercooled liquid or a superheated solid respectively, with their possible reformulation as Stefan problems with 'negative' latent heat, are of wider interest because their mathematical formulation and solution may be relevant to other physical situations. For example, if the specific heat is zero in part of the temperature range, the mathematical treatment may be relevant to certain types of Hele-Shaw flow with a free boundary surrounding a point sink (Ockendon 1978). These Hele-Shaw flows relate to the motion of a blob of Newtonian fluid sandwiched in the narrow gap between two plane parallel surfaces when further fluid is injected into or withdrawn from the blob at some fixed point (Richardson 1972). A fuller description has been given in §§2.12.1 and 2.12.2. Similar mathematical problems also arise in certain models of electrochemical machining (Fitz-Gerald and McGeough 1969, 1970; Crowley 1979). Further discussions of some numerical treatments of problems of this kind are to be found in §§2.12.2, 2.12.3, and 6.2.7. Crank (1975a, Chapter 13) gives Neumann-type solutions for several diffusion problems including the general approach by Danckwerts (1950) which also allows for movement of the medium on one side of the boundary relative to the other. Mikhailov (1975) obtains similarity solutions of coupled equations for temperature and moisture distributions together with the position of a moving evaporation front during the drying of a porous region $x > 0$, and he also studied (1976) the corresponding freezing problem. Tayler (1975) refers to similarity solutions describing the solidification of an alloy as formulated in §1.3.7 for some particular initial and boundary conditions. He also remarks on the limiting cases of instantaneous diffusion of heat or material and of no diffusion of material. Kamin (1976) and Peletier and Gilding (1976) obtain similarity solutions for a non-linear filtration problem obeying $\partial u/\partial t = \partial^2(u^m)/\partial x^2$, $m > 1$.

3.3. Similarity solutions in cylindrical and spherical coordinates

Similarity solutions of problems with radial symmetry, if they exist, can be expected to be functions of $r/t^{\frac{1}{2}}$ only by analogy with the linear cases discussed in §3.2. Frank (1950) presents solutions contained in a dissertation by Rieck in 1924 and quoted by Huber (1939), and applies them to spherical and cylindrical phase growth. Retaining Frank's nomenclature for ease of reference, concentration (or temperature) is denoted by $\phi = \phi(r, t) = \phi(s)$ where $s = r/(Dt)^{\frac{1}{2}}$, D is the diffusivity, and r the radial coordinate. Then the usual diffusion equation for spherical symmetry becomes

$$\frac{\partial^2 \phi}{\partial s^2} = -\left(\frac{s}{2} + \frac{2}{s}\right)\frac{\partial \phi}{\partial s}, \tag{3.16}$$

and successive integration yields

$$\phi - \phi_\infty = A\{s^{-1}\exp(-\tfrac{1}{4}s^2) - \tfrac{1}{2}\pi^{\frac{1}{2}}\operatorname{erfc}(\tfrac{1}{2}s)\}$$
$$\equiv AF_3(s), \tag{3.17}$$

where ϕ_∞ is the value of ϕ at infinity. The constant A is related by Frank (1950) to q, the amount of diffusant expelled per unit volume of new phase (analogous to latent heat) formed as growing proceeds, and he deduces from the conservation condition on $r = R$

$$A = \tfrac{1}{2}qS^3\exp(\tfrac{1}{4}S^2), \tag{3.18}$$

where $S = R/(Dt)^{\frac{1}{2}}$ and R is the outer radius of the sphere. On the surface

$r = R$, ϕ has the constant value ϕ_S, where

$$\phi_S = \tfrac{1}{2}qS^3 \exp(\tfrac{1}{4}S^2)F_3(S) + \phi_\infty$$
$$\equiv qf_3(S) + \phi_\infty. \tag{3.19}$$

For the corresponding radial cylindrical problem Frank (1950) obtains the solutions

$$\phi - \phi_\infty = -\tfrac{1}{2}A \; \mathrm{Ei}(-\tfrac{1}{4}s^2) \equiv AF_2(s), \tag{3.20}$$

where Ei is the exponential integral and on the surface of the growing cylinder the constant ϕ_S is

$$\phi_S = -\tfrac{1}{4}qS^2 \exp(\tfrac{1}{4}S^2)\mathrm{Ei}(-\tfrac{1}{4}S^2) + \phi_\infty$$
$$\equiv qf_2(S) + \phi_\infty. \tag{3.21}$$

In the linear case for comparison

$$f_1(S) = \tfrac{1}{2}S \exp(\tfrac{1}{4}S^2)F_1(S) \tag{3.22}$$

and

$$F_1(s) = \pi^{\frac{1}{2}} \mathrm{erfc}(\tfrac{1}{2}s). \tag{3.23}$$

Frank (1950) tabulates the functions $F_3(s)$, $f_3(S)$, $F_2(s)$, and $f_2(S)$ and quotes series and asymptotic expansions for $F(s)$ and $f(S)$ for both the spherical and cylindrical cases.

Carslaw and Jaeger (1959, p. 295) give similarity solutions in cylindrical and spherical coordinates where the region $r < R$ consists of solid at its melting point and $r > R$ of supercooled liquid whose temperature tends to $V <$ the melting temperature as $r \to \infty$. They also solve a cylindrical problem of freezing by an axial line source.

3.4. Similarity solutions and moving heat sources

Lightfoot (1929) developed a method suitable if the thermal properties of the solid and liquid are assumed to be the same. If the solidification front in a one-dimensional problem is at $x = s(t)$, and moves with a velocity $\dot{s}(t)$, the latent heat of solidification is liberated on the front at a rate

$$L\rho\dot{s}(t). \tag{3.24}$$

This liberation of heat may be represented by a moving source of heat at $s(t)$ of strength (3.24). The temperature at any point will be due to the superposed contributions from this moving source, having the initial and boundary conditions in mind. An integral equation for $s(t)$ can be developed which expresses the fact that the temperature on the boundary at $x = s(t)$ is always the melting temperature. In this way, Lightfoot

(1929) solved the problem of solidification in $x > 0$ with initial temperature U above the melting temperature U_m and the surface $x = 0$ maintained at zero for $t > 0$. Carslaw and Jaeger (1959, p. 293) give the following alternative presentation of Lightfoot's method. The temperature due to the plane moving heat source (3.24) may be written

$$u(x, t) = \frac{L}{2c(\pi k)^{\frac{1}{2}}} \int_0^t \frac{\dot{s}(\tau)\, d\tau}{(t-\tau)^{\frac{1}{2}}} \left[e^{-\{x-s(\tau)\}^2/4k(t-\tau)} - e^{-\{x+s(\tau)\}^2/4k(t-\tau)} \right], \quad (3.25)$$

which is obtained by summing contributions to the temperature at (x, t) from sources of strength $L\rho\dot{s}(\tau)\, d\tau$ instantaneously at $x = s(\tau)$ for the short time interval τ to $\tau + d\tau$, by integrating over the whole period of time $\tau = 0$ to t.

In addition, the temperature at (x, t) is contributed to by the heat originally in the region $x > 0$ associated with the initial uniform temperature U. This contribution, having in mind the surface condition of zero temperature at $x = 0$, is

$$w(x, t) = U \operatorname{erf} \frac{x}{2(kt)^{\frac{1}{2}}}, \quad (3.26)$$

which also satisfies the initial condition $w(x, 0) = U$, $x > 0$. Thus, the condition that the temperature at $x = s(t)$ is the melting temperature U_m is, on combining (3.25) and (3.26),

$$u\{s(t), t\} + w(s(t), t) = U_m, \quad (3.27)$$

which is an integral equation for $s(t)$.

Lightfoot solved the problem by adopting the similarity transformation

$$s(t) = 2\lambda(kt)^{\frac{1}{2}} \quad (3.28)$$

and expressing the integral (3.25) in terms of error functions. In terms of new variables

$$y = x/s(t), \qquad \tau = \left(\frac{1-z^2}{1+z^2}\right)^2, \quad (3.29)$$

(3.25) becomes

$$u(x, t) = \frac{2L\lambda}{c\pi^{\frac{1}{2}}} (I_1 - I_2), \quad (3.30)$$

where the integrals I_1 and I_2 reduce to

$$I_1 = \tfrac{1}{4}\pi e^{\lambda^2} \operatorname{erfc} \lambda (1 + \operatorname{erf} \lambda y), \qquad y < 1, \quad (3.31)$$

$$= \tfrac{1}{4}\pi e^{\lambda^2} \operatorname{erfc} \lambda y (1 + \operatorname{erf} \lambda), \qquad y > 1, \quad (3.32)$$

$$I_2 = \tfrac{1}{4}\pi e^{\lambda^2} \operatorname{erfc} \lambda y \operatorname{erfc} \lambda, \quad (3.33)$$

where clearly from the first of (3.29), $y \gtrless 1$ correspond to the liquid and solid regions respectively, and $y = 1$ is the solidification front where (3.31) and (3.32) are identical. Substituting (3.31) and (3.33) with $y = 1$, and (3.26), the integral equation (3.27) becomes

$$\frac{L\lambda\pi^{\frac{1}{2}}}{c} e^{\lambda^2} \operatorname{erfc} \lambda \operatorname{erf} \lambda + U \operatorname{erf} \lambda = U_m, \qquad (3.34)$$

which is the Neumann expression for λ in this problem when $K_1 = K_2$ and $k_1 = k_2$. Lightfoot (1929) extends (3.25) so as to include an infinite number of image terms instead of just the two given and obtains an approximate solution for the solidification of a finite plane sheet.

Rathjen and Jiji (1971) extend Lightfoot's idea of treating the solidification front as a moving heat source to obtain an analytical solution of the two-dimensional problem of the solidification of a liquid, initially at a uniform temperature and occupying the quarter-space $x, y > 0$, subject to a constant temperature on the boundaries $x = 0$, $y = 0$. They express the temperature as the sum of two parts. The first is a standard solution of the heat flow equation in a rectangular corner, without phase changes, satisfying the initial and boundary conditions of the present problem; the second part represents the contribution from the moving heat source by using Green's functions for the rectangular corner with zero initial and surface temperatures. After some manipulation and use of similarity variables $x/t^{\frac{1}{2}}$, $y/t^{\frac{1}{2}}$, an integro-differential equation for the movement of the solidification boundary is obtained. Numerical solutions are computed by using superhyperbolae to approximate the position of the moving boundary. Values of the temperature field follow for all times. Comparisons are made with finite-difference calculations and with experimental results.

By a convenient choice of temperature variables, Lightfoot's restriction of equal conductivities and specific heats in the two phases is replaced by the requirement that only the ratios of conductivity to specific heat need be the same. Even this restriction is unnecessary if the liquid is initially at the solidification temperature.

Budhia and Kreith (1973) proceed very much in the same way as Rathjen and Jiji (1971) to obtain analytical solutions for melting or freezing of materials in a wedge-shaped space with wedge angle between 0 and 360 degrees. The material is initially at a uniform temperature within the wedge and the surface temperatures are kept constant but not necessarily equal. The solutions are expressed in radial coordinates (r, ϕ) and time t, and a similarity transformation $r/t^{\frac{1}{2}}$ is justified. To solve their integro-differential equation Budhia and Kreith represent the moving boundary by a one-parameter hyperbola.

Computed results are given for three examples: (i) liquid at the fusion temperature inside a 90° wedge with the surface temperatures equal and below the fusion temperature; (ii) the same problem except that the liquid is initially at a temperature higher than the fusion temperature and wedge angles of 60° and 270° are considered as well as 90° (a way of improving a first analytic approximation is used when the diffusivities in the two phases differ); and (iii) the same problem but the surfaces of the 90° wedge are maintained at unequal temperatures lower than the fusion temperature.

3.5. Integral-equation formulations

The use of moving heat sources in §3.4 is a special way of deriving an integral equation and likewise the similarity solution is a particular form. Several other more general ways of formulating moving-boundary problems in terms of integral equations and their solutions have been found useful. One advantage of introducing integral equations in linear heat-flow problems generally is that only the values of the unknowns on the boundaries of the domain enter into the formulation. Although Stefan-type problems are essentially non-linear because of the condition on the moving boundary, nevertheless linearity frequently exists in each of the domains on either side of the boundary and the integral equation method is still useful.

3.5.1. Integral transforms

Evans *et al.* (1950) used Laplace transforms to express a one-phase moving boundary problem of the kind described in §1.2.1 but with a constant heat source applied to the outer surface of the domain. In the nomenclature of Evans *et al.* (1950) the problem is to find $x = x(t)$ where

$$\frac{\partial u}{\partial t} = \alpha^2 \frac{\partial^2 u}{\partial x^2}, \qquad \alpha^2 = K/c\rho, \tag{3.35}$$

$$u = 0, \qquad x \geqslant x(t), \tag{3.36}$$

$$A \frac{\mathrm{d}x(t)}{\mathrm{d}t} = \frac{\partial u}{\partial x}(x(t), t), \qquad A = L\rho/K, \tag{3.37}$$

$$x(0) = 0, \tag{3.38}$$

$$\partial u/\partial x = g(t), \qquad x = 0, \qquad t \geqslant 0, \tag{3.39}$$

where $g(t)$ is a time-dependent heat input and L here is the latent heat of recrystallization from α to β crystals when iron is heated through this phase-change temperature.

In order to obtain the integral equation satisfied by the moving boundary, $x = x(t)$, we rewrite this relationship as $t = f(x)$ and reformulate the problem as

$$\frac{\partial u}{\partial t} = \alpha^2 \frac{\partial^2 u}{\partial x^2}, \qquad t > f(x) \tag{3.40}$$

$$u(x, t) = 0, \qquad t \leqslant f(x), \tag{3.41}$$

$$\partial u/\partial x = g(t), \qquad x = 0, \qquad t \geqslant 0, \tag{3.42}$$

$$f(0) = 0, \tag{3.43}$$

$$\frac{\partial u}{\partial x}(x, f(x)) = \frac{A}{f'(x)}, \qquad f' = df/dx. \tag{3.44}$$

Define the Laplace transform of $u(x, t)$ as

$$\bar{u}(x, p) = \int_0^\infty u(x, t)e^{-pt}\, dt = \int_{f(x)}^\infty ue^{-pt}\, dt, \tag{3.45}$$

and hence from (3.40) obtain

$$\frac{d^2\bar{u}}{dx^2} - \frac{p\bar{u}}{\alpha^2} = -Ae^{-pf(x)}, \tag{3.46}$$

with the boundary conditions

$$\partial\bar{u}/\partial x = \bar{g}, \qquad x = 0; \qquad \lim_{x\to\infty} \bar{u}(x, p) = 0. \tag{3.47}$$

The solution of (3.46) is

$$\bar{u}(x, p) = \frac{A\alpha}{2p^{\frac{1}{2}}} \int_0^\infty \exp\left\{-pf(\xi)\right\}\left[\exp\left\{-\frac{p^{\frac{1}{2}}}{\alpha}|x-\xi|\right\} + \exp\left\{-\frac{p^{\frac{1}{2}}}{\alpha}(x+\xi)\right\}\right] d\xi$$

$$-\frac{\alpha\bar{g}}{p^{\frac{1}{2}}} \exp\left\{\frac{-p^{\frac{1}{2}}}{\alpha}x\right\}, \tag{3.48}$$

and the inverse transform is

$$u(x, t) = \frac{A\alpha}{2} \int_0^\infty \frac{1}{[\pi\{t-f(\xi)\}]^{\frac{1}{2}}} \left[\exp\left\{\frac{-(x-\xi)^2}{4\alpha^2[t-f(\xi)]}\right\} + \right.$$

$$\left. + \exp\left\{\frac{-(x+\xi)^2}{4\alpha^2[t-f(\xi)]}\right\}\right] d\xi$$

$$- \alpha \int_0^t \frac{g(t-\tau)}{(\pi t)^{\frac{1}{2}}} \exp\left(\frac{-x^2}{4\alpha^2\tau}\right) d\tau. \tag{3.49}$$

Use of the condition $u(x, f(x)) = 0$ from (3.41) in 3.49) gives

$$2 \int_0^{f(x)} \frac{g(t-\tau)}{\tau^{\frac{1}{2}}} \exp\left(\frac{-x^2}{4\alpha^2 \tau}\right) d\tau$$

$$= A \int_0^\infty \frac{1}{\{f(x)-f(\xi)\}^{\frac{1}{2}}} \left[\exp\left\{\frac{-(x-\xi)^2}{4\alpha^2[f(x)-f(\xi)]}\right\} + \right.$$

$$\left. + \exp\left\{\frac{-(x+\xi)^2}{4\alpha^2[f(x)-f(\xi)]}\right\}\right] d\xi, \qquad f(x) > f(\xi), \quad (3.50)$$

which is an integral equation for the determination of $t = f(x)$. In order to solve equation (3.50) we return to the specification of the moving boundary as $x = x(t)$ and put $t = f(x)$, $\tau = f(\xi)$, $x = x(t)$, $\xi = \xi(\tau)$ which is permissible provided $t = f(x)$, is monotonically increasing. After these changes on the right-hand side of (3.50) we replace the new $t - \tau$ by τ' and finally drop the superscript and write $\tau' = \tau$ so that (3.50) becomes

$$2 \int_0^t \frac{g(t-\tau)}{\tau^{\frac{1}{2}}} \exp\left\{\frac{-x^2(t)}{4\alpha^2 \tau}\right\} d\tau$$

$$= A \int_0^t \frac{x(t-\tau)}{\tau^{\frac{1}{2}}} \left[\exp\left\{\frac{-[x(t)-x(t-\tau)]^2}{4\alpha^2 \tau}\right\} + \right.$$

$$\left. + \exp\left\{\frac{-[x(t)+x(t-\tau)]^2}{4\alpha^2 \tau}\right\}\right] d\tau. \qquad \tau > 0. \quad (3.51)$$

For the case of $g(t) = g = $ constant, Evans *et al.* (1950) obtain the first two terms in a power series solution for $x = x(t)$ i.e.

$$x(t) = C_1 t + C_2 t^2 + \dots \qquad (3.52)$$

where

$$C_1 = g/A, \qquad C_2 = -\frac{1}{2} \frac{g^3}{A^3 \alpha^2}, \qquad (3.53)$$

which agrees with the first two terms of a Taylor series obtained by the same authors for $x(t)$ about $t = 0$. When $g(t)$ is an analytic function of t, the first two coefficients C_1, C_2 in the solution of (3.51) are also shown to agree with the corresponding terms of the appropriate Taylor series. They are

$$C_1 = g(0)/A, \qquad C_2 = \frac{-g^3(0)}{2A^3 \alpha^2} + \frac{\dot{g}(0)}{2A}. \qquad (3.54)$$

Ockendon (1975) shows how an integral formulation of the oxygen diffusion problem described by equations (1.63) in §1.3.10 follows by a similar use of Laplace transforms. The integral equations are found to be those derived by Hansen and Hougaard (1974) and they are obtained in

§3.5.2 by using Green's functions. Again it is necessary that $x(t)$ should be monotonic.

Ockendon (1975) also draws attention to the usefulness of Fourier transforms in infinite or semi-infinite domains and cites the Fourier transform derivation of Lightfoot's (1929) integral equation based on a moving heat source as in equations (3.25), (3.26), and (3.27).

3.5.2. Green's functions

The use of Green's functions for the solution of heat flow problems subject to conditions prescribed on fixed boundaries is well known. Carslaw and Jaeger (1959, Chapter 14) give the basic theory and solutions to a selection of standard problems in terms of appropriate Green's functions. Kolodner (1956, 1957) formulated Stefan problems in terms of integral equations, including the problem of the freezing of a lake of finite depth, by using simple Green's functions, though he does not explicitly say so. It is interesting that in such an early paper Kolodner remarked that the equations in cases of physical interest are all of Volterra type of the second kind and hence are amenable to numerical treatment. Rubinstein (1971) based an analysis of the existence, uniqueness, and stability of solutions of Stefan problems on integral equation formulations incorporating Green's functions. He also briefly examined the numerical evaluation of integral solutions coupled with estimates of accuracy. More recently Rubinstein (1980*a*) discussed the application of integral equation techniques to several Stafan problems including a one-dimensional problem with analytical input data, the solidification of a binary alloy, an axially symmetric problem with concentrated thermal capacity (see §1.3.11), and a problem with a hyperbolic instead of the more usual parabolic heat conduction equation. Collatz (1978) too used integral equations to find practical error bounds. Chuang and Szekely (1971) employed Green's functions to solve, in integral equation form, the problem of a solid slab, symmetrically placed in its own melt. There is a prescribed initial temperature, not necessarily uniform, and on the moving, outer surfaces $x = \pm X(t)$ the temperatures, u, are the melting temperature, u_M, and there is a second condition of the form

$$\alpha \frac{\partial u}{\partial x} + \beta u = f(t), \qquad x = \pm X(t),$$

with $X(t) = \ell$ at $t = 0$. The numerical evaluation of the integral solution proceeds by a process of successive approximation in which the locus of the melting boundary is approximated by linear segments in the (x, t) plane. Later papers by Chuang and Szekely (1972) and Chuang and Ehrich (1974) deal with corresponding cylindrical and spherical problems.

An extension to the solidification of iron–carbon alloys is outlined by Chuang *et al.* (1975), who give some numerical results.

Hansen and Hougaard (1974) applied Green's functions to obtain solutions of the oxygen diffusion problem prescribed by equations (1.57) to (1.60) in §1.3.10. This paper contains useful details of the analysis and also numerical comparisons with those obtained by Crank and Gupta (1972a,b). It has the added interest that Hansen and Hougaard treat the problem in its implicit form (Sackett 1971) in which the moving boundary condition does not contain the velocity of the boundary explicitly. Their solution proceeds along the following lines:

A Green's function G defined by

$$G(x, x', t' - t) = \frac{1}{2\pi\sqrt{\pi}(t' - t)} \left\{ \exp\left(\frac{-(x - x')^2}{4(t' - t)}\right) + \exp\left(\frac{-(x + x')^2}{4(t' - t)}\right) \right\} \quad (3.55)$$

for $t' > t$ and

$$G(x, x', t' - t) = 0 \quad (3.56)$$

for $t > t'$, satisfies

$$\frac{\partial^2 G}{\partial x^2} + \frac{\partial G}{\partial t} = -\delta(x - x')\delta(t' - t), \quad (3.57)$$

where δ is the Dirac delta function, for x and $x' \geqslant 0$ and for all t. It also satisfies the boundary condition

$$G_x(0, x', t' - t) = 0 \quad (3.58)$$

for all t and $x' > 0$.

Following the standard treatment of Green's functions the integral

$$\int_0^{t'} \int_0^{x_0(t)} \left[G\left(\frac{\partial^2 c}{\partial x^2} - \frac{\partial c}{\partial t}\right) - c\left(\frac{\partial^2 G}{\partial x^2} + \frac{\partial G}{\partial t}\right) \right] dx \, dt \quad (3.59)$$

reduces to

$$\int_0^{t'} \int_0^{x_0(t)} G \, dx \, dt + c(x', t'), \quad (3.60)$$

having in mind the oxygen diffusion equation (1.57) and (3.57). But also, integrating (3.59) by parts and using (3.56), the integral can be written

$$\int_0^{t'} \left[G\frac{\partial c}{\partial x} - c\frac{\partial G}{\partial x} \right]_{x=0}^{x=x_0(t)} dt - \int_0^1 [Gc]_{t=0}^{t=t_0(x)} dx, \quad (3.61)$$

where $t = t_0(x)$ is the inverse of $x = x_0(t)$, the path of the moving boundary in the (x, t) plane. The first integral in (3.61) and the contribution from $t = t_0(x)$ in the second one vanish because of the two boundary conditions (1.59) and (1.58). On inserting the initial condition (1.60) and then equating (3.61) and (3.60) we find

$$c(x', t') = -\int_0^{t'} \int_0^{x_0(t)} G(x, x', t'-t)\, dx\, dt + \frac{1}{2}\int_0^1 G(x, x', t')(1-x)^2\, dx.$$

(3.62)

It follows from (3.56) and (3.57) that

$$G(x, x', t') = \int_0^\infty [\delta(x-x')\delta(t-t') + \frac{\partial^2 G}{\partial x^2}(x, x', t'-t)]\, dt, \qquad (3.63)$$

and so (3.62) becomes

$$c(x', t') = \tfrac{1}{2}(1-x')^2 - \int_0^{t'} \int_0^{x_0(t)} G(x, x', t'-t)\, dx\, dt$$

$$+ \frac{1}{2}\int_0^\infty \int_0^1 \frac{\partial^2 G}{\partial x^2}(x, x', t'-t)(1-x)^2\, dx\, dt. \qquad (3.64)$$

Integrating the last integral of (3.64) twice by parts with respect to x and using (3.58) and (3.56) we obtain

$$c(x', t') = \tfrac{1}{2}(1-x')^2 - \int_0^{t'} G(0, x', t'-t)\, dt +$$

$$+ \int_0^{t'} \int_{x_0(t)}^1 G(x, x', t'-t)\, dx\, dt. \qquad (3.65)$$

We can now substitute the chosen form of $G(x, x', t'-t)$ from (3.55) and obtain

$$c(x', t') = \tfrac{1}{2}(1-x')^2 - 2\left(\frac{t'}{\pi}\right)^{\frac{1}{2}} \exp\left(-\frac{x'^2}{4t'}\right) +$$

$$+ x' \operatorname{erfc}\left(\frac{x'}{2\sqrt{t'}}\right) + R(x', t'), \qquad (3.66)$$

where

$$R(x', t') = \frac{1}{2}\int_0^{t'} \left\{ \operatorname{erfc}\left[\frac{x_0(t)-x'}{2(t'-t)^{\frac{1}{2}}}\right] - \operatorname{erfc}\left[\frac{1-x'}{2(t'-t)^{\frac{1}{2}}}\right] + \right.$$

$$\left. + \operatorname{erfc}\left[\frac{x_0(t)+x'}{2(t'-t)^{\frac{1}{2}}}\right] - \operatorname{erfc}\left[\frac{1+x'}{2(t'-t)^{\frac{1}{2}}}\right] \right\}\, dt. \qquad (3.67)$$

The first integral in (3.65) leads to the second and third terms in (3.66), after suitable integration by parts, and the second integral in (3.65) leads to the quantity $R(x', t')$.

If now we were to let x' in (3.66) approach $x_0(t')$ and remember from the first of (1.59) that $c(x_0(t'), t') = 0$ we would obtain an integral equation for the function $x = x_0(t)$. Hansen and Hougaard (1974) point out, however, that alternative integral equations can be obtained by differentiating (3.66) with respect to x' or t and in each case putting $x' = x_0(t')$. They decide to use the first of these forms as being simpler than the second or than (3.66). Writing $x_0(t)$, $x_0(t')$ as x, x' respectively for convenience they obtain

$$x' = 1 - 2\,\mathrm{erfc}\left(\frac{x'}{2\sqrt{t'}}\right) + (1-x')\mathrm{erf}\left(\frac{1-x'}{2\sqrt{t'}}\right) + (1+x')\mathrm{erfc}\left(\frac{1+x'}{2\sqrt{t'}}\right)$$

$$-2\left(\frac{t'}{\pi}\right)^{\frac{1}{2}}\left\{1 - \exp\left[-\frac{(1-x')^2}{4t'}\right] + \exp\left[\frac{-(1+x')^2}{4t'}\right]\right\}$$

$$-\frac{1}{\sqrt{\pi}}\int_0^{t'}\frac{1}{(t'-t)^{\frac{1}{2}}}\left\{\exp\left[-\frac{(x-x')^2}{4(t'-t)}\right] - 1 - \exp\left[-\frac{(x+x')^2}{4(t'-t)}\right]\right\}dt. \quad (3.68)$$

Once the function $x = x_0(t)$ has been determined from this integral equation, c can be evaluated from (3.66) everywhere in the range $0 < x < x_0(t)$ where $c(x, t) > 0$. Hansen and Hougaard (1974) adopt a simple iterative numerical procedure to find $x = x_0(t)$ from (3.68). Writing $t_i = ih$, where h is a suitable time step, they evaluate the integral by a standard numerical procedure, actually Simpson's formula coupled with Newton's three-eighths rule. They express (3.68) as

$$f_i = N(f_i, t_i)/D(f_i, t_i) \quad (3.69)$$

where $f_i = 1 - x_0(ih)$, and, writing generally $f_j = 1 - x_0(jh)$, $i \leqslant j \leqslant i$,

$$N(f_i, t_i) = \frac{1}{\sqrt{\pi}}\sum_{j=1}^{i}\frac{\alpha_j}{(t_i - t_j)^{\frac{1}{2}}}\left\{\exp\left(\frac{-(f_i - f_j)^2}{4(t_i - t_j)}\right) - 1 - \exp\left(\frac{-(2 - f_i - f_j)^2}{4(t_i - t_j)}\right)\right\}$$

$$+ 2\,\mathrm{erfc}\left(\frac{1 - f_i}{2\sqrt{t_i}}\right) - 2\,\mathrm{erfc}\left(\frac{2 - f_i}{2\sqrt{t_i}}\right) + 2\left(\frac{t_i}{\pi}\right)^{\frac{1}{2}}$$

$$\times\left\{1 + \exp\left(\frac{-(2 - f_i)^2}{4t_i}\right) - \exp\left(\frac{-f_i^2}{4t_i}\right)\right\}, \quad (3.70)$$

$$D(f_i, t_i) = 1 + \mathrm{erf}\left(\frac{f_i}{2\sqrt{t_i}}\right) - \mathrm{erfc}\left(\frac{2 - f_i}{2\sqrt{t_i}}\right). \quad (3.71)$$

The coefficients α_j, of course, depend on the quadrature formulae adopted. The iterative procedure based on (3.69) is

$$f_i^{(n+1)} = N(f_i^n, t_i)/D(f_i^n, t_i), \quad (3.72)$$

and the starting value $f_i^{(0)}$ is obtained by quadratic extrapolation from the values $f_{i-1}, f_{i-2}, f_{i-3}$ calculated at the three immediately preceding times.

Crank and Gupta (1972a) derived a formula for $c(x', t')$ for small times

TABLE 3.1

Position of the moving boundary as a function of time

Time	x	Time	x
0.000	1.000 00	0.1800	0.501 09
0.0200	1.000 00	0.1900	0.345 37
0.0400	0.999 18	0.1950	0.206 52
0.0600	0.991 80	0.1955	0.187 08
0.0800	0.971 55	0.1960	0.162 66
0.1000	0.935 01	0.1965	0.132 84
0.1200	0.879 16	0.1970	0.091 75
0.1400	0.798 91	0.1972	0.067 08
0.1600	0.683 37	0.1974[a]	0.000 00

[a] Estimated by extrapolation.

which is identical with the solution (3.66) above except for the term $R = R(x', t')$. Hansen and Hougaard (1974) used the first three terms in (3.66) to compute $x_0(t_i)$ for $t_i < 0.02$. For larger times they used (3.72). Finally, they computed $c(x, t)$ for various values of x and t from (3.66) and showed that the results are in good agreement with those obtained from the small-time solution of Crank and Gupta (1972a) given by (3.66) without the term $R(x', t')$. Positions of the moving boundary are reproduced from Hansen and Hougaard's paper (1974) in Table 3.1, and Table 3.2 shows a selection of their values of $c(x, t)$.

These results are compared with those of other authors in Tables 4.1, 4.2, and 6.6–8.

TABLE 3.2

Values of $10^6 c$

t					x					
	0.0	0.1	0.2	0.3	0.4	0.5	0.6	0.7	0.8	0.9
0.000	500 000	405 000	320 000	245 000	180 000	125 000	80 000	45 000	20 000	5 000
0.002	449 537	401 927	319 973	245 000	180 000	125 000	80 000	45 000	20 000	5 000
0.010	387 162	365 072	309 949	243 275	179 804	124 986	79 999	45 000	20 000	5 000
0.050	247 687	240 175	218 841	186 952	148 990	109 634	72 961	42 029	18 856	4 629
0.100	143 177	139 294	128 082	110 787	89 295	65 892	43 018	23 059	8 232	603
0.150	63 087	60 845	54 503	44 503	32 353	19 583	8 251	1 005	0	0
0.160	48 823	46 840	41 136	32 434	21 927	11 304	2 890	0	0	0
0.180	21 781	20.287	16 066	9 942	3 523	0	0	0	0	0
0.190	9 021	7 817	4 578	799	0	0	0	0	0	0
0.195	2 884	1 914	32	0	0	0	0	0	0	0

3.5.3. Embedding technique

A simple example of the meaning of 'embedding' is provided by the case of a slab of ice melting under the action of heat falling on its surface and with the water being instantaneously removed on formation. The ice phase, which decreases in size as melting proceeds, is considered as part of a larger slab which retains the dimensions of the original slab of ice before melting started and therefore has fixed boundaries. The ice phase is said to be "embedded" in the larger slab of constant dimensions in a mathematical sense. Clearly, only those parts of the embedding slab which coincide with the actual ice phase at any time have physical meaning, and other parts are fictitious. In particular, the outer boundaries which mark the extent of the initial slab of ice have fictitious boundary conditions associated with them. Such conditions are not prescribed directly by the physical problem but must be constructed in a way such that the physical conditions prescribed on the moving, melting interface are satisfied. This is the example that Boley (1961) used in order to introduce his embedding technique.

Consider a slab of thickness ℓ with surface heating conditions

$$K\frac{\partial T_1}{\partial x} + hT_1 = f_1(t), \qquad x = 0, \qquad 0 < t \leq t_m, \tag{3.73}$$

$$K_1\frac{\partial T_1}{\partial x} + h_1 T_1 = 0, \qquad x = \ell, \qquad 0 < t \leq t_m, \tag{3.74}$$

where K, K_1, h, h_1 are constants, $T_1(x, t)$ is the temperature before melting commences, and $f_1(t)$ is a prescribed function. The conditions (3.73) and (3.74) together with the heat flow equation

$$\frac{\partial T_1}{\partial t} = k\frac{\partial^2 T_1}{\partial x^2}, \qquad 0 < x < \ell, \qquad 0 < t \leq t_m, \tag{3.75}$$

where k is heat diffusivity, and some prescribed initial condition, e.g.

$$T_1 = 0, \qquad 0 < x < \ell, \qquad t = 0, \tag{3.76}$$

suffice to define $T_1(x, t)$ as long as no melting occurs. If the surface reaches the melting temperature T_m at time t_m then

$$T_1(0, t_m) = T_m. \tag{3.77}$$

For $t \geq t_m$ a new temperature $T_2(x, t)$ is defined which satisfies the equation

$$\frac{\partial T_2}{\partial t} = k\frac{\partial^2 T_2}{\partial x^2}, \qquad s(t) < x < \ell, \qquad t \geq t_m, \tag{3.78}$$

and the following conditions

$$T_2\{s(t), t\} = T_m, \qquad s(t_m) = 0 \tag{3.79}$$

$$K \frac{\partial T_2}{\partial x} + hT_m = \rho L \frac{ds}{dt} + f_1(t), \qquad x = s(t), \tag{3.80}$$

$$K_1 \frac{\partial T_2}{\partial x} + h_1 T_2 = 0, \qquad x = \ell, \tag{3.81}$$

$$T_2(x, t_m) = T_1(x, t_m), \tag{3.82}$$

where $s(t)$ denotes the thickness of the melted (and removed) water i.e. the distance the melting interface has penetrated the original slab of ice, ρ is density, and L the latent heat of melting.

The solution $T_2(x, t)$, which has physical significance only in the region $s(t) < x < \ell$, is now extended over the whole region $0 < x < \ell$ and this extended solution is denoted by $T(x, t)$; thus the solid phase $s(t) < x < \ell$ is 'embedded' in the fictitious space $0 < x < \ell$. Clearly, $T(x, t)$ will satisfy conditions (3.79–82) and equation (3.78) will be satisfied in $0 < x < \ell$. Condition (3.73) on $x = 0$, however, must be replaced by

$$K \frac{\partial T}{\partial x} + hT = f_1(t) + f(t), \tag{3.83}$$

where $f(t)$ is an unknown function which, together with $s(t)$, is to be determined so that the conditions (3.79) and (3.80) are satisfied. Boley (1961) invokes the appropriate form of Duhamel's theorem (Carslaw and Jaeger 1959, p. 31), which states:

If $v = F(x, t)$ represents the temperature at the point x and time t in a solid in which the initial temperature is zero, while radiation takes place into a medium at temperature unity, then the solution of the problem when radiation takes place into a medium at temperature $\phi(t)$ is given by

$$v = \int_0^t \phi(\lambda) \frac{\partial}{\partial t} F(x, t - \lambda) \, d\lambda. \tag{3.84}$$

In the present problem we write $F(x, t) = T_0(x, t)$ which therefore satisfies the equations:

$$\frac{\partial T_0}{\partial t} = \frac{K \partial^2 T_0}{\partial x^2}, \qquad 0 < x < \ell, \qquad t > 0, \qquad T_0(x, 0) = 0, \tag{3.85}$$

$$K \frac{\partial T_0}{\partial x} + hT_0 = h, \qquad x = 0, \qquad t > 0, \tag{3.86}$$

$$K_1 \frac{\partial T_0}{\partial x} + h_1 T_0 = 0, \qquad x = \ell, \qquad t > 0. \tag{3.87}$$

The condition (3.86) expresses radiation into a medium at temperature unity on the surface $x = 0$. A standard solution for $T_0(x, t)$, satisfying (3.85–7), is easily deduced, for example, from the results of Carslaw and Jaeger (1959, p. 118), and substitution of F by T_0 in (3.84) gives

$$T(x, t) = \int_0^{t_m} f_1(t_1) \frac{\partial T_0}{\partial t} (x, t - t_1)\, dt_1$$

$$+ \int_{t_m}^t \{f_1(t_1) + f(t_1)\} \frac{\partial T_0}{\partial t} (x, t - t_1)\, dt_1 \quad (3.88)$$

or, on rearranging,

$$T(x, t) = \int_0^t f_1(t - t_1) \frac{\partial T_0}{\partial t_1} (x, t_1)\, dt_1 + \int_0^{t-t_m} f(t - t_1) \frac{\partial T_0}{\partial t_1} (x, t_1)\, dt_1.$$

$$(3.89)$$

Substitution of $T(x, t)$ into the conditions (3.79) and (3.80) leads to

$$\int_0^t f_1(t - t_1) \frac{\partial T_0}{\partial t_1} \{s(t), t_1\}\, dt_1 + \int_0^{t-t_m} f(t - t_1) \frac{\partial T_0}{\partial t_1} \{s(t), t_1\}\, dt_1 = T_m, \qquad t \geq t_m,$$

$$(3.90)$$

$$k \int_0^t f_1(t - t_1) \frac{\partial^2 T_0}{\partial x\, \partial t_1} \{s(t), t_1\}\, dt_1 + k \int_0^{t-t_m} f(t - t_1) \frac{\partial^2 T_0}{\partial x\, \partial t_1} \{s(t), t_1\}\, dt_1$$

$$= \rho L \frac{ds}{dt} (t) + f_1(t) - h T_m, \qquad t \geq t_m. \quad (3.91)$$

Solution of these two simultaneous integro-differential equations, subject to the initial condition $s(t_m) = 0$, yields the two unknown functions $s(t)$ and $f(t)$. After this, the temperature $T(x, t)$ follows from (3.88) or (3.89).

As an illustration, Boley (1961) considered the problem of a semi-infinite solid, $x > 0$, with a prescribed constant heat flux Q_0 at $x = 0$. In this case

$$h = 0, \qquad f_1(t) = -Q_0, \qquad T_0 = -\frac{2}{K} \sqrt{(kt)} \operatorname{ierfc}\left(\frac{x}{2\sqrt{(kt)}}\right), \quad (3.92)$$

and equations (3.90), (3.91) become

$$-\int_0^{t-t_m} f(t - t_1) e^{-s^2(t)/(4kt_1)} \frac{dt}{t_1^{\frac{1}{2}}} = K T_m \sqrt{\frac{\pi}{k}} - 2 Q_0 (\pi t)^{\frac{1}{2}} \operatorname{ierfc}\left(\frac{s}{2\sqrt{(kt)}}\right),$$

$$(3.93)$$

$$-s(t) \int_0^{t-t_m} f(t - t_1) e^{-s^2(t)/(4kt_1)} \frac{dt_1}{t_1^{\frac{3}{2}}} = 2(k\pi)^{\frac{1}{2}} \left\{ Q_0 \operatorname{erf}\left(\frac{s}{2\sqrt{(kt)}}\right) - \rho L \frac{ds}{dt} \right\}.$$

$$(3.94)$$

The pre-melting solution, T_1, in this example is

$$T_1(x, t) = \frac{2Q_0}{K} (kt)^{\frac{1}{2}} \text{ierfc}\left(\frac{x}{2\sqrt{(kt)}}\right). \tag{3.95}$$

In terms of the non-dimensional parameters

$$y = \frac{t}{t_m} - 1, \qquad \xi(y) = \frac{Q_0 s(t)}{\pi^{\frac{1}{2}} K T_m}, \qquad m = \frac{\pi^{\frac{1}{2}} c T_m}{2L}, \qquad F(y) = -\frac{f(t)}{Q_0}, \tag{3.96}$$

the relation between t_m and T_m, which from (3.95) is given by

$$T_m = \frac{2Q_0}{K} \left(\frac{kt_m}{\pi}\right)^{\frac{1}{2}},$$

becomes

$$t_m = k\left(\frac{\rho L m}{Q_0}\right)^2, \tag{3.97}$$

and equations (3.93), (3.94) may be written

$$\int_0^y F(y - y_1) e^{-\xi^2(y)/y_1} \frac{\text{d}y_1}{y_1^{\frac{1}{2}}} = 2[1 - \{\pi(1+y)\}^{\frac{1}{2}} \text{ierfc}\left(\frac{\xi(y)}{\sqrt{(1+y)}}\right)], \tag{3.98}$$

$$\xi(y) \int_0^y F(y - y_1) e^{-\xi^2(y)/y_1} \frac{\text{d}y_1}{y_1^{\frac{3}{2}}} = \sqrt{\pi} \left\{\text{erf}\left(\frac{\xi(y)}{\sqrt{(1+y)}}\right) - \frac{2}{m} \frac{\text{d}\xi(y)}{\text{d}y}\right\}, \tag{3.99}$$

for $y \geqslant 0$ and the initial condition becomes

$$\xi(0) = 0. \tag{3.100}$$

Boley (1961) obtained series solutions of these equations valid for short times and then proceeded to larger times by using numerical methods. His short-time solutions are

$$F(y) = -\frac{2}{\pi} y^{\frac{1}{2}} + \frac{m}{\sqrt{\pi}} y + \frac{4}{3\pi}\left(\frac{1}{2} - m^2 - \frac{16m}{3\pi^{\frac{3}{2}}}\right) y^{\frac{3}{2}}$$

$$+ \frac{m}{2\sqrt{\pi}}\left\{\left(\frac{16}{3\pi} + \frac{35}{8}\right)\frac{m}{\sqrt{\pi}} + m^2 - \frac{1}{2}\right\} y^2 + \dots, \tag{3.101}$$

$$\xi(y) = \frac{2m}{3\pi} y^{\frac{3}{2}} - \frac{m^2 y^2}{4\sqrt{\pi}} - \frac{4m}{15\pi}\left(\frac{1}{2} - m^2 - \frac{16m}{3\pi^{\frac{3}{2}}}\right) y^{\frac{5}{2}}$$

$$+ \frac{m^2}{12\sqrt{\pi}}\left(-\frac{35m}{8\sqrt{\pi}} - m^2 + \frac{1}{2}\right) y^3 + \dots, \tag{3.102}$$

and the associated temperature is given by

$$\frac{T}{T_m} = \sqrt{\{\pi(1+y)\}} \operatorname{ierfc}\left(\frac{\bar{x}}{\sqrt{(1+y)}}\right) + \bar{x}\sqrt{y}\,\operatorname{ierfc}\left(\frac{\bar{x}}{\sqrt{y}}\right) - \frac{y}{2}\operatorname{erfc}\left(\frac{\bar{x}}{\sqrt{y}}\right)$$

$$0 \leqslant y \ll 1, \quad (3.103)$$

where $\bar{x} = Q_0 x/(\sqrt{\pi K T_m})$ and (3.103) has physical meaning only for $\bar{x} \geqslant \xi$.

In the same paper, Boley (1961) formulated the problem of the solidification of a semi-infinite liquid under arbitrary cooling rates. Both solid and liquid phases are present and so four simultaneous integro-differential equations are needed to determine two unknown heat-flow functions, together with the movement of the solidification boundary and the temperature distribution. Series solutions are developed for the particular case of constant surface cooling. It is found that the first three terms of the series expressions for this two-phase problem are identical with those of (3.101) and (3.102). This means that for short times the motion of the interface is virtually the same whether ablation occurs or not.

In a later paper, Boley (1968) developed general short-time solutions in one space dimension for melting or solidifying slabs which include arbitrary heating conditions, arbitrary rates of liquid removal (including the limiting cases of stationary and instantaneously removed liquid), temperature-dependent properties, and slabs of finite thickness.

Boley and Yagoda (1971) extended their embedding technique to obtain three-dimensional starting solutions for a melting slab or for the companion problem of the solidification of a liquid. A half-space is subjected to a general heat input on its surface, such that the change of phase starts at a point of the surface and then spreads both towards the interior of the body and along its surface. Mathematically they considered the half-space $z > 0$, initially solid at a temperature lower than the melting temperature T_m, and heated on the $z = 0$ surface by a prescribed heat input in the z-direction $Q(x, y, t)$ which has a maximum at a prescribed point of the surface, chosen as the origin. Thus they write

$$Q(x, y, t) = a(t) - b(t)x^2 - c(t)y^2 + \dots, \qquad b, c \geqslant 0. \quad (3.104)$$

The liquid phase is assumed to be removed instantaneously on formation but Boley (1961) and Wu and Boley (1966) showed that the starting solution is the same for arbitrary rates of ablation including zero and infinite. This problem was solved earlier by Boley and Yagoda (1969) for the special case of a plane heat input ($c = 0$ in (3.104)) and by Yagoda and Boley (1970) for a radially symmetric heat input ($b \equiv c$ in (3.104)).

Boley and Yagoda (1971) showed that for very short times the shape of the phase-change surface, when suitably normalized, is independent of the distribution and the rate of the heat input, and of the thermal properties of the slab. The magnitude of the melt depth does, of course, depend on these parameters. The initial extents of melting in directions normal to and along the surface are respectively proportional to $(t-t_m)^{\frac{3}{2}}$ and $(t-t_m)^{\frac{1}{2}}$, where t_m is the time at which melting starts at the origin. In fact, they established that for short times the melt depth (X, Y, τ), where all quantities are rendered suitably dimensionless, is given by

$$(X, Y, \tau) = A_2 \tau^{\frac{3}{2}} \left\{ 1 - \left(\frac{X}{X_0(\tau)} \right)^2 - \left(\frac{Y}{Y_0(\tau)} \right)^2 \right\}^{\frac{3}{2}}, \qquad \tau \ll 1, \quad (3.105)$$

where $X_0(\tau)$, $Y_0(\tau)$ indicate the extent of melting along the surface in the X, Y coordinate directions, τ is the time after melting commences, and $X_0(\tau) = A\tau^{\frac{1}{2}}$, $Y_0(\tau) = B\tau^{\frac{1}{2}}$ subject to $(X/X_0)^2 = (Y/Y_0)^2 \leq 1$. The constants A, B, and A_2 are complicated expressions of C_1, C_{2x}, and C_{2y}, where

$$C_1 = \frac{\partial}{\partial t} \left(\frac{T}{T_m} \right), \qquad C_{2x} = \frac{\partial^2}{\partial X^2} \left(\frac{T}{T_m} \right), \qquad C_{2y} = \frac{-\partial^2}{\partial Y^2} \left(\frac{T}{T_m} \right),$$

all evaluated at $X = Y = Z = \tau = 0$.

Boley (1975) summarized investigations of the uniqueness of various embedding solutions. There is a review by Boley (1978) of important practical problems requiring further study in Wilson *et al.* (1978). The subject of upper and lower bounds for solutions is included. Boley (1963) in discussing an ablation problem proposed a comparison theorem which allows upper and lower bounds to be constructed by imposing heat fluxes on the surface of the embedding body which are respectively always larger or smaller than the desired flux. An analogous result for a melt that remains in position was established by Boley (1964). The lack of comparison theorems and of solution bounds for problems with variable heat parameters is noted by Boley (1978) coupled with tentative suggestions for examining cases of limited variations of heat properties. Bounds and approximate solutions for linear heat equations with non-linear boundary conditions are established by Glasser and Kern (1978) who insert physically sensible inequalities into zero and first-order moments of the heat equation.

Ferriss and Hill (1974) used an embedding technique to solve the oxygen diffusion problem set out in equations (1.57–60) of §1.3.10. They introduced a fictitious condition

$$\partial c/\partial x = f(t), \qquad x = 1, \qquad (3.106)$$

which replaces conditions (1.59) of the original problem and solved the linear problem defined by (3.106) together with (1.57), (1.58), and (1.60) in the embedding region $0 \leq x \leq 1$ by using a cosine integral transform. Substitution of the two conditions (1.59) into the transform solution yields a pair of simultaneous non-linear, integral equations for $f(t)$ and $s(t)$. Ferriss and Hill employed a standard numerical method of solving the integral equations starting from a small-time solution

$$c(x, t) = \frac{2}{\pi^2} \sum_{p=\frac{1}{2},\frac{3}{2}} \frac{e^{-p^2\pi^2 t}}{p^2} \cos p\pi x + \tfrac{1}{2}(x^2 - 1), \qquad (3.107)$$

which is simpler than the series of complementary error functions used by Crank and Gupta (1972a). They tabulated numerical values of $c(x, t)$ and of $s(t)$ and made comparisons with a finite-difference method of their own in the same paper and with the results of Crank and Gupta.

Comments by Ferriss and Hill (1974) on the embedding technique are probably true in general. The amount of algebra involved in developing the embedding formulation is by no means trivial and can be very considerable. They found that the computing time required by a finite-difference method was less than for the embedding solution, due mainly to the tri-diagonal form of the finite-difference equations. On the other hand, the embedding approach permitted larger time-steps for a pre-scribed accuracy and, furthermore, a variable time-step could be conveniently introduced such that successive time-steps decreased in geometric progression. The embedding solution also concentrates complete information about c and its derivative into the two integral equations, once $f(t)$ is known. It is worth noting at this point that Boley and collaborators have only solved problems in which the solutions in the fixed embedding domain can be written down immediately and analytically.

3.5.4. Heat-balance integral method: Goodman

By integrating the one-dimensional heat flow equation with respect to the space variable x, and inserting the boundary conditions, Goodman (1958) produced an integral equation which expresses the overall heat balance of the system. A review article by Goodman (1964) is a useful introduction to integral methods in heat flow problems generally, including changes of phase. Earlier papers by Goodman from 1958 onwards and by other authors are cited.

Successive steps in Goodman's method are:

(i) Assume a particular form for the dependence of the temperature on the space variable which is consistent with the boundary conditions, e.g. assume a polynomial relationship.

(ii) Integrate the heat flow equation with respect to the space variable over the appropriate interval and substitute the assumed temperature distribution to obtain the heat-balance integral.

(iii) Solve the integral equation to obtain the motion of the phase-change boundary and then the time dependence of the temperature distribution.

The method can conveniently be illustrated by solving the one-phase melting-ice problem in one space dimension as defined in non-dimensional terms by equations (1.21–26) with $\lambda = 1$. Integration of (1.21) with respect to x, from $x = 0$ to $s(t)$, gives

$$\frac{d}{dt}\int_0^{s(t)} u \, dx = -\left(\frac{ds}{dt} + \frac{\partial u}{\partial x}(0, t)\right). \tag{3.108}$$

Assume a temperature distribution in the water phase given by

$$u = a(t)\{x - s(t)\} + b(t)\{x - s(t)\}^2, \tag{3.109}$$

which automatically satisfies (1.25). It is convenient to modify condition (1.26) to the form

$$\left(\frac{\partial u}{\partial x}\right)^2 = \frac{\partial^2 u}{\partial x^2}, \qquad x = s(t), \tag{3.110}$$

by using the standard formula

$$\frac{du}{dt} = \frac{\partial u}{\partial x}\frac{ds}{dt} + \frac{\partial u}{\partial t} = 0, \qquad x = s(t)$$

to replace ds/dt in (1.26). Remembering that $u = 1$, $x = 0$, (3.109) gives $1 = -as + bs^2$ and, on differentiation, $\partial u/\partial x = a$, $\partial^2 u/\partial x^2 = 2b$ at $x = s(t)$, so that the substitution into (3.110) and some manipulation yield

$$a = (1 - \sqrt{3})/s(t), \qquad b = (as + 1)/s^2. \tag{3.111}$$

Substitution of (3.109) incorporating $a(t)$, $b(t)$ from (3.111), and integrating the left side of (3.108) gives

$$\tfrac{1}{6}(-1 + \sqrt{3} + 2)\frac{ds}{dt} = -\frac{ds}{dt} - (a - 2bs).$$

Further substitution gives $a - 2bs = -(1 - \sqrt{3} + 2)/s$ and so finally

$$s\frac{ds}{dt} = \frac{6(3 - \sqrt{3})}{7 + \sqrt{3}} = \tfrac{1}{2}\alpha^2, \tag{3.112}$$

say, from which it immediately follows that

$$s(t) = \alpha t^{\frac{1}{2}}, \tag{3.113}$$

since $s(0) = 0$ from (1.24). The temperature distribution follows readily by substituting in (3.109) from (3.113) and (3.111). Instead of using the heat-balance integral (3.108) after finding the temperature profile, Goodman (1958) suggested use of (1.26) as an alternative which leads to the differential equation $s \, ds/dt = (\sqrt{3} - 1)$.

Goodman and Shea (1960) applied the heat-balance method to the two-phase problem of the melting of a finite slab which is initially at a uniform temperature below the melting point. Initial sub-cooling was also considered by Yuen (1980). The relevant equations are conveniently written (Fox 1975)

$$\frac{\partial u_1}{\partial t} = k_1 \frac{\partial^2 u_1}{\partial x^2}, \qquad 0 < x < s(t), \tag{3.114}$$

$$\frac{\partial u_2}{\partial t} = k_1 \frac{\partial^2 u_2}{\partial x^2}, \qquad s(t) < x < \ell, \tag{3.115}$$

where the suffixes 1 and 2 denote the liquid and solid regions respectively. Certain conditions are also specified on $x = 0$ and ℓ, and on the melting boundary we have

$$\left. \begin{array}{c} K_1 \dfrac{\partial u_1}{\partial x} - K_2 \dfrac{\partial u_2}{\partial x} = \rho L \dfrac{ds}{dt}, \\[2mm] u_1 = u_2 = u_{\mathrm{m}}. \end{array} \right\} \quad x = s(t). \tag{3.116}$$

Goodman and Shea (1960) solved the pre-melting problem by the heat-balance method though, of course, standard solutions are readily available (Carslaw and Jaeger 1959) for their boundary conditions on $x = 0$ and ℓ.

Following the steps outlined in (i–iii) above, (3.114) is integrated from $x = 0$ to $s(t)$ and (3.115) from $s(t)$ to ℓ to obtain the integral equations

$$k_1 \left\{ \left(\frac{\partial u_1}{\partial x} \right)_{s(t)} - \left(\frac{\partial u_1}{\partial x} \right)_0 \right\} = \int_0^{s(t)} \frac{\partial u_1}{\partial t} \, dx = \frac{d}{dt} \int_0^{s(t)} u_1 \, dx - u_{\mathrm{m}} \frac{ds}{dt}, \tag{3.117}$$

$$k_2 \left\{ \left(\frac{\partial u_2}{\partial x} \right)_{\ell} - \left(\frac{\partial u_2}{\partial x} \right)_{s(t)} \right\} = \int_{s(t)}^{l} \frac{\partial u_2}{\partial t} \, dx = \frac{d}{dt} \int_{s(t)}^{l} u_2 \, dx + u_{\mathrm{m}} \frac{ds}{dt}. \tag{3.118}$$

Elimination of $\partial u_1/\partial x$ and $\partial u_2/\partial x$ at $x = s(t)$ from (3.116–118) now yields

$$\left(\frac{K_2}{k_2} u_{\mathrm{m}} - \frac{K_1}{k_1} u_{\mathrm{m}} - \rho L \right) \frac{ds}{dt} + \frac{K_1}{k_1} \frac{d\theta_1}{dt} + \frac{K_2}{k_2} \frac{d\theta_2}{dt} = K_2 \left(\frac{\partial u_2}{\partial x} \right)_{\ell} - K_1 \left(\frac{\partial u_1}{\partial x} \right)_0, \tag{3.119}$$

where

$$\theta_1 = \int_0^{s(t)} u_1(x, t) \, dx, \qquad \theta_2 = \int_{s(t)}^{\ell} u_2(x, t) \, dx. \tag{3.120}$$

The two heat-balance equations (3.117) and (3.118) have been coupled together through the melting condition (3.116) to yield (3.119).

Now suitable forms for u_1 and u_2 are chosen, e.g.

$$\left.\begin{array}{l} u_1(x, t) - u_m = a_1(t)(x - s) + b_1(t)(x - s)^2, \\ u_2(x, t) - u_m = a_2(t)(x - s) + b_2(t)(x - s)^2, \end{array}\right\} \tag{3.121}$$

and the parameters $a_1(t)$ etc. determined by satisfying all the boundary conditions and the melting condition. Proceeding as in the single-phase problem discussed above the parameters $a_1(t)$, $a_2(t)$, etc. are determined by satisfying the boundary conditions and preserving the heat balance. Thus, for example, with boundary conditions

$$K_1 \frac{\partial u_1}{\partial x} = -H, \qquad x = 0; \qquad \frac{\partial u_2}{\partial x} = 0, \qquad x = \ell; \qquad u_m = 0,$$
$$\tag{3.122}$$

Goodman and Shea (1960) produced the differential equations

$$-\rho L \frac{ds}{dt} + \frac{K_1}{k_1} \frac{d\theta_1}{dt} + \frac{K_2}{k_2} \frac{d\theta_2}{dt} = H, \tag{3.123}$$

$$\theta_1 = s^2 \left(\frac{H}{2K_1} - \frac{1}{3k_1} \frac{d\theta_1}{dt} \right), \tag{3.124}$$

$$\theta_2 = -\frac{(\ell - s)^2}{3k_2} \frac{d\theta_2}{dt}. \tag{3.125}$$

Here, (3.123) comes immediately from (3.119) on substituting from (3.122); (3.124) is derived by first evaluating $a_1(t)$, $b_1(t)$ from the conditions (3.122) and (3.117) and finally using the first of (3.120). Equation (3.125) is determined similarly from the corresponding relations for the solid phase.

The initial conditions needed in order to solve (3.123–125) are

$$s(t_m) = 0, \qquad \theta_1(t_m) = 0, \qquad \theta_2(t_m) = -H\ell^2/(3K_2). \tag{3.126}$$

The last of (3.126) comes by integration with respect to x of the appropriate pre-melting solution at $t = t_m$, the time when melting commences, which is given by Goodman and Shea (1960) as

$$u_2(x, t_m) = \frac{-H\ell}{2K_2} \{1 - (1 - x/\ell)^2\}, \tag{3.127}$$

and $\theta_2(t_m) = \int_0^\ell u_2(x, t) \, dx$ from the second of (3.120) and the first of (3.126).

Goodman and Shea obtained solutions in series of equations (3.123–5) together with (3.126) by a complicated analysis. But numerical solutions are easily obtainable by standard methods such as the Runge–Kutta method.

For an isothermal condition on $x = \ell$, so that (3.122) are replaced by

$$K_1 \frac{\partial u_1}{\partial x} = -H, \qquad x = 0; \qquad u_2 = -V, \qquad x = \ell; \qquad u_m = 0,$$

$$(3.128)$$

Goodman and Shea (1960) derived the following set of ordinary differential equations

$$\rho L \frac{ds}{dt} + \frac{K_1}{k_1} \frac{d\theta_1}{dt} + \frac{1}{2} \frac{K_2}{k_2} \frac{d\theta_2}{dt} + \frac{K_2 V}{\ell - s} = H, \qquad (3.129)$$

$$\theta_1 = s^2 \left(\frac{H}{2K_1} - \frac{1}{3k_1} \frac{d\theta_1}{dt} \right), \qquad (3.130)$$

$$\theta_2 = -\frac{V}{2}(\ell - s) - \frac{(\ell - s)^2}{12k_2} \frac{d\theta_2}{dt}, \qquad (3.131)$$

$$s(t_m) = 0, \qquad \theta_1(t_m) = 0, \qquad \theta_2(t_m) = -\frac{V\ell}{3} - \frac{H\ell^2}{6K}. \qquad (3.132)$$

Poots (1962b) applied the heat-balance integral method to single-phase melting problems in the semi-infinite region, the cylinder, and the sphere, by using a two-parameter quadratic profile. He also developed a short-time series solution. Lardner and Pohle (1961) also considered cylindrical problems.

Goodman (1961, 1964) and Boley and Estenssoro (1977) suggested ways of incorporating variable heat properties in heat-balance integral methods, though not specifically in relation to problems of phase change. Imber and Huang (1973) also studied temperature-dependent thermal properties and when the dependence is linear they concluded that the freezing of a semi-infinite liquid is well approximated by the constant heat-properties solution when the properties at the fusion temperature are used. Clearly, the mathematical manipulation in heat-balance methods for anything other than relatively simple problems can be very lengthy. The choice of a satisfactory approximation to the temperature profile is acknowledged to be a major difficulty in the heat-balance approach. The use of a higher-order polynomial, for example, does not necessarily improve the accuracy of the solution. Goodman himself (1964), Bell (1978), Langford (1973), and others quote illustrations. Boley (1973) suggested that improvement is obtained when approximate solutions, such as the simple heat-balance method produces, are superimposed by using convolution or Duhamel integrals, or by the method of images. Boley and Estenssoro (1977) likewise introduced the idea of 'multiple penetration depths' and expressed the temperature as a series, each term of which was taken to vanish beyond an unknown penetration depth below the surface. Thus for a half-space, initially at temperature zero and with its surface raised to a constant temperature T_0, the temperature $T(x, t)$ is written as

$$T(x, t)/T_0 = \sum_{i=0}^{m-1} d_i x^i \left(1 - \frac{x}{q_i}\right)^2 H(x, q_i)$$

where

$$H(x, x_i) = \begin{cases} 1, & x < x_i, \\ 0, & x > x_i, \end{cases}$$

and $d_0 = 1$, $d_i = d_i(t)$, $q_i = q_i(t)$ for $i \geq 0$. It is assumed that $q_i(t) > 0$, $i = 0, 1, \ldots, m-1$. Integration of the weighted-moments expression

$$\int_0^\infty \left(k \frac{\partial^2 T}{\partial x^2} - \frac{\partial T}{\partial t}\right) x^n \, dx = 0, \qquad n = 0, 1, 2, \ldots, 2(m-1);$$

using $T = T(x, t)$ from above leads to $(2m-1)$ equations to be solved for the $(2m-1)$ unknown d_i and q_i. Solutions were given for the cases $m = 1$, 2, 3 and found to provide improved approximations to the exact solution as more penetration depths were included. No numerical solutions were obtained for variable thermal properties but the fundamental relationships were set out.

3.5.5. *Heat-balance method with spatial subdivision*

Noble (1975) suggested that the most promising way of systematically improving accuracy would be by repeated spatial subdivision, using quadratic profiles in each sub-region, as in finite-element treatments. Bell (1978) solved the single-phase melting problem by adopting a modified form of Noble's suggestion in which equal subdivision of the dependent variable, temperature, is used rather than the more usual equal division of the space variable. The solution of a system of differential equations provides the positions of each isotherm identified by the subdivision, including automatically that of the melting boundary, at successive times.

Bell (1978) used the equations

$$\frac{\partial u}{\partial t} = k \frac{\partial^2 u}{\partial x^2}, \qquad 0 < x < s_0(t), \qquad t > 0, \tag{3.133}$$

$$u(x, t) = u_0, \qquad s_0(t) \leq x, \qquad t \geq 0, \tag{3.134}$$

$$u(0, t) = u_s, \qquad t > 0, \tag{3.135}$$

$$-K \frac{\partial u}{\partial x} = L\rho \frac{ds}{dt}, \qquad x = s_0(t), \qquad t > 0. \tag{3.136}$$

The temperature range u_0 to u_s is subdivided into n equal intervals as in

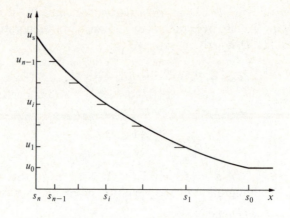

FIG. 3.1. Sub-division of region

Fig. 3.1. The depth of penetration of the isotherm u_i at time t is denoted by $s_i(t)$ where

$$u_i = u_0 + \frac{i}{n}(u_s - u_0), \qquad i = 0, 1, \ldots, n,$$

and $u_n = u_s$. Integration of (3.133) over the sub-region $s_{i+1}(t) \leq x \leq s_i(t)$ gives

$$\frac{d}{dt}\left[\int_{s_{i+1}}^{s_i} u(\xi, t)\, d\xi\right] - u_i \frac{ds_i}{dt} + u_{i+1} \frac{ds_{i+1}}{dt} = k\left(\frac{\partial u}{\partial x}\right)_i - k\left(\frac{\partial u}{\partial x}\right)_{i+1},$$

more conveniently written as

$$\frac{d}{dt}\left[\int_{s_{i+1}}^{s_i} u(\xi, t)\, d\xi + u_{i+1}s_{i+1} - u_i s_i\right] = -k\left(\frac{\partial u}{\partial x}\right)_{i+1} + k\left(\frac{\partial u}{\partial x}\right)_i, \quad (3.137)$$

which, remembering (3.136), takes the special form

$$\frac{d}{dt}\left[\int_{s_1}^{s_0} u(\xi, t)\, d\xi + u_1 s_1 - (u_0 - L/c)s_0\right] = -k\left(\frac{\partial u}{\partial x}\right)_1, \qquad s_1(t) \leq x \leq s_0(t).$$

$$(3.137a)$$

Clearly, (3.137) expresses the heat balance of the sub-region $s_{i+1}(t) \leq x \leq s_i(t)$. Considering each sub-region (s_{i+1}, s_i), $i = 0, 1, \ldots, n-1$, in turn, n equations of the form (3.137) are produced in terms of the n unknowns s_i $(i = 0, 1, \ldots, n-1)$ and the corresponding known temperatures u_i, recalling that $s_n = 0$. The heat-balance equation commonly used in the Goodman integral method is simply the sum of these n equations. The temperature $u(x, t)$ is approximated by a series of profiles v_i which simultaneously satisfy the n equations (3.137), $i = 0, 1, \ldots, n-1$. Thus $v_i(x, t) = u(x, t)$ in the sub-region $s_{i+1} \leqslant x \leqslant s_i$, $i = 0, 1, \ldots, n-1$, and hence $v_i(s_i) = u_i$ and $v_i(s_{i+1}) = u_{i+1}$. If v_i is chosen to be a quadratic function in x then one extra condition is needed to specify v_i in each sub-region. These can be the conditions that express the continuity of heat flux between each subdivision and its neighbours, together with the special condition (3.136) at $x = s_0$ for the end subdivision. For convenience, let

$$x = s_{i+1} + y_i, \qquad s_{i+1} \leqslant x \leqslant s_i, \qquad \ell_i = s_i - s_{i+1},$$

so that a suitable quadratic form for v_i is

$$v_i(s_{i+1} + y_i) = u_{i+1} - y_i(u_{i+1} - u_i)/\ell_i + a_i y_i (1 - y_i/\ell_i), \qquad (3.138)$$

where the coefficient a_i is determined from the condition for continuity of heat flux

$$(\partial v_i/\partial x)_{s_i} = (\partial v_{i-1}/\partial x)_{s_i}, \qquad 1 \leqslant i \leqslant n-1. \qquad (3.139)$$

The recurrence relation

$$a_i = \frac{u_s - u_0}{n} \left(\frac{1}{\ell_{i-1}} - \frac{1}{\ell_i} \right) - a_{i-1}, \qquad 1 \leqslant i \leqslant n-1 \qquad (3.140)$$

follows immediately from (3.138) and (3.139), so that, once a_0 is determined from (3.136) the a_i can be generated. After differentiating (3.138) and putting $i = 0$, $y_0 = \ell_0$ to obtain $(\partial u/\partial x)_{s_0}$ and $(\partial^2 u/\partial x^2)_{s_0}$, use of the condition (3.136), rewritten in the form of (3.110), namely

$$\frac{L}{c} \left(\frac{\partial^2 u}{\partial x^2} \right)_{s_0} = \left(\frac{\partial u}{\partial x} \right)_{s_0}^2, \qquad (3.141)$$

leads to the result

$$a_0 = -(u_s - u_0)\beta/(\ell_0 n), \qquad (3.142)$$

where

$$\beta = 1 + \gamma - (\gamma^2 + 2\gamma)^{\frac{1}{2}}, \qquad \gamma = Ln/\{(u_s - u_0)c\}, \qquad (3.143)$$

and the negative square root is chosen to fit the physical fact that as the latent heat L becomes larger, the melting boundary moves more slowly and the temperature distribution approaches a linear steady state. This requires, from (3.138) that $a_i \to 0$, $i = 0, 1, \ldots, n-1$, and in particular $a_0 \to 0$ as $L \to \infty$ which is satisfied by (3.142) and (3.143).

Combination of (3.138), (3.142), and (3.137a) gives the equation

$$\frac{d}{dt} \left\{ (3 - \beta + 6\gamma)\ell_0 + 6(1 + \gamma) \sum_{j=1}^{n-1} \ell_j \right\} = \frac{6k}{\ell_0}(1 + \beta), \qquad (3.144)$$

where

$$\sum_{j=1}^{n-1} \ell_j = s_1.$$

Similarly, after generating the as from (3.140) and substituting (3.138) into (3.137), the remaining $(n-1)$ equations are found to be:

$$\frac{d}{dt}\left[\ell_i^2\left\{2\sum_{j=1}^{i-1}\frac{(-1)^{i-j-1}}{\ell_j}+\frac{(-1)^{i-1}(1+\beta)}{\ell_0}\right\}+2\ell_i+6\sum_{j=i+1}^{n-1}\ell_j\right]$$
$$=\frac{12k}{\ell_i}-12k\left\{2\sum_{j=1}^{i-1}\frac{(-1)^{i-j-1}}{\ell_j}+\frac{(-1)^{i-1}(1+\beta)}{\ell_0}\right\}, \qquad 1\leqslant i\leqslant n-1, \quad (3.145)$$

which together with (3.144) form a system of n first-order non-linear differential equations for the n penetration parameters $\ell_i(t)$, $0\leqslant i\leqslant n-1$.

Bell (1978) suggested that the initial singularity in the equations can be dealt with by use of a series solution, e.g. that of Poots (1962b), but refers his own numerical calculations to the situation in which the penetration of the melting interface is proportional to $(kt)^{\frac{1}{2}}$. Solutions of the form $\ell_i=\lambda_i(kt)^{\frac{1}{2}}$, $\lambda_i>0$, are defined by reduced forms of (3.144) and (3.145), namely

$$(3-\beta+6\gamma)\lambda_0^2+6(1+\gamma)\lambda_0\left\{\sum_{j=1}^{n-1}\lambda_j\right\}-12(1+\beta)=0 \qquad (3.144a)$$

and

$$\lambda_i^3\left(\sum_{j=1}^{i-1}\frac{(-1)^{i-j-1}}{\lambda_j}+\frac{(-1)^{i-1}(1+\beta)}{2\lambda_0}\right)+\lambda_i^2$$
$$+3\lambda_i\left(\sum_{j=i+1}^{n-1}\lambda_j+8\sum_{j=1}^{i-1}\frac{(-1)^{i-j-1}}{\lambda_j}+\frac{4(-1)^{i-1}(1+\beta)}{\lambda_0}\right)-12=0,$$
$$1\leqslant i\leqslant n-1. \quad (3.145a)$$

Bell (1978) started a Newton-type iterative process of solution using initial estimates of $\lambda_1,\ldots,\lambda_{n-1}$ taken from Goodman's integral solution (1964), which corresponds to the case $n=1$ in Bell's method. Table 3.3 compares Bell's results for the position of the melting interface $s_0/(kt)^{\frac{1}{2}}$, for different subdivisions, with those for the exact solution (Carslaw and Jaeger 1959) for the case $L/(u_s-u_0)c=1$.

TABLE 3.3
Position of melting interface/$(kt)^{\frac{1}{2}}$ *i.e.* $s_0(t)/(kt)^{\frac{1}{2}}$; $L/(u_s-u_0)c=1$

n	1	2	4	8	32	Exact solution
$s_0(t)/(kt)^{\frac{1}{2}}$	1.320 03	1.254 52	1.242 91	1 240 70	1.240 15	1.240 13

TABLE 3.4

Percentage error in position of melting front/$(kt)^{\frac{1}{2}}$, i.e. in $s_0(t)/(kt)^{\frac{1}{2}}$, compared with exact solution. Two sub-regions $n = 2$: different latent heat parameters; $L/(u_s - u_0)c$

$L/(u_s - u_0)c$	1	10	50	100
% error in $s_0(t)/(kt)^{\frac{1}{2}}$	+1.16	+0.24	+0.05	+0.02

Table 3.4 shows results obtained when only two sub-regions are used for different values of the latent heat parameter $L/(u_s - u_0)c$. This method, based on equal subdivision of the dependent variable temperature, seems a promising one to apply to other heat-flow problems with or without phase changes.

In a later paper, Bell (1979) applied his spatial subdivision method to the solidification of a liquid around a cylindrical pipe but did not assume a similarity solution. The more general equations were solved by a Runge–Kutta method, and also by a simpler numerical integration algorithm, starting from a short-time series based on Poots's (1962b) solution.

3.5.6. Multi-dimensional problems

Poots (1962a) extended the heat-balance integral method to study the movement of a two-dimensional solidification front in a liquid, initially at the fusion temperature, contained in a uniform prism of square cross-section of side a. The surface is maintained at a constant temperature below the fusion temperature of the liquid. Constant thermal properties are assumed.

The prism is bounded by the isothermal surface

$$C_0(x, y) = (x^2 - a^2)(y^2 - a^2) = 0$$

at temperature T_0, where $T_0 < T_F$, the fusion temperature of the liquid. At any time, while the process of inward solidification is occurring, the two-dimensional solidification front is defined to be the surface $C_F(x, y, t) = 0$, an isothermal on which $T = T_F$.

After introducing the dimensionless variables

$$\left. \begin{array}{ccc} \theta = \dfrac{T - T_0}{T_F - T_0}, & \beta = \dfrac{\rho L k}{K(T_F - T_0)}, & \tau = \dfrac{kt}{a^2}, \\[2mm] \ell = s/a, & X = x/a, & Y = y/a, \end{array} \right\} \tag{3.146}$$

the definitive equations may be written

$$\frac{\partial^2 \theta}{\partial X^2} + \frac{\partial^2 \theta}{\partial Y^2} = \frac{\partial \theta}{\partial \tau},$$

(3.147)

$$\left. \begin{array}{ll} \theta = 0 \quad \text{on} \quad C_0(X, Y) = (X^2 - 1)(Y^2 - 1) = 0, \quad & \tau > 0 \\ \theta = 1, \quad\quad C_F(X, Y, \tau) = 0, \end{array} \right\}$$

(3.148)

and

$$\oint_{C_F = 0} \frac{\partial \theta}{\partial \nu} \, d\ell = -\beta \frac{dv}{d\tau},$$

(3.149)

where

$$v(C_F, C_0) = \int\int_{C_F = 0}^{C_0 = 0} dX \, dY.$$

The initial condition is

$$\theta = 1, \quad\quad \tau = 0.$$

(3.150)

In these equations, ℓ is the non-dimensional distance along $C_F(x, y, t) = 0$ in an anti-clockwise direction and ν is the outward normal to this solidification front which is assumed to have the shape

$$C_F(X, Y, \tau) = (X^2 - 1)(Y^2 - 1) - \varepsilon(\tau) = 0$$

(3.151)

where ε is an unknown function of time τ. The initial condition (3.150) now becomes

$$\varepsilon = 0, \quad\quad \tau = 0,$$

(3.152)

and (3.151) implies that $\varepsilon = 1$ at the instant solidification is completed.

The heat balance integral is now obtained by integrating both sides of (3.147) over the solid phase bounded by the contours $C_0 = 0$ and $C_F = 0$. Application of Green's theorem to the result and use of (3.149) leads to the relationship

$$\beta \frac{dv}{d\tau} - \oint_{C_0 = 0} \frac{\partial \theta}{\partial \nu} \, d\ell = \int\int_{C_F = 0}^{C_0 = 0} \frac{\partial \theta}{\partial \tau} \, dX \, dY.$$

(3.153)

A second integral

$$\beta \frac{dv}{d\tau} = \int\int_{C_F = 0}^{C_0 = 0} \left\{ \left(\frac{\partial \theta}{\partial X}\right)^2 + \left(\frac{\partial \theta}{\partial Y}\right)^2 + \theta \frac{\partial \theta}{\partial \tau} \right\} dX \, dY$$

(3.154)

follows by applying Green's theorem to the divergence theorem

$$\text{div}(\theta \, \nabla\theta) = \nabla\theta \cdot \nabla\theta + \theta \, \nabla^2\theta = \nabla\theta \cdot \nabla\theta + \theta \frac{\partial\theta}{\partial\tau},$$

on substituting from (3.147), and using (3.148) and (3.149). Poots (1962a) completed the analysis for an assumed one-parameter temperature distribution

$$\theta = (X^2-1)(Y^2-1)/\varepsilon \tag{3.155}$$

and also for a two-parameter form

$$\theta = \frac{(X^2-1)(Y^2-1)}{\varepsilon}(1-g) + g\left(\frac{(X^2-1)(Y^2-1)}{\varepsilon}\right)^2. \tag{3.156}$$

Ivantsov (1947) showed that for an isothermal phase-change surface which moved with a constant velocity, the shape of a paraboloid of revolution satisfied the differential equation and boundary conditions. Horvay and Cahn (1961) extended the same approach to other assumed surface shapes.

Riley and Duck (1971) considered the three-dimensional freezing of a cuboid.

3.6. Approximate solutions

3.6.1. Steady-state approximation and improvements

In Neumann's similarity solution discussed in §3.2, the equation (3.14) describing the motion of the solidification boundary in the single-phase problem was seen to reduce to (3.15) for λ small, which corresponds to a latent heat L that is large compared with the specific heat capacity of the ice.

Stefan (1889a) and sundry authors later obtained the result (3.15) by using the physical approximation that for large L, implying a slowly moving solidification boundary, the temperature distribution at any instant behind the moving boundary is the steady-state distribution which would be set up if the boundary were to be fixed in its position at that instant. In effect, the heat flow is rapid enough to maintain this pseudo 'steady state'.

Thus, putting $\partial u_1/\partial t = 0$ in (3.1), the steady-state solution which satisfies (3.4) and the first of (3.6) is

$$u_1 = U_1(1 - x/s). \tag{3.157}$$

Substitution of (3.157) in the moving-boundary condition, $-K_1 \partial u_1/\partial x = L\rho \, ds/dt$, for the single-phase problem yields

$$s^2 = 2K_1 U_1 t/L\rho \qquad (3.158)$$

which is equivalent to (3.15) remembering that in (3.14) $\lambda = \alpha/(2k_1^{\frac{1}{2}}) = \frac{1}{2}s/(k_1 t)^{\frac{1}{2}}$ from (3.10), where $k_1 = K_1/c_1\rho$.

Crank (1975, p. 310) used the steady-state approximation to solve the diffusion version of the one-phase melting problem for a surface boundary condition

$$\ell \, \partial u/\partial t = K \, \partial u/\partial x, \qquad x = 0, \qquad t > 0, \qquad (3.159)$$

with ℓ and K constants. Crank (1957b) gave the corresponding approximations for the cylinder and the sphere together with estimates of accuracy based on numerical calculations.

The essential simplifying feature of the above method is that the assumption of the steady-state temperature profile avoids the need to solve the heat-flow equation. Various authors have exploited this advantage by assuming other more sophisticated profiles, of which the steady-state solution is usually the leading term in a series. The higher terms may therefore provide an estimate of the errors associated with the steady-state assumption.

Thus, Solomon (1978) described an approximate solution, following Megerlin (1968), for a constant surface heat flux. The method follows the same pattern as that of Goodman (§3.5.4) but no attempt is made to preserve the heat balance. Take the problem defined by the following:

$$\frac{\partial u}{\partial t} = \frac{\partial^2 u}{\partial x^2}, \qquad 0 < x < s(t), \qquad (3.160)$$

$$\partial u/\partial x = -1, \qquad x = 0, \qquad t > 0, \qquad (3.161)$$

$$\partial u/\partial x = -ds/dt, \qquad u = 0, \qquad x = s(t), \qquad t > 0, \qquad (3.162)$$

$$u = 0, \qquad x > s(t); \qquad s(0) = 0. \qquad (3.163)$$

Consider

$$u(x, t) = a_1(x - s) + a_2(x - s)^2 \qquad (3.164)$$

where a_1, a_2, s are functions of t to be determined and the second of (3.162) is automatically satisfied. Using (3.162) and its alternative form (see equation (3.110)) coupled with differentiation of (3.164) it immediately follows that

$$u = -\frac{ds}{dt}(x - s) + \frac{1}{2}\left(\frac{ds}{dt}\right)^2 (x - s)^2. \qquad (3.165)$$

Substitution of (3.161), after differentiation of (3.165), leads to

$$\frac{ds}{dt} = \frac{-1 + \sqrt{(1 + 4s)}}{2s},$$

the negative root being discarded on physical grounds, and finally to the solution

$$t = \tfrac{1}{2}s + \tfrac{1}{12}(1 + 4s)^{\frac{3}{2}} - \tfrac{1}{12}. \tag{3.166}$$

We note that, in this problem, the steady-state approximation is $u = s - x$ and $ds/dt = 1$, which corresponds to the first term in (3.165). For this term, therefore, a heat balance over the whole solid region is preserved.

Solomon (1978) compared his results computed from (3.166) with those obtained by Rose (1960) for the same problem based on a weak solution. Table 3.5 is an extract from Solomon's table (1978).

Solomon, Wilson, and Alexiades (1981) showed that the solution to the Stefan problem with a convective boundary condition tends to the quasi-steady approximation as the specific heat tends to zero.

Pekeris and Slichter (1939) study the formation of ice around a cold cylinder embedded in water initially at $0\,°C$, starting from an assumed temperature profile given by

$$u = A + B \ln x + \left(\frac{a^2}{K}\right)\{\tfrac{1}{4}(A - \dot{B})x^2 + \tfrac{1}{4}\dot{B}x^2 \ln x\}$$

$$+ \left(\frac{a^4}{k^2}\right)\left\{\frac{x^4}{128}(2\ddot{A} - 3\ddot{B}) + \frac{\ddot{B}}{64}x^4 \ln x\right\} + \dots, \tag{3.167}$$

where A, B are functions of time, $\dot{B} = dB/dt$ etc., a is the radius of the cylinder, and $x = r/a$. They finally arrive at the solution, for $y = \{s(t)/a\}^2$,

$$y = y_0 + \frac{cU}{L}f(y_0) + \dots, \tag{3.168}$$

TABLE 3.5

Comparison of $t(s)$ from (3.166) and Rose (1960)

$s(t)$	t[Eqn (3.166)]	t(Rose)[a]
0	0	0
0.4	0.450	0.466
0.8	1.030	1.034
1.2	1.690	1.681
1.6	2.440	2.394

[a] The agreement is stated to be within the limits of accuracy of Rose's values.

where the surface of the cylinder, $r = a$, is maintained at temperature $-U(t)$, y_0 is the solution of

$$y_0 \ln y_0 + 1 - y_0 = \frac{4K}{L\rho a^2} \int_0^t U \, dt', \qquad (3.169)$$

and

$$f(y_0) = \frac{2(y_0 - 1)}{(\ln y_0)^2} - \frac{1 + y_0}{\ln y_0}. \qquad (3.170)$$

The first term in (3.168) is again the steady-state approximation to y, since the appropriate steady-state solution is

$$u = \frac{U \ln(s/r)}{\ln(s/a)},$$

from which (3.169) follows on using the usual conditions $\partial u/\partial r = (L\rho/K) \, ds/dt$, $u = 0$ on the solidification boundary and integrating the resulting differential equation, $(L\rho/K) \, ds/dt = -U/\{s \ln(s/a)\}$, remembering that $y = (s/a)^2$.

Pekeris and Slichter (1939) show graphs of $y_0 \ln y_0 + 1 - y_0$ and of $f(y_0)/y_0$ for y_0 ranging between 0 and 700, together with a plot of the growth of ice according to the steady-state approximation for a choice of physical conditions.

The idea of postulating temperature profiles was extended to multi-dimensional problems of crystal growth by Horie and Chehl (1975). They chose the profiles either by introducing a 'penetration depth' or by using the heat-flow equation itself.

3.6.2. Spatial subdivision

Solomon (1978) extends the Megerlin method in a manner analogous to Bell's (1978) extension of the heat-balance method by spatial subdivision. Whereas Bell divided the temperature range into equal intervals, Solomon uses equal subdivisions of the space range and postulates an approximate temperature profile in each subdivision.

Taking the one-phase Stefan melting problem he denotes the jth subdivision by

$$x_{j-1} < x < x_j, \qquad j = 1, 2, \ldots, N, \qquad (3.171)$$

where

$$x_j(t) = (j/N)s(t), \qquad j = 0, 1, 2, \ldots, N,$$

and lets

$$u^i(x, t) = u_j(t) + a_j(t)(x - x_j) + b_j(t)(x - x_j)^2. \qquad (3.172)$$

Continuity of u^j and of its first derivative are required at $x = x_j$, where u^j also satisfies the heat-flow equation. Also u^1 and u^N satisfy the conditions $u(0, t) = u_0 > u_f$, $-K \, \partial u(s, t)/\partial x = \rho L \, ds/dt$ respectively, where u_f is the freezing temperature. From a series of recurrence relations for the a_j, b_j, and the assumptions that the u_j are a constant for each j, and that $s = 2\lambda \sqrt{(kt)}$, Solomon derives the following transcendental equation for a root λ, which depends on N:

$$\frac{c}{L}(u_0 - u_f) = \frac{\lambda^2}{N}\left[1 + 3\left\{1 + \frac{2\lambda^2}{3N^2}\right\}\prod_{\beta=2}^{N}\left(1 + \frac{2\beta\lambda^2}{N^2}\right)\right.$$

$$\left. + 2\sum_{\beta=2}^{N-1}\prod_{\ell=\beta+1}^{N}\left(1 + \frac{2\ell\lambda^2}{N^2}\right)\right]. \quad (3.173)$$

Writing $c(u_0 - u_f)/(L\sqrt{\pi}) = F_N(\lambda)$ from (3.173) Solomon tabulates F_N, $N = 0, 1, 2, 3$, for $0 \leq \lambda \leq 2$ at intervals of 0.2 and notes slow convergence but concludes that limits of 10 per cent relative accuracy are given by

$$\frac{c}{L\sqrt{\pi}}(u_0 - u_f) \sim \begin{cases} 4 & \text{for} \quad N = 1, \\ 4.6 & \text{for} \quad N = 2, \\ 5 & \text{for} \quad N = 3. \end{cases}$$

However, this range of apparently up to 1000 °F even for $N = 1$ is not a serious practical limit for the melting of ice or a typical wax. However, caution is advised for the recrystallization of iron, say from α to γ type, where the allowable temperature range would be around 100 °F for $N = 1$ and 130 °F for $N = 3$.

This approach, based on spatial subdivision, is closely similar to the use of cubic splines by Crank and Gupta (1972b).

Solomon applied spatial subdivision to the constant boundary-flux problem for which results obtained by the Megerlin method without subdivision have been given in Table 3.5. In fact, he used equation (3.166) as a starting solution for a Runge–Kutta scheme for handling the spatial-subdivision equations. Some of his results are shown in Table 3.6

TABLE 3.6

Moving-boundary position
$s(t)$

t	$N = 1$	$N = 2$	$N = 5$
0	0	0	0
4	2.404	2.244	2.241
8	4.094	3.708	3.723
12	5.5549	4.928	4.961
16	6.865	6.009	6.058
20	8.085	6.995	7.058
24	9.233	7.911	7.987
28	10.324	8.772	8.860

for various values of N. Solomon concluded that $N = 2$ produces acceptable accuracy, though a heat-balance check reveals heat discrepancies for
$N = 2$ of around 16 per cent for $t = 4$ rising to 35 per cent for $t = 28$, but
reducing to 3.6 per cent for $N = 4$ and only 2 per cent for $N = 6$ and 10.

3.6.3. The inverse Stefan problem

Kreith and Romie (1955) described an iterative method for improving
the steady-state approximation for the inward solidification of a cylinder
and a sphere when the rate of movement of the interface is constrained to
be constant. The problem is to determine the necessary temperature–time
dependence which must be imposed on the surface of the cylinder or
sphere to maintain a constant rate of solidification. The physical condition
on the interface is $-K\, \partial T/\partial R = \rho L\, dS/d\tau$ with $\partial T/\partial R = g = \text{constant}$, $\tau =$
time, and S is the thickness of the frozen zone.

Expressed in non-dimensional terms, the cylinder problem requires the
solution of the following equations with $G = gcr_0/L$, where r_0 is the radius
of the cylinder, c is specific heat, and

$$u = \frac{T - T_f}{r_0 g}, \qquad r = \frac{R}{r_0}, \qquad t = \frac{Kg\tau}{\rho L r_0}, \qquad s = \frac{S}{r_0}, \qquad G = \frac{gcr_0}{L},$$

where T_f is the fusion temperature and T_0 is the temperature on the outer
surface $R = r_0$:

$$\frac{1}{r} \frac{\partial}{\partial r}\left(r \frac{\partial u}{\partial r}\right) = G \frac{\partial u}{\partial t}, \tag{3.174}$$

$$\partial u/\partial r = -ds/dt = +1, \qquad u = 0, \qquad r = 1 - s, \tag{3.175}$$

$$u = 0, \qquad s = 0, \qquad t = 0. \tag{3.176}$$

It is clear from (3.175) and (3.176) that the position of the solidification
interface at time t is given by $r = 1 - s = 1 + t$. Kreith and Romie express
the temperature u by the series

$$u = u_1 + u_2 + u_3 + \ldots + u_n + \ldots, \tag{3.177}$$

each term of which is related to the preceding term by the equation

$$\frac{1}{r} \frac{\partial}{\partial r}\left(r \frac{\partial u_n}{\partial r}\right) = \frac{G\, \partial u_{n-1}}{\partial t} \tag{3.178}$$

and hence, after integration, by

$$u_n = G \int_{1+t}^{r} \frac{dr}{r} \int_{1+t}^{r} r \frac{\partial u_{n-1}}{\partial t}\, dr, \qquad n > 1. \tag{3.179}$$

The first term u_1 of the series is obtained by setting $G = 0$ in (3.178),
which leads immediately to $r\, \partial u/\partial r = \text{constant} = 1 - s$, since on $r = 1 - s$,

$\partial u/\partial r = 1$, and so, remembering that $u = 0$ on $1 - s$, it follows that

$$u_1 = \phi \ln(r/\phi), \qquad \phi = 1 + t. \qquad (3.180)$$

This term is the steady-state approximation and the new variable ϕ is the radius of the solidification front. The next four terms given by Kreith and Romie are

$$\left.\begin{aligned}
u_2 &= -G\left\{\left(\frac{r^2}{4} + \tfrac{3}{4}\phi^2\right)\ln\frac{r}{\phi} + \frac{\phi^2}{2} - \frac{r^2}{2}\right\} \\[2mm]
U_3 &= G^2\left\{(\tfrac{5}{16}\phi^3 + \tfrac{3}{8}\phi r^2)\ln\frac{r}{\phi} + \tfrac{21}{64}\phi^3 - \tfrac{5}{16}\phi r^2 - \frac{r^4}{64\phi}\right\} \\[2mm]
u_4 &= -G^3\left\{(\tfrac{35}{384}\phi^4 + \tfrac{15}{64}\phi^2 r^2 + \tfrac{3}{128}r^4)\ln\frac{r}{\phi}\right. \\[2mm]
&\qquad \left. + \tfrac{278}{2304}\phi^4 - \tfrac{17}{256}\phi^2 r^2 - \tfrac{7}{128}r^4 + \frac{r^6}{2304}\,\phi^2\right\} \\[2mm]
u_5 &= G^4\left\{(\tfrac{198}{9216}\phi^5 + \tfrac{35}{384}\phi^3 r^2 + \tfrac{15}{512}r^4\phi)\ln\frac{r}{\phi}\right. \\[2mm]
&\qquad + \frac{2325}{73728}\phi^5 + \frac{31}{4608}\phi^3 r^2 - \frac{77}{2048}r^4\phi \\[2mm]
&\qquad \left. - \frac{1}{1536}\frac{r^6}{\phi} - \frac{1}{73728}\frac{r^8}{\phi}\right\}.
\end{aligned}\right\} \qquad (3.181)$$

For values of $G \geqslant -1.5$ the first five terms are stated by Kreith and Romie to give the temperature u to an accuracy within one or two per cent of the exact value.

They plot the required temperature–time histories on the cylinder surface for values of G from zero to -1.5, i.e. a range of constant speeds of the solidification front. These are reproduced in Fig. 3.2. They show that in order to solidify the cylinder completely in such a way that the interface maintains a constant speed to the end, the surface temperature has to be raised above the melting temperature. Such a condition is not allowed for in the problem as formulated in equations (3.174–6). This behaviour, of course, is a consequence of the speeding up of the interface as it approaches the centre when the outer surface temperature is maintained constant.

By the same method, Kreith and Romie (1955) obtain the temperature profile for a constant-speed solidification front in a sphere expressed as

$$u = -\sum_{n=1}^{\infty} G^n r^{n+1}\frac{n}{(n+2)!} + \sum_{i=-1}^{\infty}\sum_{j=0}^{\infty}(-1)^{i-1}G^{i+1}r^{i-1}\phi^{i+2}\frac{2i+j+2}{(i+2)!\,j!}.$$

$$(3.182)$$

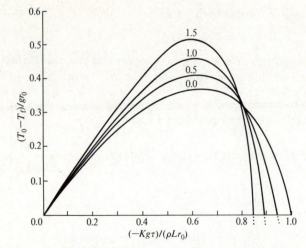

FIG. 3.2. Temperature–time history which must be imposed on the surface of a cylinder solidifying at a constant rate. Numbers on curves are values of $-gcr_0/L$

They plot the temperature–time history on the outer surface of the sphere (Fig. 3.3) needed to maintain a constant speed of the solidification boundary. Comments about the final stages of solidification are the same for the sphere as for the cylinder.

For reference, Kreith and Romie quote the corresponding solution for a constant-speed interface in a semi-infinite solid obtained by Stefan (1891), namely

$$\frac{c}{L}(T-T_{\mathrm{f}}) = 1 - \exp\left(\frac{g^{2}cKt}{\rho L^{2}} - \frac{gcx}{L}\right), \qquad (3.183)$$

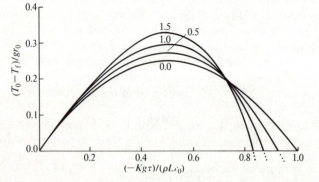

FIG. 3.3. Temperature–time history which must be imposed on the surface of a sphere solidifying at a constant rate. Numbers on curves are values of $-gcr_0/L$

where u_f is the melting temperature and τ is time. The temperature on the outer surface, $x = 0$, clearly has to be

$$T_0 = T_f + \frac{L}{c}\left(1 - \exp\frac{g^2 cKt}{\rho L^2}\right),$$

i.e. it takes negative values which increase exponentially with time.

3.6.4. *Fictitious diffusivity approximations*

Churchill and Gupta (1977) based approximate solutions of Stefan problems on classical solutions of the heat-flow equation in the absence of phase changes. They introduced fictitious heat diffusivities. As an example, the two-phase Neumann problem set out in equations (3.1–6) in §3.2 reduces to a simple heat-flow problem in the special case of zero latent heat and uniform properties. The solidification boundary, as such, no longer exists but the motion of the melting temperature is derivable from the well-known solution of the simple heat-flow problem satisfying all the boundary conditions for $L = 0$. It is convenient to rewrite the problem of §3.2 in terms of a melting temperature U_m and so to rewrite the solutions (3.12) and (3.13) as

$$\frac{U_1 - u_1}{U_1 - U_m} = \frac{\text{erf}\{x/2(k_1 t)^{\frac{1}{2}}\}}{\text{erf}\,\lambda}, \tag{3.184}$$

$$\frac{u_2 - U_2}{U_m - U_2} = \frac{\text{erfc}\{x/2(k_2 t)^{\frac{1}{2}}\}}{\text{erfc}\{\lambda(k_1/k_2)^{\frac{1}{2}}\}}, \tag{3.185}$$

where the condition (3.5) has been replaced by $u = U_2$, $x \to \infty$, $t \geq 0$ and $U_1 > U_m > U_2$; also $\lambda = \alpha/(2k_1^{\frac{1}{2}})$ so that (3.10) has become $s = 2\lambda(k_1 t)^{\frac{1}{2}}$ and hence the equation for λ is now

$$\frac{e^{-\lambda^2}}{\lambda\,\text{erf}\,\lambda} - \left(\frac{U_2 - U_m}{U_m - U_1}\right)\left\{\frac{K_2\rho_2 c_2}{K_1\rho_1 c_1}\right\}^{\frac{1}{2}}\frac{e^{-\lambda^2 k_1/k_2}}{\lambda\,\text{erfc}\{\lambda(k_1/k_2)^{\frac{1}{2}}\}} = \frac{\pi^{\frac{1}{2}}L}{c_1(U_m - U_1)} \tag{3.186}$$

instead of (3.11). From (3.184) we find that the flux F on $x = 0$ is

$$F = -K_1\frac{\partial u_1}{\partial x} = \frac{K_1(U_1 - U_m)}{(\pi k_1 t)^{\frac{1}{2}}\,\text{erf}\,\lambda}. \tag{3.187}$$

For $L = 0$, $\lambda = \infty$, subscripts can be dropped and instead of (3.184) and (3.185) we have the standard solution (Crank 1975, p. 32)

$$\frac{U_1 - u}{U_1 - U_2} = \text{erf}\left\{\frac{x}{2(kt)^{\frac{1}{2}}}\right\} \tag{3.188}$$

and (3.187) is replaced by

$$F = K(U_1 - U_2)/(\pi k t)^{\frac{1}{2}}. \tag{3.189}$$

Putting $u = U_m$ in (3.188) gives for the motion of the U_m isotherm, which is the solidification boundary, in the hypothetical case of zero latent heat

$$s(t) = 2(kt)^{\frac{1}{2}} \operatorname{erf}^{-1}\{(U_1 - U_m)/(U_1 - U_2)\}. \qquad (3.190)$$

Comparison of (3.187) and (3.189) shows that the latter equation would give the exact surface heat flux for a finite latent heat if

$$K\rho c = K_1 \rho_1 c_1 \{(U_1 - U_m)/(U_1 - U_2)\operatorname{erf} \lambda\}^2,$$

which can be secured by letting $K = K_1$ and taking a fictitious value k_f for the thermal diffusivity k given by

$$k_f = \left(\frac{k_1 (U_1 - U_2)\operatorname{erf} \lambda}{(U_1 - U_m)}\right)^2. \qquad (3.191)$$

An alternative interpretation is that a fictitious heat capacity c_f is to be used where

$$c_f = \frac{c_1 (U_1 - U_m)}{\{(U_1 - U_2)\operatorname{erf} \lambda\}^2}, \qquad (3.192)$$

coupled with putting $K\rho = K_1 \rho_1$.

These procedures cannot yield the complete temperature distribution correctly. However, the exact value for the position $s(t)$ of the solidification boundary can be obtained from equation (3.190), for $U_1 \neq U_m$, by choosing

$$k_s = k_1 \left(\frac{\lambda}{\operatorname{erf}^{-1}\{(U_1 - U_m)/(U_2 - U_1)\}}\right)^2. \qquad (3.193)$$

Churchill and Gupta (1977) proposed that these fictitious properties can be used to approximate the behaviour in other geometries and for different boundary conditions. In this way, the solutions for non-linear Stefan problems can be deduced from the well-known analytical and numerical solutions of simple heat-flow problems. The choice of k_f or k_s depends on whether accuracy in the surface heat flux or in the location of the phase-change interface is desired. If both quantities are important, both fictitious values can be employed. Churchill and Gupta (1977) cited the problems of the freezing of a liquid outside a cylinder and the freezing of a quarter-space as examples of satisfactory agreement between results obtained using the fictitious heat parameters defined in (3.191) and (3.193) and those that other authors have calculated by numerical methods.

Churchill and Evans (1971) tabulated and plotted values of λ and $\operatorname{erf} \lambda$ for a wide range of the parameters $L/\{c_1(U_1 - U_m)\}$, $\{(U_2 - U_m)/(U_m - U_1)\}\{(K_2 c_2 \rho_2)/(K_1 c_1 \rho_1)\}^{\frac{1}{2}}$, and k_1/k_2 which facilitate the use of the fictitious heat parameters. Their values for $10^4 \lambda$ are tabulated in Table 3.7 to

TABLE 3.7

Values of $10^4\,\lambda$, the solution of (3.186), for values of $L/c_1(U_1-U_m)$, k_1/k_2, and $p = \{(U_2-U_m)/(U_m-U_1)\}\{(K_2c_2\rho_2)/(K_1c_1\rho_1)\}^{\frac{1}{2}}$

$\dfrac{L}{c_1(U_1-U_m)}$	k_1/k_2	$p=0.5$	$p=1.0$	$p=1.5$	$p=2.0$	$p=3.0$	$p=5.0$	$p=10.0$
0.0	0.0	9451	6584	4991	3982	2803	1731	
0.0	0.5	7371	5137	3973	3245	2376	1547	
0.0	1.0	6841	4769	3708	3046	2253	1488	
0.0	1.5	6504	4536	3538	2917	2172	1447	
0.0	2.0	6256	4365	3413	2821	2110	1415	
0.5	0.5	5686	4403		3010			
0.5	1.0	5456	4179	3392	2855	2168	1461	803
0.5	1.5	5295	4028	3266	2752	2096	1422	789
0.5	2.0	5168	3912	3170	2673	2041	1392	778
0.5	3.0	4972	3738		2555			
0.5	4.0	4822	3607		2467			
0.5	5.0	4700	3502		2397			
0.5	6.0	4596	3415		2338			
1.0	0.5	4835	3933		2826			
1.0	1.0	4699	3778	3152	2701	2092	1435	793
1.0	1.5	4599	3669	3054	2615	2029	1399	780
1.0	2.0	4519	3584	2977	2548	1981	1371	770
1.0	3.0	4391	3452		2448			
1.0	4.0	4290	3351		2371			
1.0	5.0	4205	3268		2310			
1.0	6.0	4133	3198		2258			
1.5	1.0	4198	3480	2961	2571	2026	1411	793
1.5	1.5	4129	3396	2881	2498	1969	1377	780
1.5	2.0	4072	3330	2818	2442	1925	1350	777
2.0	0.5	3903	3338		2550			
2.0	1.0	3834	3246	2804	2461	1966	1389	789
2.0	1.5	3782	3180	2737	2398	1915	1356	776
2.0	2.0	3739	3126	2683	2348	1875	1331	766
2.0	3.0	3612	3041	2271				
2.0	4.0	3669	2974	2212				
2.0	5.0	3564	2918	2163				
2.0	6.0	3521	2870	2121				
3.0	1.0	3329	2899	2557	2280	1863	1347	780
3.0	1.5	3296	2853	2507	2232	1821	1318	768
3.0	2.0	3268	2815	2467	2193	1787	1295	758
5.0	1.0	2732	2455	2220	2021	1702	1276	764
5.0	1.5	2715	2428	2189	1988	1671	1252	753
5.0	2.0	2700	2405	2163	1962	1646	1233	744
10.0	1.0	2037	1891	1761	1643	1442	1143	728
10.0	1.5	2030	1880	1746	1627	1424	1127	719
10.0	2.0	2024	1870	1734	1613	1410	1113	711

three or four significant figures with the headings adjusted to the present nomenclature.

Ockendon (1975) drew attention to the possible non-uniformity of the limit for zero latent heat in one-phase problems.

3.6.5. *Perturbation methods*; *asymptotic solutions*; *stability*

There appear to be very few solutions of Stefan problems in the form of series expansions derived by perturbation methods. Even for one-dimensional problems, the analysis is very complicated and often prohibitive beyond the first approximation. Fox (1975) quotes Jiji (1970), who succeeded in obtaining only the zero-order approximation in a problem of the solidification of a fluid surrounding a cold solid cylinder. Fleishman, Gingrich, and Mahar (1978) and Fleishman and Mahar (1977) report applications of perturbation linearization methods to obtain approximate solutions of two-dimensional steady-state free-boundary problems. Elliptic partial differential equations of the form $\mathrm{div}(K \, \mathrm{grad} \, u) + f = 0$ are considered in which K and/or f depend discontinuously on u and a small parameter ε appears in a boundary condition. References are given by Fleishman *et al.* (1978) to earlier papers on the one-dimensional versions of similar problems.

Friedman (1964), Ockendon (1975), and Elliott and Ockendon (1982) draw attention to the value of asymptotic methods in which either the space or time variables or some physical parameters are taken to be large or small. They can sometimes illuminate aspects of behaviour not revealed by weak and variational methods. Indeed, numerical methods in general frequently need to be supplemented by small or large time or distance solutions near singularities at zero time, for example, or when approaching a boundary at infinity. Rubinstein (1971), Evans *et al.* (1950), Cannon and Hill (1967), Cannon, Douglas, and Hill (1967), and others have studied asymptotic expansions for large and small independent variables. Howarth and Poots (1976), in considering the solidification of a half-space of liquid cooling by black-body radiation, obtained both large and small-time series solutions. They then applied a process of renormalization in which certain coefficients in the large-time series are chosen so that this solution, when formally expanded for small times, is made to agree as far as possible with the small-time solution. A useful uniformly valid expansion is obtained.

Ockendon (1980) and Elliott and Ockendon (1982) explored certain instabilities of moving boundaries in a selection of problems by perturbation and asymptotic techniques. Chadam and Ortoleva (1982) adapted Rubinstein's (1982b) stability analysis to show that planar melting is stable but planar solidification is unstable unless surface tension is included. Small amplitude perturbations are also analysed by Turland and

Peckover (1980), who refer to several earlier papers including an extensive non-linear stability analysis of the freezing of a dilute binary alloy by
Wollkind and Segel (1970).

Rasmussen and McGeough (1981) examined non-linear stability in
electro-forming problems by perturbation and numerical techniques.

We discuss now in some detail the application of perturbation techniques to two problems in which the ratio of the latent heat to the sensible
or specific heat is large, as it often is in practice.

(i) *An evaporation ablation problem.* Andrews and Atthey (1975*a,b*)
considered the mathematical problems posed by the use of high-power
lasers and electron beams for deep-penetration welding and cutting of
thick materials. The intense heating may cause a part of the material
surface to evaporate so that a hole is formed (Fig. 3.4) and there is a
moving boundary. Part of the energy of the laser beam, for example, is
absorbed at the moving surface as the latent heat of evaporation and part
is conducted into the material below.

Andrews and Atthey (1975*a*) considered first the simplest case in
which all the energy supplied by the beam is assumed to be absorbed as
latent heat. If a uniformly distributed power W is applied normally over

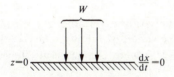

Pre-heating of surface
to evaporation temperature

Early development of
boundary motion

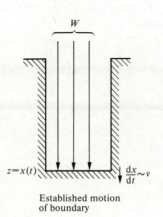

Established motion
of boundary

FIG. 3.4. Various stages of hole formation

an area A of the surface, a cylindrical hole will be formed, and if in a time interval δt the volume of material evaporated is $A \, \delta x$ then

$$h \rho A \, \delta x = W \, \delta t,$$

where h is the heat required to vaporize unit mass of the material of density ρ, i.e.

$$dx/dt = W/(Ah\rho). \tag{3.194}$$

This speed, dx/dt, is an upper limit for the rate of penetration which will be approached after a finite time if some conduction of heat into the material occurs. The coupled process constitutes a Stefan problem in radial coordinates. First, Andrews and Atthey (1975a) obtained a perturbation solution for a planar evaporating boundary using an expansion parameter ε defined by

$$\varepsilon = \bar{c} T_v / L, \tag{3.195}$$

where $\bar{c}$ and L are average specific heat and latent heat respectively and T_v is the boiling point. They quote physical data to justify taking ε small ($\varepsilon \sim 0.2$). Non-dimensional variables

$$\theta = S/S_v, \qquad \zeta = z/\ell, \qquad \tau = vt/\ell, \qquad \xi = x/\ell,$$

are introduced where S is the Kirchhoff temperature variable, z is normal to the evaporating surface $z = x(t)$, v is the speed of the boundary in the evaporation-controlled limiting case, and ℓ is a characteristic length defined by

$$\ell = \bar{K}/(\rho \bar{c} v), \tag{3.196}$$

where $\bar{K}$ is an average conductivity over the temperature range $(0, T_v)$. The problem is then specified by the equations

$$\left. \begin{array}{ll} \partial \theta^2 / \partial \zeta^2 = \partial \theta / \partial \tau, & \\ d\xi/d\tau - 1 - \varepsilon (\partial \theta / \partial \zeta + 1) = 0, \qquad \theta = 1, \qquad \zeta = \xi(\tau) & \\ \theta \to 0, \qquad \zeta \to \infty. & \end{array} \right\} \tag{3.197}$$

Substituting the trial perturbation series for $\theta, \xi, \eta = d\xi/d\tau$,

$$\theta(\zeta, \tau) = \theta_0(\zeta, \tau) + \varepsilon \theta_1(\zeta, \tau) + \dots,$$

$$\xi(\tau) = \xi_0(\tau) + \varepsilon \xi_1(\tau) + \dots,$$

$$\eta(\tau) = \eta_0(\tau) + \varepsilon \eta_1(\tau) + \dots,$$

into (3.197) gives to zero order

$$\partial^2 \theta_0 / \partial \zeta^2 = \partial \theta_0 / \partial \tau, \qquad \eta_0 = 1, \qquad \theta_0 = 1. \tag{3.198}$$

Ignoring the time for the boundary to reach the evaporation temperature (which can be shown to be $O(\varepsilon^2)$) the zero-order solution,

$$\xi_0 = \tau, \qquad\qquad (3.199)$$

comes by integrating the second of (3.198). Taking the Laplace transform in a frame of reference moving with the boundary and putting $\theta_0 = 0$, $\tau = 0$ gives

$$\theta_0 = \tfrac{1}{2}e^{-(\zeta-\tau)}\,\text{erfc}\{(\tfrac{1}{2}\zeta-\tau)/\tau^{\frac{1}{2}}\} + \tfrac{1}{2}\,\text{erfc}\{\tfrac{1}{2}\zeta/\tau^{\frac{1}{2}}\}. \qquad (3.200)$$

The first-order equations

$$\partial^2\theta_1/\partial\zeta^2 = \partial\theta_1/\partial\tau, \qquad \eta_1 = 1 + \partial\theta_0/\partial\zeta, \qquad \theta_1 = 0$$

lead to

$$\eta_1 = \tfrac{1}{2}\,\text{erfc}(\tfrac{1}{2}\tau^{\frac{1}{2}}) - (\pi\tau)^{-\frac{1}{2}}e^{-\tau/4}$$

and hence

$$\eta = 1 + \varepsilon\{\text{erfc}(\tfrac{1}{2}\tau^{\frac{1}{2}}) - (\pi\tau)^{-\frac{1}{2}}e^{-\tau/4}\} + O(\varepsilon^2). \qquad (3.201)$$

This solution is a good approximation for $\tau = O(1)$ but for $\tau = O(\varepsilon^2)$ its second term is $O(1)$ rather than $O(\varepsilon)$. For small times (3.201) provides an outer solution which needs to be supplemented by an inner solution that allows for the pre-evaporation heating of the surface, initially at some temperature below the boiling point.

During the pre-heating period the heat conduction equation is still the first of (3.197), which is to be solved for conditions

$$\partial\theta/\partial\zeta = -1 - 1/\varepsilon, \qquad \zeta = 0$$

$$\theta = 0, \qquad \zeta > 0, \qquad \tau = 0; \qquad \theta \to 0, \qquad \zeta \to \infty, \qquad \tau \geqslant 0.$$

The standard solution (Carslaw and Jaeger, 1959) is

$$\theta = (1 + 1/\varepsilon)\{2(\tau/\pi)^{\frac{1}{2}}\exp(-\tfrac{1}{4}\zeta^2/\tau) - \zeta\,\text{erfc}(\tfrac{1}{2}\zeta/\tau^{\frac{1}{2}})\},$$

which gives a pre-heating time

$$\tau_{\text{p}} = \tfrac{1}{4}\pi\varepsilon^2/(1+\varepsilon)^2 \qquad\qquad (3.202)$$

that is $O(\varepsilon^2)$. The temperature $\theta_{\text{p}}(\zeta, \tau_{\text{p}})$ is

$$\theta_{\text{p}} = \exp[-\{\zeta(1+\varepsilon)/\pi^{\frac{1}{2}}\varepsilon\}^2] - \{\zeta(1+\varepsilon)/\varepsilon\}\text{erfc}\{\zeta(1+\varepsilon)/\pi^{\frac{1}{2}}\varepsilon\}. \qquad (3.203)$$

For the early stages of evaporation (3.202) and (3.203) suggest the new variables

$$\zeta' = \zeta/\varepsilon, \qquad \tau' = \tau/\varepsilon^2 - \tfrac{1}{4}\pi, \qquad \xi' = \xi/\varepsilon^2, \qquad \theta' = \theta, \qquad \eta' = \eta.$$

The heat conduction equation becomes

$$\partial^2\theta'/\partial\zeta'^2 = \partial\theta'/\partial\tau'$$

with the boundary conditions

$$\eta' = 1 + \partial\theta'/\partial\zeta' + \varepsilon, \qquad \theta' = 1, \qquad \zeta' = \varepsilon\xi'(\tau'),$$
$$\theta' \to 0, \qquad \zeta' \to \infty,$$

and the initial condition (3.203) becomes

$$\theta' = \exp[-\{\zeta'(1+\varepsilon)/\pi^{\frac{1}{2}}\}^2] - \zeta'(1+\varepsilon)\operatorname{erfc}\{\zeta'(1+\varepsilon)/\pi^{\frac{1}{2}}\}$$

at $\tau' = -\frac{1}{4}\pi\varepsilon(2+\varepsilon)/(1+\varepsilon)^2$.

The zero-order perturbation solution based on

$$\theta'(\zeta', \tau') = \theta_0'(\zeta', \tau') + \varepsilon\theta_1'(\zeta', \tau') + \dots,$$
$$\eta'(\tau') = \eta_0'(\tau') + \varepsilon\eta_1'(\tau') + \dots$$

must satisfy the equation

$$\partial^2\theta_0'/\partial\zeta'^2 = \partial\theta_0'/\partial\tau' \tag{3.204}$$

and the conditions

$$\left.\begin{array}{c}
\eta_0' = 1 + \partial\theta_0'/\partial\zeta', \qquad \theta_0' = 1, \qquad \zeta' = 0, \\[4pt]
\theta' \to 0 \quad \text{as} \quad \zeta' \to \infty, \qquad \tau' > 0 \\[4pt]
\theta_0' = \exp(-\zeta'^2/\pi) - \zeta'\operatorname{erfc}(\zeta'/\pi^{\frac{1}{2}}), \qquad \tau' = 0.
\end{array}\right\} \tag{3.205}$$

For very small times, i.e. $\tau' = O(\varepsilon^2)$, the evaporating boundary will have moved only a small distance from $\zeta' = 0$. We therefore temporarily discard the condition for η_0' on $\zeta' = 0$ and solve the resulting problem for θ_0' on fixed boundaries. Afterwards, the discarded condition can be used to determine the motion of the evaporating boundary. Thus by taking the Laplace transform of (3.204) and using the conditions (3.205) except the one for η_0' we obtain for the transform, $\bar{\theta}_0'(p)$,

$$d^2\bar{\theta}_0'/d\zeta'^2 = p\bar{\theta}_0' - \exp(-\zeta'^2/\pi) + \zeta'\operatorname{erfc}(\zeta'/\pi^{\frac{1}{2}}), \tag{3.206}$$
$$\bar{\theta}_0' = 1/p, \qquad \zeta' = 0, \qquad \bar{\theta}_0' \to 0, \qquad \zeta' \to \infty.$$

The solution of (3.206) obtained by using Green's functions and then integrating by parts is

$$\bar{\theta}_0' = \tfrac{1}{2}\exp(\tfrac{1}{4}\pi p)[2\exp(-p^{\frac{1}{2}}\zeta')\operatorname{erf}(\tfrac{1}{2}p\pi)^{\frac{1}{2}} +$$
$$+ \exp(p^{\frac{1}{2}}\zeta')\operatorname{erfc}\{\zeta'/\pi^{\frac{1}{2}} + \tfrac{1}{2}(p\pi)^{\frac{1}{2}}\}$$
$$- \exp(-p^{\frac{1}{2}}\zeta')\operatorname{erfc}\{\zeta'/\pi^{\frac{1}{2}} - \tfrac{1}{2}(p\pi)^{\frac{1}{2}}\}]p^{-\frac{3}{2}} +$$
$$+ (1/p)\exp(-\zeta'^2/\pi) - (\zeta'/p)\operatorname{erfc}(\zeta'/\pi^{\frac{1}{2}}). \tag{3.207}$$

The Laplace transform applied to the discarded condition for η_0' gives $\bar{\eta}_0' = 1/p + \partial\bar{\theta}_0'/\partial\zeta'$ at $\zeta' = 0$ and substitution for $\partial\bar{\theta}_0'/\partial\zeta'$ from (3.207) finally yields

$$\bar{\eta}_0 = (1/p)e^{\frac{1}{4}\pi p}\operatorname{erfc}\{\tfrac{1}{2}(p\pi)^{\frac{1}{2}}\}.$$

Inversion by the convolution theorem gives an inner solution

$$\eta' = (2/\pi)\text{arc sin}\{(1+\tfrac{1}{4}\pi/\tau')^{-\frac{1}{2}}\}+\text{O}(\varepsilon). \tag{3.208}$$

Andrews and Atthey (1975a) use a standard matching principle to construct the uniformly valid approximate solution

$$\eta = [1 + \varepsilon\{\tfrac{1}{2}\,\text{erfc}(\tfrac{1}{2}\tau^{\frac{1}{2}}) - (\pi\tau)^{-\frac{1}{2}}e^{-\frac{1}{4}\tau}\}][(2/\pi)/\{1+\varepsilon/(\pi\tau)^{\frac{1}{2}}\}]$$
$$\times[\text{arc sin}\{(1-\tfrac{1}{4}\pi\varepsilon^2/\tau)^{\frac{1}{2}}\}], \tag{3.209}$$

which has errors of $\text{O}(\varepsilon)$ when $\tau = \text{O}(\varepsilon^2)$ and of $\text{O}(\varepsilon^2)$ when $\tau = \text{O}(1)$. Typical results were shown graphically by Andrews and Atthey (1975a). Figures 3.5 and 3.6 show the velocity and position of the evaporating boundary. They also examined typical orders of magnitude for the non-planar radial problem of actual hole formation and concluded that the one-dimensional solution gives a good approximation except possibly very close to the sides of the hole. Andrews and Atthey (1975b) described a modified form of the Douglas and Gallie (1955) finite-difference method to obtain numerical solutions for a source of variable power density because the perturbation technique is unwieldy.

It is the presence of the power term W/A in the moving-boundary condition $W/A = L\rho\,dx/dt - \partial S/\partial n$ that makes this problem more amenable to a perturbation technique than the classical Stefan problem is, without a power term. In the hole-formation problem, the heat conduction term can be neglected to a first approximation compared with the power and latent heat terms.

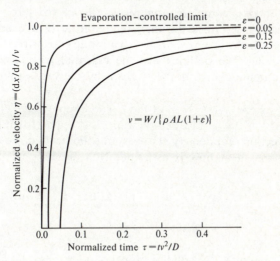

FIG. 3.5. Variation of normalized velocity of evaporating boundary with normalized time: heat diffusivity $D = \bar{K}/\rho\bar{c}$

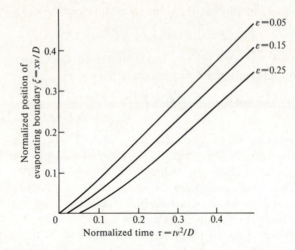

FIG. 3.6. Variation of normalized position of evaporating boundary with normalized time: heat diffusivity $D = \bar{K}/\rho\bar{c}$

(ii) *Stefan's problem for a sphere.* The determination of the time taken to complete the inward solidification of a molten sphere, initially at the fusion temperature, presents mathematical difficulties. For example, the analyses by Pedroso and Domoto (1973) and by Riley, Smith, and Poots (1974) break down before the centre freezes. Stewartson and Waechter (1976) developed a complete asymptotic theory for large latent heat which adequately describes the final temperature profile and they also determined the first four terms of the asymptotic expansion of the time to complete freezing. We outline here the essential features of their approach. More manipulative detail is given by Stewartson and Waechter (1976).

The liquid sphere $0 \leqslant r^* \leqslant a$ is at the fusion temperature T_F^* throughout and at time $t^* = 0$ the surface temperature is dropped to and maintained at T_W^*. Convection and density changes on solidification are neglected as usual, and radial heat flow is assumed. Non-dimensional variables are defined to be

$$r = r^*/a, \qquad S = S^*/a, \qquad t = \{k(T_F^* - T_W^*)/a^2\rho L\}t^*,$$

$$\theta(r, t) = \frac{T^* - T_F^*}{T_F^* - T_W^*} \frac{r^*}{a},$$

where $S^*(t^*)$ denotes the solidification boundary and K, ρ, c are heat conductivity, density, and specific heat respectively. The problem is then defined by the equations

$$\left.\begin{array}{ll}
\partial^2 \theta/\partial r^2 = (1/\lambda)\, \partial \theta/\partial t, & S(t) < r < 1, \\
S(0) = 1, \qquad \theta(1, t) = -1, & t > 0, \\
\theta(r, t) \to 0, \qquad \partial \theta/\partial r \to S\, dS/dt, & r \to S(t)
\end{array}\right\} \qquad (3.210)$$

in the solid phase, and

$$\lambda = L/\{c(T_F^* - T_W^*)\}.$$

Here L is the latent heat and we take $\lambda \gg 1$. As in the earlier papers Stewartson and Waechter put

$$\theta(r, t) = -1 + \sum_{n=0}^{\infty} \frac{\alpha^{(n)}(t)(1-r)^{2n+1}}{\lambda^n (2n+1)!} \qquad (3.211)$$

where $\alpha^0(t) = \alpha(t)$ and $\alpha^{(n)}$, $n = 1, 2, \ldots$, are the successive derivatives of a function α to be determined. Use of the boundary conditions on $r = S(t)$ leads to equations

$$\sum_{n=0}^{\infty} \frac{\alpha^{(n)}(t)(1-S)^{2n+1}}{\lambda^n (2n+1)!} = 1, \qquad \sum_{n=0}^{\infty} \frac{\alpha^n(t)(1-S)^{2n}}{\lambda^n (2n)!} = -SS^{(1)} \qquad (3.212)$$

and by writing

$$S = \sum_{n=0}^{\infty} \lambda^{-n} S_n(t), \qquad \alpha = \sum_{n=0}^{\infty} \lambda^{-n} \alpha_n(t), \qquad (3.213)$$

where $S_0(0) = 1$, $S_n(0) = 0$, $n \geq 1$, substituting into (3.212) and equating successive powers of λ^{-1} to zero we find

$$\left.\begin{array}{c}
-\tfrac{1}{2} S_0^2 + \tfrac{1}{3} S_0^3 = t - \tfrac{1}{6}, \\[6pt]
S_1 = \dfrac{1}{6S_0}(1 - S_0), \qquad S_2 = -\dfrac{1}{72S_0^3} - \dfrac{13}{360S_0^2} + \dfrac{1}{20S_0}, \\[10pt]
S_3 = \dfrac{1}{432S_0^5} + \dfrac{607}{45360S_0^4} - \dfrac{29}{11340S_0^3} - \dfrac{149}{11340S_0^2},
\end{array}\right\} \qquad (3.214)$$

with corresponding formulae for α_n The first of (3.124) shows that near $t = \tfrac{1}{6}$, $S_0 = \{2(\tfrac{1}{6} - t)\}^{\frac{1}{2}}$ approximately and hence the first approximation to the time $t = t_E$ at which $s = 0$ and the sphere is completely frozen is $t_E = \tfrac{1}{6}$. The other relationships in (3.214), however, show that the expansions (3.213) break down when $S_0^2 \sim \lambda^{-1}$ and an inner rescaled solution is needed. The form of S_n near $t = \tfrac{1}{6}$ needed to match inner and outer solutions is seen, in terms of $\eta = 2(\frac{1}{6} - t)$, to be

$$S_0 = \eta^{\frac{1}{2}} + \frac{1}{3}\eta + O(\eta^{\frac{3}{2}}), \qquad S = \frac{1}{6\eta^{\frac{1}{2}}} - \frac{2}{9} + O(\eta^{\frac{1}{2}}),$$
$$S_2 = -\frac{1}{72\eta^{\frac{3}{2}}} - \frac{1}{45\eta} + O(\eta^{-\frac{1}{2}}). \tag{3.215}$$

To obtain an inner solution Stewartson and Waechter (1976) introduce

$$\alpha = 1 + \lambda^{\frac{1}{2}}A(\lambda, x), \qquad x = \lambda(t_E - t), \qquad S = \lambda^{-\frac{1}{2}}Y(\lambda, x) \tag{3.216}$$

and express the conditions (3.212) as

$$Y = A(1 - \lambda^{-\frac{1}{2}}Y) - \frac{A^{(1)}}{3!}(1 - \lambda^{-\frac{1}{2}}Y)^3 + \frac{A^{(2)}}{5!}(1 - \lambda^{-\frac{1}{2}}Y)^5 + \ldots, \tag{3.217a}$$

where the differentiation is with respect to x, and

$$YY^{(1)} = 1 + \lambda^{-\frac{1}{2}}\left\{A - \frac{A^{(1)}}{2!}(1 - \lambda^{-\frac{1}{2}}Y)^2 + \frac{A^{(2)}}{4!}(1 - \lambda^{-\frac{1}{2}}Y)^4 + \ldots\right\}. \tag{3.217b}$$

In terms of intermediate variables $\Phi(\lambda, x)$, $\Psi(\lambda, x)$ defined as

$$\Phi(\lambda, x) = A - \frac{A^{(1)}}{3!} + \frac{A^{(2)}}{5!} + \ldots + \frac{(-1)^n A^{(n)}}{(2n+1)!} + \ldots,$$
$$\Psi(\lambda, x) = A - \frac{A^{(1)}}{2!} + \frac{A^{(2)}}{4!} + \ldots + \frac{(-1)^n A^{(n)}}{(2n)!} + \ldots, \tag{3.218}$$

(3.217) reduce to

$$Y(\lambda, x) = \Phi - \lambda^{\frac{1}{2}}Y\Psi - \lambda^{-1}\frac{Y^2}{2!}\Phi^{(1)} + \lambda^{-\frac{3}{2}}\frac{Y^3}{3!}\Psi^{(1)} + \lambda^{-2}\frac{Y^4}{4!}\Phi^{(2)} + \ldots,$$
$$YY^{(1)} = 1 + \lambda^{-\frac{1}{2}}\Psi + \lambda^{-1}Y\Phi^{(1)} - \lambda^{-\frac{3}{2}}\frac{Y^2}{2!}\Psi^{(1)} - \lambda^{-2}\frac{Y^3}{3!}\Phi^{(2)} + \ldots. \tag{3.219}$$

Starting with the second of (3.219) and putting all terms $O(\lambda^{-\frac{1}{2}})$ to zero gives $Y \sim (2x)^{\frac{1}{2}}$; then from the first equation $\Phi = (2x)^{\frac{1}{2}}$ and Ψ follows from (3.218). The expression for Y is then corrected to include terms $O(\lambda^{-\frac{1}{2}})$ and the cycle repeated. By taking Laplace transforms of (3.218), Stewartson and Waechter deduce after some manipulation that

$$\Psi(\lambda, x) = \Phi(\lambda, x) - 2\sum_{n=1}^{\infty}\int_x^{\infty}\Phi^{(1)}(\lambda, z)\exp[-n^2\pi^2(z - x)]\,dz. \tag{3.220}$$

They then substitute the expansions

$$\left.\begin{array}{l} \Phi = \Phi_0 + \lambda^{-\frac{1}{2}}\Phi_1 + \lambda^{-1}\Phi_2 + \dots, \\ \Psi = \Psi_0 + \lambda^{-\frac{1}{2}}\Psi_1 + \lambda^{-1}\Psi_2 + \dots, \\ Y = Y_0 + \lambda^{-\frac{1}{2}}Y_1 + \lambda^{-1}Y_2 + \dots, \end{array}\right\} \tag{3.221}$$

into (3.217) to obtain

$$Y_0 Y_0^{(1)} = 1, \qquad Y_0 = (2x)^{\frac{1}{2}}, \tag{3.222a}$$

$$\Phi_0 = Y_0 = (2x)^{\frac{1}{2}}, \tag{3.222b}$$

$$\frac{\mathrm{d}}{\mathrm{d}x}(Y_0 Y_1) = \Psi_0, \tag{3.222c}$$

$$\Phi_1 = Y_1 + Y_0\Psi_0, \tag{3.222d}$$

$$\frac{\mathrm{d}}{\mathrm{d}x}(\tfrac{1}{2}Y_1^2 + Y_0 Y_2) = \Psi_1 + Y_0\Phi_0^{(1)}, \tag{3.222e}$$

$$\Phi_2 = Y_2 + \Psi_0 Y_1 + Y_0\Psi_1 - \tfrac{1}{2}Y_0^2\Phi_0^{(1)}, \text{ etc.} \tag{3.222f}$$

By using the Laplace transform of (3.220) with $\Phi_0 = (2x)^{\frac{1}{2}}$, the asymptotic expansion of the Ψ_0 as $x \to 0$ is found to be

$$\Psi_0(x) \sim -(2\pi)^{-\frac{1}{2}}\{\gamma + \ln(1/x\pi^2)\}, \tag{3.223}$$

where $\gamma = 0.5772 \dots$ is Euler's constant. Then (3.222c) gives

$$Y_1 \sim -\tfrac{1}{2}(x/\pi)^{\frac{1}{2}}\{\gamma + 1 + \ln(1/x\pi^2)\} \tag{3.224}$$

as $x \to 0$, and

$$\Phi_1 \sim -\tfrac{1}{2}(x/\pi)^{\frac{1}{2}}\{1 + 3\gamma + 3\ln(1/x\pi^2)\} \tag{3.225}$$

follows from (3.222d) as $x \to 0$. Inspection reveals that $Y_1/Y_0 \to \infty$ as $x \to 0$ and it is likely that the expansion (3.221) breaks down when $\xi = O(1)$, where $\xi = \lambda^{-\frac{1}{2}}\ln(1/x)$, which implies through the second of (3.216) that the inner solution fails within a time $O(\lambda^{-1}e^{-\lambda^{\frac{1}{2}}})$ of the time for complete freezing. Subsequent analysis confirms these limits but meanwhile we can write

$$\Phi \sim (2x)^{\frac{1}{2}}\left\{1 - \frac{3\xi}{2(2\pi)^{\frac{1}{2}}} + \dots - \frac{1}{2(2\pi\lambda)^{\frac{1}{2}}}(1 + \gamma - 6\ln\pi + \dots) + O(\lambda^{-1})\right. \tag{3.226}$$

as $x \to 0$ for fixed ξ, where higher powers of ξ have been omitted.

By studying the behaviour of Y, Φ, and Ψ as $x \to \infty$, Stewartson and Waechter (1976) arrive at results for the inner solution which agree with the two-layer analysis by Riley *et al.* (1974) with a trivial correction. They find that as $x \to \infty$

$$Y_1 \sim 2x - \frac{1}{3} + \frac{\delta_1}{(2x)^{\frac{1}{2}}} - \frac{1}{90x} + \frac{1}{1890x^2} + \dots, \qquad (3.227)$$

where $\delta_1 = (2/\pi^5)^{\frac{1}{2}}\zeta(3)$ and ζ is the Riemann zeta-function, and develop an expansion for t_E in the form

$$t_E = \tfrac{1}{6} + \sum_{n=2}^{\infty} b_n \lambda^{-n/2} \qquad (3.228)$$

with

$$b_2 = \tfrac{1}{6}, \qquad b_3 = -\delta_1, \qquad b_4 = -E_1 = -0.0607 \pm 1\%,$$

where

$$E_1 = 2 \sum_{n=1}^{\infty} \frac{1}{n^2 \pi^2} \int_0^{\infty} e^{-\alpha} \Phi_1 \left(\frac{\alpha}{n^2 \pi^2} \right) d\alpha$$

and Φ_1 is defined by (3.222d). A further stage of matching yielded

$$b_5 = -2 \sum_{n=1}^{\infty} \frac{1}{n^2 \pi^2} \int_0^{\infty} e^{-\alpha} \Phi_2 \left(\frac{\alpha}{n^2 \pi^2} \right) d\alpha$$

but a numerical value was not determined. The terminal temperature distribution T_E^* was found to be given by

$$\frac{T_E^* - T_F^*}{T_F^* - T_W^*} \sim -1 + \frac{1}{r}(2\pi/\lambda)^{\frac{1}{2}} \sum_{n=1}^{\infty} \frac{\sin n\pi r}{n^2 \pi^2}$$

$$\sim -1 - \frac{1}{r\pi}(2/\pi\lambda)^{\frac{1}{2}} \int_0^{\pi r} \ln(2 \sin \tfrac{1}{2}y) \, dy. \quad (3.229)$$

This expression predicts that T_E^* has a logarithmic singularity as $r \to 0$ whereas $T_E^* = T_F^*$ from the boundary conditions. This non-uniformity stems from the behaviour of Y_1/Y_0 as $x \to 0$ and the breakdown of (3.221) referred to above.

Stewartson and Waechter (1976) examine further the properties of a terminal core of exponentially small radius by using $\xi = \lambda^{-\frac{1}{2}} \ln(1/x) = \lambda^{-\frac{1}{2}} \ln\{1/\lambda(t_E - t)\}$ as independent variable instead of x. In order to obtain a uniformly valid expression for T_E^* correct to order $\lambda^{-\frac{1}{2}}$ they write

$$\Phi = (2x)^{\frac{3}{2}}[F_0'(\xi) + \lambda^{-\frac{1}{2}} F_1'(\xi) + \dots],$$
$$Y = (2x)^{\frac{3}{2}}[G_0(\xi) + \lambda^{-\frac{1}{2}} G_1(\xi) + \dots], \qquad (3.230)$$

where F_n, G_n are functions of ξ to be found and differentiation is with respect to ξ. Some analysis leads to a parallel expression for Ψ which is

$$\Psi \sim -\frac{\lambda^{\frac{1}{2}}}{(2\pi)^{\frac{1}{2}}} F_0(\xi) + (2\pi)^{-\frac{1}{2}}\{F_1(\xi) + (2\ln 2 - 2)F_0'(\xi) - \gamma - 2\ln 2\pi\} + O(1).$$

(3.231)

The unknown functions F_n, G_n are now found by substitution of (3.230) and (3.231) into (3.219) and solving the ensuing differential equations. On writing $Z = 1 + \xi/2(2\pi)^{\frac{1}{2}}$ their expressions for $n = 0$, $n = 1$ become

$$G_0(\xi) = Z^{-1}, \qquad F_0'(\xi) = G_0^3,$$

$$G_1(\xi) = \frac{-1}{2(2\pi)^{\frac{1}{2}}Z^2}[\gamma + 1 - 2\ln \pi + (5 - 6\ln 2)\ln Z],$$

$$F_1'(\xi) = \frac{-3}{2(2\pi)^{\frac{1}{2}}Z^4}[\gamma + \tfrac{1}{3} - 2\ln \pi + (5 - 6\ln 2)\ln Z].$$

These are used in the expression

$$\theta = -r + \lambda^{-\frac{1}{2}}\left[(1 - r)\Phi + 2\sum_{n=1}^{\infty} n\pi \sin n\pi r \int_x^{\infty} \Phi(y)\exp\{-n^2\pi^2(y - x)\}\,\mathrm{d}y\right],$$

obtained by them previously in the derivation of (3.229), to give in particular

$$\theta(r, t_E) = -r + \left(\frac{2\pi}{\lambda}\right)^{\frac{1}{2}} \sum_{n=1}^{\infty} \frac{\sin n\pi r}{n^2\pi^2}$$

$$\times [F_0'(N) + \lambda^{-\frac{1}{2}}\{F_1'(N) - (2 - \gamma - 2\ln 2)F_0''(N)\}], \quad (3.232)$$

where $N = \lambda^{-\frac{1}{2}}\ln n^2\pi^2$. The evaluation of (3.232) is carried out by using a theorem by Zygmund (1959) and shows that the terminal temperature $\theta_E = \theta(r, t_E)$ satisfies

$$\frac{\partial\theta_E}{\partial r} \sim -\frac{1}{(1 + R)^2} + \frac{1}{(2\pi\lambda)^{\frac{1}{2}}}\left[\frac{(5 - 6\ln 2)\ln(1 + R)}{(1 + R)^3}\right] + \dots, \quad (3.233)$$

where $R = (2\pi\lambda)^{-\frac{1}{2}}\ln(1/\pi r)$. When $R \ll 1$

$$\frac{\partial\theta_E}{\partial r} \sim -1 + \frac{2}{(2\pi\lambda)^{\frac{1}{2}}}\ln\left(\frac{1}{\pi r}\right),$$

which is in agreement with (3.229) when $r \ll 1$, so that a composite expression for θ_E, uniformly valid for all $r < 1$, is now available. The boundary condtions which imply $\theta_E = 0$ at $r = 0$ and, from the definition of θ, $\partial\theta/\partial r = 0$ at $r = 0$ are seen to be satisfied by (3.232) since the right-hand side of (3.233) approaches zero as $R \to \infty$.

Soward (1980) obtained the same key equations for the sphere as Stewartson and Waechter (1976) by using a method of images which by-passed some of their complicated analysis. The minor errors he referred to are corrected in the above equations (personal communication from K.S.). Soward also obtained results for the freezing of the terminal cores of both sphere and cylinder by a new and simpler method based on a similarity variable.

Davis and Hill (1982) developed a series solution of the sphere problem after first using a boundary-fixing transformation.

4. Front-tracking methods

4.1. Numerical techniques

FINITE-DIFFERENCE methods have been used extensively for the numerical solution of moving boundary problems and, in recent years, finite-element techniques have been introduced. In Ockendon and Hodgkins (1975) accounts by Crank (1975b) and by Fox (1975) are to be found. More recent work is described in Wilson, Solomon, and Boggs (1978), by Fox (1979), and by Crank (1981). Furzeland (1977a, 1979a) surveys many methods and presents an extensive bibliography. In a later paper (1980) he examines a selection of methods for computational efficiency and accuracy, programming complexity, and ease of generalization to more than one space dimension and to more complicated problems.

This chapter deals with numerical methods which compute, at each step in time, the position of the moving boundary. When the solution is computed at points on a fixed grid in the space–time domain following the usual methods for obtaining a numerical solution of the simple heat-flow equation, the boundary will in general be between two grid points at any given time. Therefore, special formulae are needed to cope with terms like $\partial u/\partial x$ and ds/dt, as well as with the partial differential equation itself, in the neighbourhood of the moving boundary. These formulae must allow for unequal space intervals. The difficulty is compounded if implicit finite-difference formulae are used, since the position of the moving boundary is not known at the new time and some iterative procedure is usually inevitable. Alternatively, the grid itself has to be deformed in some way, or some transformation of variables adopted, so that the moving boundary is always on a grid line or is fixed in the transformed domain. An early paper on finite elements (Fisher and Medland 1974) used special averaging procedures in the element containing the moving boundary.

4.2. Fixed finite-difference grid

Suppose the heat-flow equation is to be solved by using finite-difference replacements for the derivatives in order to compute values of temperature, u_{ij}, at discrete points $(i\,\delta x, j\,\delta t)$ on a fixed grid in the (x, t)

plane. At any time $j\,\delta t$, the phase-change boundary will usually be located between two neighbouring grid points, say $i\,\delta x$ and $(i+1)\,\delta x$. This can be allowed for by using modified finite-difference formulae which incorporate unequal space intervals near the moving boundary. Interpolation formulae of Lagrangian type can be used (Crank 1957a). Confining attention to three-point formulae, we have for the general function $f(x)$ which takes known values $f(a_0)$, $f(a_1)$, $f(a_2)$ at the three points $x = a_0, a_1, a_2$ respectively

$$f(x) = \sum_{j=0}^{2} l_j(x)f(a_j), \tag{4.1}$$

where

$$\ell_j(x) = \frac{p_2(x)}{(x-a_j)p_2'(a_j)}, \qquad p_2(x) = (x-a_0)(x-a_1)(x-a_2),$$

and $p_2'(a_j)$ is its derivative with respect to x at $x = a_j$. It follows that

$$\frac{df}{dx} = \ell_0'(x)f(a_0) + \ell_1'(x)f(a_1) + \ell_2'(x)f(a_2), \tag{4.2}$$

where

$$\ell_0'(x) = \frac{(x-a_1)(x-a_2)}{(a_0-a_1)(a_0-a_2)},$$

and similarly for $\ell_1'(x)$, $\ell_2'(x)$. Furthermore,

$$\frac{1}{2}\frac{d^2f}{dx^2} = \frac{f(a_0)}{(a_0-a_1)(a_0-a_2)} + \frac{f(a_1)}{(a_1-a_2)(a_1-a_0)} + \frac{f(a_2)}{(a_2-a_0)(a_2-a_1)}. \tag{4.3}$$

Figure 4.1 shows the moving boundary at time $t = j\,\delta t$, when it is a fractional distance $p\,\delta x$ between the grid lines $i\,\delta x$ and $(i+1)\,\delta x$. The points a_0, a_1, a_2 are identified with the grid lines $(i-1)\,\delta x$, $i\,\delta x$, and the moving boundary itself, and correspondingly $f(a_0)$, $f(a_1)$, and $f(a_2)$ with

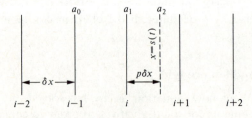

FIG. 4.1. Fixed grid

$u_{i-1,j}$, $u_{i,j}$, and u_B on the boundary. Then for $x < s(t)$ we have

$$\frac{\partial^2 u}{\partial x^2} = \frac{2}{(\delta x)^2} \left(\frac{u_{i-1}}{p+1} - \frac{u_i}{p} + \frac{u_B}{p(p+1)} \right), \qquad x = i\,\delta x, \qquad (4.4)$$

and

$$\frac{\partial u}{\partial x} = \frac{1}{\delta x} \left(\frac{p u_{i-1}}{p+1} - \frac{(p+1)u_i}{p} + \frac{(2p+1)u_B}{p(p+1)} \right), \qquad x = s(t). \qquad (4.5)$$

Similarly for $x > s(t)$ we have

$$\frac{\partial^2 u}{\partial x^2} = \frac{2}{(\delta x)^2} \left(\frac{u_B}{(1-p)(2-p)} - \frac{u_{i+1}}{1-p} + \frac{u_{i+2}}{2-p} \right), \; x = (i+1)\,\delta x, \qquad (4.6)$$

and

$$\frac{\partial u}{\partial x} = \frac{1}{\delta x} \left(\frac{2p-3}{(1-p)(2-p)} u_B + \frac{2-p}{1-p} u_{i+1} - \frac{1-p}{2-p} u_{i+2} \right), \; x = s(t). \qquad (4.7)$$

These formulae for the space derivatives are used in conjunction with the usual explicit or implicit replacements of time derivatives in the heat-flow equation and in the conditons on the phase-change boundary, $x = s(t)$. For points other than $i\,\delta x$, $s(t)$, and $(i+1)\,\delta x$, the usual finite-difference formulae for equal space intervals are used.

The numerical solution of the one-phase problem, defined by equations (1.21–6) in Chapter 1, provides a simple example. At the points $\delta x, 2\,\delta x, \ldots, n\,\delta x, \ldots, (i-1)\,\delta x$, eqn (1.21) is replaced by the simple explicit formula

$$u_{n,j+1} = u_{n,j} + \frac{\delta t}{(\delta x)^2}(u_{n-1,j} - 2u_{n,j} + u_{n+1,j}), \qquad n = 1, 2, \ldots, i-1,$$
$$(4.8)$$

and, from (1.22), $u_{0j} = 1$. At the point $i\,\delta x$, instead of (4.8) we write, using (4.4),

$$u_{i,j+1} = u_{i,j} + \frac{2\,\delta t}{(\delta x)^2} \left(\frac{u_{i-1,j}}{p_j + 1} - \frac{u_{ij}}{p_j} \right), \qquad (4.9)$$

since $u_B = 0$ from (1.25). Similarly substitution of (4.5) into (1.26) and writing $s_j = (i + p_j)\,\delta x$ yields

$$p_{j+1} = p_j - \frac{\delta t}{\lambda(\delta x)^2} \left(\frac{p_j u_{i-1,j}}{p_j + 1} - \frac{(p_j + 1)u_{i,j}}{p_j} \right). \qquad (4.10)$$

The steps in the numerical solution, starting from known values of all variables at each grid point at time $j\,\delta t$, are:

(i) Calculate $u_{n,j+1}$, $n = 1, 2, \ldots, (i-1)$, from (4.8).
(ii) Calculate $u_{i,j+1}$ from (4.9).
(iii) Calculate p_{j+1} from (4.10).
(iv) Repeat the steps (i)–(iii) but substituting variables $u_{n,j+1}$, $u_{i,j+1}$, p_{j+1}, etc. on the right-hand sides of (4.8–10) to obtain $u_{n,j+2}$, $u_{i,j+2}$, p_{j+2}. When p exceeds unity, the special equations (4.9) and (4.10) are applied to the points $i\,\delta x$ and $(i+1)\,\delta x$ and so on.

Crank and Gupta (1972a) in solving the oxygen diffusion problem defined by equations (1.57–60) in §1.3.10 used the Lagrangian formula (4.4) to approximate the diffusion equation (1.57) in the neighbourhood of the moving boundary. In the absence of an explicit expression, such as the Stefan condition, for the velocity of the moving boundary, they used the double boundary condition (1.59) to deduce higher space derivatives for substitution in a Taylor series.

Thus, differentiation of the first of (1.59) with respect to t gives

$$\frac{dc}{dt} = \left(\frac{\partial c}{\partial x}\right)_{x=s}\left(\frac{ds}{dt}\right) + \left(\frac{\partial c}{\partial t}\right)_{x=s} = 0. \tag{4.11}$$

By using (1.57) and the second of (1.59) in (4.11) we obtain

$$\frac{\partial^2 c}{\partial x^2} = 1, \qquad x = s. \tag{4.12}$$

Differentiation of (1.57) with respect to x yields

$$\frac{\partial^2 c}{\partial x\,\partial t} = \frac{\partial^3 c}{\partial x^3}. \tag{4.13}$$

Again, differentiating the second of (1.59) gives

$$\frac{d}{dt}\left(\frac{\partial c}{\partial x}\right) = \left(\frac{\partial^2 c}{\partial x^2}\right)_{x=s}\frac{ds}{dt} + \left(\frac{\partial^2 c}{\partial t\,\partial x}\right)_{x=s} = 0,$$

and hence using (4.12) and (4.13), assuming the order of differentiation by x and t can be interchanged, we obtain

$$\frac{\partial^3 c}{\partial x^3} = -\frac{ds}{dt}, \qquad x = s.$$

Similarly, it follows that

$$\frac{\partial^4 c}{\partial x^4} = \left(\frac{ds}{dt}\right)^2, \qquad \frac{\partial^5 c}{\partial x^5} = -\frac{d^2 s}{dt^2} - \left(\frac{ds}{dt}\right)^3 \quad \text{etc.}$$

Now, the Taylor's series for c_i, the oxygen concentration at the grid point

$i\,\delta x$, next to the moving boundary as in Fig. 4.1, can be written as

$$c_i = c(s) - p\,\delta x\left(\frac{\partial c}{\partial x}\right)_{x=s} + \tfrac{1}{2}(p\,\delta x)^2\left(\frac{\partial^2 c}{\partial x^2}\right)_{x=s} - \tfrac{1}{6}(p\,\delta x)^3\left(\frac{\partial^3 c}{\partial x^3}\right)_{x=s} + \ldots$$

$$= \tfrac{1}{2}(p\,\delta x)^2 + \tfrac{1}{6}(p\,\delta x)^3\frac{\mathrm{d}s}{\mathrm{d}t} + \ldots. \quad (4.14)$$

Provided the boundary is not moving too quickly the first term of the series provides a reasonable approximation and gives

$$p = \frac{\sqrt{(2c_i)}}{\delta x}. \quad (4.15)$$

Once $c_{i,j+1}$ has been computed from the equivalent of (4.9), p_{j+1} follows immediately from (4.15) and the step-by-step solution can proceed.

Various finite-difference schemes have been proposed for approximating both the Stefan condition on the moving boundary and the partial differential equation at the neighbouring grid point. Murray and Landis (1959) use a lumped-parameter interpretation of the temperatures at points on a fixed grid. In particular, in the lump containing the fusion front at any time, they introduce two fictitious temperatures, one obtained by quadratic extrapolation from temperatures in the solid region and the other from temperatures in the liquid region. The fusion temperature and the current position of the interface are incorporated in the fictitious temperatures, which are then substituted into the standard approximation such as (4.8) to compute temperatures $u_{i,j+1}$, $u_{i+1,j+1}$, instead of using special formulae like (4.9). For the motion of the fusion front an expression analogous to (4.10) is used but based on Taylor expansions for $\partial u/\partial x$. To avoid loss of accuracy associated with the extrapolation formulae, Furzeland (1977b) suggested using an approximation for $\partial u/\partial x$ centred on the moving boundary and containing a fictitious value of u to be eliminated by use of (4.8), for example.

The advantages of implicit finite-difference replacements of the time derivatives have been explored by Ehrlich (1958), Koh *et al.* (1969), Saitoh (1972), and Bonnerot and Jamet (1974), for example. Meyer (1976) obtained an $O((\delta t)^2)$ approximation for the movement of the fusion boundary by using the three-point backward formula

$$-\frac{\mathrm{d}s}{\mathrm{d}t} = \frac{1}{2\,\delta t}(3s_j - 4s_{j-1} + s_{j-2}) \qquad \text{at} \qquad t = j\,\delta t.$$

Ciment and Guenther (1974) in a study of problems typified by the immiscible displacement of one fluid by another in a loosely packed, porous medium employed a method of local, spatial, mesh-refinement on both sides of the moving front, previously analysed by Ciment and Sweet

(1973). They also incorporated the idea used earlier by Douglas and Gallie (1955), described in §4.3.1, of adjusting the time step so that the moving boundary coincided with one of the refined mesh points.

Huber (1939) used an implicit scheme for approximating the heat-flow equation in each time step, but he tracked the moving boundary explicitly using the Stefan condition. Rubinstein (1971) simplified Huber's numerical scheme and Fasano and Primicerio (1973) proved convergence of order $\frac{1}{2}$.

In a monumental work, Lazaridis (1970) used explicit finite-difference approximations on a fixed grid to solve two-phase solidification problems in both two and three space dimensions. He wrote the heat-balance condition on the solidification boundary in Patel's form (see eqn (1.56), §1.3.9) and developed numerical schemes based on an auxiliary set of differential equations which express the fact that the moving boundary is an isotherm. These auxiliary forms are compatible with Patel's expression (1.56) in enabling the boundary movement to be computed in the directions of the coordinate axes rather than along the normals to itself. Standard finite-difference approximations to the heat flow equation were used at grid points far enough from the solidification boundary. Near the boundary, formulae for unequal intervals such as (4.5) and (4.7) were incorporated into the auxiliary equations. To avoid loss of accuracy associated with singularities which can arise when the moving boundary is too near a grid point, localized quadratic temperature profiles were used. The manipulation is very complicated indeed. In fact, Lazaridis (1970) merely outlined the derivation of his equations for the two-dimensional case only and referred to his thesis (1969) for details and for the development of the three-dimensional relationships. Results were presented for solidification problems in a prism of square cross-section and for a three-dimensional corner.

4.3. Modified grids

Various ways of modifying the grid have been proposed, all with the aim of avoiding the increased complication and loss of accuracy associated with unequal space intervals near the moving boundary.

4.3.1. Variable time step

Douglas and Gallie (1955) rather than using a fixed time step and searching for the boundary decided to determine, as part of the solution, a variable time step such that the moving boundary coincides with a grid line in space at each time level. They treated the problem defined by the system

$$\frac{\partial u}{\partial t} = \frac{\partial^2 u}{\partial x^2}, \qquad 0 < x < s(t), \qquad t > 0, \tag{4.16}$$

$$\partial u / \partial x = -1, \qquad x = 0, \qquad t > 0, \tag{4.17}$$

$$dx/dt = -\partial u / \partial x, \qquad u = 0, \qquad x = s(t), \tag{4.18}$$

$$s(0) = 0. \tag{4.19}$$

By integrating (4.16) with respect to x and then t and using (4.17), (4.18), (4.19), an alternative form of boundary condition is seen to be

$$s(t) = t - \int_0^{s(t)} u(x, t) \, dx. \tag{4.20}$$

They used the notation

$$x_i = i \Delta x, \qquad t_n = \sum_{k=0}^{n-1} \Delta t_k, \qquad f_{i,n} = f(x_i, t_n),$$

and took $\Delta t_0, \ldots, \Delta t_{n-1}, u_{i,k}, k = 0, \ldots, n$ to be known. Then, for $\Delta t_n^{(0)} > 0$ chosen arbitrarily, they defined $u_{i,n+1}^{(0)}$ as the solution of the finite-difference representation of (4.16–20), namely the case $r = 0$ of the general scheme

$$\frac{u_{i-1,n+1}^{(r)} - 2u_{i,n+1}^{(r)} + u_{i+1,n+1}^{(r)}}{(\Delta x)^2} = \frac{u_{i,n+1}^{(r)} - u_{i,n}^{(r)}}{\Delta t_n^{(r)}}, \qquad i = 1, \ldots, n, \qquad r \geq 0, \tag{4.16a}$$

$$u_{0,n+1}^{(r)} - u_{1,n+1}^{(r)} = \Delta x, \tag{4.17a}$$

$$\Delta t_n^{(r+1)} = \left(n + 1 + \sum_{i=1}^{n} u_{i,n+1}^{(r)} \right) \Delta x - t_n, \tag{4.20a}$$

together with $u_{n+1,n+1}^{(r)} = 0$ from the second of (4.18). The solution proceeds iteratively by using (4.20a) to correct the assumed time step. The simplest special forms were used for the interval $x = 0$ to $x = \Delta x$. Douglas and Gallie established convergence and stability of their implicit scheme. In a footnote they pointed out that the iteration of Δt would be avoided by using $u_{i,n}$ in place of $u_{i,n+1}^{(r)}$ in (4.20a) to obtain Δt_n.

Vasilev (1964) and Nogi (1974) extended this method to more general one-dimensional problems but there is no obvious way of applying it to two or three dimensions.

Gupta and Kumar (1980) formulated the same set of finite-difference equations as Douglas and Gallie but they used a difference form of the boundary condition (4.18) to update Δt instead of (4.20a), i.e. they took $\Delta t_n^{(r+1)} = (\Delta x)^2 / u_{n,n+1}^{(r)}$ based on both conditions (4.18). They thus avoided the instability which develops as x increases and (4.20a) enters a closed loop because it becomes very sensitive to rounding errors. Goodling and Khader (1974) incorporated their finite-difference form of (4.18) into the system of equations to be solved for an arbitrary value of $u_{n,n+1}^{(r)}$ which is then updated by (4.17a). Gupta and Kumar (1981a) in a study of a convective boundary condition at the fixed end found this latter method fails to converge as x increases because it is too sensitive to a small change in $u_{n,n+1}$. Gupta and Kumar's (1981a) results are in good agreement with those obtained by the other variable time-step methods and by a Goodman's (1958) integral method (see §3.5.4).

Yuen and Kleinman (1980) used another way of updating Δt by successive approximation.

Gupta and Kumar (1981b) adapted the method of Douglas and Gallie (1955) to solve the oxygen diffusion problem (see §1.3.10). Other authors have to introduce interpolation or extrapolation procedures for calculating the total absorption time at which the moving boundary, marking the innermost penetration of oxygen, reaches the outer fixed surface. In Gupta and Kumar's method, the total absorption time emerges from the final step in the normal computing procedure. Their results for the position of the moving boundary are compared with others in Table 4.1. Hansen and Hougaard (1974) predicted the total absorption time to be between 0.1972 and 0.1977. Gupta and Kumar (1981b) obtained 0.1973 which is close to the accurate value of 0.197 434 obtained by Dahmardah and Mayers (1983) (see §5.1).

Kumar (1982) discussed in detail various variable time-step methods and their application to problems mentioned above and to others including phase-change with non-uniform initial temperatures, time-dependent boundary conditions, the dissolution of a spherical gas bubble in a liquid, and the freezing of liquid inside and outside a cylinder. For most problems, numerical results obtained by several methods were compared.

4.3.2. Variable space grid

Murray and Landis (1959) kept the number of space intervals between $x = 0$ and $x = s(t)$, i.e. between a fixed and a moving boundary, constant and equal to I, say, for all time (Fig. 4.2). Thus, for equal space intervals, $\delta x = s(t)/I$ is different in each time step. The moving boundary is always on the Ith grid line. They differentiated partially with respect to time t, following a given grid line instead of at constant x. Thus, for the line $i\,\delta x$ they had

$$\left(\frac{\partial u}{\partial t}\right)_i = \left(\frac{\partial u}{\partial x}\right)_t \left(\frac{\mathrm{d}x}{\mathrm{d}t}\right)_i + \left(\frac{\partial u}{\partial t}\right)_x. \tag{4.21}$$

Murray and Landis assumed that a general grid point at x moved

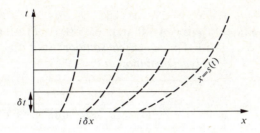

FIG. 4.2. Murray and Landis variable grid

according to the expression

$$\frac{\mathrm{d}x_i}{\mathrm{d}t} = \frac{x_i}{s(t)}\frac{\mathrm{d}s}{\mathrm{d}t}. \tag{4.22}$$

The one-dimensional heat equation becomes

$$\left(\frac{\partial u}{\partial t}\right)_i = \frac{x}{s(t)}\frac{\mathrm{d}s}{\mathrm{d}t}\frac{\partial u}{\partial x} + \frac{\partial^2 u}{\partial x^2}, \tag{4.23}$$

and $s(t)$ is updated at each time step by using, for example, a suitable finite-difference form of the boundary condition $\mathrm{d}s/\mathrm{d}t = -\partial u/\partial x$ on $x = s(t)$. The method was extended by Heitz and Westwater (1970) to convection problems and by Tien and Churchill (1965) to cylindrical geometry.

Crank and Gupta (1972b) avoided the complications due to the unequal grid size near the moving boundary by moving the whole uniform grid system with the velocity of the moving boundary. They used this technique to solve the system of equations (1.57–60) describing the oxygen diffusion problem. With reference to Fig. 4.3, the moving boundary is at $x = 1$ at $t = 0$ and at time Δt it has moved a distance ε to the left. The whole grid system is moved a distance ε to the left, represented by the broken grid lines in Fig. 4.3. Thus, the unequal interval is transferred to the $x = 0$ end, which is found to be a more tractable region, and the irregularities in the boundary motion, noted by Crank and Gupta (1972a)

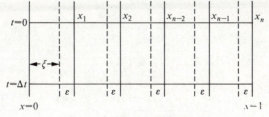

FIG. 4.3. Moving grid system

TABLE 4.1

Free boundary position $10^4 s(t)$: $\delta x = 0.01$, $\delta t = 0.001$
*except for MVTS. The numerical solutions start from the
analytical solution at* $t = 0.025$

Method	Time							
	0.04	0.06	0.10	0.12	0.14	0.16	0.18	0.185
FGL	9988	9905	9312	8747	7912	6756	4849	4014
MG	9988	9903	9301	8719	7882	6682	4766	4048
ML	9988	9904	9309	8740	7930	6776	4974	4308
MVTS	9950	9899	9249	8703	7916	6825	4768	—
MMB	9993	9920	9356	8796	7992	6832	5085	—
HH	9992	9918	9350	8792	7989	6834	5011	4334

FGL = fixed-grid Lagrange (Crank and Gupta, 1972a); MG = moving grid (Gupta, 1973, 1974); ML = Murray and Landis (1959), started from $x = 0.1$; MVTS = modified variable time step (Gupta and Kumar, 1981b); MMB = Miller, Morton, and Baines (1978), extrapolated values, $\delta x = 0.05$, $\delta t = 0.0002$; HH = Hansen and Hougaard (1974).

previously, are smoothed out. Proceeding in time intervals Δt as usual, an approximation to the value of u at time $t = \Delta t$ one step Δx to the left of the moving boundary is first calculated explicitly. Then, from the equation corresponding to (4.15) above, the new position of the boundary at $t = \Delta t$ is obtained, i.e. $1 - \varepsilon$ in Fig. 4.3. The whole grid is now moved a distance ε to the left and values of u calculated explicitly at time $t = \Delta t$ for the new grid points. Crank and Gupta (1972b) described two ways of obtaining the interpolated values of u at the new grid points, to be used for the next time step, based on cubic splines or polynomials. Later Gupta (1973, 1974) avoided the interpolations by using a Taylor's expansion in space and time variables and derived an equation which is essentially a particular case of the Murray and Landis equation (4.23). Since Gupta has every grid point moving with the velocity of the moving boundary, the

TABLE 4.2

Values of $10^4 u(0, t)$ *at fixed sealed surface*

Method	Time							
	0.04	0.06	0.10	0.12	0.14	0.16	0.18	0.185
FGL	2745	2238	1434	1094	781	490	220	156
MG	2745	2238	1434	1093	780	490	219	155
ML	2745	2238	1434	1093	780	489	218	154
HH	2743	2236	1432	1091	779	488	218	153

[a] See footnote to Table 4.1.

relationship (4.22) does not apply. Instead, $dx_i/dt = ds/dt$ everywhere and the first term on the right-hand side of (4.23) is simply $(ds/dt)(\partial u/\partial x)$. The second term becomes $(\partial^2 u/\partial x^2 - 1)$ in Gupta's oxygen diffusion example. Tables 4.1 and 4.2 compare different results. Other solutions are given in Tables 3.1, 3.2, 6.6–8.

4.3.3. Finite elements: adaptive meshes

Bonnerot and Jamet (1974, 1975) used a space grid which was adapted at each time step to construct quadrilateral finite elements in space and time for the non-rectangular (x, t) grid. They solved an integral or weak form of the one-dimensional heat-flow equation using bilinear, isoparametric test-functions and numerical quadrature. Their implicit, iterative formulation was shown to be a generalization of the Crank–Nicolson method. They dealt with a boundary moving in a prescribed way and also with a Stefan problem.

Wellford and Ayer (1977) used this one-phase Stefan problem to test their finite-element method applicable to multi-phase problems. They used a fixed grid of standard space–time finite elements but elements which contained the free boundary had special features incorporating discontinuous interpolation. The position of the free boundary was assumed to vary linearly within a special element, and a temperature distribution $T = a + bx + ct + dxt$ was assumed on one side of the free boundary and $T = e + fx + gt + hxt$ on the other. The unknowns in the finite-element approximation were the temperatures at the corners of the space–time element, the positions of the free boundary at t and $t + \Delta t$, together with the heat flux jumps across the interface at t and $t + \Delta t$. A Galerkin formulation of the problem was evaluated. Good agreement was obtained with the results of Bonnerot and Jamet (1975) by using a relatively sparse grid of 20 elements.

Later, Bonnerot and Jamet (1977) extended their method to a simple, one-phase problem in two dimensions, specified as follows:

$$\frac{\partial u}{\partial t} = \frac{\partial^2 u}{\partial x^2} + \frac{\partial^2 u}{\partial y^2}, \qquad 0 \leqslant x \leqslant 1, \qquad 0 \leqslant y \leqslant s(x, t), \qquad t > 0 \quad (4.24)$$

$$\partial u/\partial x = 0, \qquad x = 0, \qquad x = 1, \qquad t > 0, \qquad\qquad (4.25)$$

$$u = 1, \qquad y = 0, \qquad 0 \leqslant x < 1, \qquad t > 0, \qquad\qquad (4.26)$$

$$\left. \begin{aligned} s(x, 0) &= 2 + \cos \pi x \\[2mm] u(x, y, 0) &= 1 - \frac{y}{2 + \cos \pi x} \end{aligned} \right\} \qquad \begin{aligned} & 0 \leqslant x \leqslant 1, \\[2mm] & 0 \leqslant y \leqslant s(x, 0), \end{aligned}$$

$$(4.27)$$

$$\partial u/\partial n = -\lambda v_n, \qquad u = 0, \qquad y = s(x, t), \qquad 0 < x < 1, \qquad t \geqslant 0, \qquad \lambda > 0.$$
$$(4.28)$$

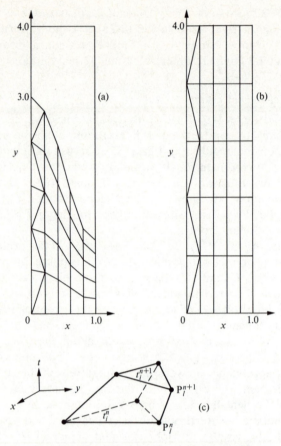

FIG. 4.4. Two-dimensional variable space grid

Figure 4.4(a) shows a simplified picture of the bases of their triangular elements at time $t = 0$ and Fig. 4.4(b) shows how the number of elements remains the same but each is stretched in the y-direction as the phase-change boundary moves in the direction of y increasing. Figure 4.4(c) shows a typical isoparametric finite element in space and time, in the time interval from t^n to t^{n+1}. Jamet (1979) established the stability and convergence of his schemes together with error estimates. He also referred to other similar contemporary schemes devised to solve the same problem.

Jamet (1978) introduced a new class of Galerkin-type approximations which are continuous with respect to the space variables but allow discontinuities with respect to the time variable between each time step.

Thus the elements may be chosen arbitrarily at each time step with no connection with the elements corresponding to the previous step. This is a more flexible procedure than the one mentioned above (Bonnerot and Jamet 1974, 1975), but the general mathematical theory was developed only for a prescribed moving boundary and no numerical experiments were reported. Subsequently, Bonnerot and Jamet (1979) proposed a third-order-accurate discontinuous finite-element method for a one-dimensional Stefan problem, based on biquadratic finite elements. This work will now be described in more detail.

Bonnerot and Jamet (1979) solve the following problem:

$$\frac{\partial u}{\partial t} = \frac{\partial^2 u}{\partial x^2}, \tag{4.29}$$

$$u(x, 0) = u^0(x), \qquad 0 \leq x \leq a^0, \tag{4.30}$$

$$u(0, t) = g(t), \qquad 0 < t \leq T, \tag{4.31}$$

$$u(a(t), t) = 0, \qquad 0 < t \leq T, \tag{4.32}$$

$$\frac{da}{dt} = -c\frac{\partial u}{\partial x}, \qquad x = a(t), \qquad 0 < t \leq T, \tag{4.33}$$

$$a(0) = a^0, \tag{4.34}$$

where $a(t)$ is an unknown, positive function specifying the position of the Stefan phase-change boundary which starts at $a(0) = a^0$. The partial differential equation is replaced by the integral relation

$$-\int_{t^n}^{t^{n+1}} \int_0^{a(t)} u\frac{\partial \phi}{\partial t} \, dx \, dt + \int_{t^n}^{t^{n+1}} \int_0^{a(t)} \frac{\partial u}{\partial x}\frac{\partial \phi}{\partial x} \, dx \, dt$$

$$+ \int_0^{a(t^{n+1})} u(x, t^{n+1})\phi(x, t^{n+1}) \, dx - \int_0^{a(t^n)} u(x, t^n)\phi(x, t^n) \, dx = 0, \tag{4.35}$$

which relates to the strip $t^n \leq t \leq t^{n+1}$ depicted in Fig. 4.5. If (4.35) is satisfied for all differentiable functions ϕ which vanish at $x = 0$ and $x = a(t)$ for all t, u is a unique solution of (4.29). The relation (4.35) comes from the usual integration by parts of (4.29) with introduction of the boundary conditions through Green's theorem. In the numerical treatment of (4.35) the fourth integral contains the known values of u at the time t^n, which have been calculated from the lower strip in the previous time interval $t^{n-1} < t \leq t^n$. The unknown values for $t^n < t \leq t^{n+1}$ appear in the first three integrals of (4.35).

Each strip $t^n < t \leq t^{n+1}$, $0 \leq n \leq N$ say, is now discretized into biquadratic finite elements as in Fig. 4.5, where the two ends are arcs of parabolas mapped from the reference finite-element shown in (ξ, η) coordinates. The family of nine points $P_{i+\mu}^{n+\nu}$, where μ and ν are two indices which each

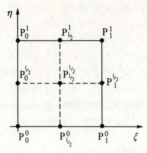

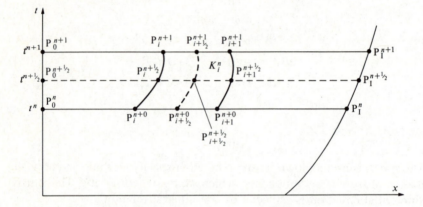

FIG. 4.5. Biquadratic finite element in strip $t^n < t < t^{n+1}$

take the values 0, $\frac{1}{2}$, and 1, are located at $(x_{i+\mu}^{n+\nu}, t^{n+\nu})$, where i is an integer, $0 \le i \le I - 1$, and $x_0^{n+\nu} = 0$,

$$x_I^{n+\nu} = a(t^{n+\nu}), \qquad x_{i+\frac{1}{2}}^{n+\nu} = \tfrac{1}{2}(x_i^{n+\nu} + x_{i+1}^{n+\nu}).$$

An attractive feature of this method lies in the fact that the discretization of each strip $t^n < t \le t^{n+1}$ can be completely independent of the discretization of the previous strip $t^{n-1} < t \le t^n$. For example, the nodes P_i^{n+0} in $t^n < t \le t^{n+1}$ can be different from the nodes P_i^n of the strip $t^{n-1} < t \le t^n$. In fact, since the line $t = t^n$ is not in the range $t^n < t \le t^{n+1}$ by definition, the notation u_i^{n+0}, for example, implies $u_i^{n+0} = \lim u_i(t^{n+\varepsilon})$, $\varepsilon > 0$, $\varepsilon \to 0$. The number of elements for any strip can also be chosen independently of other strips. However, in their paper Bonnerot and Jamet (1979) simply took $P_i^{n+0} = P_i^n$. By using Simpson quadrature formulae within each finite element to evaluate the integrals in the discretized form of (4.35), Bonnerot and Jamet (1979) obtained a linear system of equations with a square matrix. They give details for computing the coefficients.

In order to gain one order of accuracy, Bonnerot and Jamet (1979) approximated the derivative $\partial u/\partial x = Du(t)$, say, in the moving boundary condition (4.33), as follows. Denote by $v_h^{n+\nu}$ the function of x obtained by cubic interpolation of the values $u_{i+\mu}^{n+\nu}$ at the nodes $P_{i+\mu}^{n+\nu}$ for $i+\mu = I - \frac{3}{2}$, $I-1$, $I-\frac{1}{2}$, and I. Let $(Du)_h^{n+\nu}$ denote $\partial v^{n+\nu}/\partial x$ at the end-point $P_I^{n+\nu}$ and approximate $Du(t)$ in the interval $t^n < t \leqslant t^{n+1}$ by the function $(Du)_h(t)$ obtained by quadratic interpolation of the three values of $(Du)_h^{n+\nu}$ for $\nu = 0, \frac{1}{2}, 1$. The moving boundary condition (4.33) can now be discretized. The curve $x = a(t)$ is approximated by a continuous curve $x = a_h(t)$ which in each interval $t^n \leqslant t \leqslant t^{n+1}$ is a section of a parabola $x = q_2^{(n)}(t)$, which denotes a polynomial of degree $\leqslant 2$ with respect to the variable t. Let $a^{n+\nu} = a_h(t^{n+\nu})$. At each time step, the values of $a^{n+\frac{1}{2}}$ and a^{n+1} are computed from the integrated form of (4.33) with respect to t;

$$a^{n+\nu} = a^n - c \int_{t^n}^{t^{n+\nu}} (Du)_h(t)\, dt \tag{4.36}$$

for $\nu = \frac{1}{2}$ and 1, where $(Du)_h(t)$ denotes the approximation of $Du(t)$ obtained by quadratic interpolation above.

An iterative procedure is now needed since the values of $(Du)_h(t)$ in the interval $t^n < t \leqslant t^{n+1}$ depend on the computed values of u_h in this strip which in turn depend on $a^{n+\frac{1}{2}}$ and a^{n+1}. Bonnerot and Jamet (1979) adopted the following iterative procedure:

   (i) Given values at the ℓth stage of the iteration denoted by $a^{n+\frac{1}{2},\ell}$, $a^{n+1,\ell}$, $u_h(x, t)^\ell$, calculate $\{(Du)_h(t)\}^\ell$ as above;
   (ii) Calculate $a^{n+\frac{1}{2},\ell+1}$ $a^{n+1,\ell+1}$ by inserting $\{(Du)_h(t)\}^\ell$ in (4.36).

The iterations are started from.

$$a^{n+\nu,0} = q_2^{n-1}(t^{n+\nu}) \quad \text{for} \quad \nu = \frac{1}{2} \quad \text{and} \quad 1, n \geqslant 1,$$

i.e. from the quadratic extrapolation of a^{n-1}, $a^{n-\frac{1}{2}}$, and a^n. For $n = 0$, $a^{1,0} = a^{\frac{1}{2},0} = a^0$ is used.

Bonnerot and Jamet (1979) claimed that their method has the further advantage that it is specially well suited to handle problems with singularities, e.g. discontinuities in the initial functions $u^0(x)$, or in the boundary values including an initial discontinuity on the free boundary which then moves with infinite speed at $t = 0$. They quoted numerical results to establish that the convergence is of order 3 even in computing a discontinuity, and that their method is unconditionally stable. They also demonstrated that initial singularities generate no irregularity of the computed values at later times, such as the oscillations which are often produced by other methods, and that the accuracy remains very satisfactory.

Bonnerot and Jamet (1981) applied a modified and extended form of their third-order-accurate discontinuous finite-element method to solve the one-dimensional Stefan problem involving three phases which appear and disappear described in §1.3.6. Their scheme is conservative. One interesting feature of their results is that when they appear the moving boundaries start with zero initial speed. For mathematical results concerning appearing and disappearing phases the reader is referred to Cannon and Primicerio (1973); Cannon *et al.* (1967); and Sherman (1971).

Miller, Morton, and Baines (1978) described a method which is closely similar to that of Bonnerot and Jamet (1974, 1975) but they used finite differences in time with finite elements in space which are of standard length apart from the last two, which are adapted at each time step to fit the new position of the moving boundary. Miller *et al.* (1978) studied the oxygen diffusion problem defined by equations (1.57–60).

They represented the approximate solution at the discrete times t_n, $n = , 1, 2, \ldots$, as

$$U^n(x) = \sum_{j=1}^{J} Q_j^n \phi_j^n(x), \tag{4.37}$$

The coefficients Q_j^n are to be calculated so as to give the best representation of $u(x, t_n)$ in eqn (1.57) in terms of the basis functions $\phi_j^n(x)$, in the least-squares sense. An appropriate weak form of the differential equation is obtained in the usual way by multiplying (1.57) by a suitable test function $\psi(x)$, integrating over x and t, and using the boundary conditions (1.58) and (1.59). The weak form is

$$\int_0^{x_0(t_{n+1})} u(x, t_{n+1})\psi(x)\,dx - \int_0^{x_0(t_n)} u(x, t_n)\psi(x)\,dx$$

$$= -\int_0^{t_{n+1}} dt \left[\int_0^{x_0(t)} \psi_x u_x\,dx + \int_0^{x_0(t)} \psi(x)\,dx \right]. \tag{4.38}$$

When u is replaced by U and ψ by ϕ_i^{n+1}, and the θ-method quadrature is applied to the right-hand side of (4.38), the Galerkin equations which determine the Q_i^n in (4.37) are

$$\int \{U^{n+1}(x) - U^n(x)\}\phi_i^{n+1}(x)\,dx = -\theta\Delta t \left[\int_0^{x_0(t_{n+1})} \frac{dU^{n+1}}{dx} \frac{d\phi_i^{n+1}}{dx}\,dx \right.$$

$$\left. + \int_0^{x_0(t_{n+1})} \phi_i^{n+1}\,dx \right] - \Delta t(1-\theta) \left[\int_0^{x_0(t_n)} \frac{dU^n}{dx} \frac{d\phi_i^{n+1}}{dx}\,dx + \int_0^{x_0(t_n)} \phi_i^{n+1}\,dx \right].$$

Substitution of (4.37) into the approximated form of (4.38) produces the system of equations

$$[\mathbf{M} + \theta(t_{n+1} - t_n)\mathbf{K}]\mathbf{Q}^{n+1} = \mathbf{R} - (t_{n+1} - t_n)\mathbf{S}, \tag{4.39}$$

where

$$R_i = \int_0^{x_0(t_n)} U^n(x)\phi_i^{n+1}(x)\,\mathrm{d}x$$

$$S_i = (1-\theta)\int_0^{x_0(t_n)} \frac{\mathrm{d}U^n}{\mathrm{d}x}\frac{\mathrm{d}\phi_i^{n+1}}{\mathrm{d}x}\,\mathrm{d}x + \theta\int_0^{x_0(t_{n+1})} \phi_i^{n+1}(x)\,\mathrm{d}x$$

$$+ (1-\theta)\int_0^{x_0(t_n)} \phi_i^{n+1}(x)\,\mathrm{d}x,$$

and $\mathbf{M}$ and $\mathbf{K}$ are the usual mass and stiffness matrices at time level t_{n+1} given by

$$\left.\begin{aligned}M_{ij} &= \int_0^{x_0(t_{n+1})} \phi_i^{n+1}(x)\phi_j^{n+1}(x)\,\mathrm{d}x,\\[2mm] K_{ij} &= \int_0^{x_0(t_{n+1})} \frac{\mathrm{d}\phi_i^{n+1}}{\mathrm{d}x}\frac{\mathrm{d}\phi_j^{n+1}}{\mathrm{d}x}\,\mathrm{d}x.\end{aligned}\right\}$$

Linear basis functions are used throughout.

In the interests of computing economy, Miller *et al.* (1978) adapted only the two elements neighbouring the moving boundary. After calculating a new boundary point at time t_{n+1} as explained below, they positioned the penultimate node halfway between this point and the last fixed node (Fig. 4.6). The calculation proceeds through successive time steps until the two non-standard elements become less than half the size of a standard element. The total number of elements is then decreased by one by omitting the penultimate node and allowing the next one along to be repositioned so as to preserve two non-standard elements of equal length next to the moving boundary. In order to evaluate the integrals in $\mathbf{R}$ and $\mathbf{S}$, Miller *et al.* extended the last basis function $\phi_J^{n+1}(x)$ by the constant value unity. They found by experimenting with other possibilities that this gave the best results.

A novel procedure was adopted for locating the moving boundary. With reference to (1.59) the condition $\partial u/\partial x = 0$ was treated as a natural

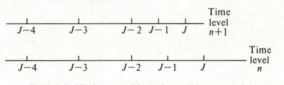

FIG. 4.6. Halfway position of penultimate mode

condition by including in the system (4.39) the Galerkin equation obtained from using ϕ_J^{n+1} as a test function. The condition $u = 0$ was then used to estimate the boundary position $x_0(t_{n+1})$. The boundary conditions (1.59) show that the solution in the neighbourhood of the moving boundary $x_0(t)$ is of the form

$$f(x) = \lambda(x_0 - x)^2, \qquad x_0 - h \leqslant x \leqslant x_0,$$

where h is the length of each non-standard element. In order to be consistent with the piecewise linear nature of the approximate solution obtained by using linear basis functions, Miller *et al.* used the best straight line approximation to $f(x)$ in the least-squares sense, which is

$$y = -\lambda h(x - x_0 + h) + \tfrac{5}{6}\lambda h^2.$$

This gives $x = x_0 - h/6$ when $y = 0$, i.e. when the last node is correctly positioned at the boundary the zero of the solution, U, should be taken to be at P in Fig. 4.7 such that BP/BA $= \tfrac{5}{6}$. This '$\tfrac{5}{6}$ rule' is used in positioning the adapted mesh. From the similar triangles in Fig. 4.7, this is equivalent to finding the mesh (with non-standard elements of length h) such that the solution on it satisfies

$$Q_{J-1}^{n+1} + 5Q_J^{n+1} = 0,$$

where Q_{J-1}^{n+1}, Q_J^{n+1} are the last two nodal values at time t_{n+1}. Miller *et al.* described an iterative procedure which they applied to the last p equations only, keeping Q_{J-p}^{n+1} fixed, since slight changes near the moving boundary have little effect on the solution remote from the boundary. Their results compare favourably with those of Hansen and Hougaard (1974) (see Tables 4.1, 4.2 for one comparison).

Gelinas, Doss, and Miller (1981) describe the automatic solution of general partial differential equations in one space dimension by a moving finite-element method. The nodes migrate continuously and systematically to regions where they are most needed to preserve accuracy, e.g.

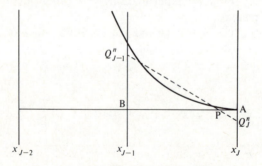

FIG. 4.7. Positioning last mode by 5/6 rule

regions of high gradients. Extension to two dimensions is mentioned. Adaptable, non-uniform grids are used by Jones and Thompson (1980) to obtain accuracy in finite-difference calculations.

4.4. Method of lines

For problems in one space dimension, the time variable is discretized in this method which is a well-known one for solving fixed-boundary problems. The partial differential equation is replaced by a sequence of ordinary differential equations at discrete time levels. Meyer (1970, 1977a,b,c, 1978a,b,c) developed the method in relation to moving boundary problems in a consistent and generalized way. The 1977c paper contains references to earlier solutions of sundry isolated problems by various authors.

We introduce the method of lines through the one-dimensional two-phase Stefan problem specified by

$$\frac{\partial}{\partial x}\left(k_1\frac{\partial u_1}{\partial x}\right)-c_1\frac{\partial u_1}{\partial t}=F_1, \qquad b_1<x<s(t), \qquad t>0 \qquad (4.41)$$

$$\frac{\partial}{\partial x}\left(k_2\frac{\partial u_2}{\partial x}\right)-c_2\frac{\partial u_2}{\partial t}=F_2, \qquad s(t)<x<b_2, \qquad t>0, \qquad (4.42)$$

$$u_1=\beta_1(t)\frac{\partial u_1}{\partial x}+\alpha_1(t), \qquad x=b_1, \qquad t>0, \qquad (4.43)$$

$$\frac{\partial u_2}{\partial x}=\beta_2(t)u_2+\alpha_2 t, \qquad x=b_2, \qquad t>0, \qquad (4.44)$$

where $k_1, c_1, F_1, k_2, c_2, F_2$ may be functions of x and t, e.g. $k_1=k_1(x,t)$, etc.

On the free boundary $s(t)$ we have the conditions

$$u_1=\mu_1(s(t),t): \qquad u_2=\mu_2(s(t),t), \qquad (4.45)$$

$$k_1\frac{\partial u_1}{\partial x}-k_2\frac{\partial u_2}{\partial x}+\lambda(s(t),t)\frac{ds}{dt}=\mu_3(s(t),t), \qquad x=s(t), \qquad t>0,$$

$$(4.46)$$

Meyer (1978b) showed that the introduction of convection terms $\partial u_1/\partial x, \partial u_2/\partial x$ causes no further difficulty.

At the nth time level we can approximate the system (4.41–46) by

$$(k_i u_i')' - c_i \frac{u_i - v_i(x)}{\delta t} = F_i, \qquad i = 1, 2.$$

$$u_1(b_1) = \beta u_1'(b_1) + \alpha_1,$$

$$u_2'(b_2) = \beta_2 u_2(b_2) + \alpha_2, \tag{4.47}$$

$$u_1(s) = \mu_1, \; u_2(s) = \mu_2, \qquad k_1 u_1'(s) - k_2 u_2'(s) + \lambda\left(\frac{s - s_{n-1}}{\delta t}\right) = \mu_3$$

where the prime indicates d/dx, $\delta t = t_n - t_{n-1}$, and where all parameters are evaluated at $t = t_n$. The functions v_i denote u_i at the previous time level or their linear extensions beyond their domain of definition in the neighbourhood of the moving interface. Thus if the boundary is moving in the direction of x increasing, so that $s(t_n) > s(t_{n-1})$, we use

$$v_1(x) = u_1(s_{n-1}) + (x - s_{n-1})u_1'(s_{n-1}), \qquad x \geqslant s_{n-1}. \tag{4.48}$$

A full description of the method of lines applied to the discretized system (4.47) was given by Meyer (1978c). We require to find the positions of the moving boundary at successive times $t = t_n$ and to solve the sequence of second-order, ordinary differential equations. The discretization of the time derivative does not depend on the linear structure of the equations (4.41–6), and if these equations were non-linear the usual 'shooting technique' could be used to solve them. For linear equations and conditions on the fixed boundaries, Meyer (1973, 1978c) developed a convenient method of solution by introducing the well-known Riccati transformations, to be used at $t = t_n$;

$$\left.\begin{aligned}
u_1(x) &= U(x)\phi_1(x) + w(x), \\
\phi_2(x) &= R(x)u_2(x) + z(x), \\
\phi_i &= k_i(x, t_n)u_i', \qquad i = 1, 2.
\end{aligned}\right\} \tag{4.49}$$

We obtain U, w, R, and z by solving the initial value problems

$$\left.\begin{aligned}
\frac{dU}{dx} &= \frac{1}{k_1} - \frac{c_1 U^2}{\delta t}, & U(b_1) &= \beta_1/k_1, \\
\frac{dw}{dx} &= -\frac{c_1}{\delta t} U(w - v_1) - UF_1, & w(b_1) &= \alpha_1, \\
\frac{dR}{dx} &= \frac{c_2}{\delta t} - \frac{R^2}{k_2}, & R(b_2) &= \beta_2 k_2, \\
\frac{dz}{dx} &= \frac{R}{k_2} z - \frac{c_2 v_2}{\delta t} + F_2, & z(b_2) &= \alpha_2 k_2,
\end{aligned}\right\} \tag{4.50}$$

where all parameters take their values at $t = t_n$.

By substituting $k_1 u_1'$ and $k_2 u_2'$ from the Riccati relations (4.49) into the discretized boundary condition in the last of (4.47), we see that the position of the interface at $t = t_n$ is the root $x = s_n$ of the equation

$$\frac{\mu_1(x, t_n) - w(x)}{U(x)} - \{R(x)\mu_2(x, t_n) + z(x)\}$$

$$+ \lambda(x, t_n)\left(\frac{x - s_{n-1}}{\delta t}\right) - \mu_3(x, t_n) = 0. \quad (4.51)$$

Once $s(t_n)$ is known, the functions u_i are obtained by backward integration of

$$\frac{du_2}{dx} = \frac{1}{k_2}(Ru_2 + z), \qquad u_2(s) = \mu_2(s), \qquad s < x < b_2, \quad (4.52)$$

from the second of (4.49), and

$$\frac{d\phi_1}{dx} = \frac{c_1 U\phi_1 + w - v_1}{\delta t} + F_1,$$

with

$$\phi_1(s) = \frac{\mu_1(s) - w(s)}{U(s)}, \qquad b_1 < x < s,$$

from the first equations of (4.47) and (4.49), and $u_1(x)$ follows from the first of (4.49).

Meyer (1978c) suggested integration of the differential equations by the simple trapezoidal rule coupled with linear interpolation between nodal values to locate the $x = s_n$ of (4.51). He discussed the relative merits of this and more elaborate methods.

Meyer (1977a,b) treated two- and three-dimensional problems by coupling his one-dimensional algorithms with the locally one-dimensional methods familiar in fixed boundary problems, e.g. fractional-step methods in alternating directions (see, for example, Williams 1979).

Thus, Meyer (1977a) considered the problem defined by

$$\nabla \cdot \{K(x, t)\nabla u\} + a(x, t) \cdot \nabla u + b(x, t)u - c(x, t)\frac{\partial u}{\partial t} = f(x, t) \quad (4.53)$$

for $x = (x_1, x_2)$ in the domain $D(t)$, $t > t_0$, with conditions $u = \alpha(x, t)$, for points x on those parts of the boundary $\partial D_1(t)$, $t > t_0$, which coincide with the coordinate axes, and $u = g_0(x, t)$ for points x on the free boundary $\partial D_2(t)$, together with the second free boundary condition

$$\nabla u = \left(g_1 - \lambda_1 \frac{\partial s_1}{\partial t}, g_2 - \lambda_2 \frac{\partial s_2}{\partial t}\right). \quad (4.54)$$

The initial value $u(x, 0) = u_0(x)$ within the initial domain $D(t_0)$ is also given. One half of the equation (4.53) is discretized in time and integrated in alternate directions, i.e. first

$$L_1 u = \frac{\partial}{\partial x_1}\left(K(x, t)\frac{\partial u}{\partial x_1}\right) + a_1(x, t)\frac{\partial u}{\partial x_1} + \tfrac{1}{2}b(x, t)u$$

$$-\tfrac{1}{2}c(x, t)\frac{\partial u}{\partial t} = \tfrac{1}{2}f(x, t), \qquad x_2 > 0, \quad (4.55)$$

is integrated from t_n to $t_{n+\frac{1}{2}}$ with (4.54) replaced by $\partial u/\partial x_1 = g_1(x, t) - \lambda_1\,\partial s_1/\partial t$. The initial value is $u(x, t_n)$ on $D(t_n)$, x_2 is considered as a free parameter and the other boundary conditions are satisfied. This one-dimensional problem is solved by the method of lines and invariant imbedding (Meyer 1977a) to yield values of $u(x, t_{n+\frac{1}{2}})$ and $D(t_{n+\frac{1}{2}})$ which then serve as initial values for the second, one-dimensional problem defined by

$$L_2 u = \frac{\partial}{\partial x_2}\left(K(x, t)\frac{\partial u}{\partial x_2}\right) + a_2(x, t)\frac{\partial u}{\partial x_2} + \tfrac{1}{2}b(x, t)u$$

$$-\tfrac{1}{2}c(x, t)\frac{\partial u}{\partial t} = \tfrac{1}{2}f(x, t), \qquad x_1 > 0, \quad (4.56)$$

to be discretized in time and integrated over the interval $t_{n+\frac{1}{2}}$ to t_{n+1}, and where x_1 is now a free parameter. Equation (4.55) is solved N times for N different values of the parameter $x_2 = x_2^k$, $k = 1, 2, \ldots, N$. These are uncoupled equations and can be solved simultaneously, followed by the solution of N equations (4.56) for N different values of $x_1 = x_1^k$, $k = 1, 2, \ldots, N$.

Meyer (1977a,b) pointed out that the disadvantage of this alternating direction method is its inability to track free boundaries which move essentially along one of the coordinate axes, so that the boundary cannot be described as an invertible function with respect to two orthogonal axes. Point sources may also lead to difficulties.

Meyer (1978a) therefore proposed a second method based on the method of lines combined with an SOR type iteration, where the sub-sidiary, one-dimensional problems are again solved by invariant imbed-ding. The free boundary at any given time is visualized as a curve or surface over a base domain, e.g. $z = z(x, y, t)$, $r = r(\theta, \phi, t)$ in three dimensions or $y = y(x, t)$, $r = (\theta, t)$ in two dimensions. The spacial variables of the base domain as well as time are then discretized and all derivatives with respect to these variables are replaced by appropriate difference ratios. A system of coupled free-boundary problems for second-order equations in the remaining space variable results. This system is solved by cycling through it using a Gauss–Seidel or SOR algorithm and using invariant imbedding to solve the one-dimensional free-boundary problem on each line.

As an example, consider the problem defined by the following system

$$\nabla^2 u - \frac{\partial u}{\partial t} = f(x, y, t) \qquad \text{in the time dependent} \tag{4.57}$$

$$\text{domain } D(t),$$

$$u = \alpha(x, y, t) \quad \text{on the fixed boundary } \partial D_1(t), \tag{4.58}$$

$$\left. \begin{aligned} g_1\left(x, y, t, u, \frac{\partial u}{\partial n}, \frac{\partial s}{\partial t}\right) &= 0 \\ g_2\left(x, y, t, u, \frac{\partial u}{\partial n}, \frac{\partial s}{\partial t}\right) &= 0 \end{aligned} \right\} \quad \text{on the free boundary } \partial D_2(t). \tag{4.59}$$

It is assumed that the free boundary $\partial D_2(t)$ can be expressed as $y = s(x, t)$ and that any line $x = $ constant cuts $\partial D_2(t)$ at most once. In contrast to the fractional step method, however, it is not necessary for $\partial D_2(t)$ to be expressible as the inverse function $x = \sigma(y, t)$, say. After discretizing time and x the following equations are to be solved at time level t_n:

$$\left. \begin{aligned} & \frac{\partial^2 u_i}{\partial y^2} + \frac{1}{(\Delta x)^2}(u_{i+1} - 2u_i + u_{i-1}) - \frac{1}{\Delta t}(u_i - u_{i,n-1}) = f(x_i, y, t_n) \quad \text{in} \quad D(t_n), \\ & u_i(y) = \alpha(x_i, y, t_n) \quad \text{on} \quad \partial D_1(t_n), \\ & g_j\left(x_i, s_i, t_n, u_i(s_i), \frac{\partial u_i}{\partial n}, \frac{s_i - s_{i,n-1}}{\Delta t}\right) = 0 \quad \text{on} \quad \partial D_2(t_n), \qquad j = 1, 2 \end{aligned} \right\}$$

$$\tag{4.60}$$

where

$$\frac{\partial u_i}{\partial n} = \left\{ \left(\frac{u_{i+1} - u_{i-1}}{2\Delta x}\right)\left(\frac{s_{i+1} - s_{i-1}}{2\Delta x}\right) - \frac{\partial u_i}{\partial y}(s_i) \right\} \bigg/ \left\{ 1 + \left(\frac{s_{i+1} - s_{i-1}}{2\Delta x}\right)^2 \right\}^{\frac{1}{2}}. \tag{4.61}$$

This is a multi-point system of ordinary differential equations with free boundary points. A successive over-relaxation scheme of solution, starting with an initial guess $(u_i^0, s_i^0)_{i=1}^N$, proceeds by solving in the kth iteration the equations (4.60) for $\bar{u}_i$ where $u_{i+1} = u_{i+1}^{k-1}$, $u_{i-1} = u_{i-1}^k$, $s_{i+1} = s_{i+1}^{k-1}$, $s_{i-1} = s_{i-1}^k$, i.e. new values are used as soon as they are available in the usual SOR fashion. Invariant embedding can conveniently be used to solve the sequence of one-dimensional problems and the location of the free boundary is essentially uncoupled from the solution of the second-order equation in (4.60). Once $(\bar{u}_i, s_i^k)$ are found the solution for the next line is obtained by incorporating the values

$$u_i^k = u_i^{k-1} + \omega(\bar{u}_i - u_i^{k-1}),$$

where ω is a relaxation factor chosen by experience to maximize the speed of convergence with hope of an improvement from one time level to the next.

The attraction of using the invariant imbedding method of solution, as in the one-dimensional problems discussed earlier in this section, is that it is little affected by the complicated form of the free-boundary condition involving (4.61).

Meyer (1977b) dealt similarly with an elliptic problem in radial coordinates. He mentions matters of the existence and uniqueness of his solutions and anticipates satisfactory convergence on the evidence of computed examples.

Meyer (1978a) solved a Stefan and an ablation problem by the method of lines with SOR. Furzeland (1979b) produced a general-purpose computer package, incorporating Meyer's ideas, for one-dimensional, nonlinear moving-boundary problems and tested it on a large variety of problems. George and Damle (1975) considered non-uniform initial temperatures.

The method of lines with a Riccati transformation was applied to the binary alloy solidification problem by Meyer (1981).

5. Front-fixing methods

5.1. One-dimensional problems

AN alternative to tracking the moving front is to fix it by a suitable choice of new space coordinates. For example, in the simple one-dimensional melting problem specified by equations (1.21–6), the transformation

$$\xi = x/s(t) \tag{5.1}$$

fixes the melting boundary at $\xi = 1$ for all t. By using the standard relationships

$$\frac{\partial u}{\partial x} = \frac{1}{s(t)} \frac{\partial u}{\partial \xi}, \qquad \frac{\partial^2 u}{\partial x^2} = \frac{1}{\{s(t)\}^2} \frac{\partial^2 u}{\partial \xi^2},$$

$$\left(\frac{\partial u}{\partial t}\right)_x = \frac{\partial u}{\partial \xi}\frac{\partial \xi}{\partial t} + \left(\frac{\partial u}{\partial t}\right)_\xi = -\frac{x}{\{s(t)\}^2}\frac{ds}{dt}\frac{\partial u}{\partial \xi} + \left(\frac{\partial u}{\partial t}\right)_\xi$$

to transform from $u(x, t)$ to $u(\xi, t)$, the heat equation (1.21) can be written

$$\frac{\partial^2 u}{\partial \xi^2} = s^2 \frac{\partial u}{\partial t} - s\xi \frac{ds}{dt}\frac{\partial u}{\partial \xi}, \qquad 0 < \xi < 1, \qquad t > 0, \tag{5.2}$$

where $s = s(t)$, and (1.26) becomes, for $\lambda = 1$,

$$-\frac{1}{s}\frac{\partial u}{\partial \xi} = \frac{ds}{dt}, \qquad \xi = 1, \qquad t > 0. \tag{5.3}$$

The transformation (5.1) was proposed by Landau (1950) and first applied to a finite-difference scheme by Crank (1957a). Lotkin (1960) used unequal intervals in ξ and t and divided differences to obtain an economic improvement in accuracy. Temperature-dependent thermal properties were included by Citron (1962) and Mastanaiah (1976). Ferriss and Hill (1974) solved the implicit oxygen diffusion with absorption problem (see §1.3.10) by using the same transformation. Some of their results are included in Table 6.3 for convenience. Dahmardah and Mayers (1983) used the variable $x/s(t)$ in a Fourier-series solution of the oxygen problem which supports the results of Hansen and Hougaard (1974) in Tables 4.1 and 4.2 except in the final stages. Further examples are discussed in Hoffmann (1977, Volume III, pp. 4, 49, 91). Höhn (1978) proved second-order convergence of an explicit finite-difference scheme based on (5.1).

If a fully implicit numerical scheme is used to solve (5.2) together with the necessary conditions on a fixed boundary, $\xi = 0$, and say $u = 0$ on $\xi = 1$, the additional condition (5.3) on $\xi = 1$ can be incorporated in the way eqn (5.15) is treated later in a two-phase problem. Alternatively, and necessarily for an implicit type of moving boundary condition, in which ds/dt does not appear, some estimated value of s must be iteratively improved in each time step. Thus the oxygen diffusion problem has conditions $u = 0$, $\partial u/\partial x = 0$ on the moving boundary and so Ferriss and Hill (1974), having introduced a variable $\phi = u + t$, used $\partial \phi/\partial \xi = 0$ on $\xi = 1$ in order to solve the boundary-value problem equivalent to (5.2) for two estimated values of s, denoted by s_1 and s_2. Since $\phi = t$ on $\xi = 1$, s must satisfy $F(s) \equiv (\phi/t) - 1 = 0$ on $\xi = 1$. Thus if $F(s_1)$ and $F(s_2)$ are evaluated from the solution of the equivalent of (5.2) the method of Regula-Falsi yields an improved estimate s_3 given by

$$s_3 = \{s_1 F(s_1) - s_2 F(s_2)\}/\{F(s_2) - F(s_1)\},$$

and so on. When advancing the time to the next step, the initial estimate for s is taken to be the final value at the previous time level.

In attempting to obtain quadratic convergence, Baumeister *et al.* (1980) developed a Newton-type iterative scheme. Following the ideas of Bonnerot and Jamet (1974) and Fasano and Primicerio (1977), the Newton step requires the determination of the zero of an integral form of the moving boundary condition $\partial u/\partial x = -ds/dt$, namely

$$F(s) = s - b + \int_0^t g(\tau)\, d\tau + \int_0^{s(t)} u(x, t)\, dx - \int_0^b f(x)\, dx,$$

where $s(0) = b$ and g and f are boundary flux and initial temperature functions. Although quadratic convergence is secured, the cost in computing time is considerable because an auxiliary heat flow problem must be solved on a given but time-variable domain in each step of the iteration in order to obtain the necessary Frechet derivative.

Nitsche (1980) used the boundary-fixing transformation (5.1) coupled with a new time variable, $\tau = \int_0^t s^{-2}(t)\, dt$, to obtain a weak formulation of a single-phase Stefan problem with zero-flux condition on the fixed boundary. A finite-element discretization was effected and used to establish existence and regularity results but no numerical solutions were obtained. Several generalizations of the method, including an application to the oxygen diffusion problem, were mentioned together with an extensive list of references to papers on finite-element formulations and free-boundary problems in general.

Furzeland (1980) considered the more general, linear equation with constant coefficients for the unknown $u_i(x, t)$ in each phase $i = 1$ and 2,

$$k_i \frac{\partial^2 u_i}{\partial x^2} + a_i \frac{\partial u_i}{\partial x} + b_i u_i - c_i \frac{\partial u_i}{\partial t} = f_i(x, t), \qquad 0 < t < t^*, \qquad (5.4)$$

with phase 1 defined as $\ell_1 \leqslant x \leqslant s(t)$ and phase 2 as $s(t) \leqslant x \leqslant \ell_2$, where $s(t)$ denotes the moving boundary. Initial and boundary conditions of the form

$$\left. \begin{array}{ll} u_1 = u_{10}(x), & \ell_1 \leqslant x \leqslant s(0) \\ u_2 = u_{20}(x), & s(0) \leqslant x \leqslant \ell_2 \end{array} \right\} \quad t = 0, \quad s(0) \text{ given}, \qquad (5.5)$$

$$\alpha_1(i, t) u_i + \alpha_2(i, t) \frac{\partial u_i}{\partial x} = \alpha(i, t), \qquad x = \ell_i, \qquad 0 < t < t^*, \qquad (5.6)$$

are allowed for, together with the following general form of the moving boundary conditions:

$$\left. \begin{array}{l} u_1 = u_2 = \mu(s, \dot{s}, t) \\ F\left(u, \dfrac{\partial u_1}{\partial x}, \dfrac{\partial u_2}{\partial x}, \dfrac{\partial u}{\partial t}, s, \dot{s}, t\right) = 0 \end{array} \right\} \quad x = s(t), \quad 0 < t < t^*. \qquad (5.7)$$

The function F need not be a polynomial in its variables, though it commonly is, and the $\dot{s}$ term will be absent in an implicit problem. In phase 1, $\ell_1 \leqslant x \leqslant s(t)$, the moving boundary can be fixed by the coordinate transformation

$$\xi_1 = \frac{x - \ell_1}{s(t) - \ell_1}, \qquad (5.8)$$

so that $x = \ell_1$ becomes $\xi_1 = 0$ and $x = s(t)$ is $\xi_1 = 1$. The corresponding transformation for phase 2 is

$$\xi_2 = \frac{x - \ell_2}{s(t) - \ell_2}. \qquad (5.9)$$

Considering phase 1 and dropping the subscript $i = 1$ in (5.4) the equation becomes

$$k \frac{\partial^2 u}{\partial \xi^2} + (s - \ell_1)(a + c\xi\dot{s}) \frac{\partial u}{\partial \xi} + (s - \ell_1)^2 bu - (s - \ell_1)^2 c \frac{\partial u}{\partial t}$$

$$= (s - \ell_1)^2 f(x, t), \qquad 0 < \xi < 1, \qquad t > 0, \qquad (5.10)$$

after multiplying through by $(s - \ell_1)^2$ for numerical convenience.

The usual finite-difference discretization in time at $t^{n+\theta} = (n + \theta) \Delta t$, where $\theta = \frac{1}{2}$ for Crank–Nicolson and $\theta = 1$ for a fully implicit scheme, is combined with central differences for space derivatives to give the tridiagonal system

$$\theta[A_j - B_j]u_{j-1}^{n+1} + [cS^2 + \theta(2B_j - \Delta tS^2 b)]u_j^{n+1} - \theta[A_j + B_j]u_{j+1}^{n+1}$$
$$= (1-\theta)[B_j - A_j]u_{j-1}^n + [cS^2 - (1-\theta)(2B_j - \Delta tS^2 b)]u_j^n$$
$$+ (1-\theta)[A_j + B_j]u_{j+1}^n - \Delta tS^2 f(x_j, t^{n+1}), \quad (5.11)$$

where

$$S = s^{n+\theta} - \ell_1, \qquad s^{n+\theta} = \theta s^{n-1} + (1-\theta)s^n, \qquad \dot{s}^{n+\theta} = \frac{s^{n+1} - s^n}{\Delta t},$$

$$A_j = \frac{\Delta t}{2\Delta \xi} S[a + c(\xi_j \dot{s})^{n+\theta}], \qquad B_j = \frac{\Delta t}{(\Delta \xi)^2} k, \qquad x_j = j\,\Delta\xi.$$

The fixing of the moving boundary at $\xi = 1$ allows

$$\frac{\partial u_i}{\partial x} = \frac{1}{(s - \ell_i)} \frac{\partial u_i}{\partial \xi}, \qquad i = 1, 2, \tag{5.12}$$

to be expressed as standard finite-difference approximations. Thus, Furzeland (1980) examines three of the possible choices for approximating $\partial u_1 / \partial \xi$, i.e.

$$(u_N - u_{N-1})/\Delta\xi, \qquad (3u_N - 4u_{N-1} + u_{N-2})/2\Delta\xi, \qquad \text{and}$$

$$(u_{N+1} - u_{N-1})/2\,\Delta\xi, \tag{5.13}$$

where $j = N$ is $\xi = 1$ and u_{N+1} is a fictitious point, and compares them numerically for a test problem of Stefan type. The first of (5.13) was inferior to the other two forms. Moving-boundary conditions with an explicit relationship for $\dot{s}$ such as

$$\left. \begin{aligned} u &= F_2(s, t) \\ \dot{s} &= F_1\left(s, t, \frac{\partial u_1}{\partial x}, \frac{\partial u_2}{\partial x}\right) \end{aligned} \right\} \quad x = s(t), \qquad \xi = 1, \tag{5.14} \tag{5.15}$$

are easily incorporated into an iterative scheme for the solution of (5.11) by approximating (5.15) at time $t^{n+\theta}$ as

$$s^{n+1} = s^n + \Delta t[\theta F_1^{n+1} + (1-\theta)F_1^n], \tag{5.16}$$

and using one of (5.13) for the space derivatives. Moving boundary conditions of implicit type, with $\dot{s}$ absent, as in

$$\left. \begin{aligned} \frac{\partial u_i}{\partial x} &= F_1(i, s, t) \\ F_2(s, t, u_1, u_2) &= 0 \end{aligned} \right\} \quad \text{or} \quad \left. \begin{aligned} u_i &= F_2(i, s, t) \\ F_1\left(s, t, \frac{\partial u_1}{\partial x}, \frac{\partial u_2}{\partial x}\right) &= 0 \end{aligned} \right\} \tag{5.17} \tag{5.18}$$

can be incorporated by using one of (5.17) as the known boundary condition for (5.11) and estimating s from a *regulia falsi* solution of the corresponding eqn (5.18) as Ferriss and Hill (1974) did.

Equation (5.11) is solved to find $u^{n+1,1}$ with an initial guess $s^{n+1,0}$ obtained, for example, by extrapolation from the two previous time levels. Then an improved estimate $s^{n+1,1}$ is obtained from the moving boundary conditions, as described above, and the scheme iterated until convergence is achieved to a prescribed accuracy (usually 2–3 iterations in Furzeland's test problems for a precision of 10^{-4}). If $s(t) = \ell_1$ for a finite length of time then (5.8) is singular. The original eqn (5.1) needs to be solved by using, for example, a Crank–Nicolson scheme on a fixed domain until $s(t) - \ell_1$ becomes non-zero. This is not necessary at $t = 0$ as long as $s(\theta \, \Delta t) > 0$ in (5.11).

Furzeland (1979*b*) developed a general-purpose computer program based on the above iterative algorithm. Mastaniah (1976) used essentially the same algorithm to solve problems with temperature-dependent thermal properties, i.e. k and c depend on u. Thus $k(u)$ was evaluated at the mid points $(j \pm \frac{1}{2}, n + \frac{1}{2})$ by using Taylor's series to extrapolate $u^{n+\frac{1}{2}}$ from previous time levels.

An alternative way of treating eqn (5.4), suggested in a private communication by C. M. Elliott to Furzeland (1979*b*, 1980), is to discretize only the space derivatives and to integrate the resulting ordinary differential equations in time along constant $\xi = \xi_j$ lines. For example, eqn (5.2) is approximated by

$$\frac{du_j}{dt} = \frac{1}{s^2} \frac{u_{j-1} - 2u_j + u_{j+1}}{(\Delta \xi)^2} + \xi_j \frac{\dot{s}}{s} \frac{(u_{j+1} - u_{j-1})}{2 \, \Delta \xi},$$

$$j = 0, 1, \ldots, N-1, \qquad 0 < \xi < 1, \qquad t > 0. \quad (5.19)$$

Moving boundary conditions containing an explicit relation for $\dot{s}$ can be approximated, for example, by

$$u_N = 0, \qquad \dot{s} = -(3u_N - 4u_{N-1} + u_{N-2})/2 \, \Delta \xi, \qquad \xi = 1, \qquad t > 0.$$
$$(5.20)$$

An attraction of this approach is that well-established algorithms, e.g. the Gear algorithm, for stiff ordinary differential equations such as (5.19) can be used. Two readily available implementations are the NAG algorithm DO2AJF or an algorithm by Aitchison (1975) adapted for sparse matrices.

The stiff-system algorithm will automatically generate du_j/dt, $j = 0, 1, \ldots, N-1$, and $\dot{s}$, and will produce a solution vector $(u_0, u_1 \ldots, u_{N-1}, s)$ at required time intervals, provided an explicit boundary condition of type (5.20) exists. This method is readily adaptable to non-linear equations and moving boundary conditions, and to multi-component systems including combined heat and mass transfer. Implicit conditions on the moving boundary that do not contain $\dot{s}$ are less amenable directly, but implicit conditions can be rendered explicit by the Schatz (1969) transformation or by repeated differentiation (Crank and Gupta 1972a).

Furzeland (1980) gave detailed comparisons of the boundary-fixing methods in one space dimension for three test problems from the points of view of accuracy, computing time, convenience of use, etc. Some of these results are given in Tables 6.2 and 6.3.

5.2. Body-fitted curvilinear coordinates

The one-dimensional coordinate transformations introduced in §5.1 are simple cases of the more general transformation of a curved-shaped region in two or more dimensions into a fixed, rectangular domain. Since the new coordinates are chosen to fit the shape of the original region, they are commonly called 'body-fitted' or 'natural' coordinates. The original coordinate system (x, y) is transformed into a new system (ξ, η), which may be orthogonal or non-orthogonal, such that the curved boundaries in the (x, y) plane become (ξ, η) coordinate lines. Corresponding to the regular straight-lined (ξ, η) mesh is a curvilinear (x, y) mesh as in Fig. 5.1.

In moving-boundary problems the region changes with time, but a fixed (ξ, η) mesh which corresponds to a moving (x, y) mesh can be used for all time. The movements of the boundary and of the mesh points in the original region appear only as changes in x and y at the corresponding, fixed, (ξ, η) points at each time step. The attractions of working in a simple, straight-lined fixed region such as a rectangle with no loss of accuracy in discretization near curved boundaries are offset to some extent by the increased complexity of the transformed partial differential equation and boundary conditions, together with the additional work of generating the curvilinear mesh at each time step. But it is an advantage to have direct control over the (x, y) mesh spacing so that, for example, a

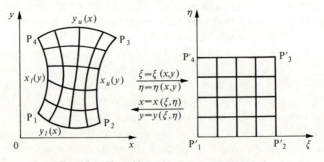

FIG. 5.1. Transformation to rectangular mesh

finer mesh can be used in regions of special interest such as near the moving boundary or a singularity.

Following a pioneer paper by Winslow (1967), extended by Chu (1971), successive authors have proposed various ways of using a curvilinear grid. Useful lists of references were given by Thompson, Thames, and Mastin (1974) and by Furzeland (1977*a*). Oberkampf (1976) discussed some useful generalized mapping functions. Furzeland (1977*b*) summarized some of the transformations, orthogonal and non-orthogonal, that have been proposed. For moving-boundary problems attention has been focussed mainly on non-orthogonal transformations, which are conveniently introduced through Laplace's equation for a general dependent variable $\phi = \phi(x, y) = \phi(\xi, \eta)$ satisfying

$$\frac{\partial^2 \phi}{\partial x^2} + \frac{\partial^2 \phi}{\partial y^2} = 0. \tag{5.21}$$

The following standard differential relationships exist:

$$\xi_x = y_\eta/J, \qquad \xi_y = -x_\eta/J, \tag{5.22}$$

$$\eta_x = -y_\xi/J, \qquad \eta_y = x_\xi/J, \tag{5.23}$$

where $\xi_x = \partial\xi/\partial x$, etc., and J is the Jacobian $x_\xi y_\eta - x_\eta y_\xi \neq 0$. For example, given

$$x = x(\xi, \eta), \qquad y = y(\xi, \eta)$$

we have

$$dx = x_\xi \, d\xi + x_\eta \, d\eta; \qquad dy = y_\xi \, d\xi + y_\eta \, d\eta. \tag{5.24}$$

By solving (5.24) for $d\xi$ and $d\eta$ we obtain

$$d\xi = \frac{y_\eta \, dx - x_\eta \, dy}{x_\xi y_\eta - x_\eta y_\xi} = \frac{y_\eta \, dx - x_\eta \, dy}{J},$$

from which the first of (5.22) follows immediately and, similarly, the other derivatives in (5.22) and (5.23). Then we have

$$d\phi = \phi_\xi \, d\xi + \phi_\eta \, d\eta$$

and so

$$\frac{\partial \phi}{\partial x} = \phi_\xi \xi_x + \phi_\eta \eta_x = (\phi_\xi y_\eta - \phi_\eta y_\xi)/J \tag{5.25}$$

on using (5.22). The second derivative follows in the standard way since

$$\frac{\partial^2 \phi}{\partial x^2} = \frac{\partial}{\partial x}\left(\frac{\partial \phi}{\partial x}\right) = \frac{1}{J^2}\left(\frac{\partial}{\partial \xi} y_\eta - \frac{\partial}{\partial \eta} y_\xi\right)^2 \phi$$

$$= (\phi_{\xi\xi} y_\eta^2 + \phi_{\eta\eta} y_\xi^2 - 2\phi_{\xi\eta} y_\eta y_\xi)/J^2, \tag{5.26}$$

and similarly for other higher derivatives. Using these derivative relations, Laplace's equation for $\phi(x, y)$ transforms into

$$A\phi_{\xi\xi} + B\phi_{\xi\eta} + C\phi_{\eta\xi} + D\phi_{\xi} + E\phi_{\eta} = 0, \qquad (5.27)$$

where

$$A = \xi_x^2 + \xi_y^2 = (x_\eta^2 + y_\eta^2)/J^2, \qquad (5.28)$$

$$B = 2(\xi_x\eta_x + \xi_y\eta_y) = -2(x_\xi x_\eta + y_\xi y_\eta)/J^2, \qquad (5.29)$$

$$C = \eta_x^2 + \eta_y^2 = (x_\xi^2 + y_\xi^2)/J^2, \qquad (5.30)$$

$$D = \xi_{xx} + \xi_{yy}, \qquad E = \eta_{xx} + \eta_{yy}. \qquad (5.31)$$

Chu (1971) quoted expressions for C and D in terms of x_ξ, y_ξ, etc. and, together with Oberkampf (1976), derived similar expressions for more general equations than Laplace's. Normal derivatives of ϕ on a boundary $y = g(x)$ are given by

$$\phi_n = \frac{g'\phi_x - \phi_y}{\{1 + (g')^2\}^{\frac{1}{2}}} = \frac{1}{J\{1 + (g')^2\}^{\frac{1}{2}}} \times \{\phi_\xi(g'y_\eta + x_\eta) - \phi_\eta(g'y_\xi + x_\xi)\}$$

$$(5.32)$$

where $g' = \mathrm{d}y/\mathrm{d}x$.

Time derivatives of $\phi(x, y, t)$ are transformed from a given point (x, y) to the corresponding (ξ, η) point by the relationship

$$(\phi_t)_{x,y} = (\phi_t)_{\xi,\eta} - \frac{1}{J}(y_\eta\phi_\xi - y_\xi\phi_\eta)(x_t)_{\xi,\eta}$$

$$-\frac{1}{J}(x_\xi\phi_\eta - x_\eta\phi_\xi)(y_t)_{\xi,\eta}. \qquad (5.33)$$

Thus all derivatives of $\phi(x, y, t)$ can be expressed in terms of derivatives at fixed points in the transformed region, whether the original (x, y) domain and mesh are time dependent or not.

Of the various transformations mentioned by Furzeland (1977b) the simplest one to use in moving-boundary problems is due to Oberkampf (1976), who chose

$$\xi = \frac{x - x_\ell(y)}{x_u(y) - x_\ell(y)}, \qquad \eta = \frac{y - y_\ell(x)}{y_u(x) - y_\ell(x)}, \qquad (5.34)$$

where x_ℓ, x_u, y_ℓ, y_u are the four curved sides in Fig. 5.1. Even if, for example, the upper boundary y_u is moving, the second of (5.34) ensures that it is the stationary coordinate line $\eta = 1$ in the transformed plane. In general, x_ℓ etc. represent sets of discrete values of boundary points and can be deliberately chosen to give any required mesh spacing, a facility which is not available if orthogonal transformations are used. It follows from (5.34) that ξ, η are then known discrete functions of x and y, and that the derivatives ξ_x, ξ_y, etc. needed in the coefficients $A-E$ are available as suitable discrete approximations. Finite elements with bivariate blending functions (Gordon and Hall 1973) provide alternative mappings to (5.34) and so do isoparametric curvilinear coordinates (Zienkiewicz and Phillips 1971).

5.3. Example of use of body-fitted coordinates

Furzeland (1977b) took the one-phase two-dimensional problem of Bonnerot and Jamet (1977) set out in equations (4.24–28) in §4.3.3 as an example, with $\lambda = 1$ in (4.28).

The region below the moving boundary in the physical (x, y) plane is transformed into the unit square of Fig. 5.2 using the transformations

$$\xi = x, \qquad \eta = \eta(x, y). \tag{5.35}$$

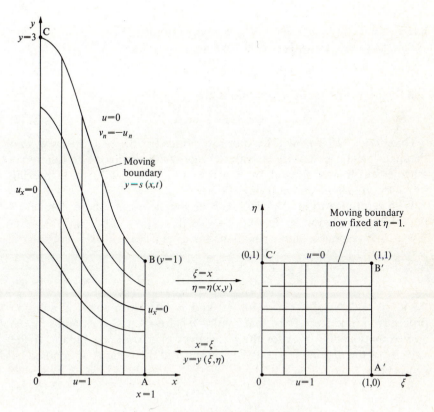

FIG. 5.2. Physical plane and transformed plane

Combining (5.27) and (5.33) the heat equation (4.24) becomes

$$u_t = Au_{\xi\xi} + Bu_{\xi\eta} + Cu_{\eta\eta} + D'u_\xi + E'u_\eta,$$

$$0 \leqslant \xi \leqslant 1, \qquad 0 \leqslant \eta \leqslant 1, \qquad t > 0, \quad (5.36)$$

where

$$D' = D + (x_\eta y_t - y_\eta x_t)/J, \qquad E' = E + (y_\xi x_t - x_\xi y_t)/J \qquad (5.37)$$

and A–E are given by (5.28–31) inclusive. If now the Oberkampf transformations (5.34) appropriate for this problem are used, i.e.

$$\xi = x, \qquad \eta = y/s(x, t), \qquad (5.38)$$

the transformed equation (5.36) becomes

$$u_t = au_{\xi\xi} + bu_{\xi\eta} + cu_{\eta\eta} + du_\xi + eu_\eta,$$

$$0 \leqslant \xi \leqslant 1, \qquad 0 \leqslant \eta \leqslant 1, \qquad t > 0, \quad (5.39)$$

where

$$a = 1, \qquad b = -2ys_\xi/s^2, \qquad c = (1/s)^2 + (b/2)^2, \qquad d = 0,$$

$$e = (y/s)_{\xi\xi} + y_t/s, \qquad (5.40)$$

having used the facts that $y_\eta = s$, $y_\xi = ys_\xi/s$ from (5.38).

Similarly, the derivative conditions are

$$su_\xi - (ys_\xi u_\eta/s) = 0, \qquad \xi = 0, \quad 1, \qquad (5.41)$$

and the moving boundary conditions are

$$u = 0, \qquad y_t = -(1 + s_\xi^2)u_\eta/s, \qquad \eta = 1. \qquad (5.42)$$

Duda *et al.* (1975) used the transformations (5.38). Hsu, Sparrow, and Patankar (1981) combined it with an energy-conserving finite-difference scheme which was applied by Sparrow and Hsu (1981) to a freezing problem outside a tube carrying coolant.

With reference to an $N_1 \times N_2$ mesh of size h in the square domain of Fig. 5.2 and denoting $\xi_k = kh$, $k = 0, 1, \ldots, N_1$, and $\eta_e = \ell h$, $\ell = 0, 1, \ldots, N_2$, suitable approximations for s_ξ and $s_{\xi\xi}$ are

$$s_\xi = (s_{k+1} - s_{k-1})/(2h), \qquad s_{\xi\xi} = (s_{k+1} - 2s_k + s_{k-1})/h^2,$$

$$k = 1, 2, \ldots, N-1. \quad (5.43)$$

At $k = 0$ and N, $s_\xi = 0$ since $u_x = 0$ and $u = 0$. The numerical solution proceeds in time steps δt so that $t = t^n = n\,\delta t$, $n = 0, 1, 2, \ldots$, and suitable approximations are used for the derivatives of u over the square. If the velocity of the moving boundary, y_t, on $\eta = 1$ at the point ξ_k is denoted by $(y_t)_k$, then

$$y_k^{n+1} = y_k^n + \delta t(y_t)_k, \qquad (5.44)$$

where $(y_t)_k$ is approximated from (5.42) by using the first of (5.43) for s_ξ

and either a three-point, end-on formula for u_η on $\eta = 1$ or the standard 'fictitious-point' approximation suggested by Furzeland (1977b) to preserve $O(h^2)$ accuracy.

The numerical algorithm becomes:

  (i) given u and s at time level n, compute (x, y) values at each (ξ, η) mesh point using (5.38);
 (ii) compute the new position of the moving boundary at time $(n+1)$ using (5.44) with a discretized form of (5.42);
(iii) use the second of (5.38) to calculate the corresponding changes in y at all mesh points (ξ, η);
 (iv) solve the heat eqn (5.39) with coefficients (5.40) by suitably discretizing the time and space variables. Return to step (ii) and repeat for each time level.

Furzeland (1977b) presented numerical results based on this algorithm and on the use of an alternative 'equipotential' transformation of Winslow (1967) instead of the Oberkampf transformation (equation (5.38)). The Winslow transformation is more expensive in computer time because an additional subsidiary equation has to be solved at each time step, but it does offer a smoother and more flexible control over the curvilinear mesh spacing. Furzeland (1977b) found that both his algorithms compared well with the results of Bonnerot and Jamet (1977). Further applications of the general method were discussed by Hoffman (1977, Volume III).

Saithoh (1978) used a version of the transformation (5.34) in radial coordinates which provides a useful way of handling a phase-change boundary problem within an arbitrarily shaped region. He referred to the earlier work of Duda *et al.* (1975). With reference to Fig. 5.3, where the coordinates are r, ϕ, the region within the fixed boundary $r = B(\phi)$ is initially occupied by liquid at the freezing temperature. The solidification

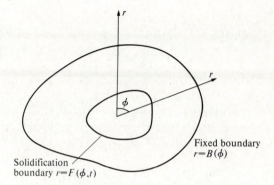

FIG. 5.3. Boundary fixing in radial coordinates

boundary is defined by $r = F(\phi, t)$. In appropriate non-dimensional terms the relevant set of equations is

$$\frac{\partial u}{\partial t} = \nabla^2 u = \frac{1}{r} \frac{\partial}{\partial r}\left(r \frac{\partial u}{\partial r}\right) + \frac{1}{r^2 \phi_0^2} \frac{\partial^2 u}{\partial \phi^2},$$

$$\frac{\partial u}{\partial \phi} = 0, \qquad \phi = 0, 1,$$

$$u = u_W(\phi, t), \qquad r = B(\phi),$$

$$u = 0, \frac{\partial F}{\partial t} = St\left[1 + \left(\frac{1}{\phi_0 F} \frac{\partial F}{\partial \phi}\right)^2\right]\frac{\partial u}{\partial r}, \qquad r = F(\phi, t),$$

where ϕ_0 is the angular width of the region and St is Stefan's number.

In two space dimensions the new variable

$$\eta = \frac{r - F(\phi, t)}{B(\phi) - F(\phi, t)}$$

fixes the moving boundary at $\eta = 0$ and the outer, fixed boundary at $\eta = 1$. The corresponding transformation in three dimensions is

$$\zeta = \frac{r - F(\phi, \theta, t)}{B(\phi, \theta) - F(\phi, \theta, t)},$$

where (r, ϕ, θ) are radial coordinates. In the two-dimensional case the transformed system is

$$\frac{\partial u}{\partial t} = \left\{\frac{1}{(B - F)^2} + \frac{1}{R}\left(\frac{\partial \eta}{\partial \phi}\right)^2\right\}\frac{\partial^2 u}{\partial \eta^2}$$

$$+ \left\{\frac{1}{r(B - F)} - \frac{\partial \eta}{\partial \phi} + \frac{1}{R} \frac{\partial^2 \eta}{\partial \phi^2}\right\}\frac{\partial u}{\partial \eta} + \frac{2}{R} \frac{\partial \eta}{\partial \phi} \frac{\partial^2 u}{\partial \phi \partial \eta} + \frac{1}{R} \frac{\partial^2 u}{\partial \phi^2},$$

$$\partial u / \partial \phi = 0, \qquad \phi = 0, 1; \qquad u = u_W(\phi, t), \qquad \eta = 1;$$

$$\frac{\partial F}{\partial t} = St \frac{\partial u}{\partial \eta} \frac{1}{B - F}\left\{1 + \left(\frac{1}{F \phi_0} \frac{\partial F}{\partial \phi}\right)^2\right\}, \qquad u = 0, \qquad \eta = 0,$$

$$R = (\phi_0 r)^2.$$

Thus, the transformed partial differential equation has to be solved in the fixed rectangular region $0 \leq \phi \leq 1$, $0 \leq \eta \leq 1$.

Any point at which the fixed boundary $r = B(\phi)$ is not smooth, such as the corner of a square, needs special consideration. For the latter case, for example, Saitoh (1978) substituted the equation

$$\frac{\partial u}{\partial r} = \frac{1}{r} \frac{\partial}{\partial r}\left(r \frac{\partial u}{\partial r}\right) + \frac{1}{r \, d\phi}\left(\frac{\partial u}{\partial \phi}\right)_-$$

where $(\partial u/\partial \phi)_-$ is the value of the derivative at $\phi = \pi/4 - \mathrm{d}\phi$; this equation expresses the heat balance of an element between r, $r + \mathrm{d}r$ and $\pi/4$, $\pi/4 - \mathrm{d}\phi$.

Saitoh (1978) compared his numerical solution for the solidification of a square region of liquid with values measured experimentally and found good agreement. He also presented calculated values for some triangular regions.

5.4. Isotherm migration method

A particular case of the curvilinear transformation depicted in Fig. 5.1 is one in which the dependent variable u is interchanged with one of the space variables, e.g. $u = u(x, y)$ becomes $x = x(u, y)$.

In a one-dimensional time-dependent heat-flow problem the temperature u can be interchanged with the space variable x, so that the solution evaluates $x(u, t)$ instead of the more traditional $u(x, t)$. The latter expresses the time dependence of temperature at chosen values of x; the former denotes how a specified temperature u moves through the medium: in other words how isotherms migrate. This so-called Isotherm Migration Method (IMM) was proposed by Chernousko (1970), independently by Dix and Cizek (1970), and subsequently developed and extended to two space dimensions by Crank *et al.* (1973, 1975, 1978, 1979) and by Turland and Wilson (1977) and Turland (1979).

The IMM is particularly suited to melting and freezing problems since the phase boundary is itself an isotherm, provided the phase change takes place at a constant temperature. The IMM is, in fact, a boundary-fixing method and, in one space dimension, x is evaluated at successive times in the fixed domain $0 \le u \le 1$, for example. As the governing equations are discretized in temperature u, mesh lines are isotherms along which temperature-dependent heat parameters are known in advance and so can be handled simply and efficiently. In other methods they are coupled with the evaluation of u, say, in the (x, t) plane.

5.4.1. IMM in one space dimension

Consider as an illustration the one-phase Stefan problem

$$\frac{\partial u}{\partial t} = \frac{\partial^2 u}{\partial x^2}; \frac{\mathrm{d}s}{\mathrm{d}t} = -\lambda \frac{\partial u}{\partial x}, \qquad u = 0, \qquad x = s(t);$$

$$u = 1, \qquad x = 0, \qquad t > 0; \qquad u = 0, \qquad x > 0, \qquad t = 0. \qquad (5.45)$$

Instead of writing $u = u(x, t)$ we take $x = x(u, t)$ as a new dependent variable. By the usual partial-derivative relations we have

$$\frac{\partial u}{\partial x} = \left(\frac{\partial x}{\partial u}\right)^{-1}, \qquad \left(\frac{\partial x}{\partial t}\right)_u = -\left(\frac{\partial u}{\partial t}\right)\left(\frac{\partial x}{\partial u}\right), \qquad (5.46)$$

where $(\partial x/\partial t)_u$ signifies the rate of change of x with t at constant u, i.e. denotes the movement of an isotherm, and finally

$$\frac{\partial^2 u}{\partial x^2} = \frac{\partial}{\partial x}\left(\frac{\partial x}{\partial u}\right)^{-1} = -\left(\frac{\partial^2 x}{\partial u^2}\right)\left(\frac{\partial x}{\partial u}\right)^{-3}. \tag{5.47}$$

Use of (5.46) and (5.47) transforms (5.45) to

$$\frac{\partial x}{\partial t} = \left(\frac{\partial x}{\partial u}\right)^{-2}\frac{\partial^2 x}{\partial u^2}, \qquad 0 < u < 1, \qquad t > 0, \tag{5.48}$$

$$\frac{ds}{dt} = -\lambda\left(\frac{\partial x}{\partial u}\right)^{-1}, \qquad u = 0, \qquad t > 0, \tag{5.49}$$

$$x = 0, \qquad u = 1, \qquad t > 0, \tag{5.50}$$

and a starting solution is needed for small times as with most finite-difference methods when there is a singularity at $x = 0$, $t = 0$.

The derivatives in (5.48) can be approximated by finite differences to give, for example, an explicit expression for x_i^{n+1}, the value of x at $u = i\,\delta u$, $t = (n+1)\,\delta t$ in terms of values already available at $(i\,\delta u, n\,\delta t)$. The computation proceeds on a mesh of size δu, δt in the (u, t) plane (Fig. 5.4) and values of x_i^n are calculated at the mesh points using

$$x_i^{n+1} = x_i^n + 4\,\delta t\,\frac{(x_{i+1}^n - 2x_i^n + x_{i-1}^n)}{(x_{i-1}^n - x_{i+1}^n)^2} \tag{5.51}$$

in general, and from (5.49) on $u = 0$, $i = 0$,

$$s^{n+1} = s^n - \frac{\lambda\,\delta t\,\delta u}{x_0^n - x_1^n}. \tag{5.52}$$

A rigorous analysis of the stability of the non-linear difference scheme has not been attempted. Dix and Cizek (1970) applied the heuristic argument that since a virtual increase in x_i^n must produce an increase in

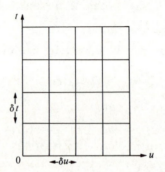

FIG. 5.4. IMM transformed plane

x_i^{n+1} so the coefficient of x_i^n in (5.51) must be positive and hence

$$\delta t < \tfrac{1}{8}(x_{i-1} - x_{i+1})^2. \tag{5.53}$$

Inclusion of the criterion (5.53) in the computer algorithm gives a simple way for the time step δt to be adjusted automatically as the solution proceeds. Dix and Cizek (1970) showed the truncation error of this IMM is proportional to δt and $(\delta u)^2$. Crank and Phahle (1973) showed satisfactory agreement between isotherm positions at selected times obtained by the IMM with those obtained from the well-known analytical solution (see (3.12), (3.14)).

Conditions on a fixed boundary, different from $u = 1$ on $x = 0$ in (5.45) for example, would call for some but not a major modification of the method. Thus a boundary condition $u = g(t)$ on $x = 0$, $t > 0$ would result in a known but curved boundary in the (u, t) plane on which $x = 0$. Conditions involving derivatives can also be handled. The condition $\partial u/\partial x = 0$ on $x = 0$ clearly causes difficulties in (5.48) but the value of u at which $x = 0$ at each time level can easily be found by expressing u as the parabola $u = ax^2 + b$, where a and b are obtained by inserting the computed values of x at the two mesh points neighbouring the boundary. This u value is then used in the solution of (5.48) at the next time level, albeit with an unequal interval in u.

5.4.2. IMM *in multi-dimensional problems*

(i) *Partial transformation of variables.* Crank and Gupta (1975) extended the IMM to the solution of general heat-flow problems in two space dimensions with particular reference to Stefan problems. As an example, they computed numerical results for the solidification of a square prism of fluid initially at the melting temperature, $u = 1$, throughout and the surface of which is subsequently maintained at $u = 0$ below the melting temperature.

Solutions are required of the equation

$$\frac{\partial u}{\partial t} = \frac{\partial^2 u}{\partial x^2} + \frac{\partial^2 u}{\partial y^2}, \tag{5.54}$$

subject to the boundary conditions

$$u = 0 \quad \text{on } g(x, y) = (x^2 - 1)(y^2 - 1) = 0, \qquad t \geq 0, \tag{5.55}$$

where the prism occupies the space $-1 \leq x \leq 1$, $-1 \leq y \leq 1$, and

$$u = 1 \quad \text{on} \quad f(x, y, t) = 0, \qquad t > 0 \tag{5.56}$$

where $f(x, y, t) = 0$ is the phase-change surface at time t. Initially we have

$$f(x, y, 0) \equiv g(x, y) = 0, \qquad t = 0. \tag{5.57}$$

Because of the symmetry of this problem, only the quadrant of the prism enclosed between $x = 1$, $y = 1$ and the axes $x = y = 0$ need be considered, and there are symmetry conditions

$$\left.\frac{\partial u}{\partial x}\right)_{x=0} = \left.\frac{\partial u}{\partial y}\right)_{y=0} = 0. \tag{5.58}$$

Crank and Gupta (1975) applied the IMM transformation to the y variable only in equation (5.54) and obtained

$$\frac{\partial y}{\partial t} = -\left\{\frac{\partial^2 u}{\partial x^2} - \frac{\partial^2 y}{\partial u^2}\left(\frac{\partial y}{\partial u}\right)^{-3}\right\}\frac{\partial y}{\partial u}, \tag{5.59}$$

using the relationships (5.47) with y written for x. The IMM form of the appropriate Patel expression (see equation (1.56) in §1.3.9) for the usual Stefan boundary condition $\partial u/\partial n = -\beta v_n$ is

$$\frac{\partial y}{\partial t} = \frac{1}{\beta}\left\{1 + \left(\frac{\partial y}{\partial x}\right)^2\right\}\left(\frac{\partial y}{\partial u}\right)^{-1}. \tag{5.60}$$

Values of y are calculated on a (u, x) mesh for successive time steps δt and $y_{i,j}^k$ signifies $y(i\,\delta u, j\,\delta x, k\,\delta t)$ in the usual way. Crank and Gupta replaced (5.60) by the explicit form

$$\frac{y_{N,j}^{k+1} - y_{N,j}^k}{\delta t} = \frac{1}{\beta}\left\{1 + \left(\frac{y_{N,j}^k - y_{N,j-1}^k}{\delta x}\right)^2\right\}\frac{\delta u}{y_{N,j}^k - y_{N-1,j}^k}, \tag{5.61}$$

where $N\,\delta u = 1$.

To be consistent with this first-order backward-difference approximation to $(\partial y/\partial u)^{-1}$ they approximated (5.59) also explicitly by

$$\frac{y_{i,j}^{k+1} - y_{i,j}^k}{\delta t} = -\left(\frac{y_{i,j}^k - y_{i-1,j}^k}{\delta u}\right)\frac{\partial^2 u}{\partial x^2} + \frac{y_{i-1,j}^k - 2y_{i,j}^k - y_{i+1,j}^k}{(y_{i,j}^k - y_{i-1,j}^k)^2}. \tag{5.62}$$

In order to discretize $(\partial^2 u/\partial x^2)_y$ at the point $(i\,\delta u, j\,\delta x)$ values of u are required, at three equally spaced values of x, for which $y = y_{i,j} = y(i\,\delta u, j\,\delta x)$. Crank and Gupta (1975) interpolated linearly the values of u corresponding to $y_{i,j}$ at x_{j-1} and x_{j+1}, e.g. at x_{j-1} they used

$$u = \frac{u_{i+1}(y_{i,j-1} - y_{i,j}) - u_i(y_{i+1,j-1} - y_{i,j})}{y_{i,j-1} - y_{i+1,j-1}}. \tag{5.63}$$

In order to take maximum advantage of the symmetry of their problem they used special interpolation and extrapolation procedures on the axes and on the line $y = x$. They used the one-parameter integral method of Poots (1962a) (see §3.5.6) to calculate the positions of some isotherms and the interface a short time after the start of solidification, and continued from there with the IMM. Reasonable agreement with earlier

TABLE 5.1

Values of the y coordinate on the solid–liquid interface for fixed values of x at various times. Solution starts from the values taken from the Poots (1962a) one-parameter method at $t = 0.0461$. $\delta x = \delta u = 0.1$; $\delta t = 0.0001$, $\beta = 1.561$

	x						
t	0.0	0.1	0.2	0.3	0.4	0.5	0.6
0.05	0.8125	0.8106	0.8048	0.7940	0.7764	0.7476	0.6904
0.10	0.6979	0.6965	0.6921	0.6836	0.6683	0.6392	0.5606
0.15	0.6157	0.6141	0.6095	0.6000	0.5810	0.5201	
0.20	0.5473	0.5453	0.5394	0.5268	0.4789		
0.25	0.4865	0.4838	0.4755	0.4567	0.3894		
0.30	0.4302	0.4263	0.4146	0.3654			
0.35	0.3766	0.3708	0.3534	0.2859			
0.40	0.3337	0.3158	0.2623				
0.45	0.2816	0.2585	0.1893				
0.495	0.2376	0.2056	0.1097				

results by Allen and Severn (1962) and Lazaridis (1970) was demonstrated. Their values for the movement of the freezing front are given in Table 5.1 and their isotherms are shown in Fig. 5.5.

Crank and Gupta (1975) discussed how the fixed boundary conditions influence the choice of $\partial^2 u/\partial y^2$ or $\partial^2 u/\partial x^2$ as the term to which the IMM transformation should best be applied. They also indicated how alternative conditions on the fixed boundaries can be handled.

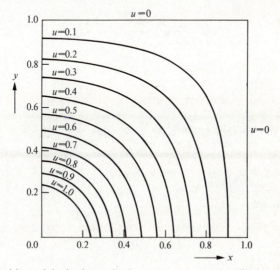

FIG. 5.5. Positions of the isotherms having temperatures $u = 0.0(0.1)1.0$ at $t = 0.495$

A similar extension of IMM to two dimensions was used by Durack and Wendroff (1977).

(ii) *Complete transformation of variables.* The interpolation used to evaluate $\partial^2 u/\partial x^2$ in (5.59) becomes more burdensome in three space dimensions and, in any case, approximations such as (5.63) are of lower accuracy than the replacement of the other second derivative. An alternative is to generalize the IMM for multi-dimensional problems by using a complete transformation, in which one space variable is chosen as the new dependent variable and the independent variables are the remaining space coordinates and temperature.

The complete transformation is conveniently performed following, for example, Boadway's (1976) treatment of fluid flow problems. The equation for heat flow in a homogeneous medium in which the heat conductivity K and specific heat c may be functions of temperature u, and ρ is the density can be written

$$c\rho \frac{\partial u}{\partial t} = \frac{\partial}{\partial x}\left(K\frac{\partial u}{\partial x}\right) + \frac{\partial}{\partial y}\left(K\frac{\partial u}{\partial y}\right) + \frac{\partial}{\partial z}\left(K\frac{\partial u}{\partial z}\right). \tag{5.64}$$

Here, temperature u is a function of the space coordinates (x, y, z) at a given time i.e.

$$u = u(x, y, z), \tag{5.65}$$

and hence

$$du = \frac{\partial u}{\partial x}dx + \frac{\partial u}{\partial y}dy + \frac{\partial u}{\partial z}dz. \tag{5.66}$$

We introduce two dummy variables ϕ and ζ with

$$\phi = \phi(x, y, z), \qquad \zeta = \zeta(x, y, z) \tag{5.67}$$

and relations equivalent to (5.66) for $d\phi$ and $d\zeta$. We require new functions

$$x = x(\phi, \zeta, u), \qquad y = y(\phi, \zeta, u), \qquad z = z(\phi, \zeta, u) \tag{5.68}$$

for which we have

$$dx = \frac{\partial x}{\partial \phi}d\phi + \frac{\partial x}{\partial \zeta}d\zeta + \frac{\partial x}{\partial u}du, \tag{5.69}$$

and similarly for dy, dz. Then by using (5.69) together with (5.66) and the corresponding expressions for $d\phi$ and $d\zeta$ we obtain

$$
dx = \frac{\partial x}{\partial u}\left(\frac{\partial u}{\partial x}dx + \frac{\partial u}{\partial y}dy + \frac{\partial u}{\partial z}dz\right)
$$
$$
+ \frac{\partial x}{\partial \phi}\left(\frac{\partial \phi}{\partial x}dx + \frac{\partial \phi}{\partial y}dy + \frac{\partial \phi}{\partial z}dz\right)
$$
$$
+ \frac{\partial x}{\partial \zeta}\left(\frac{\partial \zeta}{\partial x}dx + \frac{\partial \zeta}{\partial y}dy + \frac{\partial \zeta}{\partial z}dz\right), \tag{5.70}
$$

together with two equivalent expressions, one for dy and one for dz.

By collecting the dx terms in the three expressions we obtain

$$
\begin{pmatrix}
\dfrac{\partial x}{\partial \phi} & \dfrac{\partial x}{\partial \zeta} & \dfrac{\partial x}{\partial u} \\[2ex]
\dfrac{\partial y}{\partial \phi} & \dfrac{\partial y}{\partial \zeta} & \dfrac{\partial y}{\partial u} \\[2ex]
\dfrac{\partial z}{\partial \phi} & \dfrac{\partial z}{\partial \zeta} & \dfrac{\partial z}{\partial u}
\end{pmatrix}
\begin{pmatrix}
\dfrac{\partial \phi}{\partial x} \\[2ex]
\dfrac{\partial \zeta}{\partial x} \\[2ex]
\dfrac{\partial u}{\partial x}
\end{pmatrix}
=
\begin{pmatrix}
1 \\[2ex]
0 \\[2ex]
0
\end{pmatrix}
\tag{5.71}
$$

and hence, on solving,

$$
\frac{\partial \phi}{\partial x} = \left(\frac{\partial y}{\partial \zeta}\frac{\partial z}{\partial u} - \frac{\partial z}{\partial \zeta}\frac{\partial y}{\partial u}\right)\Big/ A, \tag{5.72}
$$

where A is the determinant of the matrix in (5.71), and similarly for $\partial \zeta/\partial x$ and $\partial u/\partial x$.

Collecting dy terms yields corresponding expressions for $\partial\phi/\partial y$, $\partial\zeta/\partial y$, $\partial u/\partial y$, and the dz terms give $\partial\phi/\partial z$, $\partial\zeta/\partial z$, $\partial u/\partial z$. Proceeding to second derivatives we require, for example,

$$
\frac{\partial}{\partial x}\left(K\frac{\partial u}{\partial x}\right) = \frac{\partial}{\partial \phi}\left(K\frac{\partial u}{\partial x}\right)\frac{\partial \phi}{\partial x} + \frac{\partial}{\partial \zeta}\left(K\frac{\partial u}{\partial x}\right)\frac{\partial \zeta}{\partial x} + \frac{\partial}{\partial u}\left(K\frac{\partial u}{\partial x}\right)\frac{\partial u}{\partial x}, \tag{5.73}
$$

into which expressions of type (5.72) are to be substituted for all the first derivatives. Similar general expressions can be obtained for $\partial(K \, \partial u/\partial y)/\partial y$ and $\partial(K \, \partial u/\partial z)/\partial z$.

Finally, for the IMM we wish to interchange temperature u with one space variable, say z, i.e. to write $z = z(x, y, u)$, and reference to the last of (5.68) shows that we must now identify ϕ with x and ζ with y in the above analysis. When this is done in the development of (5.73) and corresponding expressions for $\partial(K \, \partial u/\partial y)/\partial y$ and $\partial(K \, \partial u/\partial z)/\partial z$ noting that $A = \partial z/\partial u$ and also that

$$
\left(\frac{\partial z}{\partial t}\right)_u = -\left(\frac{\partial z}{\partial u}\right)\left(\frac{\partial u}{\partial t}\right), \tag{5.74}
$$

the heat eqn (5.64) can be written finally as

$$c\rho \frac{\partial z}{\partial t} = \left(\frac{\partial z}{\partial u}\right)\left\{\frac{\partial}{\partial x}\left(k\frac{\partial z}{\partial x}\right)+\frac{\partial}{\partial y}\left(k\frac{\partial z}{\partial y}\right)\right\}$$
$$-\frac{\partial k}{\partial u}-\left(\frac{\partial z}{\partial x}\right)\frac{\partial}{\partial u}\left(k\frac{\partial z}{\partial x}\right)-\frac{\partial z}{\partial y}\frac{\partial}{\partial u}\left(k\frac{\partial z}{\partial y}\right), \tag{5.75}$$

where for convenience k denotes $K/(\partial z/\partial u)$. In the special case of K, c, ρ constant, (5.75) reduces to

$$\frac{c\rho}{K}\frac{\partial z}{\partial t} = \left[\left(\frac{\partial^2 z}{\partial x^2}+\frac{\partial^2 z}{\partial y^2}\right)\left(\frac{\partial z}{\partial u}\right)^2 + \frac{\partial^2 z}{\partial u^2}\left\{1+\left(\frac{\partial z}{\partial x}\right)^2+\left(\frac{\partial z}{\partial y}\right)^2\right\}\right.$$
$$\left. -2\frac{\partial^2 z}{\partial x\,\partial u}\left(\frac{\partial z}{\partial x}\right)\left(\frac{\partial z}{\partial u}\right)-2\frac{\partial^2 z}{\partial y\,\partial u}\left(\frac{\partial z}{\partial y}\right)\left(\frac{\partial z}{\partial u}\right)\right]\left(\frac{\partial z}{\partial u}\right)^{-2}. \tag{5.76}$$

An iterative finite-difference solution of the steady-state form of this equation in a three-dimensional free-boundary problem is developed for eqn (8.70) in §8.4.2.

In a Stefan moving-boundary problem, a three-dimensional extension of Patel's condition (5.60), to be used in conjunction with (5.76) and other fixed-boundary conditions, is

$$\frac{\partial z}{\partial t}=\frac{1}{\beta}\left\{1+\left(\frac{\partial z}{\partial x}\right)^2+\left(\frac{\partial z}{\partial y}\right)^2\right\}\left(\frac{\partial z}{\partial u}\right)^{-1}. \tag{5.77}$$

Turland (1979) derived (5.75) by performing the complete transformation in a manner directly relevant to the present problem. He then regrouped the terms in (5.75) by using the identities, in subscript notation,

$$z_u\frac{\partial}{\partial x}\left(\frac{K}{z_u}z_x\right)=\frac{\partial}{\partial x}(Kz_x)-\frac{z_{ux}Kz_x}{z_u} \tag{5.78}$$

and

$$z_x\frac{\partial}{\partial u}\left(\frac{K}{z_u}z_x\right)=\frac{\partial}{\partial u}\left(\frac{K}{z_u}z_x^2\right)-\frac{K}{z_u}z_x z_{ux}, \tag{5.79}$$

where K/z_u has been reinserted for k. When (5.78) and (5.79) are substituted into (5.75) together with an equivalent grouping of the corresponding y terms, (5.75) can be written in the form

$$c\rho\frac{\partial z}{\partial t}=-\left(\frac{\partial\Phi^u}{\partial u}+\frac{\partial\Phi^x}{\partial x}+\frac{\partial\Phi^y}{\partial y}\right), \tag{5.80}$$

where

$$\Phi^u=K(1+z_x^2+z_y^2)/z_u, \qquad \Phi^x=-Kz_x, \qquad \Phi^y=-Kz_y. \tag{5.81}$$

Physically, the Φ terms represent heat fluxes and Turland (1979) preferred the form (5.81) because it leads readily to difference schemes in which the heat content of the system is conserved, which may not be true of numerical schemes based on (5.75). Conservative difference schemes of this kind were derived by Potter (1973) for the heat-flow equation in its usual form, by considering the heat balance of a volume element. Turland (1981) derived (5.80) directly in the same way by considering the fluxes into and out of the volume bounded by the planes in which u and $u + du$, α_1 and $\alpha_1 + d\alpha_1$, α_2 and $\alpha_2 + d\alpha_2$ are constants where, for generality, α_1, α_2, α_3 are taken to be curvilinear coordinates and u still denotes temperature. The thermal properties of the medium are allowed to be temperature dependent, but otherwise the medium is homogeneous.

In the orthogonal curvilinear system the metric is defined by

$$ds^2 = h_1^2 \, d\alpha_1^2 + h_2^2 \, d\alpha_2^2 + h_3^2 \, d\alpha_3^2, \tag{5.82}$$

and a volume of coordinate space by

$$V = \int_{\alpha_2}^{\alpha_2 + d\alpha_2} \int_{\alpha_3}^{\alpha_3 + d\alpha_3} Z \, d\alpha_2' \, d\alpha_3' \tag{5.83}$$

where

$$Z = \int_{\alpha_1 = 0}^{\alpha_1(u)} h_1 h_2 h_3 \, d\alpha_1'. \tag{5.84}$$

The intention is to interchange the curvilinear coordinate α_1 with temperature u. The functional integral $I(u, u + \Delta u, \alpha_2, \alpha_2 + \Delta\alpha_2, \alpha_3, \alpha_3 + \Delta\alpha_3, t)$ is therefore defined by

$$I = \int_u^{u + \Delta u} \rho c V \, du' = [hV]_u^{u + \Delta u} - \int_{V(u)}^{V(u + \Delta u)} h \, dV' \tag{5.85}$$

where $h = \int_0^{u(\alpha_1)} \rho c \, du'$. Thus, correspondingly,

$$\frac{\partial I}{\partial t} = \iiint \rho c \frac{\partial Z}{\partial t} \, d\alpha_2 \, d\alpha_3 \, du = -\int \rho c \frac{\partial u}{\partial t} \, dV'. \tag{5.86}$$

The partial derivatives of I and Z with respect to time are taken with u and $u + \Delta u$ constant, though the corresponding value of α_1 may change. The appropriate heat flow equation is

$$\rho c \frac{\partial u}{\partial t} = \nabla \cdot (K \nabla u) \tag{5.87}$$

and hence Gauss's theorem coupled with the second form of $\partial I / \partial t$ in (5.86) yields

$$\frac{\partial I}{\partial t} = -\oiint K \nabla u \cdot dS, \tag{5.88}$$

where the integral is taken over the surface of the volume ΔV formed by the intersections of the surfaces u and $u + \Delta u$, α_2 and $\alpha_2 + \Delta \alpha_2$, α_3 and $\alpha_3 + \Delta \alpha_3$. This volume is not a cuboid in shape because its specification includes the pair of surfaces of constant u values rather than constant α_1. The heat flow per unit area at a given point, perpendicular to planes of constant u, is

$$-K \nabla u = -K \left\{ \frac{1}{h_1} \frac{\partial u}{\partial \alpha_1} \hat{\boldsymbol{\alpha}}_1 + \frac{1}{h_2} \frac{\partial u}{\partial \alpha_2} \hat{\boldsymbol{\alpha}}_2 + \frac{1}{h_3} \frac{\partial u}{\partial \alpha_3} \hat{\boldsymbol{\alpha}}_3 \right\}, \qquad (5.89)$$

where $\hat{\boldsymbol{\alpha}}_1, \hat{\boldsymbol{\alpha}}_2, \hat{\boldsymbol{\alpha}}_3$ are unit vectors. The element of area $dS^{(u)}$ of the constant u surfaces is

$$dS^{(u)} = h_1 h_2 h_3 \, \Delta \alpha_2 \, \Delta \alpha_3 \frac{\nabla u}{\partial u / \partial \alpha_1},$$

where $\partial u / \partial \alpha_1$ is taken at constant α_2, α_3, and t. Thus, the net contribution to the heat flux from the sides u and $u + \Delta u$ of the volume element is, to first order,

$$\frac{\partial I^{(u)}}{\partial t} = -\frac{\partial}{\partial u} \left\{ (K \nabla u) h_1 h_2 h_3 \, \Delta \alpha_2 \, \Delta \alpha_3 \frac{\nabla u}{\partial u / \partial \alpha_1} \right\} \Delta u. \qquad (5.90)$$

For the constant α_2 surfaces

$$dS^{(\alpha_2)} = h_1 h_3 \, \Delta \alpha_1 \, \Delta \alpha_3 \hat{\boldsymbol{\alpha}}_2 = \frac{h_1 h_3}{\partial u / \partial \alpha_1} \Delta u \, \Delta \alpha_3 \hat{\boldsymbol{\alpha}}_2$$

which, when coupled with (5.89), gives the net contribution

$$\frac{\partial I^{(\alpha_2)}}{\partial t} = -K \frac{\partial}{\partial \alpha_2} \left\{ \frac{1}{h_2} \frac{\partial u}{\partial \alpha_2} \frac{h_1 h_3}{\partial u / \partial \alpha_1} \right\} \Delta u \, \Delta \alpha_3 \, \Delta \alpha_2. \qquad (5.91)$$

The expression for $\partial I^{(\alpha_3)} / \partial t$ follows similarly. Finally, by collecting terms we find

$$\frac{\partial I}{\partial t} = \frac{\partial I^{(u)}}{\partial t} + \frac{\partial I^{(\alpha_2)}}{\partial t} + \frac{\partial I^{(\alpha_3)}}{\partial t}$$

$$= -\left[\frac{\partial}{\partial u} \left(\frac{h_1 h_2 h_3 K |\nabla u|^2}{\partial u / \partial \alpha_1} \right) + K \frac{\partial}{\partial \alpha_2} \left(\frac{h_1 h_3}{h_2} \frac{\partial u / \partial \alpha_2}{\partial u / \partial \alpha_1} \right) \right.$$

$$\left. + K \frac{\partial}{\partial \alpha_3} \left(\frac{h_1 h_2}{h_3} \frac{\partial u / \partial \alpha_3}{\partial u / \partial \alpha_1} \right) \right] \Delta u \, \Delta \alpha_2 \, \Delta \alpha_3. \qquad (5.92)$$

To complete the interchange of u and α_1, which is to become the new dependent variable, we write

$$\partial u / \partial \alpha_1 = 1 / (\partial \alpha_1 / \partial u) \qquad \text{and} \qquad (\partial u / \partial x)_{\alpha_1} = -(\partial \alpha_1 / \partial x) / (\partial \alpha_1 / \partial u). \quad (5.93)$$

Then, in the limit $\Delta u, \Delta\alpha_2, \Delta\alpha_3$ all approaching zero the first of (5.86) becomes

$$\frac{\partial I}{\partial t} = \rho c \frac{\partial Z}{\partial t} \Delta\alpha_2 \, \Delta\alpha_3 \, \Delta u \qquad (5.94)$$

and differentiation of (5.84) gives

$$\partial Z/\partial t = h_1 h_2 h_3 \, \partial\alpha_1/\partial t. \qquad (5.95)$$

Inserting (5.89) and (5.93), (5.94), and (5.95) in (5.92) we obtain the partial differential equation for $\alpha_1(\alpha_2, \alpha_3, u, t)$ which can be written in the conservative form of (5.80), i.e.

$$-\rho c h_1 h_2 h_3 \frac{\partial\alpha_1}{\partial t} = \frac{\partial\Phi^{(u)}}{\partial u} + \frac{\partial\Phi^{(\alpha_2)}}{\partial\alpha_2} + \frac{\partial\Phi^{(\alpha_3)}}{\partial\alpha_3}, \qquad (5.96)$$

where

$$\Phi^{(u)} = \frac{h_2 h_3}{h_1} \frac{K}{\partial\alpha_1/\partial u} \left\{ 1 + \frac{h_1^2}{h_2^2} \left(\frac{\partial\alpha_1}{\partial\alpha_2}\right)^2 + \frac{h_1^2}{h_3^2} \left(\frac{\partial\alpha_1}{\partial\alpha_3}\right)^2 \right\},$$

$$\Phi^{(\alpha_2)} = \frac{-h_1 h_3}{h_2} K \frac{\partial\alpha_1}{\partial\alpha_2} \qquad \Phi^{(\alpha_3)} = -\frac{h_1 h_2}{h_3} K \frac{\partial\alpha_1}{\partial\alpha_3}. \qquad (5.97)$$

Special cases of practical interest can easily be extracted from (5.96) and (5.97). For example, on putting $h_1 = h_2 = h_3 = 1$ and identifying α_1 with z, α_2 with x, and α_3 with y we regain (5.80) and (5.81) which were seen to be equivalent to (5.75). The usual heat-flow equation in one space variable can be written in the form

$$\rho c \frac{\partial u}{\partial t} = \frac{1}{r^p} \frac{\partial}{\partial r} \left(K r^p \frac{\partial u}{\partial r} \right),$$

where for plane geometry $p = 0$ and $r = x$, and for cylindrical and spherical symmetry r is the radial coordinate and $p = 1$ or 2 respectively. For plane geometry $h_1 = h_2 = h_3$, for cylindrical coordinates $h_1 = 1$, $h_2 = r$, $h_3 = 1$, and for spherical polar coordinates $h_1 = 1$, $h_2 = r$, $h_3 = r \sin\theta$. Thus IMM expressions in three dimensions for the plane, cylindrical, and spherical geometries are easily deduced from (5.96) and (5.97) by appropriate choice of h_1, h_2, h_3. In particular, the IMM expressions in one space variable become

$$\rho c \frac{\partial r}{\partial t} = -\frac{1}{r^p} \frac{\partial}{\partial u} \left(\frac{K r^p}{\partial r/\partial u} \right), \qquad (5.98)$$

with $p = 0, 1, 2$ as above.

The two-dimensional case in spherical polar coordinates (of interest for example in studies of the melt-pool growth in a reactor substrate (Turland

1979)) is the case of $h_1 = 1$, $h_2 = r$, $h_3 = r \sin \theta$, and in (5.96), (5.97), $\alpha_1 = r$, $\alpha_2 = \theta$, and α_3 terms are ignored. Thus (5.96) becomes

$$-\rho c r^2 \sin \theta \frac{\partial r}{\partial t} = \frac{\partial}{\partial u} \left[\frac{K r^2 \sin \theta}{\partial r / \partial u} \left\{ 1 + \frac{1}{r^2} \left(\frac{\partial r}{\partial \theta} \right)^2 \right\} \right] - \frac{\partial}{\partial \theta} \left(K \sin \theta \frac{\partial r}{\partial \theta} \right). \quad (5.99)$$

In order to illustrate how equation (5.96), written in terms of the heat fluxes Φ, leads to a conservative form of difference equation, we consider the discretization of the one-dimensional plane case of (5.96) and its derivation as presented above. Thus the one-dimensional analogue of (5.85) may be written (Turland 1981)

$$I(u_1, u_2, t) = \int_{u_1}^{u_2} \rho c x \, du,$$

and hence, using (5.98) with $p = 0$,

$$\frac{\partial I}{\partial t} = \int_{u_1}^{u_2} \rho c \frac{\partial x}{\partial t} \, du = -\int_{u_1}^{u_2} \frac{\partial}{\partial u} \left(\frac{K}{\partial x / \partial u} \right) du = -\left[\frac{K}{\partial x / \partial u} \right]_{u_1}^{u_2}. \quad (5.100)$$

Defining the heat content per unit area, H, by $H(x_1, x_2, t) = \int_{x_1}^{x_2} h \, dx$ and $h = \int_0^{u(x)} \rho c \, du$, then the equivalent of (5.85) is

$$I = [hx]_{u_1}^{u_2} - H = (h_2 x_2 - h_1 x_1) - H, \quad (5.101)$$

where x_1, h_1 correspond to u_1, etc. It follows from (5.101) that if x_1, u_1 and x_2, u_2 are fixed (being the sides of a volume element) then conservation of the quantity I implies conservation of total heat.

In order now to set up difference equations for numerical solution we can consider the set of isotherms $u_i = u_0 + i \, \Delta u$ for $0 \leqslant i \leqslant j$ say. Then use of (5.100) gives

$$\frac{\partial I}{\partial t} (u_{i-\frac{1}{2}}, u_{i+\frac{1}{2}}, t) = -\left[\frac{K}{\partial x / \partial u} \right]_{u_{i-\frac{1}{2}}}^{u_{i+\frac{1}{2}}} \quad (5.102)$$

and

$$\frac{\partial I}{\partial t} (u_{\frac{1}{2}}, u_{j-\frac{1}{2}}, t) = -\left[\frac{K}{\partial x / \partial u} \right]_{u_{\frac{1}{2}}}^{u_{j-\frac{1}{2}}}. \quad (5.103)$$

We approximate by using

$$\left. \begin{aligned} I(u_{i-\frac{1}{2}}, u_{i+\frac{1}{2}}, t) &= \rho c_i x_i \, \Delta u = I_i^* \\ \frac{-K}{\partial x / \partial u} \right)_{i+\frac{1}{2}} &= \frac{-K_{i+\frac{1}{2}} \Delta u}{x_{i+1} - x_i} = \mu_{i+\frac{1}{2}}^*, \end{aligned} \right\} \quad (5.104)$$

and therefore (5.102) and (5.104) become

$$\frac{\partial I_i^*}{\partial t} = \mu_{i+\frac{1}{2}}^* - \mu_{i-\frac{1}{2}}^*, \quad \frac{dx_i}{dt} = \frac{1}{\rho c_i} \left\{ \frac{K_{i-\frac{1}{2}}}{x_i - x_{i-1}} - \frac{K_{i+\frac{1}{2}}}{x_{i+1} - x_i} \right\}. \quad (5.105)$$

By summing the first of (5.105) for $1 \leqslant i \leqslant j-1$, the difference form of (5.103) is

$$\frac{d}{dt} \sum_{i=1}^{j-1} I_i^* = \mu_{j-\frac{1}{2}}^* - \mu_{\frac{1}{2}}^*, \tag{5.106}$$

which is satisfied exactly and so the method is conservative.

Turland (1981) referred to numerical experiments based on a simple, explicit replacement of dx_i/dt in the second of (5.105). Use of an implicit scheme, however, has the usual advantage of removing the stability restriction on the time step. He proposed the linearized scheme in the time interval t^n to t^{n+1}

$$\frac{x_i^{n+1} - x_i^n}{t^{n+1} - t^n} = \tfrac{1}{2}\{a_i^n x_{i-1}^n + b_i^n x_i^n + c_i^n x_{i+1}^n + a_i^* x_{i-1}^{n+1} + b_i^* x_i^{n+1} + c_i^* x_{i+1}^{n+1}\}, \tag{5.107}$$

where

$$a_i = K_{i-\frac{1}{2}}/A_i, \qquad c_i = K_{i+\frac{1}{2}}/A_i, \qquad b_i = -(a_i + c_i), \tag{5.108}$$

$$A_i = \rho c_i (x_i - x_{i-1})(x_{i+1} - x_i), \tag{5.109}$$

$$x_i^* = x_i^n + (x_i^n - x_i^{n-1})(t^{n+1} - t^n)/(t^n - t^{n-1}). \tag{5.110}$$

The values of x_i^* calculated from the predictor formula (5.110) are inserted into (5.109) and (5.108) to evaluate the coefficients a_i^*, b_i^*, c_i^* in (5.107). With appropriate discretization of boundary conditions the values of x_i^{n+1} are then the solution of a linear, tridiagonal system of equations of type (5.107). The method is stable for all time steps and the criterion $|x_i^{n+1} - x_i^*| < \varepsilon$ for all i can be used to choose the time step. Turland (1982) solved a number of test problems satisfactorily.

(iii) *Isotherm migration along orthogonal flow lines.* Crank and Crowley (1978) described a novel approach to the solution of transient heat-flow problems in two dimensions. The movements of isotherms along orthogonal flow lines are tracked in successive small intervals of time by solving a locally one-dimensional IMM form of a radial heat equation. The changing shape and orientation of the orthogonal system are catered for by a geometric procedure. The procedure is analogous to the Lagrangian formulation of fluid flow, in which the motion of particular particles of fluid is calculated, rather than the velocity distribution at fixed points in space.

In non-dimensional form, the equation of heat flow in cylindrical polar coordinates (r, θ) may be written

$$\frac{\partial u}{\partial t} = \frac{\partial^2 u}{\partial r^2} + \frac{1}{r}\frac{\partial u}{\partial r} + \frac{1}{r^2}\frac{\partial^2 u}{\partial \theta^2}, \tag{5.111}$$

where u denotes temperature and constant thermal properties are assumed. The radial coordinate, r, is measured from an origin that remains fixed as time, t, changes. By analogy with (5.59) an IMM form of (5.111) is

$$\frac{\partial r}{\partial t} = \left(\frac{\partial^2 r}{\partial u^2}\right) \bigg/ \left(\frac{\partial r}{\partial u}\right)^2 - \frac{1}{r} - \frac{1}{r^2}\left(\frac{\partial r}{\partial u}\right)\frac{\partial^2 u}{\partial\theta^2}. \tag{5.112}$$

In general, flow of heat in two space dimensions can be represented by a family of isotherms and an associated family of flow lines, orthogonal to the isotherms. In an isotropic medium, any point on an isotherm moves along the flow line normal to the isotherm at that point. Heat flow is everywhere normal to the isotherms and never across flow lines. By confining attention to a small segment of an isotherm for a short interval of time, we can regard the isotherm element as part of a cylindrical system and identify the coordinate r, in (5.112), as the local radius of curvature of the isotherm measured from the local centre of curvature assumed fixed in its position at time t. Equation (5.112) yields the velocity $\partial r/\partial t$ of the selected element along the normal to itself.

Because the general system is distorting and rotating, both the centre of curvature for the element and the curvature itself may change with time as well as from point to point in the system. The flow lines will not strictly be radial lines of constant θ, and the local isotherms will not be exactly concentric circular arcs. But, because r in (5.112) is chosen to be along the local normal, the term $\partial^2 u/\partial\theta^2$ will be small in general, but non-zero. Crank and Crowley (1978) approximated it to zero and calculated the movement of a point on an isotherm along its normal, in a small time Δt, by solving the equation

$$\frac{\partial r}{\partial t} = \left(\frac{\partial^2 r}{\partial u^2}\right) \bigg/ \left(\frac{\partial r}{\partial u}\right)^2 - \frac{1}{r}. \tag{5.113}$$

They describe an explicit finite-difference method based on three typical isotherms (Fig. 5.6) on which the temperatures are $(j-1)\,\delta u$, $j\,\delta u$, and

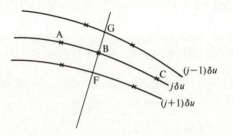

FIG. 5.6. Sketch showing relative positions of isotherms

$(j+1)\,\delta u$ and A, B, C are three points whose coordinates are known on isotherm $j\,\delta u$. The points G and F are the intersections of the radius r_m with the chords approximating the isotherms $(j-1)\,\delta u$ and $(j+1)\,\delta u$. Let $n_m = |r_m|$ be the distance of B from the centre of curvature of the arc ABC and let n_m^+ and n_m^- denote the distances of F and G from this same centre respectively. Then the usual approximations are

$$\left.\frac{\partial n}{\partial u}\right)_B = \frac{n_m^+ - n_m^-}{2\,\delta u}, \qquad \left.\frac{\partial^2 n}{\partial u^2}\right)_B = \frac{n_m^+ - 2n_m + n_m^-}{(\delta u)^2}. \tag{5.114}$$

In the interests of accuracy, it is preferable to calculate the differences $n_m^+ - n_m^-$, $n_m^+ - n_m$ and $n_m - n_m^-$ directly from the coordinates of the points B, F, and G which are known at time t. Thus, denoting by $\Delta n_{j,m}$ the movement of the point m on the isotherm $j\,\delta u$ along the normal in the small time interval from t to $t + \Delta t$, Crank and Crowley (1978) replaced (5.113) explicitly by

$$\Delta n_{j,m} = 4\,\Delta t \frac{(n_m^+ - 2n_m + n_m^-)}{(n_m^+ - n_m^-)^2} \frac{\Delta t}{n_m}. \tag{5.115}$$

If the coordinates of the point m on isotherm $j\,\delta u$ at time $t = i\,\Delta t$ are denoted by $x_{j,m}^i$, $y_{j,m}^i$, then the new coordinates at time $(i+1)\,\Delta t$ are

$$\left. \begin{aligned} x_{j,m}^{i+1} &= x_{j,m}^i + \Delta n_{j,m}^i \cos \theta_{j,m}^i, \\ y_{j,m}^{i+1} &= y_{j,m}^i + \Delta n_{j,m}^i \sin \theta_{j,m}^i, \end{aligned} \right\} \tag{5.116}$$

where $\theta_{j,m}^i$ denotes the direction of the normal to the isotherm at B measured from the x-axis, as in Fig. 5.7. It remains to calculate $\theta_{j,m}^i$ from the coordinates of the points A, B, C, and also the length of the radius of curvature, n_m, at B (x_m, y_m).

Figure 5.7 shows a section of an isotherm approximated by a circular arc ABC which is concave downwards and to which is assigned a positive

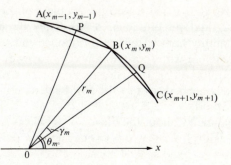

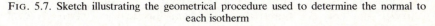

FIG. 5.7. Sketch illustrating the geometrical procedure used to determine the normal to each isotherm

curvature. The tangents at the mid-points P, Q of each of the arcs AB, BC are parallel to the corresponding chords AB, BC. Thus the change in the direction of the tangent along arc PBQ is given by $\psi_m - \psi_{m+1}$, where the ψs are the angles made by the perpendicular bisectors of the chords AB and BC respectively with the x-axis in Fig. 5.7. The arc length PBQ may be approximated by $\frac{1}{2}(s_{m+1} + s_m)$, where s_m denotes the length of the chord AB, labelled chord m. Then, the radius of curvature r_m at B (x_m, y_m) may be written

$$n_m = \left(\frac{\partial s}{\partial \psi}\right)_m = \frac{1}{2}\left(\frac{s_m + s_{m+1}}{\psi_m - \psi_{m+1}}\right). \tag{5.117}$$

By regarding BQ as a circular arc of radius $|r_m| = n_m$ and length $\frac{1}{2}s_{m+1}$, the angle marked γ_m on Fig. 5.7 may be approximated as $\frac{1}{2}s_{m+1}/n_m$. Hence the angle $\theta_{j,m}^i$ in (5.116), illustrated by θ_m in Fig. 5.7, can be calculated from

$$\theta_m = \psi_{m+1} + \gamma_m \, \mathrm{sign}(r_m) = \psi_{m+1} + s_{m+1}/(2r_m). \tag{5.118}$$

The angles ψ and distances s are readily computed from the coordinates of the points A, B, C.

If the initial data in a two-dimensional heat-flow problem are given as the coordinates of a set of points on each of a number of isotherms, the relationships (5.115–118) provide the basis of a numerical algorithm to advance each point in a succession of small time steps, Δt.

In a Stefan problem in which the phase-change boundary is itself an isothermal surface, its motion is readily calculated by the IMM form of the usual Stefan condition, e.g.

$$\frac{\partial n}{\partial t} = -\frac{1}{\beta}\left(\frac{\partial n}{\partial u}\right)^{-1}, \tag{5.119}$$

in a one-phase problem, which, suitably discretized, is used in place of (5.115).

Crank and Crowley (1978) solved the problem of the solidification of a square prism of fluid set out in §5.4.2(i). They discussed questions of accuracy and stability heuristically and obtained numerical results which are in good agreement with those obtained by Crowley (1978) (see §§6.2, 6.2.4) and by Elliott (1976) as shown in Tables 6.12, 13. The initial isotherm positions needed to start the IMM calculation were obtained by the enthalpy method.

Because of symmetry about the diagonal $y = x$ it is sufficient to work in the triangular region $0 \leqslant x \leqslant y$, $0 \leqslant y \leqslant 1$. The end points on the boundaries $x = 0$ and $y = x$ need special consideration. A circle, with its centre on the axis at $(0, y_0)$ or on the diagonal at (a, a) as appropriate, is fitted through the end point and the next point inside the region on each isotherm, in order to find the curvatures. The angles $\theta_{j,0} = \pi/2$ on the y-axis and $\theta_{j,M}$, say, $= \pi/4$ on the diagonal are already known. Thus on the axis, $x^2 + (y - y_0)^2 = r^2$, and we find from $(0, y_{j,0})$, $(x_{j,1}, y_{j,1})$ that $r = [(y_{j,0} - y_{j,1})^2 + x_{j,1}^2]/2(y_{j,0} - y_{j,1})$. On the diagonal the use of $(x - a)^2 + (y - a)^2 = r^2$ yields

$$r = \{(x_{j,M} - x_{j,M-1})^2 + (y_{j,M} - y_{j,M-1})^2\}/2(x_{j,M} + y_{j,M} - x_{j,M-1} - y_{j,M-1}),$$

and one or other of these expressions for r replaces (5.117).

Crank and Crowley (1979) described a corresponding implicit method for tracking isotherms along orthogonal flow lines, using a linearized form of equation (5.113). In general, the implicit scheme offered the usual advantage over the explicit method in enabling the use of longer time steps. However, they choose to illustrate their implicit scheme by considering the problem of Bonnerot and Jamet (1977) specified in equations (4.24–8). This problem is a more exacting test of Crank and Crowley's method for two important reasons. Firstly, the radius of curvature is positive at the left-hand end of each isotherm, and negative as the isotherm approaches $x = 1$ (Fig. 4.4(a)). Therefore, proceeding in the direction of x increasing along any isotherm the curvature increases from finite positive values, through infinity, and then from negative infinity to finite negative values, the central part of each isotherm being virtually straight. In such a situation, care is necessary to ensure that the finite-difference equations are written correctly when $r_{j,m} < 0$ and $n_{j,m} = -r_{j,m}$.

The second difficulty is associated with the initial temperature distribution. On the left boundary, $x = 0$, $\partial r/\partial t$ in (5.113) can be identified with $\partial y/\partial t$ since on $x = 0$ the isotherms are concave downwards and the centres of curvature are on $x = 0$. When the necessary derivatives are obtained from the second of the conditions (4.27) and substituted into equation (5.113) we find that on $x = 0$, $\partial y/\partial t = -\pi^2(1 - u) \leq 0$, $0 < u < 1$. However, from the first of (4.28) on the phase-change boundary, $u = 0$, we have $\partial y/\partial t = 1/3\lambda$ on $x = 0$. Thus, at $t = 0$, all the isotherms except $u = 0$ initially move in the direction of y decreasing at the left boundary $x = 0$, whereas the phase-change isotherm, $u = 0$, moves upwards. In fact, there is a discontinuity in $\partial y/\partial t$ at $t = 0$ on $x = 0$, and it is not surprising that the finite-difference solution of Crank and Crowley (1979) showed a loss of accuracy there. Accuracy can be improved by adding an extra isotherm with the temperature $\frac{1}{2}\delta u$.

On the right boundary, $x = 1$, all isotherms move upwards initially, since $\partial y/\partial t = \pi^2(1 - u)$, $0 < u \leq 1$, $x = 1$, and for $u = 0$, $\partial y/\partial t = 1/\lambda$, $x = 1$. These statements are physically consistent with the initial temperature distribution in (4.27). The negative temperature gradient from left to right along any line of constant y produces a corresponding sideways heat flow which causes the isotherms to move downward on $x = 0$ and upward on $x = 1$.

Crank and Crowley (1979) presented numerical results. Some comparisons with those of other workers for the positions of the ends of the interface are reproduced in Table 5.2.

TABLE 5.2

Comparison of positions of ends of interface
for Bonnerot and Jamet's (1977) problem
specified in equations (4.24–28)

t	JB[a]	RMF[b]	Enthalpy[c]	IMM implicit[d]	IMM explicit[e]
On $x = 0$					
0	3	3	3	3	3
1	3.021	3.015	3.031	3.073	3.057
2	3.068	3.054	3.051	3.161	3.131
On $x = 1$					
0	1	1	1	1	1
1	2.118	2.095	2.124	2.132	2.133
2	2.610	2.585	2.566	2.591	2.594

[a] JB = Bonnnerot and Jamet (1977).
[b] RMF = Furzeland, R. M. (1977b).
[c] Enthalpy results obtained by method of Crowley (1978).
[d] IMM implicit = Crank and Crowley (1979).
[e] IMM explicit results obtained by method of Crank and Crowley (1978).

6. Fixed-domain methods

6.1. Introduction

THE previous three chapters have described what are essentially 'front-tracking' methods. Even though, in Chapter 5, the phase-change boundary was 'fixed' by a change of variable which simplified the numerical work considerably, it was still necessary to satisfy the Stefan or similar derivative condition on that boundary. Furthermore, it may sometimes be difficult or even impossible to track the moving boundary directly if it does not move smoothly or monotonically with time. Quite possibly, it may not do so in more than one space dimension. The moving boundary may have sharp peaks, or double back, or it may even disappear. The possibility, therefore, of reformulating the problem in such a way that the Stefan condition is implicitly bound up in a new form of the equations, which applies over the whole of a fixed domain, is an attractive one. The position of the moving boundary appears, a posteriori, as one feature of the solution. One way of accomplishing such a reformulation is to introduce an enthalpy or total heat function.

6.2. Enthalpy method

The use of enthalpy was proposed by Eyres *et al.* (1946) and later by Price and Slack (1954) and Albasiny (1956). Rose (1960) introduced the concept of a weak solution used by Lax (1954) to obtain solutions of hyperbolic equations involving shocks. Oleinik (1960) and Kamenomostskaja (1961) provided theoretical justification. The procedure is to introduce an enthalpy function, $H(u)$, which is the total heat content, i.e. the sum of the specific or sensible heat and the latent heat required for a phase change. The heat jump at the phase-change boundary is incorporated in the definition of $H(u)$, which is

$$
\left.
\begin{aligned}
H(u) &= \int_{u_0}^{u} \rho(\theta)c(\theta)\,\mathrm{d}\theta, \qquad u < u_{\mathrm{m}}, \\[2mm]
H(u) &= \int_{u_0}^{u} \rho(\theta)c(\theta)\,\mathrm{d}\theta + \rho L, \qquad u > u_{\mathrm{m}}, \\[2mm]
\int_{u_0}^{u} \rho(\theta)c(\theta)\,\mathrm{d}\theta &\leq H(u) \leq \int_{u_0}^{u} \rho(\theta)c(\theta)\,\mathrm{d}\theta + \rho L, \qquad u = u_{\mathrm{m}}.
\end{aligned}
\right\}
\tag{6.1}
$$

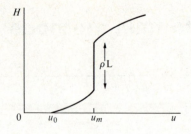

FIG. 6.1. Enthalpy function

In (6.1) u_0 is some fixed temperature less than u_m, the melting temperature: density ρ and specific heat c may both be functions of temperature u. Budak *et al.* (1965), Meyer (1973), Bonacina *et al.* (1973) incorporated the three expressions in (6.1) into the single relationship

$$H(u) = \int_{u_0}^{u} \{\rho(\theta)c(\theta) + L\rho(\theta)\delta(\theta - u_m)\} \, d\theta, \tag{6.2}$$

where δ is the Dirac impulse function. The form of $H(u)$ is shown graphically in Fig. 6.1.

Consider the two-phase, multi-dimensional problem

$$\rho_i c_i \frac{\partial u_i}{\partial t} = \nabla . (K_i \nabla u_i), \qquad i = 1, 2, \tag{6.3}$$

$$u_1 = u_2 = u_m, \qquad [K \, \partial u/\partial n]_1^2 = -\rho L v_n, \qquad x = s(t), \tag{6.4}$$

where $i = 1, 2$ denote the two phases separated by the moving phase-change boundary $x = s(t)$: $u_i(x, t)$, $i = 1, 2$, denote the temperatures u in the two phases respectively: and c_i, ρ_i, K_i may all be functions of u_i, **x**, t, where **x** is in general a three-dimensional space vector; v_n is the velocity of the moving boundary in the direction of the normal, n. For variable properties, $c(u)$, $\rho(u)$, it follows from the first two expressions of (6.1) that

$$\partial H/\partial t = \rho(u)c(u) \, \partial u/\partial t, \qquad u \gtrless u_m, \tag{6.5}$$

and hence the problem of (6.3) and (6.4) can be reformulated over the whole fixed domain occupied by the two phases together, except where $u = u_m$, as the single equation

$$\partial H(u)/\partial t = \nabla . (K \nabla u), \tag{6.6}$$

with associated initial and fixed boundary conditions.

Equation (6.6) is not obviously meaningful at $u = u_m$ since the enthalpy $H(u)$ has a jump discontinuity at $u = u_m$ of magnitude $[H(u_m)]_1^2 = L\rho$ and dH/du is infinite there. Shamsundar and Sparrow (1975), by considering the energy balance of a control volume V with surface area A, set up an integral form of the enthalpy equation, which includes fluid motion due to density changes or convection, i.e.

$$\frac{d}{dt} \int_V H \, dV + \int_A H\mathbf{v} \cdot d\mathbf{A} = \int_A K \nabla u \cdot d\mathbf{A},$$

valid throughout the domain. The second term on the left-hand side accounts for any fluid motion. By applying this integral form to an element of volume which does not include the phase-change boundary and then to one through which the interface passes, Shamsundar and Sparrow used the divergence theorem to confirm that the integral form is equivalent to (6.3) and (6.4) and to (6.6) when $\mathbf{v} = 0$.

In a one-phase problem, it follows that if the liquid is uniformly at the melting temperature initially, we have (Shamsundar and Sparrow 1976)

$$\frac{d}{dt} \int_V H_1 \, dV + \int_A H_1 \mathbf{v} \cdot d\mathbf{A} = 0,$$

where H_1 is the enthalpy of the liquid, and so on subtraction from the integral enthalpy equation above

$$\frac{d}{dt} \int_V (H - H_1) \, dV + \int_A (H - H_1)\mathbf{v} \cdot d\mathbf{A} = \int_A K \nabla u \cdot d\mathbf{A}.$$

In the solid phase $\mathbf{v} = 0$ and so an enthalpy equation which takes account of density changes in a one-phase problem is

$$\frac{d}{dt} \int_V (H - H_1) \, dV = \int_A K \nabla u \cdot d\mathbf{A},$$

which avoids explicit use of the fluid velocity.

For constant thermal properties c_1 and c_2 over each phase, constant ρ over both phases, and $u_m \equiv 0$, the enthalpy function defined in (6.1) becomes

$$H(u) = \rho c_1 u, \qquad u < 0; \qquad H(u) = \rho c_2 u + \rho L, \qquad u > 0;$$
$$0 \leqslant H \leqslant \rho L, \qquad u = 0. \quad (6.7)$$

If $K = K(u)$ in (6.6) the equation can be made linear by what is sometimes called the Kirchhoff transformation (Eyres *et al.*, 1946; Albasiny, 1956):

$$v = \int^u K(\zeta) \, d\zeta, \tag{6.8}$$

which reduces (6.6) to

$$\partial H(v)/\partial t = \nabla^2 v. \tag{6.9}$$

This transformation can sometimes simplify the numerical work in other methods too.

For constant conductivities, K_1 and K_2 in the two phases, (6.8) becomes $v = K_i u$, $i = 1, 2$, and $H(v)$ is given by

$$H(v) = \rho c_1 v/K_1, \qquad v < 0; \qquad H(v) = \rho c_2 v/K_2 + \rho L, \qquad v > 0;$$

$$0 \leqslant H(v) \leqslant \rho L, \qquad v = 0. \tag{6.10}$$

One-phase problems can be treated by extending the definition of u to a fixed, two-phase domain in which the usual two-phase equations hold but $c(u) \equiv \rho(u) \equiv 0$ in the second phase (Budak *et al.* 1965). Effectively, (6.6) is then to be solved only in the first phase.

6.2.1. Weak solutions

We have noted that $H(u)$, as defined in (6.1), is a discontinuous function of u at $u = u_m$ and this presents difficulties in interpreting equation (6.6) and its classical solutions at $u = u_m$. In such situations, the usual practice is to seek a 'weak' solution that satisfies a suitably integrated form of the equation in which derivatives of H and u do not appear (e.g. Williams, 1980, pp. 51, 142). Weak solutions for Stefan problems were introduced by Rose (1960), Oleinik (1960), and Kamenomostskaja (1961), and have been extended and analysed by many authors including Friedman (1968), Brezis (1970), and Lions (1969). Many references are given in the survey by Niezgódka (1983).

The derivation of a weak solution can be illustrated by considering a one-dimensional problem. Equation (6.6) becomes

$$\frac{\partial H}{\partial t} = \frac{\partial}{\partial x} \left\{ K(u) \frac{\partial u}{\partial x} \right\}, \qquad 0 < x < 1, \tag{6.11}$$

and we take boundary and initial conditions (Fox, 1979)

$$u(0, t) = g_1(t), \qquad u(1, t) = g_2(t), \qquad u(x, 0) = u_0(x), \tag{6.12}$$

and consider the time range $0 \leqslant t \leqslant T$. We first confine attention to this problem in the absence of any phase change so that, instead of the definitions (6.1), we assume the higher-order derivatives of H and u to exist. We now multiply (6.11) by an arbitrary test function $\phi(x, t)$ which has continuous first and second derivatives with respect to x and t and which vanishes on $x = 0$ and $x = 1$. Thus from (6.11) we obtain

$$\int_0^1 \int_0^T \left[\phi \frac{\partial}{\partial x} \left(K \frac{\partial u}{\partial x} \right) - \phi \frac{\partial H}{\partial t} \right] dx \, dt = 0,$$

and, on integrating by parts, the left-hand side becomes

$$\int_0^T \left[\phi K \frac{\partial u}{\partial x} \right]_0^1 dt - \int_0^1 \int_0^T K \frac{\partial u}{\partial x} \frac{\partial \phi}{\partial x} dx\, dt$$

$$- \int_0^1 [\phi H]_0^T\, dx + \int_0^1 \int_0^T H \frac{\partial \phi}{\partial t} dx\, dt$$

$$= - \int_0^T \left[uK \frac{\partial \phi}{\partial x} \right]_0^1 dt + \int_0^1 \int_0^T u \frac{\partial}{\partial x} \left(K \frac{\partial \phi}{\partial x} \right) dx\, dt$$

$$+ \int_0^1 H(u_0) \phi(x, 0)\, dx + \int_0^1 \int_0^T H \frac{\partial \phi}{\partial t} dx\, dt,$$

and so finally we obtain the required weak form

$$\int_0^1 \int_0^T \left\{ u \frac{\partial}{\partial x} \left(K \frac{\partial \phi}{\partial x} \right) + H(u) \frac{\partial \phi}{\partial t} \right\} dx\, dt$$

$$= \int_0^T K(g_2) g_2 \frac{\partial \phi}{\partial x}(1, t)\, dt - \int_0^T K(g_1) g_1 \frac{\partial \phi}{\partial x}(0, t)\, dt$$

$$- \int_0^1 H(u_0) \phi(x, 0)\, dx. \quad (6.13)$$

We now define a weak solution of a heat conduction problem with a phase change, subject to the boundary and initial conditions (6.12), to be a pair of bounded functions u and H, related by (6.1), for which the integral form (6.13) is satisfied for all test functions ϕ as specified.

Atthey (1974) showed that the weak formulation (6.13) includes the usual Stefan jump condition (6.4) across a phase-change surface. Thus, we assume that, as in the classical Stefan formulation, there are two regions R_1 and R_2 separated by a curve $x = s(t)$ in Fig. 6.2, and in each region the one-dimensional form of the heat conduction equation, $\rho c\, \partial u/\partial t = \partial(K\, \partial u/\partial x)/\partial x$, holds. On integrating by parts over R_1 as in the

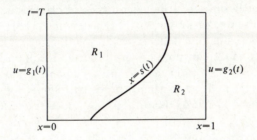

FIG. 6.2. x–t plane for weak solution containing jump discontinuities on $x = s(t)$

derivation of (6.13) we find

$$\int_0^T \int_{R_1} \left\{ u \frac{\partial}{\partial x}\left(K\frac{\partial\phi}{\partial x}\right) + H(u)\frac{\partial\phi}{\partial t} \right\} dx\, dt$$

$$= -\int_0^{s(0)} H(u_0)\phi(x,0)\, dx - \int_0^T K(g_1)\frac{\partial\phi}{\partial x}(0,t)\, dt$$

$$+ \int_{x=s(t)-0} \left[\left\{ Ku\frac{\partial\phi}{\partial x} - K\phi\frac{\partial u}{\partial x} \right\} dx - H\phi\, dt \right].$$

A similar relation holds in R_2. By adding these two relations, subtracting (6.13) and assuming u is continuous across $x = s(t)$ we obtain

$$\int_{x=s(t)} \phi\{[K\,\partial u/\partial x]\, dt + [H]\, dx\} = 0, \tag{6.14}$$

where, in general, $[F]$ denotes the jump discontinuity in the function F from $x = s(t) - 0$ to $x = s(t) + 0$. Since (6.14) holds for an arbitrary test function ϕ and an arbitrary time, it follows that

$$ds/dt = -[K\,\partial T/\partial x]/[H]. \tag{6.15}$$

In the classical one-dimensional Stefan problem, the phase change occurs on the boundary $x = s(t)$ on which $u = u_m$ and from (6.1) we know that $[H] = \rho L$. Thus (6.15) is seen to be the usual Stefan condition

$$[K\,\partial T/\partial x]_1^2 = -\rho L\, ds/dt. \tag{6.16}$$

We have thus established, through the derivations of (6.13) and of (6.16), that both the classical solution of a heat conduction problem with no change of phase and the classical solution of a one-dimensional Stefan problem satisfy the criterion for a weak solution. Rubinstein (1971) showed that the classical two-phase Stefan problem (equations (1.7–9) inclusive) is well posed with a unique solution provided the initial condition and the conditions on the fixed boundaries are twice differentiable. Similar statements can be made about the one-phase problem (Sherman 1970; Tayler 1975) subject to further restrictions on certain parameters. Furthermore, various authors have proved the existence and uniqueness of the weak solution (6.13) (Kamenomostskaja 1961; Oleinik, 1960) and other suitably defined weak solutions for multi-phase and higher-dimensional Stefan problems with derivative boundary conditions (Friedman 1968). Rubinstein (1971) presented the uniqueness proof by Kamenomostskaja and the existence proof by Oleinik for a simple case. A relevant survey by Niezgódka (1983) and more references are discussed in §6.4.7.

From all the various studies of classical and weak solutions we can conclude that for any heat-conduction problem for which a classical solution exists, the unique weak solution which exists will also be the unique classical solution. In §6.2.2 it will be seen that certain finite-difference schemes can be proved to converge to the unique weak solution (Oleinik 1960) despite the discontinuities in the derivatives of u across the phase-change boundary. The existence of the weak solution is thus established through the proof of convergence. The discontinuities make it difficult to prove directly the convergence of finite-difference solutions to the classical solutions. The weak solution, therefore, has important properties both theoretical and numerical. Indeed Rubinstein (1979) claimed that it was only for the weak solution of the multi-phase Stefan problem in several space variables that existence had been proved and saw this as a general position in the theory of partial differential equations. He was critical of attempted proofs by Datzev (1970) and Budak and Moskal (1970, 1971). Niezgódka (1983) however, considered that the existence of classical solutions for some multi-dimensional problems had been established by Meirmanov (1981a,b—all in Russian). Considerable progress has certainly been made in theoretical studies of one and two-phase multi-dimensional problems by formulating them as variational inequalities (see §§6.4–6.4.7).

6.2.2. *An explicit finite-difference scheme*

As a simple example, we describe an explicit finite-difference scheme for the solution of the one-dimensional form of (6.9), where the variable v is related to temperature, u, through (6.8). Standard modifications will allow the scheme to be adapted to include the form (6.11) with constant or variable K, for example, and also alternative conditions on the fixed boundaries. The derivative conditions

$$\partial v/\partial x = \alpha v, \qquad x = 0, \tag{6.17}$$

$$\partial v/\partial x = 0, \qquad x = \ell, \tag{6.18}$$

and the initial condition,

$$H(x, 0) = h_0(x), \tag{6.19}$$

studied by Atthey (1974) provide a convenient introduction to his more general problem with body heating discussed in §6.2.6. Also, the numerical scheme will require (6.10) to be rewritten

$$v = H/d_1, \qquad H < 0; \qquad v = (H - L\rho)/d_2, \qquad H > L\rho;$$

$$v = 0, \qquad 0 \leq H \leq \rho L, \tag{6.20}$$

where, for convenience, $d_i = \rho c_i/K_i$, $i = 1, 2$. The simplest explicit replacement of (6.9) is

$$H_m^{n+1} = H_m^n + \lambda(v_{m+1}^n - 2v_m^n + v_{m-1}^n), \tag{6.21}$$

where $\lambda = \delta t/(\delta x)^2$, $\delta x = \ell/M$, $\delta t = T/N$ for positive integers $M, N; T > 0$ is an arbitrary upper time boundary; and $H_m^n = H(m \, \delta x, \, n \, \delta t)$ etc., with $1 \le m \le M - 1$, $0 \le n \le N$. The conditions (6.17–19) are taken as

$$v_0^n = v_1^n(2 - \alpha \, \delta x)/(2 + \alpha \, \delta x), \qquad v_{M-1}^n = v_M^n, \qquad 0 \le n \le M,$$
$$\text{(6.22)}$$

$$H_m^0 = h_0(m \, \delta x), \qquad 0 \le m \le M. \tag{6.23}$$

The first of (6.22) comes by approximating (6.17) as $(v_1^n - v_0^n)/\delta x = \frac{1}{2}\alpha(v_1^n + v_0^n)$, where v_0 is the value at $x = 0$.

Assuming the numerical solution has progressed as far as $t = n \, \delta t$, so that values of v_m^n, H_m^n are known for $0 \le m \le M$, we proceed to $(n+1) \, \delta t$ as follows:

  (i) compute H_m^{n+1} from (6.21) for $1 \le m \le M - 1$.
 (ii) derive v_m^{n+1} from one of the three relationships in (6.20) appropriate to the value of H_m^{n+1}.
(iii) compute v_0^{n+1} and v_m^{n+1} from (6.22).

Clearly this enthalpy scheme avoids the difficulties of front-tracking. Instead, the position of the moving boundary has to be determined in retrospect by inspection of the computed values of H and v. Since (6.21) is a conservative integrated form of (6.9) for the element δx in the interval δt, the acute problem of the infinite derivative $\partial H/\partial t$ is avoided. Nevertheless, the numerical solution is likely to exhibit a stepwise behaviour and this may impair the accuracy with which the boundary can be located. Ways of improving this accuracy are considered in §6.2.5, together with comments on implicit finite-difference schemes and extended applications of the enthalpy method.

Before leaving the simple explicit scheme (6.21) we discuss the convergence of the numerical solution to the unique weak solution. The details of the outline of the convergence proof which follows are to be found in papers by Kamenomostskaja (1961), Oleinik (1960), Atthey (1972, 1974, 1975).

Piecewise-constant functions $\tilde{v}$ and $\tilde{H}$ are defined by

$$\tilde{v}(k, t) = v_m^n, \qquad \tilde{H}(x, t) = H_m^n, \tag{6.24}$$

for $(m-1) \, \delta x \le x < m \, \delta x$, $n \, \delta t \le t \le (n+1) \, \delta t$, for each m, n in the numerical solution. Kamenomostskaja (1961), having established the uniform boundedness of v_m^n, applied relevant theorems of functional analysis to prove that there are sequences of the integers $M_r, N_r \to \infty$ as $r \to \infty$ such that the corresponding subsequence of the sequence $\{\tilde{v}\}$ of functions v, defined in (6.24), converges strongly to a function v. Also, the corresponding subsequence of $\{\tilde{H}\}$ converges weakly to a function H as $r \to \infty$.

We can now show that these functions v and H form a weak solution of the phase-change problem. We define, for any test function ϕ, any choice of integers M, N, and corresponding step-lengths δx, δt,

$$\phi_m^n = \phi(\{m - \tfrac{1}{2}\}\,\delta x, n\,\delta t), \qquad m = 1, \ldots, M, \qquad n = 0, \ldots, N, \quad (6.25)$$

$$\left.\begin{array}{l} \phi_0^n = \phi_1^n(2 - \alpha\,\delta x)/(2 + \alpha\,\delta x), \\ \phi_{M+1}^n = \phi_M^n, \end{array}\right\} \; n = 0, \ldots, N, \qquad (6.26)$$

together with associated piecewise-constant functions

$$\tilde{\phi} = \tilde{\phi}_m^n, \qquad \tilde{\phi}_{xx} = (\phi_{m+1}^n - 2\phi_m^n + \phi_{m-1}^n)/(\delta x)^2, \qquad \tilde{\phi}_t = (\phi_m^n - \phi_m^{n-1})/\delta t,$$
$$(6.27)$$

for $(m-1)\,\delta x \leqslant x < m\,\delta x$, $n\,\delta t \leqslant t < (n+1)\,\delta t$. We can verify that $\tilde{\phi}$, $\tilde{\phi}_{xx}$, and $\tilde{\phi}_t$ converge strongly to ϕ, $\partial^2\phi/\partial x^2$, and $\partial\phi/\partial t$ respectively as $M, N \to \infty$. Now we proceed as in the development of the weak solution (6.13) and multiply (6.21) by ϕ_m^n and sum over m and n to obtain

$$\delta t\,\delta x \sum_{m=1}^{M} \sum_{n=0}^{N-1} \frac{H_m^{n+1} - H_m^n}{\delta t}\,\phi_m^n = \delta t\,\delta x \sum_{m=1}^{M} \sum_{n=0}^{N-1} \frac{v_{m+1}^n - 2v_m^n + v_{m-1}^n}{(\delta x)^2}\,\phi_m^n. \quad (6.28)$$

On summing by parts, inserting the initial and boundary conditions, and finally replacing the sums by integrals we find

$$\int_0^T\!\!\int_0^\ell \tilde{H}\tilde{\phi}_t\,\mathrm{d}x\,\mathrm{d}t + \int_0^T\!\!\int_0^\ell \tilde{v}\tilde{\phi}_{xx}\,\mathrm{d}x\,\mathrm{d}t = -\int_0^\ell \tilde{h}_0(x)\tilde{\phi}(x, 0)\,\mathrm{d}x, \quad (6.29)$$

where $\tilde{h}_0(x)$ is a piecewise-constant approximation to the function $h_0(x)$ in (6.19). In the limit as $M \to N \to \infty$ through the values M_r and N_r, with which subsequences of $\{\tilde{v}\}$ and $\{\tilde{H}\}$ converging to v and H respectively have been associated above, (6.29) becomes

$$\int_0^T\!\!\int_0^\ell \{H\,\partial\phi/\partial t + v\,\partial^2\phi/\partial x^2\}\,\mathrm{d}x\,\mathrm{d}t = -\int_0^\ell h_0(x)\phi(x, 0)\,\mathrm{d}x. \quad (6.30)$$

Thus the functions v and H satisfy a weak solution.

The final requirement is to show that the solution of the finite-difference scheme (6.21) converges to the weak solution as M and $N \to \infty$ through any sequence of integers for which the relation $\lambda = \delta t/(\delta x)^2 \leqslant \lambda_0 < \tfrac{1}{2}$ is satisfied. If we suppose this is not so then for some $\varepsilon > 0$ and for some function, ψ, which is twice continuously differentiable, we may find subsequences of the sequences $\{\tilde{v}\}$ and $\{\tilde{H}\}$ denoted by $\{\tilde{\tilde{v}}\}$ and $\{\tilde{\tilde{H}}\}$ for which

$$\left|\int_0^T\!\!\int_0^\ell (\tilde{\tilde{v}} - v)\psi\,\mathrm{d}x\,\mathrm{d}t\right| + \left|\int_0^T\!\!\int_0^\ell (\tilde{\tilde{H}} - H)\psi\,\mathrm{d}x\,\mathrm{d}t\right| > \varepsilon. \quad (6.31)$$

But the discussion above about the properties of the sequences $\{\tilde{v}\}$ and $\{\tilde{H}\}$ apply equally to the sequences $\{\tilde{\tilde{v}}\}$ and $\{\tilde{\tilde{H}}\}$. Thus we may find a subsequence of $\{\tilde{\tilde{v}}\}$ which converges strongly to a function v_1 and a subsequence of $\{\tilde{\tilde{H}}\}$ which converges weakly to some function H_1. Now the arguments used to establish (6.30) can be used to show that T_1 and H_1 form a weak solution. But we know the weak solution to be unique and so we have $\|v - v_1\| = 0, \|H - H_1\| = 0$. Thus the above subsequences of $\{\tilde{\tilde{v}}\}$ and $\{\tilde{\tilde{H}}\}$ must converge to v and H, which contradicts (6.31). We have therefore shown that the solution of the finite-difference scheme (6.21) converges to the unique weak solution (6.30).

Convergence proofs for both explicit and implicit finite-difference schemes (see §6.2.4) and for discontinuous and smoothed enthalpy functions (see §6.2.3) have been given by other authors including Ciavaldini (1975), Jerome (1977), Schäfer (1977), and Meyer (1973).

6.2.3. Alternative forms of the enthalpy function

One definition of the enthalpy relation $H(u)$ has been given in (6.1). For some substances, however, the phase change occurs within a wide band of temperatures and the relationship between $H(u)$ and temperature u is known from experimental data or standard physical tables. This information can then be used directly in the numerical solution (see, for example, Albasiny 1956; Longworth 1975; Hodgkins and Waddington 1975).

Because of the difficulties and inaccuracies associated with the discontinuous jump in $H(u)$ at the melting temperature in the definition (6.1), which is likely to occur for pure substances, various authors have proposed smoothing over a small temperature zone. With reference to the definition (6.8) over the range $-\varepsilon \leqslant v \leqslant \varepsilon$ where $\varepsilon > 0$ is small Furzeland (1977b) lists three examples of linear smoothing defined by

$$H(v) = \begin{cases} \alpha_2 v - L_2, & v < -\varepsilon, \\ \alpha_3 v + L_3, & -\varepsilon \leqslant v \leqslant \varepsilon, \\ \alpha_1 v + L_1, & v > \varepsilon, \end{cases} \tag{6.32}$$

where the smoothing parameters $\alpha_i, L_i, i = 1, 2, 3$, are chosen by Meyer (1973) to be

$$\left. \begin{array}{lll} \alpha_1 = d_1, & \alpha_2 = d_2, & \alpha_3 = \lambda/2\varepsilon, \\ L_1 = \lambda - (c_1 + c_2)\varepsilon, & L_2 = 0, & L_3 = \tfrac{1}{2}\lambda - c_2\varepsilon, \end{array} \right\} \tag{6.33}$$

and by Elliott (1976) to be

$$\begin{array}{lll} \alpha_1 = d_1, & \alpha_2 = d_2, & \alpha_3 = \tfrac{1}{2}\{\varepsilon(\alpha_1 + \alpha_2) + \lambda\}/\varepsilon, \\ L_1 = \lambda, & L_2 = 0, & L_3 = \tfrac{1}{2}\{\varepsilon(\alpha_2 - \alpha_1) + \lambda\}, \end{array} \tag{6.34}$$

where $\lambda = \rho L$, the latent heat per unit volume. For $\varepsilon = 0$ the smoothing tends to

$$\alpha_1 = d_1, \qquad \alpha_2 = d_2, \qquad \alpha_3 = \tfrac{1}{2}(\alpha_1 + \alpha_2),$$
$$L_1 = \lambda, \qquad L_2 = 0, \qquad L_3 = \tfrac{1}{2}\lambda. \tag{6.35}$$

Friedman (1968) proved existence of a weak solution to the Stefan problem by showing that solutions for smoothed enthalpy converge to the solution of the Stefan problem as $\varepsilon \to 0$.

No theoretical analysis has been made of the effect of the smoothing parameters on the accuracy of the solution. The influence of the width of the smoothing zone has been demonstrated numerically in selected cases. Thus Bonacina *et al.* (1973), whose findings were later confirmed by Furzeland (1980), found their results for a one-dimensional, two-phase Stefan problem to be appreciably dependent on the zone width, 2ε, in (6.32). In fact, Bonacina *et al.* (1973) replaced the enthalpy jump at the phase-change by an equivalent heat capacity, $\tilde{C}(u)$, i.e. density $\times$ specific heat. Instead of an enthalpy $H(u)$ defined by

$$H(u) = \int_{u_r}^{u} C(u)\,\mathrm{d}u + \lambda\eta(u - u_f), \qquad \eta(u) = \begin{cases} 1, & u \geq u_f, \\ 0, & u < u_f, \end{cases}$$

where $u_r < u_f$ (freezing temperature) is arbitrarily chosen, they defined $\tilde{C}(u)$ by

$$\tilde{C}(u) = \frac{\mathrm{d}H(u)}{\mathrm{d}u} = C(u) + \lambda\,\delta(u - u_f),$$

$$C(u) = \begin{cases} C_1(u), & u < u_f, \\ C_2(u), & u > u_f, \end{cases}$$

where $\delta(u - u_f)$ is the Dirac function. The two-phase Stefan problem is then expressed by the single equation

$$\tilde{C}(u)\frac{\partial u}{\partial t} = \frac{\partial}{\partial x}\left\{K(u)\frac{\partial u}{\partial x}\right\}, \qquad K(u) = \begin{cases} K_1(u), & u < u_f, \\ K_2(u), & u > u_f. \end{cases}$$

Bonacina *et al.* (1973) then smoothed the Dirac function in the definition of $\tilde{C}(u)$ and also in $K(u)$ and adapted a linear form for both $\tilde{C}(u)$ and $K(u)$ over the range $u_f - \varepsilon \leq u \leq u_f + \varepsilon$.

In a similar comparison, Shamsundar (1978) found the heat–flux-time curve to be relatively smooth but displaced from the true curve for a large value of ε; as ε was reduced the smoothed-enthalpy calculated curve approached the correct result but developed a stepwise wavy character. Elliott (1976) quoted an example in which an optimum choice of ε can improve the numerical solution as much as a reduction in mesh size. Furzeland (1977*b*) quotes results but points out that the optimum ε has to be determined by experiment. Meyer (1973) on the other hand claims his calculated values for the temperature record at one point in the region for a two-dimensional melting problem around a square duct are independent of ε in the range $10^{-6} \leqslant \varepsilon \leqslant 0.5$. Incidentally, Meyer (1973) mentions a precipitation problem (Cannon and Hill 1970) in which the enthalpy actually is continuous and piecewise linear.

Budak *et al.* (1965) and Moiseynko and Samarskii (1965) used higher-order smoothings in which, for example, the $H(u)$ function near the melting temperature is approximated by a parabola which ensures continuity of the first derivative of the approximating smoothed enthalpy function as well as of the function itself. Brauner *et al.* (1983) discuss the smoothing function $H(u) = \frac{1}{2} L\rho\{1 + u/(\eta + |u|)\}$, where $\eta > 0$ is a small parameter.

6.2.4. *Other numerical schemes and multi-dimensional problems*

The discontinuous form of the enthalpy function expressed as $u = u(H)$ was used by Atthey (1974) together with the one-dimensional explicit form (6.21) though he did not use the transformation (6.8). The scheme was extended to two-dimensional problems by Crowley (1978).

She solved two problems of the inward solidification of a square cylinder of liquid initially at its freezing temperature for two different surface conditions. This problem was described in §5.4.2(i). In the first case, the surface temperature is lowered at a constant rate, corresponding to the conditions under which Saitoh (1976) carried out his experiments. Crowley's (1978) enthalpy calculations using an explicit finite-difference scheme agree well with Saitoh's experimental results and with a perturbation solution in inverse powers of the latent heat, L, until the freezing front has moved about halfway to the centre.

In the second case, the surface temperature is dropped discontinuously at the initial instant. The solution obtained by Rathjen and Jiji (1971) for solidification in an infinite corner provides a good approximation while the front remains parallel to the sides of the square cylinder away from its corners. Crowley also compared her enthalpy results graphically with those of Allan and Severn (1962), Lazaridis (1970), Crank and Gupta (1975).

Implicit finite-difference schemes with discontinuous enthalpy functions were described by Furzeland (1974, 1977*b*), Federenko (1975), Longworth (1975), and Shamsundar and Sparrow (1975). Wood, Ritchie, and Bell (1981) used a hopscotch finite-difference scheme (Gourlay, 1970). Ciavaldini (1975) used explicit and implicit finite-element schemes to solve a discretized weak form of an enthalpy formulation. Hodgkins and Waddington (1975) also introduced finite elements.

Li (1983) adopted the finite-element discretization used by Bonnerot and Jamet (1979), shown in Fig. 4.5, but applied it to a weak enthalpy formulation. This is a front-tracking method which avoids explicit treatment of the jump condition and so overcomes the difficulties of an accurate, a posteriori location of the moving boundary discussed in §6.2.5. The method is applied to a one-dimensional, two-phase problem, a two-dimensional, one-phase problem, the spot-welding problem (§6.2.6), and to alloy solidification (§6.2.7(ii)).

Longworth (1975) considered an equation of the form

$$\partial H/\partial t + \mathbf{u} \cdot \nabla H - \nabla^2 \phi = Q(\mathbf{x}, t), \tag{6.36}$$

which describes a Stefan melting problem with a heat source $Q(\mathbf{x}, t)$ moving with a velocity $\mathbf{u}$, and ϕ is the Kirchhoff variable defined in eqn (6.8). The equation refers to a frame of reference fixed in the heat source, and Longworth integrated over individual, elementary, volume cells to obtain for the ith cell of volume V_i and surface area S_i

$$\partial H_i/\partial t = Q_i - \int_{S_i} H\mathbf{u} \cdot d S + \int_{S_i} \nabla\phi \cdot d\mathbf{S},$$

where $Q_i = (1/V_i) \int_{V_i} Q \, dV$. The surface integrals are expressed in terms of H_i, ϕ_i and H_j, ϕ_j for cells j adjacent to cell i, but Longworth expresses all the ϕs in terms of the corresponding Hs through the enthalpy relationship and thus arrives at a set of non-linear finite-difference equations of the form

$$\partial \mathbf{H}/\partial t = \mathbf{F}(\mathbf{H}), \tag{6.37}$$

where $\mathbf{H}$ is a vector with components H_i and $\mathbf{F}$ is a non-linear function whose ith component F_i is a function of H_i and the values H_j for the cells j adjacent to cell i. An implicit Crank–Nicolson scheme is evaluated iteratively.

Smoothed enthalpy functions were first used in multi-dimensional problems by Budak *et al.* (1965) and Moiseynko and Samarskii (1965), with locally one-dimensional finite-difference methods. Two-dimensional, implicit finite-difference and finite-element schemes were suggested by Couch *et al.* (1970), Elliott (1976), and Meyer (1973, 1975, 1976, 1978*c*). Meyer had a particular interest in the efficient numerical solution of the non-linear equations which the implicit formulations require. Bonacina *et al.* (1973), Fisher and Medland (1974), and Comini *et al.* (1974) used three-level schemes. The latter authors aimed particularly to include simultaneous temperature dependence of thermal conductivity, heat capacity, rate of internal heat generation, and surface heat transfer coefficients. Latent heat effects were approximated by a large heat capacity over a small temperature interval enclosing the melting temperature. An improved procedure was described by Morgan *et al.* (1978).

For the implicit schemes, there arises the complication of having to decide in which of the three enthalpy intervals a given point lies at the next time level in order to use the correct relation from (6.1) or (6.32), for example. The authors cited above determined which interval was appropriate by a trial and error process, and this was time consuming especially as it was an additional part of an iterative scheme of solution. Furzeland (1977b) described a way of avoiding the trial and error process that he attributed to Shamsundar and Sparrow (1975).

A fully implicit, finite-difference scheme approximating the two-dimensional case of (6.9) is

$$H_{i,j}^{n+1} + 4rv_{i,j}^{n+1} = H_{i,j}^n + r\{v_{i,j-1}^{n+1} + v_{i-1,j}^{n+1} + v_{i+1,j}^{n+1} + v_{i,j+1}^{n+1}\}, \qquad (6.38)$$

where $h = \delta x = \delta y$, $r = \delta t/h^2$, $v_{i,j}^n = v(ih, jh, n\,\delta t)$, $H_{i,j}^n = H(v_{i,j}^n)$. The Gauss–Seidel iterative scheme, working along the rows j and up the columns i, can be written

$$H_{i,j}^{(s+1)} + 4rv_{i,j}^{(s+1)} = b_{i,j}, \qquad (6.39)$$

where the notation $n+1$ has been dropped for the unknowns and

$$b_{i,j} = r\{v_{i,j-1}^{(s+1)} + v_{i-1,j}^{(s+1)} + v_{i+1,j}^{(s)} + v_{i,j+1}^{(s)}\}, \qquad (6.40)$$

so that b_{ij} is known at the iteration $(s+1)$.

With reference to the temperature–enthalpy relations (6.20), $H<0$ implies $v<0$ and so $b_{i,j}<0$. Conversely, if $b_{i,j}<0$, then v is given by the first of (6.20) and using this in (6.39) gives

$$H_{i,j}^{(s+1)} = \frac{b_{i,j}}{1 + 4r/d_1} \qquad \text{if} \qquad b_{i,j}<0. \qquad (6.41)$$

Similarly

$$H_{i,j}^{(s+1)} = b_{i,j}, \qquad 0 \leq b_{i,j} \leq \lambda, \qquad (6.42)$$

$$H_{i,j}^{(s+1)} = \frac{b_{i,j} + 4r\lambda/d_2}{1 + 4r/d_2}, \qquad b_{i,j} > \lambda, \qquad (6.43)$$

where $\lambda = L\rho$. Thus, the known value of $b_{i,j}$ at any point in the computational scheme determines which expression in (6.20) is to be used. The argument is only valid as long as the dependence on K has been removed, e.g. by the transformation (6.8), and provided both H and u are negative in one phase and positive in the other. This second requirement can easily be satisfied for cases where the melting temperature is $u_m \neq 0$ by adopting new variables $u' = u - u_m$, $H' = H - \rho c_2 u_m$. Finite elements can be used instead of finite differences and Furzeland (1977b) includes a body-heating term $Q(x, y)$ which, by the simple modification of including Q on the left-hand side of (6.39), can be extended to $Q(u)$.

The same arguments can be applied (Furzeland, 1977b) for the smoothed enthalpy function (6.32). If $v < -\varepsilon$, then $H < -(\alpha_2\varepsilon + L_2)$ and so $H + 4rv < -\{\varepsilon(\alpha_2 + 4r) + L_2\}$. This corresponds to the test (6.41) for $L_2 \geqslant 0$ and, in fact, (6.41), (6.42), and (6.43) are to be replaced by

$$v_{i,j}^{(s+1)} = \frac{b_{i,j} + L_2}{\alpha_2 + 4r}, \qquad b_{i,j} < -b_2, \tag{6.44}$$

$$v_{i,j}^{(s+1)} = \frac{b_{i,j} - L_3}{\alpha_3 + 4r}, \qquad -b_2 \leqslant b_{i,j} \leqslant b_1, \tag{6.45}$$

$$v_{i,j}^{(s+1)} = \frac{b_{i,j} - L_1}{\alpha_1 + 4r}, \qquad b_{i,j} > b_1, \tag{6.46}$$

where $b_1 = \varepsilon(\alpha_1 + 4r) + L_1$, $b_2 = \varepsilon(\alpha_2 + 4r) + L_2$.

Furzeland (1977b) applies this iteration-avoiding modification to obtain fully implicit, finite-difference solutions of the constructed problem in two space dimensions (x, y):

$$C_i \, \partial u/\partial t = K_i \, \nabla^2 u + Q_i, \qquad i = 1, 2,$$

with

$$Q_i = Q_i(t) = 2(C_i e^{-2t} - 2K_i),$$

and Q_1 is used if $x^2 + y^2 > e^{-2t}$, Q_2 if $x^2 + y^2 < e^{-2t}$. The initial conditions are

$$u = x^2 + y^2 - 1, \qquad s(x, y, 0) = x^2 + y^2 - 1, \qquad t = 0,$$

where $s(x, y, t) = 0$ is the moving boundary on which $u = 0$ and the usual Stefan jump condition holds. The region $x \geqslant 0$, $y \geqslant 0$, $x^2 + y^2 \leqslant 1$ is considered with the boundary conditions

$$\partial u/\partial x = 0, \qquad x = 0; \qquad \partial u/\partial y = 0, \qquad y = 0.$$

The solution of this problem is $u = x^2 + y^2 - e^{-2t}$ with the moving boundary given by $s(x, y, t) \equiv x^2 + y^2 - e^{2t} = 0$. Numerical solutions obtained for both discontinuous and smoothed enthalpy with the parameters (6.33) and (6.34) are reproduced in Table 6.1.

Furzeland (1980) outlined the basis of a general-purpose program by Elliott and Furzeland (Furzeland, 1979b) for the finite-difference solution of the non-linear, one-dimensional problem

$$C_i(u) \frac{\partial u_i}{\partial t} = \frac{\partial}{\partial x} \left(K_i(u) \frac{\partial u_i}{\partial x} \right) + f_i(u_i), \qquad i = 1, 2, \tag{6.47}$$

$$\left. \begin{array}{l} u_1 = u_2 = u_M \\ \lambda \dot{s} = K_1(u_M) \, \partial u_1/\partial x - K_2(u_M) \, \partial u_2/\partial x \end{array} \right\} \quad x = s(t), \tag{6.48}$$

TABLE 6.1

Values of $10^3 u$ at $x = 0.5$, $y = 0$ for model problem (Furzeland 1977b): $c_1 = 4$, $c_2 = 1$, $k_1 = 2$, $k_2 = 1$, $\lambda = 2$; $v(H)$ signifies a discontinuous enthalpy function and $H(v)$ a smoothed function

	Time				
	0.2	0.4	0.6	0.8	1.0
$v(H)$					
$h = 0.1$, $\delta t = 0.04$	−421	−208	−34	39	102
$H(v)$, $\varepsilon = 0.001$					
$h = 0.1$, $\delta t = 0.04$	−412	−197	−22	44	103
$H(v)$, $\varepsilon = 0.05$					
$h = 0.1$, $\delta t = 0.04$	−416	−196	−43	38	107
$v(H)$					
$h = 0.05$, $\delta t = 0.01$	−421	−202	−53	46	110
$H(v)$, $\varepsilon = 0.01$					
$h = 0.05$, $\delta t = 0.01$	−417	−191	−51	47	115
$H(v)$, $\varepsilon = 0.03$					
$h = 0.05$, $\delta t = 0.01$	−421	−199	−52	47	113
Exact solution	−420	−199	−51	48	115

rewritten as the single equation

$$\frac{\partial}{\partial t} H(u(v)) = \frac{\partial^2 v}{\partial x^2} + f(u(v)), \tag{6.49}$$

where $v(u)$ is defined by (6.8) and H is the discontinuous enthalpy function

$$H(u) = \int_{u^*}^{u} C(\theta)\, d\theta + \begin{cases} \lambda & \text{if} \quad u \geq u_M \\ 0 & \text{if} \quad u < u_M \end{cases}, \tag{6.50}$$

and u^* is a reference temperature. The parameters K, C, f represent K_1, C_1, f_1 if $u < u_M$ or K_2, C_2, f_2 if $u > u_M$, and $u(v)$ is the inverse transformation of (6.8). Equation (6.49) is approximated by

$$H_j^{n+1} + 2r\theta v_j^{n+1} - \theta\, \Delta t f_j^{n+1} = a_j^n + r\theta(v_{j-1}^{n+1} + v_{j+1}^{n+1}) \equiv b_j, \tag{6.51}$$

where $r = \Delta t/(\Delta x)^2$, θ is the implicit, time-discretization parameter, and

$$a_j^n = H_j^n + \Delta t(1-\theta)f_j^n + r(1-\theta)(v_{j-1}^n - 2v_j^n + v_{j+1}^n). \tag{6.52}$$

A point, successive over-relaxation scheme (SOR) is used to solve (6.51) for $v_j^{n+1,s+1}$ by treating the right-hand side, b_j^s, as known at the previous iteration, s. The relaxation parameter ω is varied automatically according to the proportion of phase 1 to phase 2 (Elliott 1976).

The standard discussion of ω for a linear system gives the optimum value for a matrix $c\mathbf{I} + \theta\mathbf{A}\,\delta t/(\delta x)^2$ to be

$$\omega = \frac{2}{1+(1-p^2)^{\frac{1}{2}}}, \qquad p = \frac{2\theta\,\delta t}{c}\cos\pi\,\delta x.$$

The relevant matrix in (6.51) is $\mathbf{c}(v) + \theta\mathbf{A}\,\delta t/(\delta x)^2$ with $\mathbf{c}(v)$ a diagonal matrix with entries $1/K_1$ or $1/K_2$ for the frozen or melted regions respectively. Near the solution at time $(n+1)\,\delta t$, $\mathbf{c}(v)$ is approaching $\mathbf{c}(v^{n+1})$ and so Elliott (1976) calculated ω from the usual expression with the diagonal entries in $\mathbf{c}$ given by $z/K_1 + (1-z)/K_2$, where z is the proportion of mesh points in the frozen region, phase 1.

For mesh points j where the phase change is occurring, ω is set equal to unity following Meyer (1973). At each point j, an inner iteration ($p = 1, 2, \ldots$) is performed based on a Newton linearization of (6.51), i.e.

$$v_j^{n+1,p+1} = v_j^{n+1,p} + \frac{b_j - \{H + 2r\theta v - \theta f\,\Delta t\}_j^{n+1,p}}{\{dH/dv + 2r\theta - \theta\,\Delta t\,df/dv\}_j^{n+1,p}} \tag{6.53}$$

TABLE 6.2

Moving boundary position $s(t)$: two-phase Stefan problem (Bonacina et al. 1973); $k_1 = 2.22$, $k_2 = 0.556$, $c_1 = 1.762$, $c_2 = 4.226$, $\lambda = 338$; zero melting temperature, surface temperature -20, initial temperature 10. Starting time $t = 0.0012$ with $s(t) = 0.015\,975$; $\Delta t = 0.0012$ except for method (iii) and smoothed enthalpy results

	Time					
	0.0024	0.0036	0.018	0.072	0.144	0.288
(i)	0.0224	0.0276	0.0618	0.1238	0.1750	0.2476
(ii)	(0.0220)	(0.0273)				
	0.0196	0.0256	0.0617	0.1236	0.1749	0.2474
(iii)	0.0228	0.0279	0.0619	0.1237	0.1750	0.2474
(iv)	0.0250	0.0250	0.0500	0.1250	0.1750	0.2490
(v)	Non-monotonic values near				(0.180)	(0.247)
	phase change			0.1391	0.1791	0.2389
Exact solution	0.0226	0.0278	0.0619	0.1238	0.1750	0.2475

(i) Method of lines in time with invariant imbedding (Meyer 1970; Furzeland 1979b): 81 space points, $\theta = \frac{1}{2}$.

(ii) Boundary-fixing coordinate transformation (Crank 1957a); 41 points in each phase, $\theta = \frac{1}{2}$; $\theta = 1$ for values in parentheses.

(iii) Boundary-fixing transformation and method of lines in space by Elliott reported in Furzeland (1979b): 41 points in each phase, ODE in time.

(iv) Enthalpy method: 41 points, $\theta = \frac{1}{2}$, stepwise behaviour can be corrected by Voller and Cross's (1981a) procedure.

(v) Smoothed enthalpy (Bonacina et al. 1973): 81 points, $\Delta t = 0.0004$, $\varepsilon = 0.5$, results in parentheses started at $t = 0.072$; if started at $t = 0.0012$ non-monotonic instability occurs near phase-change boundary.

is used, where the SOR superscript s has been dropped and

$$dH/dv = (dH/du)/(dv/du), \qquad df/dv = (df/du)/(dv/du).$$

Non-linear conditions on the fixed boundaries are similarly treated by Newton linearizations. Denoting the converged result of (6.53) by $\hat{v}_j^{n+1}$, the outer SOR iteration is written

$$v_j^{n+1,s+1} = (1-\omega)v_j^{n+1,s} + \omega\hat{v}_j^{n+1}. \qquad (6.54)$$

Elliott (1981) established the convergence of this SOR scheme for $\theta < \omega < 2$ if $\omega = 1$ is used when $\hat{v}_j^{n+1} \cdot v_j^{n+1,s} \leq 0$, i.e. a mesh point changes phase.

TABLE 6.3

Moving boundary position and surface concentration: oxygen diffusion problem. Upper value = $u(0, t)$; lower value = $s(t)$

	Time			
	0.04	0.10	0.18	0.19
(i)				
201 points	0.2746	0.1436	0.0222	0.0090
$\theta = \frac{1}{2}$, $\Delta t = 0.001$	0.9993	0.9361	0.5065	0.3541
(ii)				
41 points	0.2770	0.1437	0.0219	0.0091
$\theta = \frac{1}{2}$, $\Delta t = 0.005$	0.9991	0.9349	0.5023	0.3495
(iii)				
41 points	0.2745	0.1433	0.0219	0.0091
ODE time	0.9992	0.9358	0.5028	0.3477
(iv)				
41 points	0.2766	0.1439	0.0221	0.0093
$\theta = \frac{1}{2}$, $t = 0.005$	1.0000	1.0000	0.4750	0.3250
* Hansen and	0.2743	0.1432	0.0218	0.0090
Hougaard (1974)	0.9992	0.9350	0.5011	0.3454

	Time				
	0.02	0.10	0.12	0.16	0.18
* Ferriss and Hill	0.3417	0.1438	0.1097	0.0493	0.0222
(1974); boundary	1.0000	0.9354	0.8800	0.6850	0.5046
fixing; 40 points,					
$\delta t = 0.0005$					
Ferriss and Hill	0.3404	0.1437	0.1091	0.0488	0.0218
(1974); embedding	1.0000	0.9353	0.8797	0.6860	0.5097
$\delta t = 0.01$					

Methods (i–iv) as in Tables 6.2 with the amended parameters indicated. Solutions start from $t = 0.0025$ except for those bearing an asterisk which start from $t = 0$.

Furzeland (1980) compared results obtained by the enthalpy program for several test problems with those obtained by other methods. Tables 6.2, 6.3 are extracted from his paper and some other results referred to in this book are assembled there for convenience.

6.2.5. *Accurate determination of phase-change boundary*

It is a central feature of the enthalpy method that the position of the phase-change boundary is determined, *a posteriori*, from the numerical solution carried out in a fixed domain. Simple inspection of the solution reveals which mesh point is undergoing a phase-change at any time and it may take several time steps for the change to be completed there. The position of the boundary can be bracketed by the two mesh points between which temperature changes from less than to greater than the phase-change temperature or vice versa. To locate the boundary more accurately, however, can present difficulties. Extrapolation of the temperature values, u or v, from either side of the phase boundary, or interpolation across the boundary, is a possible refinement but the results tend to exhibit a physically unrealistic, stepwise behaviour in the motion of the boundary and in the time history of the temperature at a typical fixed point.

Voller and Cross (1981*a*) advanced explanations of the stepwise and oscillatory behaviour of the basic enthalpy solutions both for discontinuous and smoothed enthalpy functions. The authors proposed an interpretation of the numerical solution which leads to a more accurate evaluation of the boundary movement and the temperature history at any point.

With reference first to a one-dimensional solidification problem with a discontinuous enthalpy function, Voller and Cross (1981*a*) considered an element e_i of the discretized region associated with a mesh point or node i. The total heat in this element at any time is approximated by $H_i \, \delta x$, where H_i is the enthalpy at node i and δx is the element length (Fig. 6.3). If at any time t the freezing front is in element e_i and moving towards element e_{i+1}, the total heat in e_i is the sum of the heat in the solid and

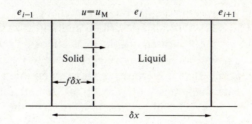

FIG. 6.3. Partially frozen element e_i

liquid parts, and so H_i can be approximated by

$$H_i = Cu_i f + (Cu_i + L)(1-f), \tag{6.55}$$

where f is the solid fraction of the element and the heat capacity C is taken to be the same in both phases. When the freezing front reaches the node i then $f = \frac{1}{2}$, $u_i = u_M$, the melting temperature, and (6.55) becomes

$$H_i = Cu_M + \tfrac{1}{2}L. \tag{6.56}$$

Thus, when any nodal enthalpy satisfies (6.56) the phase-change boundary can be located approximately on the node i. This interpretation of the state of an element undergoing a phase-change is physically more satisfying than the idea that the whole element is uniformly in some intermediate state betwen the solid and liquid phases.

Voller and Cross (1981a) concluded from (6.56) that if at any nodal point i in the basic finite-difference solution $H_i^j > Cu_M + \frac{1}{2}L$ and $H_i^{j+1} < Cu_M + \frac{1}{2}L$, for a freezing problem, the phase-change boundary had passed through node i in the time interval between $j\,\delta t$ and $(j+1)\,\delta t$. If it is further assumed that the enthalpy changes linearly in any time interval, δt, then the time t_i when the phase-change boundary is on node point i, which occurs when $H_i = Cu_M + \frac{1}{2}L$, is given by $t_i = (j+\alpha)\,\delta t$ where $\alpha < 1$ and is estimated by

$$\alpha = (\tfrac{1}{2}L + Cu_M - H_i^j)/(H_i^{j+1} - H_i^j). \tag{6.57}$$

Clearly, at time t_i the temperature at node i is u_M. The temperatures at other node points k $(k \neq i)$ are approximated by the linear interpolation

$$u_k^{i+\alpha} = \alpha(u_k^{j+1} - u_k^j) + u_k^j. \tag{6.58}$$

Voller and Cross (1981a) obtained good agreement between their finite-difference solution interpreted using the above argument and the analytical solution for a classical, two-phase Stefan problem. They extended their algorithm to include some two-dimensional problems and used it to solve a problem similar to the freezing in a square channel studied by Crowley (1978), but it was a two-phase problem as distinct from Crowley's one phase. A refinement of the method (Voller and Cross 1983) led to a more continuous tracking of the phase change.

Voller and Cross (1981a) also described an implicit scheme for numerical solution which incorporated the idea of choosing each time step so that the phase-change boundary moved from one node point to the next for each chosen time step. This was the strategy adopted by Douglas and Gallie (1955), but they did not apply it to an enthalpy equation (see §4.3.1). The advantage of this approach is that the position of the phase-change boundary is known at each time step. This means that the finite-difference formulae, and hence the nodal enthalpy and temperature distributions, have their normal accuracy and if the heat parameters are different in the two phases they can be easily and accurately accommodated in the scheme.

The essential feature of the algorithm proposed is to ensure that at each time step one and only one nodal enthalpy has the value $Cu_M + \frac{1}{2}L$. If at time t the phase-change boundary is at node k, an initial guess for the next time step, δt_k, is taken as $\delta t_k^0 = \delta t_{k-1}$. An implicit finite-difference scheme for equation (6.37), following Longworth (1975) (see equations (6.36, 37) above) is used to evaluate the nodal enthalpies for a time interval δt_k^0 and for successive estimates, δt_k^m, calculated from the iterative formula

$$\delta t_k^{m+1} = \delta t_k^m + \omega \, \delta t_k^m \left(\frac{H_k^{t+\delta t_k^m}}{Cu_M + \frac{1}{2}L} - 1 \right), \tag{6.59}$$

where ω is a relaxation factor. When $H_{k+1}^{t+\delta t_k^m}$ has converged to $Cu_M + \frac{1}{2}L$, the corresponding enthalpy and temperature values at all nodal points are adopted as the solution for the time $t + \delta t_k^m$. Voller and Cross (1981a) quoted acceptable results for a simple two-phase freezing problem and for a spot-welding problem similar to that studied by Atthey (1974). Their implicit algorithm is much faster than their explicit one because there are no stability restrictions on the time step.

Voller and Cross (1981b) confirmed that their explicit numerical scheme incorporating (6.56–8) can be satisfactorily applied to the freezing or melting of a circular cylinder. From a collection of numerical results they approximated the dimensionless solidification time t_S^* of a circular cylinder to be given by the empirical expression

$$t_S^* = (0.14 + 0.085u_0^*) + (0.252 - 0.0025u_0^*)L^*, \tag{6.60}$$

with an accuracy of 1 per cent of the numerical predictions when $0 \le u_0^* \le 2$, $2 \le L^* \le 50$. The cylinders of radius R are solidifying under the following conditions: $u = u_0 > u_M$, $0 \le r \le R$, $t < 0$; $u = u_W < u_M$, $r = R$, $t \ge 0$; u_M is the melting temperature. The non-dimensional variables in (6.60) are defined to be $t^* = Kt/(R^2 \rho c)$, $u_0^* = (u_0 - u_M)/(u_M - u_W)$, $L^* = L/\{c(u_M - u_W)\}$. The enthalpy function has a discontinuous jump L at $u = u_M$. A more tentative approximation for the melting time of general symmetrically shaped but not necessarily circular cylinders is based on an upper and a lower bound. The former is the melting time of a circular cylinder with the same cross-section as the original cylinder; the latter is based on a circular cylinder which fits totally inside the original cylinder. A reasonable approximation for the solidification time of any general, symmetrically shaped cylinder is afforded by using an approximating circular cylinder with a radius equal to the mean of the radii associated with the upper and lower bounds. The sum of the percentage errors in the upper and lower bounds is stated to be given by $E = 100(r_u^2 - r_\ell^2)/r_a$, where r_u, r_ℓ are the radii of the upper and lower bound approximating cylinders and $r_a = (r_u + r_\ell)/2$. For a square channel $E = 24$ per cent: for a pentagon $E = 14.5$ per cent.

A qualitative explanation of the stepwise time history produced by the enthalpy method was outlined by Shamsundar (1978). The detailed quantitative analysis put forward independently by Bell (1982) incorporates the same physical idea. He computed an enthalpy solution of the two-phase Stefan freezing problem posed by Goodrich (1978) and noted that the temperature distribution, particularly in the frozen phase, was effectively linear. This is because, for the physical conditions of the problem, the freezing front moves slowly and the temperature within the frozen phase is quasi-steady. Figure 6.4 shows the stepwise nature of the time history at a typical point, which is most obvious in the frozen region but exists also in the unfrozen zone. Bell noted the number of increments Δt between consecutive jumps in temperature (i.e. between the points O, A, B, C in Fig. 6.4) and found them to be in remarkably close agreement with the number of time increments taken to freeze successive Δxs of the medium. These latter were calculated from the analytic expression appropriate to this problem for the penetration of the solidification front, $s(t)$. i.e. $s(t) = 0.508\,691(K_f t)^{\frac{1}{2}}$, where K_f is the conductivity in the frozen state.

The implication is that the whole of the temperature distribution throughout the medium remains unchanged as long as the freezing front

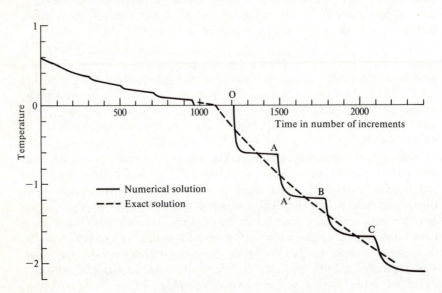

FIG. 6.4. Temperature variation at depth of 15 cm

remains within the spatial element Δx associated with a particular mesh point. A change in temperature only occurs when the front moves from the locality of one mesh point to the next. This is exactly the behaviour to be expected from the quasi-steady temperature profile. Thus, if the interval $(m - \frac{1}{2}) \Delta x \leqslant x \leqslant (m + \frac{1}{2}) \Delta x$ is changing from liquid to solid, the enthalpy computed at the point $m \Delta x$ lies between 0 and L, and the temperature at $m \Delta x$ stays at zero. Thus, there is a region with fixed boundaries, namely $x = 0$, where the temperature is $-10\,°C$ in Goodrich's example, and $x = m \Delta x$ where the temperature is $0\,°C$. The analytical expression for the temperature in this subsidiary heat-flow problem (Carslaw and Jaeger 1959, p. 100) has the form

$$-10\left(1 - \frac{x}{m \, \Delta x}\right) + \sum_{n=1}^{\infty} A_n \sin \frac{n\pi x}{m \, \Delta x} \exp\left(\frac{-n^2 \pi^2 K_f t'}{m^2 (\Delta x)^2}\right),$$

where t' is measured from the instant the temperature at $x = m \Delta x$ becomes zero. The series represents the transient part of the solution. The term $-10(1 - x/m \, \Delta x)$ is the ultimate steady state achieved theoretically after infinite time but for this problem the transient effects are very short lived. The intervals of time over which transient heat flow occurs are the short periods of rapid temperature drop in Fig. 6.4, e.g. between the points A and A'. The near-constant-temperature plateaux, e.g. between A' and B, indicate the period of the steady-state flow.

Bell also showed, with the aid of the analytical expression, that the time-scale of the transient effects increases in comparison with the time taken for an element to freeze, as m increases. This is confirmed in Fig. 6.4.

There is an implication for the linear interpolation which Voller and Cross (1981a) used as described above to determine the time at which the enthalpy is $\frac{1}{2}L$. It is that the accuracy will deteriorate when a point well below the surface freezes, because the time increments for the element to freeze will be larger, and non-linear transient effects more significant.

6.2.6. *Body heating*

There is an important class of problems in which the heat needed to produce melting comes from heat sources distributed throughout the volume of the material rather than by entering through the outer boundaries. An example is joule heating associated with electric currents within the medium. In such cases the melting does not necessarily take place across a surface as in the classical Stefan problem. Instead, there may be an extended region the whole of which is at melting temperature, and the solid and liquid phases coexist. This volume of material is called a 'mushy' region, a term used by Tien and Geiger (1967) in relation to alloys. The physical nature of a mushy region has been discussed more generally in

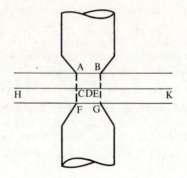

FIG. 6.5. Cross-section of a weld specimen

§1.3.1(iv). For the purpose of the present enthalpy formulation it suffices to say that within the mushy region the average enthalpy lies somewhere between its values for the solid and liquid states at the melting temperature.

Atthey (1974) described an enthalpy method to solve a welding problem in which a mushy region occurs. The spot welding of two sheets of equal thickness placed face to face is effected by passing a high electric current between two circular electrodes held together by a large force (Fig. 6.5). Atthey (1974) gave grounds for assuming that only the joule heating Q is significant. As the electrodes are cooled, the expectation that the highest temperature will occur on the central plane, $x = \ell$, of the weld specimen can be justified by the Maximum Principle before melting starts. At some time, $t = t_0$, the central plane CE in Fig. 6.5 will reach the melting temperature $u = u_M$. An amount of heat $\rho L S\, \delta x$ is required to melt a small element $\ell - \delta x \leq x \leq \ell$, where S is the area of the electrode. The joule rate of heating of the element is $QS\, \delta x$ and no other heat can enter the element since no other part of the sheets can be at a higher temperature than that at $x = \ell$. The only possibility is that the element $\ell - \delta x \leq x \leq \ell$ remains at the melting temperature for a time $L\rho/Q$. During this time, other parts of the sheets will reach the melting temperature so that during the time interval $t_0 \leq t \leq t_0 + L\rho/Q$ there will be a region, $\ell'(t) \leq x \leq \ell$ say, which is at the melting temperature and is neither fully molten nor fully solid. This is half the mushy region existing between the solidus and liquidus boundaries if the sheets are of equal thickness.

The one-dimensional heat-conduction problem for a material with constant heat capacity $C = c\rho$ can be expressed by the equation

$$K\frac{\partial^2 u}{\partial x^2} + Q = \frac{\partial H}{\partial t} \tag{6.61}$$

where $u = u(H)$ is defined by

$$\left.\begin{array}{ll} u = H/C, & H \leqslant Cu_M, \\ u = u_M, & Cu_M < H < Cu_M + L\rho, \\ u = (H-L)/C, & H \geqslant Cu_M + L\rho. \end{array}\right\} \qquad (6.62)$$

The body-heating term $Q(u)$ is taken to be a given piecewise linear function of temperature only and other boundary conditions are, for example,

$$\partial u/\partial x = \alpha u, \qquad x = 0; \qquad \partial u/\partial x = 0, \qquad x = \ell. \qquad (6.63)$$

An initial condition $H(x, 0) = h_0(x)$ is given. Atthey (1974) approximated (6.61) by the explicit finite-difference scheme

$$H_m^{n+1} = H_m^n + K\lambda\{u_{m+1}^n - 2u_m^n + u_{m-1}^n\} + Q(u_m^n)\,\delta t, \qquad (6.64)$$

where $\lambda = \delta t/(\delta x)^2$ and the conditions (6.63) become, as in (6.22), (6.23),

$$u_0^n = u_1^n(2 - \alpha\,\delta x)/(2 + \alpha\,\delta x), \qquad u_{E+1}^n = u_E^n \qquad (6.65)$$

where $E\,\delta x = \ell$. The initial condition is simply $H_m^0 = h_0(m\,\delta x)$. Proceeding as in §§6.2.1, 2, Atthey (1974) showed that the finite-difference solution converges to a unique weak solution analogous to (6.13) modified to include the body-heating term and the derivative boundary conditions (6.63).

Figure 6.6 shows a typical picture of the solid, mushy, and liquid regions from a computed solution (Atthey 1974).

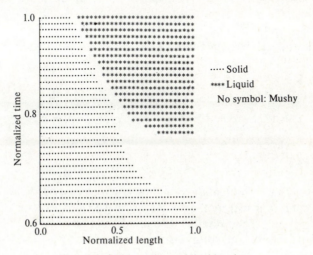

FIG. 6.6. Solid, mushy, and liquid regions

Crowley and Ockendon (1977) solved a similar problem with body heating in modelling a thermal switch. They obtained an asymptotic solution of an integral equation for large times and enthalpy-based numerical results.

6.2.7. *Other conservation forms and weak solutions*

The essential features of the enthalpy method, introduced in §6.2, are that the problem is reformulated in terms of a single differential equation which is solved over a fixed domain to obtain values of temperature and the enthalpy function. The position of the moving boundary emerges as a by-product of the computation. In heat-flow problems, enthalpy has a real physical significance and the reformulated problem is in a heat conservation form. This means that a boundary jump condition such as (6.4) is automatically satisfied. Even in problems where the concept of enthalpy has no direct relevance it is sometimes possible to secure the advantages of computing in a fixed domain by rewriting the equations in a form in which some essential physical quantity is conserved and which permits a weak solution. The following are some examples.

(i) *Cases of zero 'specific heat'.* In the classical Stefan problem, specific heat c and conductivity K are both positive and the enthalpy H is a strictly monotone function of temperature, while temperature is a single-valued function of H. Crowley (1979) formulated certain problems by introducing a 'generalized enthalpy' function incorporating a zero 'specific heat' in one or both phases. An example of the latter case is the electro-chemical machining problem described by McGeough and Rasmussen (1974) (see §2.12.3). Their quasi-steady model is defined by the equation

$$\nabla^2 \phi = 0, \tag{6.66}$$

where ϕ is the electric potential, with the boundary conditions at the anode

$$\phi = 0, \qquad M\frac{ds}{dt} = -\nabla\phi \cdot \nabla s \qquad \text{on} \qquad s(\mathbf{r}, t) = 0, \qquad M > 0, \tag{6.67}$$

and on the cathode, $r = a$,

$$\phi = -V < 0. \tag{6.68}$$

This is a one-phase problem governed by an elliptic equation and a Stefan boundary condition with non-zero latent heat, M. In order to formulate the equations on a fixed domain, the anode can be regarded as a region where $\phi \equiv 0$, $H \equiv M$, and solutions are required of

$$\partial H/\partial t = \nabla^2 \phi, \tag{6.69}$$

with boundary conditions $\phi = -v$, $H = 0$ on $r = a$, and initially

$$\phi = -v, \qquad H = 0, \qquad t = 0, \qquad \text{outside } s(\mathbf{r}, t) = 0,$$

and

$$\phi = 0, \qquad H = M, \qquad t = 0, \qquad \text{within } s(\mathbf{r}, t) = 0.$$

The conservative form of (6.69) across a surface of discontinuity $s(\mathbf{r}, t) = 0$, obtained by integrating over a small volume, is

$$[H]\, ds/dt = M\, ds/dt = [\nabla\phi \cdot \nabla s], \tag{6.70}$$

where [ ] denotes the jump between the sides where $H \leqslant 0$ and $H \geqslant M$ respectively. Since $\nabla^2\phi = 0$ inside the anode it follows that (6.67) is satisfied by a solution of (6.69). Crowley (1979) obtained an implicit, finite-difference solution for this problem by expressing (6.69) in cylindrical polar coordinates. Her results are compared with those of Christiansen and Rasmussen (1976) and Elliott (1980) in Table 8.25.

The flow of an incompressible fluid in a Hele-Shaw cell (Richardson 1972) (see §2.12.2) can be formulated in the same way (see Figs. 2.19, 20).

In Crowley's second problem the 'specific heat' is zero only in one phase in which, therefore, an elliptic differential eqution holds, but we have a parabolic equation in the other phase. This is the situation for saturated/unsaturated flow in a porous medium (Hornung 1978). In one region the flow is unsaturated, that is the medium is only partially filled with fluid and its fluid content may change. In a second region the medium is saturated and no more fluid can be added. In terms of a velocity potential, ϕ, the relevant equation is

$$\partial H/\partial t = \nabla^2\phi, \tag{6.71}$$

where H measures the air content of the medium. Hornung (1978) adopts the 'generalized-enthalpy relationships',

$$H = \begin{cases} \frac{1}{2}\phi^2, & \phi > 0, \\ 0 & \phi < 0, \end{cases} \tag{6.72}$$

and ϕ can be regarded as a generalized temperature. A sample problem is solved by Crowley (1979).

In the same paper, the uniqueness of the weak solutions for these cases in which the 'specific heat' may vanish was proved by necessary extensions of Oleinik's (1960) methods for the classical Stefan problem. It is because zero specific heat means that temperature is no longer a single-valued function of enthalpy that Oleinik's proof needs modification. Crowley (1979) introduced a function strictly monotone in both H and ϕ, defined by $F(\phi, H) = H + \gamma\phi$ where γ is any positive constant, into the uniqueness proof. This has the additional consequence that the finite-difference scheme used to solve (6.69), for example, cannot be explicit but must be implicit. Thus, an implicit scheme of the form

$$H(x, t+\delta t) - H(x, t) = \delta t \{\mu \, \nabla^2 \phi(x, t+\delta t) + (1-\mu) \, \nabla^2 \phi(x, t)\}$$

for $0 < \mu \leq 1$ can evaluate a linear combination of H and ϕ of the form $H + \gamma\phi$ used in the uniqueness proof. This follows immediately by writing the implicit scheme for a one-dimensional problem, for example, as

$$(H_m^{n+1} + 2\mu \, \delta t \phi_m^{n+1}) = H_m^n + \mu \, \delta t(\phi_{m-1}^{n+1} + \phi_{m+1}^{n+1})$$
$$+ (1-\mu) \, \delta t(\phi_{m-1}^n + \phi_{m+1}^n - 2\phi_m^n),$$

and solving by successive over-relaxation with the right-hand side treated as known from the previous iteration (see solution of (6.51)). From numerical experiments Crowley (1979) showed that the finite-difference scheme converges but no analytical proof was available because of the vanishing specific heat. Niezgódka (1980) and Visintin (1983) established the existence of weak solutions for problems with zero specific heat and examined their stability. We note in passing that Fasano and Primicerio (1979*d*) proved existence, uniqueness, and continuous dependence theorems for classical solutions for saturated/unsaturated flow in one space dimension.

Crowley's (1979) approach can be adopted in solving the oxygen diffusion with absorption problem which is formulated classically in equations (1.57–60). By introducing a generalized enthalpy function H related to oxygen concentration u by

$$H = \begin{cases} u, & u > 0, \\ 0, & u \leq 0, \end{cases} \tag{6.73}$$

and an absorption term Q defined by

$$Q = \begin{cases} -1, & u > 0, \\ 0, & u \leq 0, \end{cases} \tag{6.74}$$

the problem can be reformulated as

$$\frac{\partial H}{\partial t} = \frac{\partial^2 u}{\partial x^2} + Q, \qquad 0 < x < 1, \tag{6.75}$$

$$\partial u/\partial x = 0, \qquad x = 0; \qquad u = 0, \qquad x = 1, \qquad t > 0, \tag{6.76}$$

$$H = u = \tfrac{1}{2}(1-x)^2, \qquad 0 \leq x \leq s(0) = 1, \qquad t = 0. \tag{6.77}$$

Integration of (6.75) over a small volume element around the moving front, $x = s(t)$, for a time interval $\delta t \to 0$ shows that condition (1.59) is satisfied. Furzeland's (1980) results obtained in this way are included in

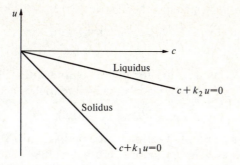

FIG. 6.7. Typical equilibrium diagram for a dilute alloy

Table 6.3. They correspond to a heat flow problem with zero latent heat and zero specific heat in the one phase where $u = 0$.

(ii) *Alloy solidification.* Crowley and Ockendon (1979) expressed equations for the solidification of an alloy in conservation form and obtained numerical solutions within a fixed domain. They adopted the simple linearized form of equilibrium or eutectic diagram (Fig. 6.7) as discussed in §1.3.7. Denoting by c the concentration of impurity and u the temperature, then the material is in the stable liquid ($i = 1$) or solid ($i = 2$) phase depending on whether $c + k_i u \gtrless 0$ respectively. Assuming that the temperature is continuous at the phase boundary and that the material on either side is in chemical equilibrium, there is a concentration discontinuity there with

$$c_S = (k_1/k_2)c_L. \tag{6.78}$$

A one-dimensional solidification process in which the impurity diffuses in both phases $i = 1, 2$ is described by the equations

$$\frac{\partial c_i}{\partial t} = \frac{\partial}{\partial x}\left(D_i \frac{\partial c_i}{\partial x}\right), \qquad i = 1, 2, \tag{6.79}$$

$$\sigma \frac{\partial u_i}{\partial t} = \frac{\partial}{\partial x}\left(K_i \frac{\partial u_i}{\partial x}\right), \qquad i = 1, 2, \tag{6.80}$$

where D is the mass diffusion coefficient and K the heat conductivity. The specific heat σ of the bulk material is taken as unity. At the phase-change interface the equilibrium conditions are

$$u_1 = u_2, \qquad c_1 = -k_1 u_1, \qquad c_2 = -k_2 u_2 \tag{6.81}$$

and the conservation of heat and mass gives

$$L\frac{ds}{dt} = -\left[K\frac{\partial u}{\partial x}\right]_1^2, \qquad [c]_1^2 \frac{ds}{dt} = -\left[D\frac{\partial c}{\partial x}\right]_1^2. \tag{6.82}$$

There are also appropriate conditions on the fixed boundaries. An enthalpy form which matches the first of (6.82) is

$$H = \begin{cases} u, & u + c/k_1 < 0, \\ u + L, & u + c/k_2 > 0. \end{cases} \tag{6.83}$$

Making the usual assumption for dilute solutions, that the concentration of impurity is proportional to its chemical potential or activity v, we have

$$c_i = bk_i v_i, \qquad i = 1, 2, \tag{6.84}$$

where b is some constant. The interface condition (6.78) becomes $v_1 = v_2$ and so v, as well as u, is continuous across the phase boundary $u + bv = 0$. Since $k_i = $ constant in this formulation, v can be defined everywhere by (6.84), though in a more general case this relation only holds on the phase boundary.

Thus, the chemical activity and concentration in the mass transfer problem are the direct analogues of temperature and enthalpy in the heat transfer and (6.79) may be written in a form analogous to (6.80), i.e.

$$\frac{\partial c_i}{\partial t} = \frac{\partial}{\partial x} \left(D_i' \frac{\partial v_i}{\partial x} \right), \qquad D_i' = D_i bk_i, \qquad i = 1, 2,$$

with

$$[v]_1^2 = 0, \qquad [c]_1^2 \frac{ds}{dt} = -\left[D' \frac{\partial v}{\partial x} \right]_1^2, \qquad u + bv = 0$$

at $x = s(t)$. Finally, therefore, we can write the conservation forms

$$\frac{\partial c}{\partial t} = \frac{\partial}{\partial x} \left(D_i' \frac{\partial v}{\partial x} \right), \qquad \frac{\partial H}{\partial t} = \frac{\partial}{\partial x} \left(K_i \frac{\partial u}{\partial x} \right) \tag{6.85}$$

coupled with

$$\begin{aligned} c = bk_1 v, & \quad H = u, & u + bv < 0, \\ bk_1 v \leqslant c \leqslant bk_2 v, & \quad u \leqslant H \leqslant u + L, & u + bv = 0, \\ c = bk_2 v, & \quad H = u + L, & u + bv > 0. \end{aligned} \tag{6.86}$$

Crowley and Ockendon (1979) were unable to prove either existence or uniqueness of a weak solution of (6.85) in general. Following Oleinik (1960) proofs can be established for the particular case of $K_i = K_i(u)$ and $D_i' = D_i'(v)$ for a pure metal where $c \equiv v \equiv 0$. Accordingly, Crowley and Ockendon (1979) proposed a numerical discretization of the general case (6.85), (6.86) with K_i and D_i constant in each phase for simplicity, and tested their solution by comparison with Rubinstein's (1971) explicit similarity solutions.

For the numerical scheme, (6.86) must be rewritten to give u, v in terms of H, c, i.e.

$$u = H, \qquad v = c/bk_1, \qquad H + c/k_1 < 0,$$
$$u = H - L, \qquad v = c/bk_2, \qquad H + c/k_2 > L. \tag{6.87}$$

The interval between $H + c/k_1 > 0$ and $H + c/k_2 < L$ needs special consideration since u is not defined there. Since the material cannot exist in equilibrium in this range, Chalmers (1964) took the values of H and c to represent an element of material of which a fraction f is solid and $1 - f$ is liquid. This is reminiscent of the procedure adopted by Voller and Cross (1981a) (see eqn (6.55)). The vaues of u, c_S, c_L for the element are compatible with local equilibrium. Thus we have

$$c = fc_S + (1 - f)c_L, \qquad H = u + (1 - f)L,$$

where $c_S = -k_1 u$, $c_L = -k_2 u$. Elimination of c_S, c_L, f yields a quadratic equation for u with the solution corresponding to taking the positive square root given by

$$u = u^* = \frac{-\{k_1 - H(k_1 - k_2)/L\} + [\{k_1 - H(k_1 - k_2)/L\}^2 - 4c(k_1 - k_2)/L]^{\frac{1}{2}}}{2(k_1 - k_2)/L} \tag{6.88}$$

so that $u = H$ on the solidus and $u = H - L$ on the liquidus boundary. The inverted form of (6.86) is therefore (6.87) together with

$$u = u^*, \qquad v = -u^*/b, \qquad H + c/k_2 - L < 0 < H + c/k_1, \tag{6.89}$$

where u^* comes from (6.88).

A simple explicit, finite-difference scheme for (6.85) if the mesh points $(n-1)\,\delta x$, $n\,\delta x$, $(n+1)\,\delta x$ are in the same phase is

$$H_n^{m+1} = H_n^m + K_n^m \frac{\delta t}{(\delta x)^2} (u_{n+1}^m - 2u_n^m + u_{n-1}^m), \tag{6.90}$$

$$c_n^{m+1} = c_n^m + D_n'^m \frac{\delta t}{(\delta x)^2} (v_{n+1}^m - 2v_n^m + v_{n-1}^m), \tag{6.91}$$

where $H_n^m = H(n\,\delta x, m\,\delta t)$, $m = 0, \ldots, M$, $n = 1, \ldots, N$, etc., and K_n^m, $D_n'^m$ take the values appropriate to that phase. In order to use (6.90) and (6.91) successfully over the whole of a fixed domain including both phases, we require them to conserve heat and mass across the phase boundary. Integration of (6.90) shows that the change in heat content in an interval δt is given by

$$\delta x \sum_1^{N-1} (H_n^{m+1} - H_n^m) = (\delta t/\delta x) \sum_1^{N-1} \{K_n^m(u_{n+1}^m - u_n^m) - K_n^m(u_n^m - u_{n-1}^m)\}$$

$$= (\delta t/\delta x) \sum_1^{N-1} (K_n^m - K_{n+1})(u_{n+1}^m - u_n^m) + K_N^m(u_N^m - u_{N-1}^m) - K_1^m(u_1^m - u_0^m).$$

Thus, there will be a heat loss or gain at the phase boundary unless $K_n^m = K_{n+1}^m$ for $n = 1, \ldots, N-1$, which implies $K_1 = K_2$. Modified forms of K and D' are therefore introduced near the phase boundary, the position of which corresponds to $u + bv = 0$ and is found by linear extrapolation on the values of $u + bv$ at the nearby mesh points. If $s(m\, \delta t) = n\, \delta x + h_1$, $0 < h_1 < \delta x$, with the solid phase in $x < s(t)$, the expressions for the space derivatives of u in (6.85) are replaced by

$$\frac{\partial}{\partial x}\left(K \frac{\partial u}{\partial x}\right)_n^m = \{K^*(u_{n+1}^m - u_n^m) - K_1(u_n^m - u_{n-1}^m)\}/(\delta x)^2, \qquad (6.92)$$

$$\frac{\partial}{\partial x}\left(K \frac{\partial u}{\partial x}\right)_{n+1}^m = \{K_2(u_{n+2}^m - u_{n+1}^m) - K^*(u_{n+1}^m - u_n^m)\}/(\delta x)^2, \qquad (6.93)$$

where K^* is the modified heat conductivity. Similar expressions are used for the space derivatives of v. Crowley and Ockendon (1979) secured the best agreement with the appropriate analytical solution by using $K^* = K_1 h_1/\delta x + K_2(1 - h_1/\delta x)$ and similarly for D'^*. At a mesh point where $u_n^m + bv_n^m = 0$, the equation

$$\frac{\partial}{\partial x}\left(K \frac{\partial u}{\partial x}\right)_n^m = \{K_2(u_{n+1}^m - u_n^m) - K_1(u_n^m - u_{n-1}^m)\}/(\delta x)^2 \qquad (6.94)$$

is used. The final, finite-difference scheme provides a conservative algorithm on a fixed domain which avoids explicit application of the classical conditions (6.82) at the phase boundary.

As a test example, Crowley and Ockendon (1979) obtained numerical solutions for the solidification of a block of molten alloy, with uniform temperature and concentration distributions, $u_0 = 1$, $c_0 = 0.1$, occupying the region $x > 0$. At time $t = 0$, the temperature at the face $x = 0$ is lowered to $u = -1$ and they assumed there is no mass flux out of the material. They obtained satisfactory agreement, for chosen values of the parameters, with Rubinstein's (1971) analytical solution, which is

$$\left. \begin{aligned}
c_1 &= -k_1 u_S, \\
c_2 &= 0.1 - (0.1 + k_2 u_S)\, \frac{\operatorname{erfc}\{x/2(D_2 t)^{\frac{1}{2}}\}}{\operatorname{erfc}\{\beta/D_2^{\frac{1}{2}}\}}, \\
u_1 &= -1 + (1 + u_S)\, \frac{\operatorname{erf}\{x/2(K_1 t)^{\frac{1}{2}}\}}{\operatorname{erf}\{\beta/K_1^{\frac{1}{2}}\}}, \\
u_2 &= 1 + (u_3 - 1)\, \frac{\operatorname{erfc}\{x/2(K_2 t)^{\frac{1}{2}}\}}{\operatorname{erfc}\{\beta/K_2^{\frac{1}{2}}\}}, \\
s(t) &= 2\beta t^{\frac{1}{2}},
\end{aligned} \right\} \qquad (6.95)$$

where β is the root of

$$L\beta\pi^{\frac{1}{2}}\,\mathrm{erf}(\beta/K_1^{\frac{1}{2}})\mathrm{erfc}(\beta/K_2^{\frac{1}{2}})\exp(\beta^2/K_1+\beta^2/K_2)$$

$$-K_1^{\frac{1}{2}}\mathrm{erfc}(\beta/K_2^{\frac{1}{2}})\exp(\beta^2/K_2)+K_2^{\frac{1}{2}}\mathrm{erfc}(\beta/K_1^{\frac{1}{2}})\exp(\beta^2/K_1)$$

$$=\{K_1^{\frac{1}{2}}\,\mathrm{erfc}(\beta/K_2^{\frac{1}{2}})\exp(\beta^2/K_2)+K_2^{\frac{1}{2}}\,\mathrm{erfc}(\beta/K_1^{\frac{1}{2}})\exp(\beta^2/K_1)\}u_S, \quad (6.96)$$

and where u_S, the constant phase-change temperature, is given by

$$u_S=\frac{0.1D_2^{\frac{1}{2}}}{k_2D_2^{\frac{1}{2}}+\beta\pi^{\frac{1}{2}}(k_1-k_2)\mathrm{erfc}(\beta/D_2^{\frac{1}{2}})\exp(\beta^2/D_2)}. \quad (6.97)$$

The error in the position of the phase boundary obtained from the finite-difference scheme compared with the analytical solution (6.95) is less than 5 per cent. Slight oscillations in the numerical solution are less noticeable the greater the number of mesh points.

A more general enthalpy-type formulation by Wilson *et al.* (1982) removes the restriction that the heat capacities in the liquid and solid phases must be equal and allows more general boundary conditions to be considered. They give a number of numerical finite-difference solutions and some comparisons with Rubinstein's solution (1971). The possible appearance of a mushy region just in front of the interface is mentioned.

(iii) *Flame-front problems.* Sometimes the equations in one part of a fixed domain are special cases of the equations in the remainder of the domain. In these cases, the latter equations are valid for the whole domain but when written in conservation form they may not satisfy the jump condition on the moving boundary, which forms part of the classical formulation. Crowley (1981) handles this situation by adding a term to the differential equation which vanishes everywhere except on the moving boundary where it has the required jump. The context is an enthalpy formulation of Buckmaster's (1979) model of flame fronts, in particular a steady-state flame tip for laminar, pre-mixed flames with large activation energy. The reaction region is considered to be confined to a flame sheet $y = h(x)$, which divides two regions where the simple diffusion and heat conduction equations are satisfied. Jump conditions hold at the flame sheet. The governing equations considered are

$$\left.\begin{array}{l}\dfrac{\partial T}{\partial x}=\dfrac{\partial^2 T}{\partial y^2}, \\[2mm] \dfrac{\partial \phi}{\partial x}=\dfrac{\partial^2 \phi}{\partial y^2}+\lambda\dfrac{\partial^2 T}{\partial y^2},\end{array}\right\} \quad 0<y<h(x), \qquad \begin{array}{r}(6.98)\\[4mm](6.99)\end{array}$$

$$T=1+T_\infty, \qquad \dfrac{\partial \phi}{\partial x}=\dfrac{\partial^2 \phi}{\partial y^2}, \qquad y>h(x). \quad (6.100)$$

The conditions on the flame sheet, $y = h(x)$, are

$$T = 1 + T_\infty, \tag{6.101}$$

$$[\phi] = 0 = [T], \tag{6.102}$$

$$\left[\frac{\partial \phi}{\partial y}\right] = -\lambda\left[\frac{\partial T}{\partial y}\right] = \lambda \, \exp\left\{\frac{\phi_{y=h}}{2(1 + T_\infty)^2}\right\}, \tag{6.103}$$

where [ ] denotes the jump from one side of the front to the other as usual. The other boundary conditions are

$$\partial T/\partial y = 0 = \partial\phi/\partial y, \qquad y = 0, \tag{6.104}$$

and on $x = 0$ we have

$$\left.\begin{array}{l} T(0, y) = T_\infty + \exp(y - h_0), \\ \phi = -\lambda(y - h_0)\exp(y - h_0), \end{array}\right\} \quad 0 < y < h_0, \qquad x = 0 \tag{6.105}$$

$$T = 1 + T_\infty, \qquad \phi = 0, \qquad y > h_0, \qquad x = 0. \tag{6.106}$$

In these expressions T denotes temperature, ϕ measures the sum of temperature and concentration of fuel related to its value far behind the flame, and $\lambda = (1 - 1/L)\theta$, where θ is the activation energy and L the Lewis number.

It is obvious on inspection that equations (6.100) are simply the constant-temperature forms of (6.98) and (6.99), which are therefore valid over the whole domain except perhaps on $y = h(x)$. The conservation form of (6.99) on the flame front can be obtained by integrating (6.99) with respect to y across the front, having replaced $\partial\phi/\partial x$ by $(\partial\phi/\partial y)(dh/dx)$. It is $[\phi]\,dh/dx = -[\partial\phi/\partial y + \lambda\,\partial T/\partial y]$, which shows that the first of the jump conditions in each of (6.102) and (6.103) are satisfied.

On the other hand, the conservative form of (6.98) gives $[T]\,dh/dx = [\partial T/\partial y]$, which does not agree with the second of each of (6.102) and (6.103). Crowley (1981) therefore modifies equation (6.98) by the addition of a term which vanishes everywhere except on $y = h(x)$, where it has the jump required in the second of (6.103). When (6.98) is rewritten as

$$\frac{\partial T}{\partial x} + \frac{\partial}{\partial y}\left\{H(1 + T_\infty - T)\exp\left(\frac{\phi_{y=h}}{2(1 + T_\infty)^2}\right)\right\} = \frac{\partial^2 T}{\partial y^2},$$

where $H(u) = 0$, $u \leqslant 0$ and $H(u) = 1$, $u > 0$, the conservation form at the discontinuity gives

$$[T]\frac{dh}{dx} - \left[H(1 + T_\infty - T)\exp\left(\frac{\phi_{y=h}}{2(1 + T_\infty)^2}\right)\right] = -\left[\frac{\partial T}{\partial y}\right],$$

which at $T = 1 + T_\infty$ implies the desired jump conditions in (6.102) and (6.103).

For numerical solution, the conditions $T = 1 + T_\infty$, $\phi = 0$ on $y = \ell$ for some sufficiently large $\ell > h_0$ are added and then the governing equations in conservation form, to be solved over the whole of the fixed domain, $0 \leq y \leq \ell$, are

$$\frac{\partial T}{\partial x} + \frac{\partial}{\partial y}\left\{ H(1 + T_\infty - T)\exp\left(\frac{\phi_{y=h}}{2(1 + T_\infty)^2}\right) \right\} = \frac{\partial^2 T}{\partial y^2}, \qquad (6.107)$$

$$\frac{\partial \phi}{\partial x} = \lambda \frac{\partial^2 T}{\partial y^2} + \frac{\partial^2 \phi}{\partial y^2}, \qquad (6.108)$$

together with the conditions (6.104–106).

A convenient numerical procedure, given ϕ, T at all mesh points $j\,\delta y$, $i\,\delta x$, $0 < y < \ell$, is

(i) find $h(i\,\delta x)$ where $T = 1 + T_\infty$ by quadratic extrapolation on T from the region $y < h(i\,\delta x)$, i.e. $T < 1 + T_\infty$;

(ii) evaluate $\phi_{y=h} \equiv \phi_h^i$ by quadratic extrapolation from both sides and taking the mean;

(iii) solve (6.107) explicitly using

$$T_j^{i+1} = T_j^i - \frac{\delta x}{2\,\delta y}(f_{j+1}^i - f_{j-1}^i) + \frac{\delta x}{(\delta y)^2}(T_{j+1}^i - 2T_j^i + T_{j-1}^i),$$

$$f_j^i = \begin{cases} \exp\left(\dfrac{\phi_h^i}{2(1 + T_\infty)^2}\right) & T_j^i < 1 + T_\infty, \\ 0 & T_j^i \geq 1 + T_\infty; \end{cases}$$

(iv) solve (6.108) as

$$\phi_j^{i+1} = \phi_j^i + \frac{\delta x}{2(\delta y)^2}\{(\phi_{j+1}^{i+1} - 2\phi_j^{i+1} - \phi_{j-1}^{i+1})$$
$$+ \lambda(T_{j+1}^{i+1} - 2T_j^{i+1} + T_{j-1}^{i+1}) + (\phi_{j+1}^i - 2\phi_j^i + \phi_{j-1}^i)$$
$$+ \lambda(T_{j+1}^i - 2T_j^i + T_{j-1}^i)\},$$

using successive over-relaxation.

Typical profiles of the flame sheet are shown in Fig. 6.8. Crowley (1981) was not able to quote existence and uniqueness proofs of weak solutions for the coupled systems of equations nor to prove the convergence of the numerical scheme just outlined. In another paper Crowley (1980) suggests enthalpy treatments of flame fronts in solid fuel and a quite different problem relating to the growth of a crystal as it is slowly withdrawn from a bath of its own melt. The movement of an air–liquid interface has to be considered as well as that of the solidification front, with the additional complication that a corner exists all the time in the interface with air.

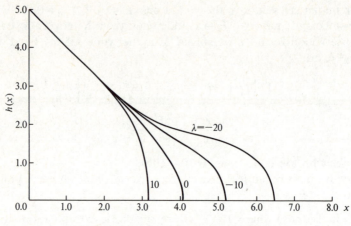

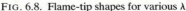

FIG. 6.8. Flame-tip shapes for various λ

6.2.8. *Conditions not amenable to enthalpy formulation*

An enthalpy formulation does not always lead to a useful method of solution. Rubinstein (1979) instances one-dimensional two-phase problems in which body-heating terms depend not only on x and t but also on $s(t)$, or in which the temperature of the phase change is not constant but is a known function of the position of the phase boundary, $u(s(t), t) = f(s(t))$. This latter condition occurs in the geophysical problem of a partially solid globe, solidifying from within, when the solidification temperature depends on the pressure which is known from the density distribution in the outer rock layer (Rubinstein 1971; Fasano and Primicerio 1977). The essence of the enthalpy method is that the position of the phase boundary is determined *a posteriori* from the numerical solution, e.g. by interpolating for the position of the level-temperature surface. Clearly this method is not applicable to the two problems proposed above. Indeed they are examples of conditions which restrict the use of weak solutions though classical solutions can be found (Gastaldi 1979).

Elliott and Ockendon (1982, p, 155) suggested a possible enthalpy formulation of an ablation problem in one space dimension. The one-phase Stefan problem in $0 < x < s(t)$ with the moving boundary condition

$$K \, \partial u / \partial x = \rho L \, ds/dt + Q \qquad \text{at} \qquad x = s(t)$$

is embedded in a fixed domain by replacing the energy supply $Q(t) > 0$ by a fictitious liquid in which

$$\partial^2 u / \partial x^2 = 0, \qquad s(t) < x < d, \qquad K \, \partial u / \partial x = Q(t), \qquad x = d,$$

where d is sufficiently large. The Stefan condition for this two-phase problem is

$$[K\ \partial u/\partial x]^{\text{liquid}}_{\text{solid}} = -\rho L\ \mathrm{d}s/\mathrm{d}t = -(K\ \partial u/\partial x)_{\text{solid}} + Q,$$

because the temperature gradient is linear in the liquid phase. Then an enthalpy function $H(u) = \rho cu$, $u < 0$, $H(u) = \rho L$, $u > 0$ can be used. However, it is not clear how such a formulation could be extended to a general ablation problem in more space dimensions.

6.3. Truncation method: alternating phase

A novel approach to moving boundary problems was originated by J. C. W. Rogers, first in connection with the Crank–Gupta oxygen diffusion and consumption problem (see §1.3.10) and later with classical Stefan problems. In the latter context, the basis of the alternating-phase truncation (APT) method of solution is to solve a heat-flow problem in either the liquid or the solid phase alternately in successive time steps. Precautions are taken to ensure that energy is conserved.

The method can conveniently be described in terms of the enthalpy formulation of the one-dimensional Stefan problem (Berger, Ciment, and Rogers 1979) defined by

$$\frac{\partial H}{\partial t} = \frac{\partial}{\partial x}\left(d_i\frac{\partial H}{\partial x}\right), \qquad H < 0, \qquad t > 0, \tag{6.109}$$

$$\frac{\partial H}{\partial t} = \frac{\partial}{\partial x}\left(d_w\frac{\partial H}{\partial x}\right), \qquad H > \lambda, \qquad t > 0, \tag{6.110}$$

$$H(a, t) = F(t), \qquad t \geq 0, \tag{6.111}$$

$$H(b, t) = G(t), \qquad t \geq 0, \tag{6.112}$$

$$H(x, 0) = H_0(x), \qquad a < x < b, \tag{6.113}$$

$$H^+ = \lambda, \qquad H^- = 0, \qquad t \geq 0, \tag{6.114}$$

$$-d_w(0)\frac{\partial H^+}{\partial x} + d_i(0)\frac{\partial H^-}{\partial x} = \lambda\frac{\mathrm{d}s}{\mathrm{d}t}, \qquad t > 0, \tag{6.115}$$

where the enthalpy $H(u)$ is given by

$$H(u) = \left\{ \begin{array}{ll} -\displaystyle\int_u^0 C_i(\xi)\ \mathrm{d}\xi, & u < 0, \\[3mm] \displaystyle\int_0^u C_w(\xi)\ \mathrm{d}\xi + \lambda, & u \geq 0. \end{array} \right\} \tag{6.116}$$

In the above, λ is the enthalpy jump, C_i, C_w the heat capacities, and K_i, K_w the thermal conductivities of ice and water respectively, with $d_i = K_i/C_i$, $d_w = K_w/C_w$, $F > \lambda$, $G < 0$ are given boundary data, and H_0 is the given initial enthalpy distribution which satisfies $H \geqslant \lambda$, $a < x < s(0)$, and $H \leqslant 0$, $s(0) < x < b$.

In the discretized problem, H_j^n denotes the value of H at time $t^n = n\,\delta t$ at the point $x_j = a + j\,\delta x$, $j = 0, 1, \ldots, J$. The APT algorithm progresses the solution from time t^n to t^{n+1} in two separate half-steps. In the first half, sufficient heat is added at each point of the solid to convert it to liquid. The heat equation is then solved for the all-liquid region and then the heat that was formerly added is subtracted away again. In the second half-step, enough heat to make the liquid region everywhere solid is subtracted away, the heat equation in the all-solid region is solved, and the subtracted heat put back again. Remembering that $H < 0$ in the solid and $H > 0$ in the liquid the APT algorithm is:

(i) Define $\bar{H}_j^n = \max(H_j^n, \lambda)$, $j = 0, 1, \ldots, J$. This means that in the solid phase $\bar{H}_j^n = \lambda$ at all points which implies an addition of heat to H^n in that phase; in the liquid phase $\bar{H}_j^n = H_j^n$.

(ii) Solve the heat flow problem defined by (6.110, 111) and with $H(b, t) = \lambda$, starting with $\bar{H}_j^n$, $0 \leqslant j \leqslant J$ as initial values, by any suitable finite-difference or finite-element method to obtain values of $\tilde{H}_j^{n+\frac{1}{2}}$ at $t = t^{n+1}$.

(iii) Whatever amounts of heat were added to the H_j^n in step (i) are now subtracted from the $\tilde{H}_j^{n+\frac{1}{2}}$ values to obtain $H_j^{n+\frac{1}{2}}$, i.e. $H_j^{n+\frac{1}{2}} = \tilde{H}_j^{n+\frac{1}{2}} - (\bar{H}_j^n - H_j^n)$, $j = 1, \ldots, J-1$, and $H_0^{n+\frac{1}{2}} = F(t^n)$, $H_J^{n+\frac{1}{2}} = G(t^n)$. This is the heat-conserving step. The solid phase is now treated similarly, i.e.

(iv) Define $\bar{H}^{n+\frac{1}{2}} = \min(H_j^{n+\frac{1}{2}}, 0)$, $j = 0, \ldots, J$, i.e. by appropriate subtraction of heat.

(v) Solve the problem defined by (6.109), (6.112) and with $H(a, t) = 0$, starting with $\bar{H}_j^{n+\frac{1}{2}}$, $0 \leqslant j \leqslant J$ as initial values at $t = t^n$, to obtain values of $\tilde{H}_j^{n+1}$ at $t = t^{n+1}$.

(vi) Whatever amounts of heat were subtracted in step (iv) are now added to the $\bar{H}_j^{n+1}$ values to obtain H_j^{n+1}, i.e. $H_j^{n+1} = \tilde{H}_j^{n+1} + (H_j^{n+\frac{1}{2}} - \bar{H}_j^{n+\frac{1}{2}})$, $j = 1, \ldots, J-1$, and $H_0^{n+1} = F(t^{n+1})$, $H_J^{n+1} = G(t^{n+1})$. This is the second heat-conserving step.

An advantage of this fixed-domain method is that if the heat parameters are taken to be constant in each phase the matrix of the simultaneous equations arising from the discretization is invariant and only one inversion is required; advancing the solution in time involves only matrix multiplication. Extension to higher space dimensions is straightforward and solutions relatively easy to obtain. An illustrative one-dimensional APT solution is shown to compare favourably with the known analytical solution.

An analytical version of the APT algorithm for a one-dimensional, one-phase Stefan problem is shown by Berger *et al.* (1979) to have $O([\delta t \ln(1 + t/\delta t)]^{\frac{1}{2}})$ rate of convergence. Numerical experiments indicate that this is a conservative estimate.

Earlier work by Rogers (1977) demonstrated that the oxygen consumption problem described in §1.3.10 is equivalent to the solution of a parabolic equation with non-linear absorption on a fixed domain. This work motivated a fixed-domain numerical method which incorporates the use of truncation (Berger, Ciment, and Rogers, 1975). These authors reformulated the oxygen diffusion problem set out in equations (1.57–60) by considering the family of non-linear problems

$$\frac{\partial c}{\partial t} = \frac{\partial^2 c}{\partial x^2} - g(c), \qquad 0 < x < 1, \qquad 0 < t < T, \tag{6.117}$$

$$\lim_{t \to 0} c(x, t) = f(x), \qquad 0 < x < 1, \tag{6.118}$$

$$\partial c/\partial x = 0, \qquad x = 0, \qquad 0 < t < T, \tag{6.119}$$

$$c = 0 \; (\text{or } \partial c/\partial x = 0), \qquad x = 1, \qquad 0 < t < T, \tag{6.120}$$

where

$$g = g_\varepsilon(c) = \begin{cases} c/\varepsilon, & 0 \le c \le \varepsilon, \\ 1, & \varepsilon \le c. \end{cases} \tag{6.121}$$

There is an associated family of solutions $c_\varepsilon(x, t)$ and a function $c(x, t) = \lim_{\varepsilon \to 0} c_\varepsilon(x, t)$, where c and $\partial c/\partial x$ are continuous functions of x and t in the relevant domain, and $c(x, t)$ satisfies (6.117–20). With

$$g(c) = \begin{cases} 0, & c = 0, \\ 1, & c > 0, \end{cases} \tag{6.122}$$

the function $c(x, t)$ satisfies (6.117) almost everywhere and is in fact a classical solution for which c is everywhere non-negative for all t. With $f(x) = \frac{1}{2}(1 - x)^2$ we have the oxygen diffusion problem. Berger *et al.* (1975) established that for reasonable physical constants and an absorption function $g(c)$ defined by (6.121) or (6.122), or a third form

$$g(c) = Vc/(K + c), \tag{6.123}$$

any standard numerical method of solution of (6.117–20) on a fixed domain could produce negative values of concentration $c(x, t)$, unless the time step were prohibitively small. This realization motivated their truncation algorithm in which any negative values that appear in the numerical solution are set equal to zero after each time step.

Thus, to advance the numerical solution of (6.117–20) with (6.122), for example, to t^{n+1}, any standard numerical scheme is used to obtain intermediate solution values, W_j^{n+1}, given the numerical solution, c_j^n, at

TABLE 6.4

Dependence of error in $10^3 c(x, t)$ on Δt: time
$t = 0.5$

				x		
Δt	0	0.1	0.2	0.3	0.4	0.5
0.05	0	2.23	4.53	6.96	4.84	
0.025	0	1.18	2.39	3.65	4.84	
0.0125	0	0.594	1.20	1.84	2.52	
analytical solution	107	70.3	40.8	18.7	4.84	0

Finite elements and Crank–Nicolson in time; $\Delta x = 0.1$; $g(c) \equiv 1$.

time t. The values for c_j^{n+1} are then obtained by truncating the negative values and taking

$$c_j^{n+1} = \max(0, W_j^{n+1}) \tag{6.124}$$

at each grid point $j\,\delta x$. The location of the moving front is taken to be the position of the first zero of c_j^{n+1}.

Berger *et al.* (1975) showed satisfactory graphical results for the Crank–Gupta oxygen problem and for a corresponding two-dimensional version. They also examined a similar oxygen consumption problem but modified the initial and fixed boundary condition so that an analytical solution was known. Thus for

$$c(x, 0) = \exp(x - 1) - x, \qquad c(0, t) = e^t - t,$$

the solution is

$$c(x, t) = \begin{cases} -x - t + \exp(x + t - 1), & x < s(t) = 1 - t, \\ 0, & x \geq s(t). \end{cases}$$

Tables 6.4 and 6.5 show the true solution at two different times and

TABLE 6.5

Dependence of error in $10^3 c(x, t)$ on form of $g(c)$ used in truncation method: time $t = 0.27$

					x				
	0	0.1	0.2	0.3	0.4	0.5	0.6	0.7	0.8
$g(c) \equiv 1$	0	0.155	0.317	0.493	0.689	0.910	1.16	0.446	0
$g(c)$ from (6.122)	0	−0.0437	−0.0890	−0.137	−0.187	−0.233	−0.251	−0.446	0
Analytical solution	212	163	119	80.5	48.9	24.5	8.10	0.446	0

Finite elements and Crank–Nicolson in time; $\Delta x = 0.01$, $\Delta t = 0.006$.

errors in numerical results obtained by the truncation method using $g(c) \equiv 1$ in Table 6.4 and also $g(c)$ given by equation (6.122) in Table 6.5 which is to be preferred. Table 6.4 indicates a convergence $O(\Delta t)$ as suggested by the authors' analysis. Comparisons between truncation results and linear complementarity solutions are made in §6.4.4.

Error estimates are examined formally by Berger and Falk (1977) for an explicit truncation method and for an implicit version by Hager (1980).

Evans and Gourlay (1977) combined the truncation technique of Berger *et al.* (1975) with the hopscotch scheme due to Gourlay (1970) in order to solve the oxygen diffusion with consumption problem in a semi-infinite, two-dimensional domain covered by a finite layer of another non-consuming material which contains an array of disc electrodes.

6.4. Variational inequalities

Another possibility is to formulate moving boundary problems in terms of certain inequalities, either of a differential or variational nature. In either case the expressions refer to a fixed domain and explicit use of the Stefan or other interface condition is avoided. Both differential and variational inequalities can lead to numerical algorithms, but rigorous analytical proofs stem mainly from the variational forms. Their dual usefulness is one attractive feature.

One of the earliest publications on theoretical properties of variational inequalities was written by Lions and Stampacchia (1967). They established the existence and uniqueness of the solution of the variational inequality

$$a(u, v-u) \geqslant (f, v-u), \tag{6.125}$$

defined as the function, u, for which the inequality is satisfied for all test functions v which satisfy certain conditions. The nomenclature with reference to a domain D in two space dimensions, for example, is that

$$a(u, v-u) = \iint_D \nabla u \cdot \nabla(v-u) \, dx \, dy$$

$$= \iint_D \{u_x(v_x - u_x) + u_y(v_y - u_y)\} \, dx \, dy \tag{6.126}$$

and

$$(f, v-u) = \iint_D f(v-u) \, dx \, dy. \tag{6.127}$$

Typically, if u is the solution of an equivalent elliptic partial differential equation with Dirichlet boundary conditions, the functions v will agree with u on the boundary ∂D of the domain D and will be square summable together with their first derivatives, which means that v, v_x, v_y exist on D and $\iint (v^2 + v_x^2 + v_y^2)\,dx\,dy < \infty$. Depending on the problem, v will be bounded above or below, e.g. $v \geqslant 0$ on D.

By analogy with the classification of partial differential equations, variational inequalities of this type are termed elliptic. Their applications to mechanics and many physical problems are described in the book by Duvaut and Lions (1976). Their application to free-boundary stationary problems has been developed by Baiocchi and his colleagues, following an approach introduced by Baiocchi (1971, 1972). This aspect is discussed more fully in Chapters 2 and 8 of the present book. Good introductions and numerous references are given by Bruch (1980) and by Elliott and Ockendon (1982).

Moving boundary problems can be expressed in terms of variational inequalities of parabolic type. A general problem can be defined in relation to Fig. 6.9, where $S(t)$ is the unknown moving boundary separating the region Ω^+, where the variable u is positive, from Ω_0, where u is identically zero. The function u satisfying conditions $u = 0$ in Ω_0, $u = u_n = 0$ on $S(t)$, $u = 0$ on Γ_2, $u = g \geqslant 0$ on Γ_1, and $u = u_0 \geqslant 0$, $t = 0$ can be defined as the solution of the variational inequality

$$(u_t, v - u) + a(u, v - u) \geqslant (f, v - u) \tag{6.128}$$

to be satisfied for all test functions v which, as in the elliptic inequality above, are square summable together with their first derivatives and which satisfy the conditions $v = g$ on Γ_1, $v = 0$ on Γ_2, and $v \geqslant 0$. Lions (1969) and Brezis (1972a,b) showed the existence and uniqueness of the solution to such parabolic inequalities and Brezis gave regularity results for the solution u.

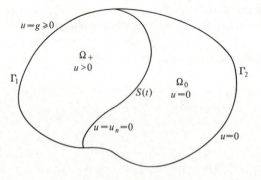

Fig. 6.9. General domain

It is easy to confirm that the inequality (6.128) is equivalent to the formulation

$$
\left.\begin{aligned}
&u_t - \nabla^2 u = f \quad \text{in} \quad \Omega_+, \\
&u = 0 \quad \text{in} \quad \Omega_0, \\
&u = u_n = 0 \quad \text{on} \quad S(t), \\
&u = 0 \quad \text{on} \quad \Gamma_2; \qquad u = g \geqslant 0 \quad \text{on} \quad \Gamma_1, \\
&u = u_0 \geqslant 0, \qquad t = 0,
\end{aligned}\right\}
\tag{6.129}
$$

again with reference to Fig. 6.9. Thus

$$
(u_t, v-u) + a(u, v-u) = \iint\limits_{\Omega_+} u_t(v-u)\, dx\, dy + \iint\limits_{\Omega_+} \nabla u \cdot \nabla(v-u)\, dx\, dy
$$

$$
= \iint\limits_{\Omega_+} (u_t - \nabla^2 u)(v-u)\, dx\, dy
$$

on integrating by parts and remembering that $v - u = 0$ on Γ_1 and grad $u = 0$ on $S(t)$. Hence

$$
(u_t, v-u) + a(u, v-u) = \iint\limits_{\Omega_+} f(v-u)\, dx\, dy \geqslant \iint\limits_{\Omega} f(v-u)\, dx\, dy
$$

which confirms (6.128) provided f is such that $u_t - \nabla^2 u - f \geqslant 0$ in the whole domain.

We shall see in §6.4.2 that the oxygen consumption problem is immediately expressible in terms of a parabolic variational inequality on a fixed domain. This is only possible, however, for Stefan problems after a Baiocchi (1972) type transformation of variable introduced by Duvaut (1973, 1975) has been applied. The new variable is continuously differentiable on the whole of a fixed domain and in particular the discontinuity in the first derivative of temperature on the freezing boundary does not appear in the new variable. Frémond (1974) used the new variable sometimes referred to as the 'freezing index'.

Various moving-boundary problems are formulated as parabolic variational inequalities in the sections that follow, their equivalence to other formulations is established, and their numerical solution is described. Theoretical aspects are dealt with by Duvaut (1973) for one-phase Stefan problems, Duvaut (1975) for two phases and by various other authors including Brezis (1972a,b), Lions (1969, 1976), Bensoussan, Lions and Papanicolau (1975), Johnson (1976), Oden and Kikuchi (1978), Aguirre-Puente and Frémond (1976), and Elliott and Ockendon (1982).

6.4.1. Introductory example

The oxygen diffusion problem defined by equations (1.57–60) is readily amenable to a variational approach (Baiocchi and Pozzi 1976). It is easy to confirm that these equations can be combined to give the differential inequalities

$$\frac{\partial c}{\partial t} - \frac{\partial^2 c}{\partial x^2} + 1 \geqslant 0, \qquad c \geqslant 0, \tag{6.130}$$

together with the equality

$$\left(\frac{\partial c}{\partial t} - \frac{\partial^2 c}{\partial x^2} + 1\right) c = 0. \tag{6.131}$$

The first of (6.130) is satisfied with equality in the region $0 < x < s(t)$ by virtue of (1.57) and the inequality holds in $s(t) \leqslant x \leqslant 1$, where $c \equiv 0$, and hence the first two terms of (6.130) vanish. The second of (6.130) follows from (1.60) and (1.58) and the fact that $c \equiv 0$ in $s(t) \leqslant x \leqslant 1$. The equality (6.131) is true because at every point in $0 \leqslant x \leqslant 1$ one of the factors vanishes.

A fully implicit finite-difference form of (6.130) is

$$\frac{c_i^{n+1} - c_i^n}{\delta t} - \frac{c_{i+1}^{n+1} - 2c_i^{n+1} + c_{i-1}^{n+1}}{(\delta x)^2} + 1 \geqslant 0, \qquad i = 1, \ldots, N-1,$$

$$\tag{6.132}$$

where $N \delta x = 1$ and, as usual, c_i^n denotes $c(i \, \delta x, n \, \delta t)$. The zero-derivative condition (1.58) on $x = 0$ can be expressed by a centralized difference or some other approximation. The set of equations (6.132) can be expressed in matrix form

$$\mathbf{Ac} + \mathbf{b} \geqslant 0, \tag{6.133}$$

where $\mathbf{A}$ is the usual tridiagonal matrix with terms $1 + 2\lambda$ on the diagonal and $-\lambda$ as the non-diagonal terms, and where $\mathbf{b} = \delta t - c_j^n$, $\lambda = \delta t / (\delta x)^2$. The vector $\mathbf{c}^{n+1} \equiv [c_0^{n+1}, c_1^{n+1}, \ldots, c_{N-1}^{n+1}]^T$. The other relationships in (6.130) and (6.131) give

$$\mathbf{c} \geqslant 0, \qquad (\mathbf{Ac} + \mathbf{b})\mathbf{c}^T = 0. \tag{6.134}$$

On introducing $\mathbf{z} = \mathbf{Ac} + \mathbf{b}$, the problem becomes that of finding a solution $\mathbf{c}^{n+1} = [c_0^{n+1}, c_1^{n+1}, \ldots, c_{N-1}^{n+1}]^T$ of

$$\mathbf{Ac} = \mathbf{z} - \mathbf{b}, \qquad \mathbf{c}^T \mathbf{z} = 0, \tag{6.135}$$

such that

$$\mathbf{c} \geqslant 0, \qquad \mathbf{z} \geqslant 0. \tag{6.136}$$

This is a standard problem in quadratic programming and several well-known methods exist for its numerical solution. One which is amenable in the present context is a generalized successive over-relaxation method due to Cryer (1971a). Sequences of vectors $\mathbf{z}^{(k+1)}$ and $\mathbf{c}^{(k+1)}$ are generated with components given by

$$z_i^{(k+1)} = b_i + \sum_{j=1}^{i-1} A_{ij}c_j^{(k+1)} - \sum_{j=i}^{N} A_{ij}c_j^{(k)} \tag{6.137}$$

$$c_i^{(k+1)} = \max\{0,\ c_i^{(k)} + \omega z_i^{(k+1)}/A_{ii}\}. \tag{6.138}$$

Clearly, without the non-negativity condition in (6.138), corresponding to the so-called complementarity condition in (6.135), this is the standard SOR algorithm with relaxation parameter ω. Cryer (1971a) proved the convergence of the method for $0 < \omega < 2$ provided the matrix $\mathbf{A}$ is positive definite. More details of this and other SOR methods are to be found in §8.5.

An alternative is the conjugate gradient method (Elliott 1976). Ichikawa and Kikuchi (1979) suggested the use of Lagrange multipliers and a penalty method. O'Leary (1977) also used the conjugate gradient technique and compared several methods.

The quadratic programming problem of (6.135) and (6.136) is equivalent to the problem

$$\text{minimize } \mathbf{b}^T\mathbf{c} + \tfrac{1}{2}\mathbf{c}^T\mathbf{A}\mathbf{c} \quad \text{for} \quad \mathbf{c} \geq 0, \tag{6.139}$$

provided $\mathbf{A}$ is symmetric and positive definite. The Cryer algorithm therefore provides a solution of the minimization formulation (6.139) and

TABLE 6.6
Free boundary position $s(t)$

Time	CS[a] $\delta x = 0.01$, $\delta t = 0.001$	FGL[b] $\delta x = 0.05$, $\delta t = 0.001$	HH[c]
0.04	0.9991	0.9992	0.9992
0.06	0.9916	0.9918	0.9918
0.10	0.9325 (9343)	0.9346	0.9350
0.12	0.8790 (8792)	0.8781	0.8792
0.14	0.7986 (7987)	0.7966	0.7989
0.16	0.6830 (6833)	0.6799	0.6834
0.18	0.5011 (5018)	0.4942	0.5011
0.185	0.4333 (4342)	0.4178	0.4334

[a] CS = complementarity solution (Elliott 1976).
[b] FGL = fixed-grid Lagrange method (Crank and Gupta 1972a, b).
[c] HH = integral equation solution (Hansen and Hougaard 1974).
Values in parentheses obtained by Elliott and Ockendon (1982) using fast algorithm (Cottle 1979).

TABLE 6.7

Values of $10^4 c(0, t)$ on the fixed sealed surface

Time	CS $\delta x = 0.01, t = 0.001$	FGL $\delta x = 0.05, t = 0.001$	*HH*
0.04	2743	2742	2743
0.06	2236	2234	2236
0.10	1432	1430	1432
0.12	1091	1089	1091
0.14	779	777	779
0.16	488	486	488
0.18	218	216	218
0.185	153	151	153

of the more general θ-type problem (6.153) with suitable extension of the vector **b** to include all terms in (6.153) of time level $n \, \delta t$.

Elliott (1976) obtained solutions of the oxygen diffusion problem by using the projected SOR algorithm (6.137) and (6.138) with z_i^{n+1} given by the Crank–Nicolson form of the left-hand side of (6.132). Some of his results are included in Tables 6.6–8, where results obtained by other workers are assembled for comparison. It is generally accepted that those of Hansen and Hougaard (1974) are the most reliable. Elliott and Ockendon (1982) presented values of $s(t)$ obtained by using the fast algorithm (Cottle 1979) described in §8.5 and their values are indicated in parentheses in Table 6.6. Other results for this problem are given in

TABLE 6.8

Values of $10^4 c(x, t)$. At each time level: top line = CS, $\delta x = 0.05$, $\delta t = 0.005$; middle line = FGL, $\delta x = 0.05$, $\delta t = 0.001$; bottom line = HH

Time	*x* 0.1	0.3	0.5	0.7	0.9
0.06	2173	1710	1019	394	41
	2173	1713	1022	396	42
	2172	1710	1019	395	41
0.10	1394	1108	659	230	5
	1394	1110	661	232	6
	1393	1108	659	231	6
0.14	755	571	284	42	
	754	571	285	42	
	754	571	284	42	

Tables 3.1, 3.2, 4.1, 4.2, and 6.3. The Fourier-series solution of Dahmardah and Mayers (1983) agrees closely with the values obtained by Hansen and Hougaard (1974) except in the final stages (see §5.1).

6.4.2. Variational form

A variational form of (6.130) and (6.131) is easily derived by multiplying the left-hand side of the first of (6.130) by $(v-c)$ using the test functions v which are non-negative in $0 < x < 1$ with $v = 0$ on $x = 1$, and for which v and $\partial v/\partial x$ are square integrable in $0 < x < 1$, i.e. $\int_0^1 v^2 \, dx$ and $\int_0^1 (\partial v/\partial x)^2 \, dx$ are bounded. Thus

$$\int_0^1 \left(\frac{\partial c}{\partial t} - \frac{\partial^2 c}{\partial x^2} + 1 \right)(v-c) \, dx = \int_0^1 \frac{\partial c}{\partial t}(v-c) \, dx$$

$$- \left[(v-c) \frac{\partial c}{\partial x} \right]_0^1 + \int_0^1 \frac{\partial c}{\partial x} \frac{\partial}{\partial x}(v-c) \, dx + \int_0^1 (v-c) \, dx.$$

Remembering that $v = c = 0$, $x = 1$, and $\partial c/\partial x = 0$, $x = 0$, together with use of (6.131), we obtain after slight rearrangement

$$\int_0^1 \frac{\partial c}{\partial t}(v-c) \, dx + \int_0^1 \frac{\partial c}{\partial x} \frac{\partial}{\partial x}(v-c) \, dx$$

$$= -\int_0^1 (v-c) \, dx + \int_0^1 \left(\frac{\partial c}{\partial t} - \frac{\partial^2 c}{\partial x^2} + 1 \right) v \, dx.$$

But the final term on the right-hand side of this equation is non-negative since $v \geq 0$, and the first of (6.130) is to be satisfied. We therefore have the variational inequality

$$\int_0^1 \frac{\partial c}{\partial t}(v-c) \, dx + \int_0^1 \frac{\partial c}{\partial x} \frac{\partial}{\partial x}(v-c) \, dx \geq -\int_0^1 (v-c) \, dx. \qquad (6.140)$$

Using the nomenclature of (6.126) and (6.127), the inequality formulation of the oxygen diffusion problem (6.140), equivalent to the differential forms (6.130) and (6.131), becomes

$$(c_t, v-c) + a(c, v-c) \geq (-1, v-c). \qquad (6.141)$$

We are required to find $c(x, t)$ for each t such that (6.141) is satisfied for all admissible test functions $v(x, t)$.

A suitable finite-element discretization of the variational inequality (6.141) leads back to the matrix condition (6.133) and hence to the algebraic problem posed by (6.135) and (6.136). This can be seen by considering the slightly more general inequality

$$(u_t, v-u) + a(u, v-u) \geq (f, v-u). \qquad (6.142)$$

Let both u and the trial functions v be expressed in terms of basis functions ϕ_i by

$$u = \sum_{i=1}^{N} u_i \phi_i = u_i \phi_i ; \qquad v = \sum_{i=1}^{N} v_i \phi_i = v_i \phi_i, \qquad (6.143)$$

where the suffix convention implying summation over all nodes of the finite element grid is used for convenience, and u_i, v_i, $i = 1, \ldots, N$, are the values at the successive nodes. As usual, ϕ_i is a basis function associated with node i, which vanishes outside all elements, for example triangular elements, that have node i as a vertex. For simplicity of explaining the alternative formulation of the oxygen diffusion problem, we confine attention for the moment to one space dimension so that the node i is common to only the two linear elements $(i-1)h \le x \le ih$ and $ih \le x \le (i+1)h$ in the line $0 \le x \le 1$ and $h = 1/N$. Recalling the variational inequality (6.142) in the expanded form of (6.140) and substituting from (6.143) we obtain

$$\int_0^1 \left(\sum_{i=1}^{N} \frac{\partial u_i}{\partial t} \phi_i \right) \left(\sum_{j=1}^{N} (v_j - u_j) \phi_j \right) dx$$

$$+ \int_0^1 \left(\sum_{i=1}^{N} u_i \frac{\partial \phi_i}{\partial x} \right) \left(\sum_{j=1}^{N} (v_j - u_j) \frac{\partial \phi_j}{\partial x} \right) dx$$

$$- \int_0^1 f \left(\sum_{j=1}^{N} (v_j - u_j) \phi_j \right) dx \ge 0. \quad (6.144)$$

Introducing a mass matrix M_{ij}, a stiffness matrix K_{ij}, and a vector f_i given by

$$M_{ij} = (\phi_i, \phi_j), \qquad K_{ij} = a(\phi_i, \phi_j), \qquad f_i = (f, \phi_j), \qquad (6.144\text{a})$$

and noting that M_{ij} and K_{ij} are symmetric, we obtain on interchanging i and j

$$M_{ij} \frac{\partial u_j}{\partial t} (v_i - u_i) + K_{ij} u_j (v_i - u_i) - f_i (v_i - u_i) \ge 0, \qquad (6.145)$$

where the shorter notation $u = u_i \phi_i$, $v = v_i \phi_i$ introduced in (6.143) is used. We can now use a θ-type finite-difference replacement for $\partial u_j / \partial t$ and let u_j^n denote the value of u_i at the point $i = j$ and time $n\, \delta t$, with u_j^{n+1} for the corresponding value at time $(n+1)\, \delta t$. The inequality (6.145) becomes

$$M_{ij} \frac{u_j^{n+1} - u_j^n}{\delta t} (v_i - u_i) + [\theta K_{ij} u_j^{n+1} + (1-\theta) K_{ij} u_j^n](v_i - u_i)$$

$$- f_i (v_i - u_i) \ge 0, \quad (6.146)$$

or alternatively, on grouping terms, we have

$$[M_{ij}/\delta t + \theta K_{ij}]u_j^{n+1}(v_i - u_i^{n+1}) \geq [f_i + \{M_{ij}/\delta t - (1-\theta)K_{ij}\}u_j^n](v_i - u_i^{n+1}).$$
(6.147)

If now we take the ϕ_i to be a set of piecewise linear basis functions satisfying

$$\phi_i = \begin{cases} \{x - (i-1)h\}/h, & (i-1)h \leq x \leq ih, \\ \{(i+1)h - x\}/h. & ih \leq x \leq (i+1)h, \end{cases}$$
(6.148)

and use the lumped integration familiar in finite-element methods, we have immediately that M_{ij} is a diagonal matrix with elements h. Noting that $\partial \phi/\partial x = \pm(1/h)$ it follows that K_{ij} is tridiagonal with diagonal elements $2/h$ and off-diagonal elements $-1/h$. If now we multiply both sides of (6.147) by $\delta t/h$, it takes the form

$$\mathbf{A}u_j^{n+1}(v_i - u_i^{n+1}) + \mathbf{b}(v_i - u_i^{n+1}) \geq 0,$$
(6.149)

where $\mathbf{A}$ is a positive definite, tridiagonal matrix with constant diagonal elements $(1+2\lambda\theta)$ and off-diagonal elements $-\lambda\theta$, and $\mathbf{b} = f_j \delta t - \{1 - (1-\theta)K_{ij}\}u_j^n$, where $\lambda = \delta t/h^2$. For $f = 1$ and $\theta = 1$ these $\mathbf{A}$ and $\mathbf{b}$ are the same as the matrices in (6.135). The solution is advanced by solving at each time step the linear complementarity equivalent of (6.149) by the Cryer algorithm of type (6.137), (6.138) or otherwise, e.g. for the oxygen diffusion problem we regain (6.135), (6.136).

6.4.3. Minimization formulation

Finally, we recall that the solution of an elliptic partial differential equation can sometimes be expressed as the function for which a certain integral takes a stationary value, often a minimum value. Correspondingly, the solution u of an elliptic variational inequality such as

$$(u, v-u) + a(u, v-u) \geq (f, v-u)$$
(6.150)

is the function which minimizes the quadratic expression

$$J(v) = (v, v) + a(v, v) - 2(f, v),$$
(6.151)

provided $a(u, v)$ is symmetric, i.e. $a(u, v) = a(v, u)$. Thus

$$J(v) - J(u) = (v, v) - (u, u) + a(v, v) - a(u, u) - \{2(f, v) - 2(f, u)\}$$
$$= (u - v, u - v) + 2(u, v - u) + 2a(u, v - u) - 2(f, v - u) \geq 0$$

if (6.150) is true. Equally, if u is known to minimize $J(v)$ over a set of values of v for which $(1-p)u + p(v)$ is contained in the same set with $0 < p < 1$, i.e. the set is convex, then immediately $(1/p)\{J((1-p)u + pv) - J(u)\} \geq 0$ and so

$$p(v-u, v-u) + pa(v-u, v-u) + 2(u, v-u) + 2a(u, v-u) - 2(f, v-u) \geq 0.$$

In the limit as $p \to 0$ we regain (6.150). The variational inequality and the minimization problem are equivalent.

With this equivalence in mind we return to the parabolic inequality (6.142) and rewrite it in terms of general discrete variables u_h^n and v_h^n of the type defined by (6.143) but without specifying the precise form of the ϕs. Thus u_h^n, for example, denotes the discrete values of the function u at the nodes of a finite-element mesh of size h at time $n \, \delta t$. The values of u_h^{n+1} are given by the solutions of a sequence of elliptic variational inequalities for successive intervals of time δt, i.e.

$$\left(\frac{u_h^{n+1} - u_h^n}{\delta t}, v_h - u_h^{n+1} \right)_h + a(u_h^{n+\theta}, v_h - u_h^{n+1})_h \geq (f_h^{n+\theta}, v_h - u_h^{n+1})_h,$$

$$(6.152)$$

which can be written as discrete forms of (6.150) by incorporating the term $u_h^n/\delta t$ with $f_h^{n+\theta}$. By using the discretized version of (6.151), we see that (6.152) is equivalent to the problem of minimizing the expression

$$(v_h, v_h)_h + \theta \, \delta t a(v_h, v_h)_h - 2 \, \delta t \left\{ \left(f_h^{n+\theta} + \frac{u_h^n}{\delta t}, v_h \right)_h - \frac{(1-\theta)}{2} a(u_h^n, v_h)_h \right\}$$

for all admissible v_h. Thus in this formulation the oxygen diffusion problem requires the minimization of the matrix expression

$$\mathbf{v}^T(\mathbf{M}/\delta t + \theta \mathbf{K})\mathbf{v} - 2 \left\{ \mathbf{f} + \left(\mathbf{M}/\delta t - \frac{(1-\theta)}{2} \mathbf{K} \right) u^n \right\}^T \mathbf{v} \qquad (6.153)$$

for $v_i \geq 0$, where the matrices $\mathbf{M}$ and $\mathbf{K}$ and the vector $\mathbf{f}$ are the $M_{i,j}$, $K_{i,j}$, and f_j of (6.144a) and we regain (6.139) for $f = 1$, $\theta = 1$, and the $\mathbf{A}$ and $\mathbf{b}$ related to $\mathbf{M}$ and $\mathbf{K}$ as in (6.149).

Lions (1969) and Brezis (1972a,b) proved that a unique solution of the parabolic inequality (6.142) exists. Elliott and Ockendon (1982) established that if the weak form, obtained by integrating (6.142) with respect to time, i.e.

$$\int_0^T [(v_t, v - u) + a(u, v - u) - (f, v - u)] \, dt \geq \tfrac{1}{2} |v(T) - u(T)|^2 - \tfrac{1}{2} |v(0) - u_0|^2,$$

has a solution in the relevant time-dependent domain, $0 \leq t \leq T$, then it is a unique solution. They also snowed that a unique solution of a discrete form of (6.142), i.e. of (6.152), exists and converges to the unique solution of the weak form and hence to the unique solution of (6.142). Stability criteria were also established for θ-forms of (6.152). Brezis (1972b), Bensoussan et al. (1975), and Elliott and Ockendon (1982) established the equivalence of the differential forms (6.130) and (6.131) with the variational inequality (6.142) and its weak form.

6.4.4. Review of equivalent forms

We have formulated the oxygen diffusion problem in three apparently different ways: (i) the system of differential equations and inequalitiies (6.130) and (6.131) and their discrete equivalents (6.135) and (6.136); (ii) the parabolic variational inequality (6.141) or (6.142) and the discretized form (6.152); and (iii) the minimization problem of (6.139) and (6.153).

The latter two forms are alternative statements of a problem in quadratic programming. In that context, the conditions for optimality, known as the Kuhn–Tucker conditions, are in fact the algebraic statements (6.135) and (6.136). They arise here from a discretization of the differential forms (6.130) and (6.131) which constitute a continuous linear complementarity problem by virtue of the equation (6.131). We have further seen that a linear, finite-element discretization of (6.141) for this particular problem leads again to the Kuhn–Tucker conditions (6.135) and (6.136).

Furthermore, Elliott (1976) draws attention to the close connection between the discretized linear complementary problem and the truncation algorithm of Berger *et al.* (1975) discussed in connection with oxygen diffusion equations (6.117–122) above. If we use the θ-type discretization of (6.130) instead of the fully implicit form (6.132) we obtain in matrix notation

$$(\mathbf{M}/\delta t)(\mathbf{c}^{n+1} - \mathbf{c}^n) + \mathbf{K}\mathbf{c}^{n+\theta} + \mathbf{1} = \mathbf{z} \geqslant 0, \qquad \mathbf{c} \geqslant 0, \qquad (6.154)$$

and from (6.131) the complementarity condition

$$\mathbf{z}^{\mathrm{T}}\mathbf{c}^{n+1} = 0. \qquad (6.155)$$

Here $\mathbf{K}$ is the usual matrix expression for the finite-difference form of the second derivative; $\mathbf{M}$ is a diagonal matrix so that in the explicit case of $\theta = 0$ the solution of (6.154) and (6.155) is simply

$$c_i^{n+1} = \max\{(\mathbf{M}^{-1}\mathbf{F})_i, 0\}, \qquad (6.156)$$

where $\mathbf{F}$ contains all terms in the first equality of (6.154) which do not involve $\mathbf{c}^{n+1}$, multiplied by δt. Berger *et al.* (1975) also solve the first equation (6.154) and project by (6.156) to ensure $c_i^{n+1} \geqslant 0$. When $\theta = 0$, therefore, or whenever $\mathbf{M}$ is the diagonal lumped mass matrix, the truncation method is identical with the solution of (6.154) and (6.155). Otherwise Elliott (1976) and Elliott and Ockendon (1982) expect truncation to be less accurate because it does not take account of the complementarity of (6.155). Instead of using the algorithm defined by (6.137) and (6.138), Berger *et al.* (1975) solve equations like

$$
\left.
\begin{aligned}
\frac{W_i^{n+1} - c_i^n}{\delta t} &= \frac{\theta}{h^2}(W_{i+1}^{n+1} - 2W_i^{n+1} + W_{i-1}^{n+1}) \\
&\quad + \frac{(1-\theta)}{h^2}(c_{i-1}^n - 2c_i^n + c_{i+1}^n) - 1
\end{aligned}
\right\}
$$

(6.157)

and then take

$$
c_i^{n+1} = \max(W_i^{n+1}, 0).
$$

The claim about the relative accuracy of the two methods is supported by Elliott's (1976) numerical solution of a model problem used by Berger (1976) to test the truncation method. It is to solve

$$
\left.
\begin{aligned}
\frac{\partial u}{\partial t} &= \frac{\partial^2 u}{\partial x^2} + f, \qquad 0 < x < 1 - t^2, \\
u &= 0, \qquad x = 0, \qquad 1, \\
f &= (40xt - 20x - 40)\exp(x + t^2 - 1) - 40xt + 44,
\end{aligned}
\right\}
$$

(6.158)

and the initial data $u(x, 0)$ are taken to conform with the exact solution

$$
u(x, t) = \begin{cases} 20x\{-x - t^2 + \exp(x + t^2 - 1)\} + \psi(x), & 0 \leqslant x \leqslant 1 - t^2, \\ \psi(x), & 1 - t^2 \leqslant x \leqslant 1 \end{cases}
$$

where $\psi(x) = 2x(1-x)$. The accuracy is expressed at any time by a relative error

$$
\left[\sum_{i=1}^{N-1} h\{u(ih) - u_i\}^2\right]^{\frac{1}{2}} \bigg/ \left[\sum_{i=1}^{N-1} h\{u(ih)\}^2\right]^{\frac{1}{2}},
$$

where u_i are the analytic values. These errors tabulated in Tables 6.9, 10 are extracted from fuller tables in Elliott (1976).

The complementarity values are clearly more accurate than the truncation ones. The Crank–Nicolson scheme ($\theta = \frac{1}{2}$) appears to offer little

TABLE 6.9

Percentage relative errors for problem (6.158) *obtained by linear complementarity method using projected* SOR

| | $\theta = 1$ | | $\theta = \frac{1}{2}$ | |
Time	$h = 1/10, \delta t = 1/100$	$h = 1/20, \delta t = 1/400$	$h = 1/10, \delta t = 1/100$	$h = 1/20, \delta t = 1/400$
0.1	0.072	0.020	0.237	0.055
0.3	0.136	0.068	0.111	0.062
0.5	0.487	0.025	0.487	0.032
0.7	0.134	0.051	0.081	0.059
0.9	0.023	0.004	0.021	0.003

TABLE 6.10

Percentage relative errors for problem (6.158) obtained by truncation method

Time	$\theta = 1$		$\theta = \frac{1}{2}$	
	$h = 1/10, \delta t = 1/100$	$h = 1/20, \delta t = 1/400$	$h = 1/10, \delta t = 1/100$	$h = 1/20, \delta t = 1/400$
0.1	0.072	0.020	0.237	0.055
0.3	1.119	0.701	0.708	0.404
0.5	4.230	1.511	2.547	0.777
0.7	5.328	1.426	2.785	0.673
0.9	0.680	0.305	0.680	0.204

improvement for the complementarity solution though it does when trunction is used. The results agree with the theoretical statement by Berger *et al.* (1975) that the difference between the two types of solution at time $t + t^*$, starting from the same data at t, is of order t^*.

Elliott and Ockendon (1982) used the fast algorithm described in §8.5 to evaluate complementarity solutions for the model problem examined in Tables 6.4, 5. This meant that the complementarity method was computationally more efficient as well as more accurate than the truncation method. Some results and errors are compared in Table 6.11 extracted from Elliott and Ockendon (1982). Their fuller table supported their theoretical conclusion that the errors are $O(h^2 + \delta t)$.

TABLE 6.11

Percentage relative errors and $s(t)$ values

t	$s = 1 - t$	$s(t):$CM[a] $h = 1/16$	% error:CM[a] $h = 1/16$	$h = 1/64$	% error:TM[a] $h = 1/64$
0.08	0.92	0.9206	0.19	0.012	0.11
0.24	0.76	0.7629	0.36	0.022	0.28
0.40	0.60	0.5972	0.39	0.022	0.47
0.56	0.44	0.4125	0.36	0.017	0.87

[a] CM = complementarity method; TM = truncation method; % error = % relative discrete L^2 error.

6.4.5. One-phase Stefan problem

An essential feature of any variational formulation is that the dependent function and certain of its derivatives must be continuous in order to permit the integrations by parts needed in the derivation of the variational inequality, for example. This requirement is satisfied by the implicit condition on the moving front $c = 0$ in the oxygen diffusion problem where $\partial c/\partial x = 0$. It is not the case, however, for the Stefan condition on a solidification boundary, where the first derivative of temperature is discontinuous for a non-zero latent heat. A variational formulation is only possible after some suitable transformation has removed the discontinuity. Such a transformation was first proposed by Duvaut (1973) for moving boundaries prompted by an idea of Baiocchi (1971) for stationary, free boundaries. Frémond (1974, 1980) called the new variable the freezing index. Subsequently, theoretical properties have been studied by Friedman and Kinderlehrer (1975), Bensoussan and Lions (1978), and Friedman (1975).

For the single-phase melting problem defined by equations (1.21–26), the Duvaut transformation, or the Frémond freezing index, is

$$w(x, t) = \int_{\ell(x)}^{t} u(x, \tau) \, d\tau, \qquad 0 \le x \le s(t); \qquad w(x, t) = 0, \qquad s(t) \le x \le 1,$$
(6.159)

where we have introduced the fixed, finite domain $0 \le x \le 1$, and where $t = \ell(x)$ defines the time when the ice at point x changes to water, i.e. $\ell(x)$ is the time when the melting boundary is at x. It is the inverse of $s(t)$ so that $\ell^{-1}(t) = s(t)$. To show that $w(x, t)$ and its first derivative $\partial w/\partial x$ are continuous on the melting boundary and throughout the range $0 \le x \le 1$, we have from (6.159) that

$$\frac{\partial w}{\partial x} = \int_{\ell(x)}^{t} \frac{\partial u}{\partial x}(x, \tau) \, d\tau, \qquad 0 \le x \le s(t), \qquad \frac{\partial w}{\partial x} = 0, \qquad s(t) \le x \le 1, \quad (6.160)$$

since the temperature u is zero on the interface $x = s(t)$ i.e. $u(x, \ell(x)) = 0$, and from (6.160) we know that $\partial w/\partial x = 0$ on $x = s(t)$. Furthermore,

$$\frac{\partial^2 w}{\partial x^2} = \int_{\ell(x)}^{t} \frac{\partial^2 u}{\partial x^2} \, d\tau - \frac{\partial \ell}{\partial x} \frac{\partial u}{\partial x}(x, \ell(x)), \qquad 0 \le x < s(t), \qquad (6.161)$$

since $\partial u/\partial x \ne 0$ on $x = s(t)$, i.e. at $(x, \ell(x))$. Noting that $\partial \ell/\partial x = -1/(\partial s/\partial t)$, so that the Stefan condition (1.26) implies $(\partial u/\partial x)(\partial \ell/\partial x) = \lambda$, equation (6.161) with use of (1.21) becomes

$$\frac{\partial^2 w}{\partial x^2} = \int_{\ell(x)}^{t} \frac{\partial u}{\partial t} \, d\tau - \lambda = \frac{\partial}{\partial t} \int_{\ell(x)}^{t} u \, d\tau - \lambda = \frac{\partial w}{\partial t} - \lambda.$$

Thus the transformed equation is

$$\frac{\partial w}{\partial t} = \frac{\partial^2 w}{\partial x^2} + \lambda, \qquad 0 \le x < s(t); \qquad w = 0, \qquad s(t) \le x \le 1, \quad (6.162)$$

with $w(x, t) > 0$ in $0 \le x < s(t)$. Thus $w(x, t)$ and its first derivative are continuous throughout the domain $0 \le x \le 1$ and so the inequalities corresponding to (6.130) and (6.131) are

$$\frac{\partial w}{\partial t} - \frac{\partial^2 w}{\partial x^2} - \lambda \geqslant 0, \qquad w \geqslant 0, \qquad 0 \leqslant x \leqslant 1, \tag{6.163}$$

$$\left(\frac{\partial w}{\partial t} - \frac{\partial^2 w}{\partial x^2} - \lambda\right)w = 0, \qquad 0 \leqslant x \leqslant 1. \tag{6.164}$$

The variational inequality is now

$$(w_t, v - w) + a(w, v - w) \geqslant (\lambda, v - w), \tag{6.165}$$

using the nomenclature adopted in (6.141).

Ichikawa and Kikuchi (1979) formulated a single-phase Stefan problem in terms of a variational inequality closely similar to (6.165) but with application to one, two, or three space dimensions and with the possibility of three sorts of condition on different segments of the boundary of the fixed domain, i.e. temperature a prescribed function of time on one segment, zero temperature on another, and a Neumann flux condition on the third segment.

Their problem is defined on an open domain D with boundary Γ in one, two, or three space dimensions and a subset Ω of D is the frozen region. On Γ_G a negative temperature $g(t)$ is prescribed and on Γ_C the temperature is zero. On Γ_F the flux is prescribed by a Neumann condition. The time-dependent interface $\Gamma_0(t)$ separates the ice and water regions and $t = S(x)$ defines the time when the water changes to ice. The formulation of the problem is: for given $g(t)$, $0 \leqslant t \leqslant T$, find $S^{-1}(t)$ and temperature $\theta(x, t)$ such that

$$\left.\begin{aligned}
&\frac{\partial \theta}{\partial t} = \nabla \cdot (k \nabla \theta) \quad \text{in } \Omega, \\
&\theta = 0 \quad \text{in } D - \Omega, \\
&\theta(x, t) = g(t) \quad \text{on } \Gamma_G, \qquad \theta(x, t) = 0 \quad \text{on } \Gamma_C, \qquad \alpha\theta = K\frac{\partial \theta}{\partial n} \quad \text{on } \Gamma_F, \\
&K \nabla \theta \cdot \nabla S(x) = \ell \quad \text{on } \Gamma_0, \qquad \theta(x, 0) = 0 \quad \text{in } D.
\end{aligned}\right\} \tag{6.166}$$

The introduction of Duvaut's transformation

$$u(x, t) = \int_{S(x)}^{t} \theta(x, \tau)\, \mathrm{d}\tau \quad \text{in } \Omega, \qquad u(x, t) = 0 \quad \text{in } D - \Omega,$$

and the reformulation of the problem follows precisely the derivation of equations (6.160) to (6.164) and leads to the equations

$$
\left.
\begin{aligned}
&u\left(\frac{\partial u}{\partial t} - \nabla \cdot (K\,\nabla u) - \ell\right) = 0 \quad \text{in } D, \\[2mm]
&u \leq 0, \qquad u(x,0) = 0 \quad \text{in } D, \\[2mm]
&\frac{\partial u}{\partial t} - \nabla \cdot (K\,\nabla u) - \ell \leq 0 \quad \text{in } D \\[2mm]
&u = \hat{g}(t) \quad \text{on } \Gamma_G, \qquad u = 0 \quad \text{on } \Gamma_C, \qquad u = K\,\partial u/\partial n \quad \text{on } \Gamma_F,
\end{aligned}
\right\}
$$

where

$$
\hat{g}(t) = \int_0^t g(\tau)\,\mathrm{d}\tau \qquad \text{for} \qquad 0 \leq t \leq \infty \qquad \text{and} \qquad \hat{g}(0) = 0.
$$

(6.167)

So far, the initial state has been taken to be water at the freezing temperature zero everywhere i.e. $\theta(x,0) = 0$ in D. Ichikawa and Kikuchi (1979) use a transformation due to Friedman and Kinderlehrer (1975) to extend their treatment to the initial state which is partially ice and partially water at zero temperature. Taking Ω_0 to denote the ice domain at the initial stage, the freezing index u is redefined as

$$
\left.
\begin{aligned}
u(x,t) &= \int_{S(x)}^t \theta(x,\tau)\,\mathrm{d}\tau \quad \text{in } \Omega - \Omega_0, \\[2mm]
&= \int_0^t \theta(x,\tau)\,\mathrm{d}\tau \quad \text{in } \Omega_0, \\[2mm]
&= 0 \quad \text{in } D - \Omega.
\end{aligned}
\right\}
$$

(6.168)

The partial differential equation for u then becomes

$$
\partial u/\partial t = \nabla \cdot (K\,\nabla u) + h,
$$

(6.169)

where $h = \ell$ in $\Omega - \Omega_0$ and $h = \theta_0(x)$ in Ω_0, where θ_0 is the initial temperature in Ω_0.

Ichikawa and Kikuchi (1979) further point out that if Γ_F is part of the boundary of the domain Ω_0 just defined, the more general boundary condition

$$
K\,\partial\theta/\partial n = \alpha\theta + \beta \quad \text{on } \Gamma_F
$$

can be considered, and that on integration with respect to time from $t = 0$ the corresponding condition for u is

$$
K\,\partial u/\partial n = \alpha u + \beta t \quad \text{on } \Gamma_F.
$$

If Γ_F is located in the boundary of Ω_0 and Ω, the generalized relation cannot be used. By multiplying the expression

$$
\partial u/\partial t - \nabla \cdot (K\,\nabla u) - \ell
$$

by $(v-u)$ and integrating by parts we obtain a generalized inequality, which incorporates the conditions on Γ_F,

$$(u_t, v-u)_D + (K\,\nabla u, \nabla(v-u))_D - (\ell, v-u)_D = (u_t - \nabla\cdot(K\,\nabla v) - \ell, v)_D$$
$$- (\alpha u, v - - u)_{\Gamma_F} \geqslant -(\alpha u, v-u)_{\Gamma_F}. \quad (6.170)$$

Here the test function $v \leqslant 0$ is used in D for all time t in $0 \leqslant t \leqslant T$ with $v = 0$ on Γ_C, $v = \hat{g}$ on Γ_G, where T is a positive real number indicating the total time of the problem. Also, the inequality

$$(u_t - \nabla\cdot(K\,\nabla u) - \ell, v)_D \geqslant 0$$

has been used and $v \leqslant 0$ so that the modified form of (6.170),

$$(u_t, v-u)_D + (K\,\nabla u, \nabla(v-u))_D + (\alpha u, v-u)_{\Gamma_F} \geqslant (\ell, v-u)_D, \quad (6.171)$$

is obtained. Ichikawa and Kikuchi (1979) quote theoretical work by Lions (1976) and Ichikawa (1978) which establishes the existence of a unique solution of (6.171) for functions v which, together with their first derivatives, are square summable on D and satisfy the boundary conditions as above.

Ichikawa and Kikuchi (1979) discretized in space using finite-elements and a generalized implicit scheme in time. They wrote (6.171) as

$$(u_t, v-u) + B(u, v-u) \geqslant (f, v-u) \quad (6.172)$$

for v as defined above and where

$$B(u, v) = (K\,\nabla u, \nabla v)_D + (\alpha u, v)_{\Gamma_F},$$
$$(u, v) = (u, v)_D, \qquad (f, v) = (\ell, v)_D.$$

The derivation corresponding to that of (6.147) shows the discretized form of (6.172) to be

$$\hat{K}_{ij} u_j^{n+1}(v_i - u_i^{n+1}) \geqslant \hat{f}_i(v_i - u_i^{n+1}), \quad (6.173)$$

where $u = \sum_{i=1}^{N} u_i \phi_i = u_i \phi_i$, $f = f_i \phi_i$, $v = v_i \phi_i$,

$$\hat{K}_{ij} = M_{ij}/\Delta t + \theta K_{ij}, \qquad \hat{F}_{ij} = M_{ij}/\Delta t - (1-\theta)K_{ij}, \qquad \hat{f}_i = f_i - \hat{F}_{ij} u_j^n,$$

and

$$M_{ij} = (\phi_i, \phi_j), \qquad K_{ij} = B(\phi_i, \phi_j), \qquad f_i = (f, \phi_i).$$

The projectional SOR algorithm becomes

$$u_{h,i}^{n+1}(k+\tfrac{1}{2}) = (1-\omega)u_{h,i}^{n+1}(k) + \omega\left(-\sum_{j=1}^{i-1} \hat{K}_{i,j} u_{h,j}^{n+1}(k+1)\right.$$
$$\left. - \sum_{j=i+1}^{N} \hat{K}_{i,j} u_{h,j}^{n+1}(k) + \hat{f}_i\right)/\hat{K}_{ii}$$
$$u_{h,i}^{n+1}(k+1) = \min(0, u_{h,i}^{n+1}(k+\tfrac{1}{2})), \quad (6.174)$$

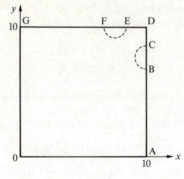

FIG. 6.10. Two-dimensional problem with coupling freezing fronts

where $u_{h,i}^{n+1}(k)$ denotes the kth iteration of $u_{h,i}^{n+1}$. Ichikawa and Kikuchi (1979) solved the variational inequality (6.173) for the one-dimensional single-phase Stefan problem using linear finite elements with different mesh sizes and found the orders of error to be exactly the step size h in the H^1 norm and $h^{1\cdot6}$ in the L^2 norm in agreement with their theoretical estimates. They also quoted results obtained by several schemes for solving the inequality for a model, one-dimensional, steady-state problem with a known exact solution and concluded that the Langrange multiplier method was unsatisfactory and that the projected SOR method was more effective than either the fixed point SOR or the penalty method.

Numerical experiments on a two-dimensional problem illustrated in Fig. 6.10, and selected because the two freezing fronts become coupled

TABLE 6.12
Coordinates of the freezing front on the axes

| | LC[a] | | IMM[b] | |
t	$\delta x = 1/20$	$\delta x = 1/80$	$\delta u = 0.2$	$\delta u = 0.1$
0.05	0.776	0.770	0.775	0.775
0.10	0.680	0.673	0.675	0.676
0.20	0.539	0.537	0.535	0.536
0.30	0.431	0.426	0.422	0.420
0.40	0.323	0.321	0.314	0.308
0.50	0.208	0.209	0.201	0.188

[a] LC = Linear complementarity method (Elliott and Ockendon 1982); $\delta t = \frac{1}{4}(\delta x)^2$.
[b] IMM = Isotherm migration method (Crank and Crowley 1978); $\delta t \equiv 0.0001$.

TABLE 6.13

Coordinates of freezing front on diagonal
$$y = x$$

t	LC[a]		IMM[a]	
	$\delta x = 1/20$	$\delta x = 1/80$	$\delta u = 0.2$	$\delta u = 0.1$
0.05	0.728	0.735	0.732	0.732
0.10	0.625	0.625	0.623	0.619
0.20	0.470	0.470	0.471	0.463
0.30	0.350	0.351	0.352	0.342
0.40	0.247	0.249	0.250	0.237
0.50	0.152	0.154	0.153	0.139

[a] See footnote to Table 6.12.

after a finite time, reveal an optimum relaxation factor of $\omega = 1$ in the projected SOR method used. Lumping of the mass matrix gives smooth freezing fronts in the first few time steps, whereas singular fronts are observed in the absence of lumping. The two solutions agree at later times.

Elliott (1976) also quoted numerical results, obtained by solving the variational-inequality formulation of one-phase Stefan problems, which agree satisfactorily with the analytical solution. His solutions for the melting of a long cylinder of square cross-section, quoted by Elliott and Ockendon (1982), are compared with those obtained by Crank and Crowley (1978) using the isotherm migration method described in §5.4.2(iii) in Tables 6.12, 13.

6.4.6. Two-phase Stefan problem

As an example of the extension of the freezing-index transformation to a two-phase Stefan problem and of differential and variational formulations in multi-space dimensions, we refer to a paper by Kikuchi and Ichikawa (1979). Their main concern was to present numerical schemes of solution based on theoretical considerations due to Duvaut (1975), Frémond (1974), and Lions (1976).

Consider a multi-dimensional domain D to be divided into a solid part, D_1, and a liquid part, D_2, with temperature $u(x, t) < 0$ in D_1 and $u(x, t) > 0$ in D_2. Here x is a space vector of position with one, two, or three components. On the phase-change interface $\Gamma_0(t)$ we have $u(x, t) = 0$. We take a problem in which there may be different conditions on three separate parts of the boundary Γ of D. The temperature $u(x, t)$ is prescribed on the boundary Γ_1; there is no heat flux on Γ_2; on Γ_3 the heat flux is taken to be proportional to the temperature. The problem can be defined by the equations and conditions

$$\left.\begin{array}{ll} C_1\,\partial u/\partial t = \nabla \cdot (K_1\,\nabla u) & \text{in } D_1, \\ C_2\,\partial u/\partial t = \nabla \cdot (K_2\,\nabla u) & \text{in } D_2, \end{array}\right\} \tag{6.175}$$

$$\left.\begin{array}{l} u(x,t)=g(x,t) \quad \text{on } \Gamma_1, \\ \partial_i u(x,t)=0 \quad \text{on } \Gamma_2 \text{ bounding } D_i, \\ \partial_i u(x,t)=-p_i u(x,t)+q_i(x,t) \quad \text{on } \Gamma_3 \text{ bounding } D_i, \end{array}\right\} \tag{6.176}$$

$$\left.\begin{array}{l} u(x,t)=0 \quad \text{on } \Gamma_0, \\ [K\,\nabla u]_1^2 \cdot \nabla \ell + L = 0 \quad \text{on } \Gamma_0, \end{array}\right\} \tag{6.177}$$

$$u(x,0)=u_0(x) \quad \text{in } D. \tag{6.178}$$

Here C_i and K_i, $i = 1, 2$, are the heat capacity and conductivity of the ith phase, g, p_i, q_i are given functions, u_0 is the initial temperature in the domain D, $\partial_i u$ is the temperature gradient normal to Γ, and $[K\,\nabla u]_1^2 = K_2(\nabla u)_2 - K_1(\nabla u)_1$. The function $t = \ell(x)$ indicates the position of Γ_0, i.e. $x = s(t)$ as in §6.4.5.

For a two-phase problem Duvaut (1975) introduced the transformation

$$w(x,t) = \int_0^t K_i u(x,\tau)\,d\tau. \tag{6.179}$$

For $u(x,t)$ continuous we have $K_i u(x,t) = \partial w/\partial t = \dot{w}(x,t)$ and so $\dot{w}(x,t)<0$ in $D_1(t)$ and $\dot{w}(x,t)>0$ in $D_2(t)$. On integrating (6.175) with respect to time and remembering that a point x may be in $D_1(0)$ initially but in $D_2(t)$ at time t we obtain relationships such as

$$\nabla \cdot \nabla w(x,t) = -C_1 u(x,0) + C_1 u(x,t), \; x \text{ in } D_1(0) \text{ and } D_1(t),$$

$$= -C_1 u(x,0) + C_2 u(x,t) + L, \; x \text{ in } D_1(0) \text{ and } D_2(t),$$

and two corresponding relations where x is initially in $D_2(0)$ but in $D_1(t)$ or $D_2(t)$ at time t. Using (6.178) the equation which holds over the whole domain D can be written

$$(C_i/K_i)\frac{\partial w}{\partial t}(x,t) - \nabla \cdot \nabla w(x,t) = C_j u_0(x) + \varepsilon_{ij} L \tag{6.180}$$

where

$$i = 1, x \quad \text{in } D_1(t), \qquad i = 2, x \quad \text{in } D_2(t), \tag{6.181}$$

$$j = 1, x \quad \text{in } D_1(0), \qquad j = 2, x \quad \text{in } D_2(0), \tag{6.182}$$

$$\varepsilon_{12} = 1, \qquad \varepsilon_{21} = -1, \qquad \varepsilon_{11} = \varepsilon_{22} = 0. \tag{6.183}$$

Now $\partial w = \int_0^t K_i\,\nabla u \cdot \mathbf{n}\,d\tau = \int_0^t \partial_i u\,d\tau$, where $\mathbf{n}$ is the outward normal unit vector on Γ, so that boundary conditions (6.176) become

$$w(x, t) = \hat{g}(x, t) \quad \text{on } \Gamma_1, \tag{6.184}$$

$$\partial w(x, t) = 0 \quad \text{on } \Gamma_2, \tag{6.185}$$

$$\partial w(x, t) = -(p_i/K_i)w(x, t) + \hat{q}(x, t) \quad \text{on } \Gamma_3, \tag{6.186}$$

where $i = 1$ for x in $D_1(t)$ and $i = 2$ for x in $D_2(t)$ and we have put

$$\hat{g}(x, t) = \int_0^t K_i g(x, \tau) \, d\tau \quad \text{with} \quad i = \begin{cases} 1 & \text{if} & g(x, \tau) < 0, \\ 2 & \text{if} & g(x, \tau) > 0, \end{cases}$$

$$\hat{q}(x, t) = \int_0^t q_i(x, \tau) \, d\tau \quad \text{with} \quad i = \begin{cases} 1 & \text{if} & x \quad \text{in } D_1(\tau) \\ 2 & \text{if} & x \quad \text{in } D_2(\tau) \end{cases}$$

$$\tag{6.187}$$

The two-phase Stefan problem is now expressed over the whole fixed domain D by the equation (6.180) with the conditions (6.184–6). The interface conditions (6.177) and the initial condition $w(x, 0) = 0$ have been included.

Under the assumption that $\dot{w}(x, t)$ is twice differentiable in D for all t in $0 \le t \le T$, Kikuchi and Ichikawa write (6.180) in the variational form

$$(d_i \dot{w}_i, v) + a_i(w, v) = (\varepsilon_{ij} L, v) + F_j(v), \tag{6.188}$$

for all t in $0 \le t \le T$, where

$$d_i = C_i/K_i, \qquad F_j(v) = \int_D C_j u_0 v \, dx + \int_{\Gamma_3} \hat{q} v \, ds,$$

$$a_i(u, v) = \int_D \nabla u \cdot \nabla v \, dx + \int_{\Gamma_3} e_i u v \, ds, \qquad e_i = p_i/K_i,$$

the test functions v and their first derivatives are square summable in D and $v(t) = 0$ on Γ_1.

Kikuchi and Ichikawa (1979) obtained finite-difference solutions based on a discretized form of (6.180) which in two space dimensions is

$$d_i^{\alpha,\beta} \dot{w}_{\alpha,\beta} + (4/\Delta^2) w_{\alpha,\beta} = (1/\Delta^2)(w_{\alpha-1,\beta} + w_{\alpha+1,\beta} + w_{\alpha,\beta-1} + w_{\alpha,\beta+1})$$
$$+ C_j^{\alpha,\beta} (u_0)_{\alpha,\beta} + \varepsilon_{ij}^{\alpha,\beta} L \tag{6.189}$$

where $d_i = C_i/K_i$, (α, β) denotes the point $(\alpha \, \Delta x, \beta \, \Delta y)$ on the rectangular net covering the domain D and $\Delta = \Delta x = \Delta y$ is adopted for simplicity, $w_{\alpha,\beta}$ denotes the value of w at the point (α, β), etc. The term $d_i^{\alpha,\beta} \dot{w}_{\alpha,\beta}$ is approximated by

$$d_i^{\alpha,\beta} \dot{w}_{\alpha,\beta}(n \, \Delta t) = d_i^{\alpha,\beta}(n \, \Delta t)\{w_{\alpha,\beta}^n - w_{\alpha,\beta}^{n-1}\}/\Delta t,$$

where $w_{\alpha,\beta}^n = w_{\alpha,\beta}(n \, \Delta t)$ and Δt is the given time interval, and either

$d_i^{\alpha,\beta}(n\,\Delta t)$ or $d_i^{\alpha,\beta}(n-1)\,\Delta t$ is used. The solution proceeds in time steps Δt by a modified SOR scheme as follows, where m denotes the mth stage of the iteration and $d_i^{\alpha,\beta}(n\,\Delta t)$ is used (a simple modification is needed if $d_i^{\alpha,\beta}(n-1)\,\Delta t$ is used instead); ω is the iteration parameter and $0<\omega<2$.

(i) Put $w_{\alpha,\beta}^n(0)=w_{\alpha,\beta}^{n-1}$.

(ii) $R_{\alpha,\beta}(m)=\tfrac{1}{4}\{w_{\alpha-1,\beta}^n(m)+w_{\alpha+1,\beta}^n(m-1)+w_{\alpha,\beta-1}^n(m)+w_{\alpha,\beta+1}^n(m-1)$

$$+\Delta^2 C_j^{\alpha,\beta}(u_0)\}$$

$$N_{\alpha,\beta}^n(m)=\varepsilon_{i,j}^{\alpha,\beta}(n,m)L-\frac{d_i^{\alpha,\beta}(n,m)}{\Delta t}\{w_{\alpha,\beta}^n(m-1)-w_{\alpha,\beta}^{n-1}\},$$

$$w_{\alpha,\beta}^n(m)=(1-\omega)w_{\alpha,\beta}^n(m-1)+\omega\left\{R_{\alpha,\beta}^n(m)+\frac{\Delta^2}{4}N_{\alpha,\beta}^n(m)\right\}.$$

(iii) Repeat till $\sum|w_{\alpha,\beta}^n(m)-w_{\alpha,\beta}^n(m-1)|/\sum|w_{\alpha,\beta}^n(m)|<\varepsilon$, where ε is the tolerance of convergence of the iterations.

The indices i,j are given by

$$i=1,\qquad \dot{w}_{\alpha,\beta}^n(m)<0,\qquad j=1,\qquad (u_0)_{\alpha,\beta}<0,$$
$$i=2,\qquad \dot{w}_{\alpha,\beta}(m)>0,\qquad j=2,\qquad (u_0)_{\alpha,\beta}>0,$$

where $\dot{w}_{\alpha,\beta}(m)=(1/\Delta t)\{w_{\alpha,\beta}^n(m-1)-w_{\alpha,\beta}^{n-1}\}$. Kikuchi and Ichikawa (1979) argue that the erm $\varepsilon_{ij}^{\alpha,\beta}(n,m)L$ varies like a step function, whereas the other terms in the iterative scheme change moderately and therefore in order to secure convergence the relationship

$$|R_{\alpha,\beta}^n(m)-\tfrac{1}{4}\Delta^2 d_i^{\alpha,\beta}(n,m)\dot{w}_{\alpha,\beta}(m)|\gg|\tfrac{1}{4}\Delta^2\varepsilon_{ij}^{\alpha,\beta}(n,m)L| \qquad (6.190)$$

must be satisfied.

Two problems are solved in one space dimension. In the first, the initial temperature is given by $u_0(x)=(x-\tfrac{1}{2})$ for $x\leq\tfrac{1}{2}$, $u_0(x)=0$ for $\tfrac{1}{2}<x<1$, and the boundary conditions are $g(0,t)=-\tfrac{1}{2}$, $g(1,t)=t$, so that for $t<1$, $\hat{g}(0,t)=-K_1 t/2$, $\hat{g}(1,t)=K_2 t^2/2$. Numerical experiments support the convergence condition $2K_1\,\Delta t\geq L\Delta$, which is an approximation based on (6.190a). It is deduced by noting that for the one-dimensional case (6.190) becomes, for $n=1$,

$$|0.5(w_{\alpha-1}^1(m)+w_{\alpha+1}^1(m-1))-0.25\,\Delta^2 d_i^\alpha(1,m)\dot{w}_\alpha^1(m)|\gg0.25L\,\Delta^2.$$
$$(6.190a)$$

and then concentrating attention on the point $x=\alpha\,\Delta=\tfrac{1}{2}$ and assuming Δt small enough for $\dot{w}_\alpha^1=\dot{w}_\alpha^0=u_{0\alpha}$ approximately. Since $u_{0\alpha}=0$ at $x=\tfrac{1}{2}$, $u_{0,\alpha-1}=-\Delta$ at $x=\tfrac{1}{2}-\Delta$, $u_{0,\alpha+1}=0$ at $x=\tfrac{1}{2}+\Delta$, and recalling that

$$w_{\alpha-1}^1=\int_0^{\Delta t}K_1 u_{\alpha-1}\,d\tau=K_1\,\Delta t u_{0,\alpha-1}=-K_1\,\Delta t\,\Delta$$

we arrive at $0.5K_1\,\Delta\cdot\Delta t\gg0.25\,\Delta^2 L$ as the convergence condition.

The second problem, also one dimensional, includes heat parameters appropriate to a silty soil with 20 per cent moisture content; initially the soil is unfrozen with temperature $u_0(x) = ax^2$, where $a = g(b, 0)/b^2$, x is the depth through the soil, and $g(b, 0)$ the initial temperature at the full depth b. Boundary conditions for $g(0, t)$, $g(b, t)$ are prescribed numerically. Numerical convergence studies support the approximate criterion $|\frac{1}{2}K_1 \Delta^2 g(0, \Delta t)| \gg \Delta^2 L$, derived as in the earlier problem.

Graphic solutions were presented by Kikuchi and Ichikawa (1979) for a two-dimensional problem defined in the square region $0 \leqslant x \leqslant 0.4$, $0 \leqslant y \leqslant 0.4$, with $C_1 = 0.5$, $C_2 = K_1 = K_2 = L = 1.0$, and zero initial temperature. Boundary conditions are $g(0, y) = (0.8 - y)(2t - 1)$, $g(x, 0.4) = (0.8 - x)(2t - 1)$, $g(0.4, y) = 0.5\sqrt{y}(1 - 3t)$, $g(x, 0) = 0.5\sqrt{x}(1 - 3t)$. In this example, the term $\varepsilon_{ij}^{\alpha,\beta}(n, m)L$ is replaced by the expression

$$\{h_1(\varepsilon^{\alpha-1,\beta} + \varepsilon^{\alpha+1,\beta} + \varepsilon^{\alpha,\beta-1} + \varepsilon^{\alpha,\beta+1}) + h_2\varepsilon^{\alpha,\beta}\}L/(4h_1 + h_2),$$

where $\varepsilon = \varepsilon_{ij}$ for convenience of writing, and the influence on the results is examined for different h_1, h_2. Also some results are obtained with a linearly smoothed phase boundary, and there is a mushy region.

Finally, Kikuchi and Ichikawa (1979) used linear finite elements to solve the variational equation (6.188) for a two-dimensional problem defined with reference to Fig. 6.11 by the conditions:

$$u(x, t) = g(t) = \begin{cases} -1.0, & 0 < t \leqslant 0.4 \quad \text{and} \quad 0.5 < t \leqslant 0.6 \\ 0.5, & 0.4 < t \leqslant 0.5 \end{cases} \quad \text{on } \Gamma_1.$$

On other parts of the boundary the normal derivative of $u(x, t)$ is zero. The heat parameters are $K_1 = K_2 = C_1 = L = 1.0$, $C_2 = 0.5$. The Cryer algorithm is used with $\omega = 1.4$ as the over-relaxation parameter. The shape of the frozen area just before it vanishes depends markedly on the width of the smoothed, phase-change region but does not seem to effect the temperature distribution in subsequent time steps.

Kikuchi and Ichikawa (1979) did not explicitly include an enthalpy

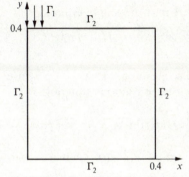

FIG. 6.11. Two-dimensional two-phase problem

function in either of their papers. Duvaut (1975) and Elliott (1976) derived a variational inequality for the two-phase Stefan problem which incorporates the enthalpy function and its integral with respect to temperature.

If heat capacity C and conductivity K are taken to be piecewise constants, a convenient definition of enthalpy $H(z)$ is

$$H(z) = \alpha_1 z, \quad z < 0; \qquad H(z) = \alpha_2 z + L, \quad z > 0;$$

$$0 \leqslant H(z) \leqslant L, \qquad z = 0, \quad (6.191)$$

where $\alpha_i = C_i/K_i$, $i = 1, 2$. Consider a domain D divided into two parts D_1, D_2 by the melting surface at $t = \ell(x)$ on which the usual Stefan condition holds. The problem is

$$\left. \begin{aligned} \alpha_i \, \partial u/\partial t &= \nabla^2 u \quad \text{in } D_i(t), \qquad i = 1, 2, \\ u &= g \quad \text{on the boundary } \partial D, \\ u &= u_0 < 0, \qquad H(u_0) = H_0, \qquad t = 0. \end{aligned} \right\} \quad (6.192)$$

In this formulation the variable u is related to temperature T by

$$u = \int_{T_M}^{T} K(\theta) \, d\theta, \quad (6.193)$$

which allows for likely discontinuities in $K(T)$, and the freezing index becomes

$$w(x, t) = \int_0^t u(x, \tau) \, d\tau \quad (6.194)$$

in place of (6.179). Proceeding as in the derivation of (6.180) above we see that

$$\nabla^2 w = \begin{cases} -\alpha_1 u_0 + \alpha_1 u, & t < \ell(x), \\ -\alpha_1 u_0 + \alpha_2 u + L, & t > \ell(x), \end{cases}$$

which can be combined into the single equation

$$\nabla^2 w + \alpha_1 u_0 = H(\partial w/\partial t) = H(w_t) \quad (6.195)$$

in place of (6.180).

Because $H(z)$ is discontinuous at $z = 0$ it is useful to introduce its integral $\phi(z)$ which is the continuous function

$$\phi(z) = \int_0^z H(z') \, dz' = \tfrac{1}{2}\alpha_1(z^-)^2 + \tfrac{1}{2}\alpha_2(z^+)^2 + Lz^+ \quad (6.196)$$

where $z^- = \min(0, z)$ and $z^+ = \max(0, z)$. It follows readily that

$$\phi(z_1) - \phi(z_2) \geqslant H(z_2)(z_1 - z_2), \quad (6.197)$$

for all real z_1 and provided $H(z_2)$ satisfies (6.191). Combining (6.195) and (6.197) yields

$$\phi(z) - \phi(w_t) \geqslant (\nabla^2 w + H_0)(z - w_t). \tag{6.198}$$

The desired variational inequality follows by substituting $z = v(x)$ and integrating (6.198). It is

$$a(w, v - w_t) + \Phi(v) - \Phi(w_t) \geqslant (H_0, v - w_t), \tag{6.199}$$

where

$$\Phi(v) = \int \phi(v(x)) \, dx$$

for all admissible v. Duvaut (1975) proved that there exists a unique solution of (6.199) for which $w = 0$, $t = 0$.

6.4.7. Mathematical results

Some references have been made in this chapter to investigations of the mathematical properties of classical, weak, and variational solutions of generalized Stefan-like problems. It is not intended in the present section to discuss these aspects in detail but simply to refer interested readers to some recent surveys and the numerous references contained in them. One up-to-date survey by Niezgódka (1983) concentrates on recent and lesser known results and extends earlier surveys by Friedman (1979) and Primicerio (1981a) with reference to the bibliography by Wilson, Solomon, and Trent (1979). A brief historical introduction is given by Pawlow (1981).

Niezgódka (1983) listed 16 applications of practical origin with source references which introduce various generalized features and have motivated research into the evaluation of solutions and their theoretical properties. His survey of mathematical results for generalized statements of Stefan-like problems covers one-dimensional and multi-dimensional problems, both single and multi-phase.

The concept of the weak solution introduced by Kamenomostskaja (1961) and extended by Oleinik (1960) was based on the Kirchhoff transformation and the enthalpy function. Kamenomostskaja considered a two-phase multi-dimensional problem and proved the existence and uniqueness of the weak solution by considering the convergence of explicit finite-difference approximations as in §6.2.2 above. Oleinik (1960) used a smoothing technique to prove existence and uniqueness for a more general one-dimensional Stefan problem. The work of both these authors was refined and extended by Friedman (1968). The more general problem with parameters dependent on space, time, and temperature, and with a derivative condition on the fixed boundary, was investigated by Budak and Gapenko (1971) for the one-dimensional case and extended to the multi-dimensional problem by Niezgódka and Pawlow (1983).

Brezis (1970) and Lions (1969) used monotonicity methods in their analyses and this approach was later developed by Jerome (1976, 1977) and Damlamian (1977, 1979).

Duvaut (1975) extended the variational-inequality analysis to two-phase problems and Frémond (1974) used the same approach independently. Various later papers are quoted by Niezgódka (1983) and by Pawlow (1981). The latter paper, together with those of Jerome (1977) and Budak and Gapenko (1971), considered domains having only piecewise-regular boundaries.

For problems in one space dimension, references are given by Niezgódka (1983) to papers on regularity and global continuity of weak solutions. Very general forms of non-linear conditions on the free boundary were examined by Fasano and Primicerio (1979a), Fasano (1980) and Niezgódka (1979), and in several papers in Russian. Papers by Fasano (1980) and Fasano and Primicerio (1979b, 1981) deal with Cauchy-type conditions on the free boundary and with critical cases for which solutions are non-existent. Other related papers by Meirmanov (1981a,b) and Pukhnachev (1980) are mentioned. Curvature of the free boundary was studied by Fasano and Primicerio (1979c), Friedman and Jensen (1977), and Primicerio (1980). A paper by Shishkin (1975) considered problems with a discontinuous free boundary.

Other sections of Niezgódka's (1983) survey refer to degenerate problems, e.g. with zero specific heat, and to global non-linearities, concentrated capacities, time and spatial periodicities, hysteresis, and non-parabolic Stefan-like problems. Finally, fundamental difficulties associated with the accurate tracking of a free boundary are mentioned. Sufficient global regularity is provided only by results for the one-dimensional problems (Wilson *et al.* 1978; Primicerio 1981a) and multi-dimensional single-phase problems (Wilson *et al.* 1978; Friedman 1975). Damlamian (1977, 1979) produced general results on the regularity of weak solutions but they do not help to determine the movement of the free boundary.

7. Analytical solution of seepage problems

7.1. Introduction

THE practical importance of free-boundary problems in porous flow has stimulated many attempts to find analytical solutions, both approximate and exact. The latter depend almost entirely on finding suitable conformal mappings of the flow region in the physical plane on to both the hodograph plane and the flow-potential plane. A comprehensive account of many ingenious transformations is given by Polubarinova-Kochina (1962). Other basic texts are by Bear (1972) and Harr (1962). In this chapter the most commonly used mappings and the ways in which they are applied to selected problems are described.

7.2. Approximate methods

7.2.1. Dupuit's approximation

Dupuit (1863) made the simplifying assumption that the slope of the free surface in unconfined flow is very small. This seems to be substantiated by observations of most ground-water flows. In steady two-dimensional flow without accretion in the vertical x–z plane, the free surface is a streamline and at every point on it the rate of flow per unit area q_s is given by Darcy's law to be

$$q_s = -K \, \mathrm{d}\phi/\mathrm{d}s = -K \, \mathrm{d}z/\mathrm{d}s = -K \, \mathrm{d}h/\mathrm{d}s, \tag{7.1}$$

where s is measured along the streamline, since $\phi = z = h$ on the free boundary. Dupuit replaced $\mathrm{d}h/\mathrm{d}s$ by $\mathrm{d}h/\mathrm{d}x$ for small gradients, which is equivalent to assuming that the equipotential surfaces are vertical, i.e. $\phi = \phi(x)$ independent of z, and the flow is assumed to be horizontal. Thus we now consider a uniform one-dimensional flow based on the expression

$$q_x = -K_x \, \mathrm{d}h/\mathrm{d}x, \qquad h = h(x), \tag{7.2}$$

and we should examine the error introduced by neglecting the vertical flow $q_z = -K_z \, \partial\phi/\partial z$.

The exact expression for the total horizontal discharge through the medium, Q, is

$$Q = -K_x \int_0^{h(x)} \frac{\partial \phi}{\partial x}(x, z)\, dz = -K_x \frac{\partial}{\partial x}\left\{ \int_0^{h(x)} \phi(x, z)\, dz - \tfrac{1}{2}h^2 \right\} = -K_x\, \partial\phi'/\partial x,$$

$$(7.3)$$

where

$$\phi' = h\bar\phi - \tfrac{1}{2}h^2, \qquad h\bar\phi = \int_0^{h(x)} \phi(x, z)\, dz.$$

By Dupuit's assumption we get

$$Q = -K_x h\, \partial h/\partial x = -K_x\, \partial(\tfrac{1}{2}h^2)/\partial x, \tag{7.4}$$

i.e. ϕ' has been replaced by $\tfrac{1}{2}h^2$. Integration by parts of the expression for ϕ' in (7.3) gives

$$\phi' = \int_0^{h(x)} \phi(x, z)\, dz - \tfrac{1}{2}h^2 = (z\phi)_0^{h(x)} - \int_0^{h(x)} z(\partial\phi/\partial z)\, dz - \tfrac{1}{2}h^2(x)$$

$$= \tfrac{1}{2}h^2(x)\left\{ 1 + \frac{2}{K_z h^2(x)} \int_0^{h(x)} zq_z(x, z)\, dz \right\}, \tag{7.5}$$

where $q_z = -K_z\, \partial\phi/\partial z$, $q_z(h) < 0$. The integral term in (7.5) measures the error introduced by replacing ϕ' by $\tfrac{1}{2}h^2$. Since on the free surface $\phi = h$, $q_z/q_x = dh/dx$, and $d\phi/dx = dh/dx = -(q_x/K_x) - (q_z/K_z)\, dh/dx$, with q_x, q_z evaluated at $z = h$, we see that

$$q_z = \frac{-K_x(dh/dx)^2}{1 + (K_x/K_z)(dh/dx)^2}, \qquad z = h.$$

The range in the error term can be expressed as

$$0 < \frac{\tfrac{1}{2}h^2 - \phi'}{\tfrac{1}{2}h^2} < \frac{(K_x/K_z)/(dh/dx)^2}{1 + (K_x/K_z)(dh/dx)^2}. \tag{7.6}$$

As an example, we apply Dupuit's formula (7.4) to the problem of the simple dam of width L separating two reservoirs with water heights h_0 and h_L (Fig. 2.1 with $x_1 = L$, $y_1 = h_0$, $y_2 = h_L$). Integrating the second eqn (7.4) we obtain, noting that Q is constant throughout the dam for steady flow and with $K_x = K$ for an isotropic homogeneous dam,

$$h^2(x) = h_0^2 - 2Qx/K, \qquad Q = K(h_0^2 - h_L^2)/(2L). \tag{7.7}$$

The correct expression for Q (i.e. without making the Dupuit assumptions) follows by integrating (7.3) and inserting $\phi = h(x) = h_0$ at $x = 0$ to obtain (Charni, 1951)

$$(Q/K)x = -\int_0^{h(x)} \phi(x, z)\, dz + \tfrac{1}{2}(h^2 + h_0^2). \tag{7.8}$$

But at $x = L$ we have $\phi(L, z) = h_L$, $0 < z \leq h_L$, and $\phi = z$, $h_L \leq z \leq h_S$, where h_S is the height of the separation point. Hence at $x = L$, using (7.8),

$$(Q/K)L = -\int_0^{h_L} h_L \, \mathrm{d}z - \int_{h_L}^{h_S} z \, \mathrm{d}z + \tfrac{1}{2}(h_S^2 + h_0^2) = \tfrac{1}{2}(h_0^2 - h_L^2),$$

and so the Dupuit formula (7.7) for the total discharge Q turns out to be exact in this case.

On the other hand, the first of (7.7) approximates the free surface by a parabola which passes through $h = h_0$ at $x = 0$ and $h = h_L$ at $x = L$. Thus the Dupuit solution does not include a seepage surface and the parabolic free surface meets both the inlet and outlet faces of the dam at finite angles given by

$$\mathrm{d}h/\mathrm{d}x = -Q/(Kh_0), \qquad x = 0; \qquad \mathrm{d}h/\mathrm{d}x = -Q(Kh_L), \qquad x = L.$$

Further discussion of the paradoxes associated with the Dupuit assumptions is to be found in Bear (1972). A rough guideline proposed for practical purposes is that the Dupuit formulae are acceptable if the definitive lengths in the direction of flow are greater than twice the depth of the flow region.

It can similarly be shown (Polubarinova-Kochina 1962) that the Dupuit expression for radial flow to a fully penetrating well of radius r_W in which $h = h_W$ is also exact and given by

$$Q = \frac{\pi K(h_R^2 - h_W^2)}{\ln(r_W/R)}, \tag{7.9}$$

where $h = h_R$ at $r = R$. Other examples of Dupuit formulae are given by Bear (1972) including cases of horizontally and vertically stratified media for which Outmans (1964) produced generalized formulae.

7.2.2. Boussinesq's equations

Dupuit's approximations can be combined with a continuity equation of flow averaged in the usual way over the vertical coordinate z in order to reduce a three-dimensional free-boundary problem to two dimensions. As an example we can consider an element of volume of the flow region bounded below by a horizontal impervious base, above by the free surface and with vertical sides at $x \pm \delta x/2$ and $y \pm \delta y/2$. If q is taken to be the flow rate per unit width in the y-direction but over the whole depth of the element $h(x, y, t)$, then in the limit the appropriate conservation expression can be written

$$-\frac{\partial}{\partial x}(q_x) - \frac{\partial}{\partial y}(q_y) + N = n\frac{\partial h}{\partial t}.$$

Here n is an effective porosity, $N(x, y, t)$ allows for accretion in the z-direction on the free surface, and any contribution associated with the increase of pressure in the element due to the change in height of the free surface is neglected. Introduction of Dupuit's approximations for q_x, q_y in an isotropic medium gives

$$\frac{1}{2}\frac{\partial}{\partial x}\left(K\frac{\partial(h)^2}{\partial x}\right)+\frac{1}{2}\frac{\partial}{\partial y}\left(K\frac{\partial(h)^2}{\partial y}\right)+N=n\frac{\partial h}{\partial t}, \tag{7.10}$$

which is known as Boussinesq's equation for unsteady porous flow with accretion on the free boundary. The steady-state form of (7.10), referred to as Forchheimer's equation, is linear in h^2. The usual special forms follow for a homogeneous medium ($K = $ constant) and in the steady state without accretion we have

$$\frac{\partial^2(h)^2}{\partial x^2}+\frac{\partial^2(h)^2}{\partial y^2}=0.$$

Suitable amendments are needed for an anisotropic medium, when $K_x \neq K_y$.

A simple extension includes a slightly sloping impervious base with height $\eta = \eta(x, y)$. Then (7.10) can be written as

$$\frac{\partial}{\partial x}\{K(h-\eta)\,\partial h/\partial x\}+\frac{\partial}{\partial y}\{K(h-\eta)\,\partial h/\partial y\}+N=n\frac{\partial h}{\partial t}, \tag{7.11a}$$

or, putting $b(x, y, t) = h(x, y, t) - \eta(x, y)$, as

$$\frac{\partial}{\partial x}\left(b\frac{\partial b}{\partial x}\right)+\frac{\partial}{\partial y}\left(b\frac{\partial b}{\partial y}\right)+\frac{\partial}{\partial x}\left(b\frac{\partial \eta}{\partial x}\right)+\frac{\partial}{\partial y}\left(b\frac{\partial \eta}{\partial y}\right)+\frac{N}{K}=\frac{n}{K}\frac{\partial b}{\partial t}, \tag{7.11b}$$

if the medium is homogeneous. If $\partial\eta/\partial x$, $\partial\eta/\partial y$ are constant, i.e. the base is plane, a change of variables to

$$x'=x+\frac{Kt}{n}\frac{\partial\eta}{\partial x}, \qquad y'=y+\frac{Kt}{n}\frac{\partial\eta}{\partial y}, \qquad t'=t,$$

reduces (7.11b) to

$$\frac{\partial}{\partial x'}\left(b\frac{\partial b}{\partial x'}\right)+\frac{\partial}{\partial y'}\left(b\frac{\partial b}{\partial y'}\right)+\frac{N}{K}=\frac{n}{K}\frac{\partial b}{\partial t'}, \tag{7.11c}$$

which has the same form as (7.10) and is sometimes also attributed to Forchheimer.

Bear (1972) presented some simple exact solutions of Forchheimer's equations and also for Boussinesq's equation for ground-water flows of practical interest. There are references to other results but they are limited in number because of the non-linearity of eqn (7.10). Various linearization techniques have been applied (Bear 1972). The first is based on the assumption that the height of the free surface above the horizontal base varies only slightly so that $h = \bar{h} + h'$ where $h' \ll h$ and $\bar{h}$ is the average height. Equation (7.10) becomes

$$K\bar{h}\{\partial^2 h'/\partial x^2 + \partial^2 h'/\partial y^2\} + N = n \, \partial h'/\partial t, \tag{7.10a}$$

which is linear in h' and some solutions can be obtained by recognizing (7.10a) as being related to the heat-flow or diffusion equation. Bear (1972) gave a number of examples and references. In the second method (7.10) is rewritten as

$$\frac{1}{2}\left(\frac{\partial^2 (h)^2}{\partial x^2} + \frac{\partial^2 (h)^2}{\partial y^2}\right) + \frac{N}{K} = \frac{n}{2Kh}\frac{\partial (h)^2}{\partial t} \tag{7.10b}$$

and variations in h are considered to be such that Kh/n is assumed constant, equal to $K\bar{h}/n$. Alternatively (7.10b) is written

$$a^2\left(\frac{\partial^2 u}{\partial x^2} + \frac{\partial^2 u}{\partial y^2}\right) + N_1 = \frac{\partial u}{\partial t}, \qquad u = \tfrac{1}{2}h^2, \qquad N_1 = \frac{N\bar{h}}{n}, \qquad a^2 = \frac{K\bar{h}}{n}. \tag{7.10c}$$

Charni (1951) wrote Boussinesq's equation for one-dimensional flow as

$$\frac{K}{n}\frac{du}{df}\frac{\partial^2 u}{\partial x^2} = \frac{\partial u}{\partial t}, \tag{7.10d}$$

where $u = \int f(h) \, dh$, and chose $du/df = 1/A = $ constant. Thus (7.10d) is a linear diffusion equation and $df/f = A \, dh$ so that $f = B \exp(Ah)$. The constants A and B are chosen so that $f \equiv h$ at the boundaries.

7.2.3. Other approximate methods

Bear (1972) referred to early work by Lembke (1887), Weber (1928), Polubarinova-Kochina (1962), and Aravin and Numerov (1965) based on a quasi-steady-state method. The free boundary is assumed always to have the shape of the Dupuit steady-state approximation for the boundary conditions, e.g. reservoir level, pertaining at any given time; the time-dependent flow pattern is approximated by a succession of steady states. Bear (1972) gave details of the unsteady solution of Aravin and Numerov (1965) for one-dimensional flow to a ditch.

Instead of reducing the dimension of the equation of flow by Dupuit's assumptions, the standard method of small perturbations can be applied to the full set of equations and boundary conditions, i.e. $\phi(x, y, z, t)$ can be determined such that $\nabla^2 \phi = 0$ in the flow region, and the free surface boundary condition,

$$n\frac{\partial \phi}{\partial t} = K\left[\left(\frac{\partial \phi}{\partial x}\right)^2 + \left(\frac{\partial \phi}{\partial y}\right)^2 + \left(\frac{\partial \phi}{\partial z}\right)^2\right] - (K + N)\frac{\partial \phi}{\partial z} + N$$

is satisfied together with appropriate conditions on the fixed boundaries.

Bear (1972) presented details based on the work of Polubarinova-Kochina (1962) and others who also discussed a shallow-flow approximation which is similar to the perturbation approach.

7.3. Hodograph method: conformal transformations

The discussion in this section will be confined to porous flow in homogeneous isotropic media and so we shall define a flow vector **q** related to a flow potential, ϕ, by

$$\mathbf{q} = -\text{grad } \phi. \tag{7.11}$$

The vector **q** is the volume flow per unit area and in two-dimensional flow **q** has components q_x, q_y in the x, y directions, where

$$q_x = -\partial\phi/\partial x, \qquad q_y = -\partial\phi/\partial y \tag{7.12}$$

and in any general direction s, $q_s = -\partial\phi/\partial s$.

7.3.1. Hodograph plane

Figure 7.1a shows any curve Γ in the physical plane x, y in which two-dimensional flow is occurring. A flow vector **q** with components q_x, q_y exists at each point of Γ and so in a second plane with cartesian axes q_x, q_y we can generate a curve Γ' (Fig. 7.1b) from the end points of the vectors **q** for all points on Γ. This curve Γ' is called the hodograph representation of Γ and Γ' is a mapping of the curve Γ in the x, y plane on to the hodograph plane q_x, q_y. Strictly, the term hodograph refers to velocities but the volume flow is a more satisfactory concept in the present context and **q** can clearly be interpreted as a suitably defined velocity.

In the various problems formulated in Chapter 2, conditions along straight or curved boundary lines were expressed in terms of flow potentials ϕ or their derivatives expressing rates of flow across the surfaces. Thus it is possible to map the various boundary segments on to the hodograph plane. Intersections of boundaries in the x, y plane map into intersections in the hodograph plane, but in general it is not possible to

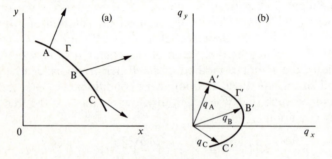

FIG. 7.1. Hodograph mapping

associate individual points in the one plane with particular points in the other plane. Fortunately, this is not necessary in the present context but we shall confine attention to cases of one-to-one correspondence between the two planes. The extension to a hodograph space with a cartesian coordinate system q_x, q_y, q_z presents no difficulties (Bear, Zaslavsky, and Irmay, 1968). The hodograph representation of the following boundary conditions are frequently required. We take the x, y plane to be vertical with y measured vertically upwards.

(i) *Impervious boundary.* This boundary is a streamline and we know that $q_y = q_x \tan \beta$, where β is the angle between the x-axis and the tangent to the streamline. In general, β varies from point to point along a curved boundary and so do q_x, q_y in an unknown way so that the hodograph mapping is not possible. However, it is clear that for a straight-line impervious boundary the hodograph is a parallel straight line passing through the origin of the hodograph plane (Fig. 7.2a,b).

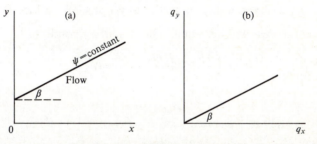

FIG. 7.2. Plane impervious boundary

(ii) *Equipotential boundary.* Here the flow is normal to the boundary and so $q_y = q_x \cot \beta = q_x \tan (\beta - \pi/2)$. Thus the hodograph line of a straight-line equipotential boundary making an angle β with the $+x$-axis is a straight line through the origin making an angle $\beta - \pi/2$ with the q_x axis, i.e. normal to the boundary in the physical plane (Fig. 7.3a,b).

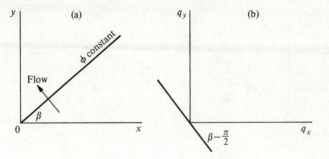

FIG. 7.3. Plane equipotential boundary

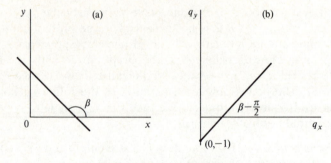

FIG. 7.4. Plane seepage face

(iii) *Seepage face.* For points along a straight seepage face as in Fig. 7.4a, since $\phi = y$ we have $\partial\phi/\partial s = -q_x \cos \beta + q_y \sin \beta = \partial y/\partial s = -\sin \beta$, i.e. $q_y = q_x \tan(\beta - \pi/2) - 1$, and the hodograph of the straight seepage face is a straight line at an angle $\beta - \pi/2$ with the q_x axis and passing through the point $(0, -1)$, i.e. perpendicular to the physical seepage face (Fig. 7.4b).

(iv) *Free surface without accretion.* The boundary condition

$$\left(\frac{\partial\phi}{\partial x}\right)^2 + \left(\frac{\partial\phi}{\partial y}\right)^2 - \frac{\partial\phi}{\partial y} = 0$$

becomes

$$q_x^2 + q_y^2 + q_y = 0 \qquad \text{or} \qquad q_x^2 + (q_y + \tfrac{1}{2})^2 = (\tfrac{1}{2})^2,$$

so that the hodograph of a free surface without accretion is a circle centred at $(0, -\tfrac{1}{2})$ with radius $\tfrac{1}{2}$ as in Figs. 7.5a,b. We note that the direction of $\mathbf{q}$ at a point P on the free surface (Fig. 7.5a) is along the tangent at P, and hence in this case the corresponding point in the hodograph plane is P′ (Fig. 7.5b) and $q_P = -\sin \alpha$.

More general examples of hodographs of these various boundaries for non-homogeneous and anisotropic media with accretion included and for

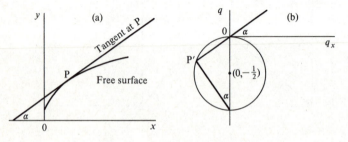

FIG. 7.5. Free surface without accretion

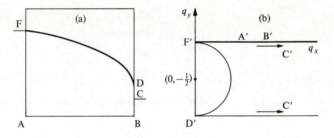

FIG. 7.6. Hodograph mapping of simple dam problem

non-steady problems are given by Bear (1972), Polubarinova-Kochina (1962), and Harr (1962).

The hodograph mapping of the simple dam problem (Fig. 7.6a,b) provides an example of a complete problem. The origin in the hodograph plane is at F′, the point corresponding to F in the physical plane: FA and BC map into F′A′ and B′C′, lines through the origin parallel to the q_x axis since in §7.3.1(ii) $\beta = \pi/2$; the impervious base has $\beta = 0$ (§7.3.1(i)) and also maps into the horizontal line A′B′; the seepage face CD maps into C′D′ through $(0, -1)$ by virtue of §7.3.1(iii); finally, by using the expression in §7.3.1(iv) the free surface FD maps into the circle centre $(0, -\frac{1}{2})$ and radius $\frac{1}{2}$ through F′D′. Since C′ is a common point of the mappings of BC and CD into parallel lines it must be at infinity.

In order to consider the mapping on to the hodograph and other planes it is convenient to introduce complex notation. Thus the potential function ϕ and the corresponding streamline function ψ can be expressed as a complex potential ζ where

$$\zeta = f(z) = \phi(x, y) + i\psi(x, y). \tag{7.13}$$

The function $w = q_x + iq_y$ defines functions in the q_x, q_y plane which has so far been referred to as the hodograph plane. However, it is more convenient to use the complex conjugate function $\tilde{w} = q_x - iq$ and to refer to the plane with axes q_x, $-q_y$ as the hodograph $\tilde{w}$-plane or the inverse hodograph plane; geometrically it is the reflection of the hodograph w-plane in the q_x axis. We note that differentiating the analytic function ζ in (7.13) gives

$$-d\zeta/dz = q_x - iq_y = \tilde{w}. \tag{7.14}$$

A simple example demonstrates the use of the $\tilde{w}$ function in solving a flow problem (Harr 1962). Figure 7.7a shows a triangular dam on a horizontal impervious base with no tail water. The whole of the side BC is a seepage face and there is no free surface. The hodograph planes w and $\tilde{w}$ (Figs. 7.7b,c) follow by the rules (i–iii) set out above. The triangular

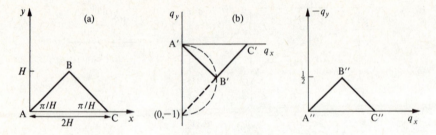

FIG. 7.7. Triangular dam with no tail water

flow region ABC in the z-plane is similar to A″B″C″ in the $\bar{w}$ plane and so $\bar{w} = C_1 z$, where C_1 is a constant. For point B, $z = H + iH$, $\bar{w} = \frac{1}{2}(1 + i) = q_x - i q_y$; hence $C_1 = 1/(2H)$ and $\bar{w} = z/(2H) = -\mathrm{d}\zeta/\mathrm{d}z$. Integration gives $\zeta = -z^2/(4H) + C_2$ and since for point B, $\phi = H$, $\psi = 0$, we find $\zeta = -z^2/(4H) + \frac{1}{2}H(2 + i)$. Finally,

$$\phi = (4H^2 - x^2 - y^2)/(4H), \qquad \psi = (H^2 - xy)/(2H),$$

so that the equipotentials are given by $x^2 - y^2 = \text{constant}$ and the streamlines by $xy = \text{constant}$.

We can identify the following successive steps in the hodograph method of solving flow problems.

(a) Map the boundary of the flow domain in the z-plane on to the hodograph w-plane by applying the rules (i–iv) or more general ones if need be.

(b) Reflect this contour in the w-plane in the real axis q_x to obtain the mapping from the z-plane to the $\bar{w}$-plane. This is a conformal mapping.

(c) Map the boundary of the flow domain in the z-plane on to the complex potential ζ-plane.

(d) Find a conformal mapping relationship $\bar{w} = \bar{w}(\zeta)$ that maps the $\bar{w}$-plane on to the ζ-plane.

(e) Use (7.14) in the integrated form

$$z = \int_{\zeta_0}^{\zeta} \frac{\mathrm{d}\zeta'}{\bar{w}(\zeta')} = x + iy, \qquad \zeta_0 = \phi_0 + i\psi_0, \qquad x = y = 0. \quad (7.15)$$

Although theoretically there is a transformation which will map any pair of simply connected regions on to each other it may not always be possible to find the expression which maps the complicated flow region directly on to the complex potential plane as required in step (c) above. However, by introducing one or more auxiliary planes and mapping functions, even complicated flow regions have been transformed into regular geometric shapes. Usually these are polygons having a finite number of vertices, one or more of which may be at infinity. The interior of a polygon can be mapped on to the upper half of another plane by the Schwarz–Christoffel transformation. Thus the flow domains in the $\bar{w}$ and ζ planes are both mapped on to an auxiliary ξ plane by functions $\bar{w} = \bar{w}(\xi)$ and $\zeta = \zeta(\xi)$ respectively. The Schwarz–Christoffel and other transformations are now described.

7.3.2. *Some useful transformations*

(i) *Schwarz–Christoffel transformation.* This transformation is central to the analytical solution of problems in porous flow: it transforms the interior of a polygon in the z-plane into the upper half of another ξ-plane, $\xi = \mu + i\nu$, in such a way that the sides of the polygon are transformed into the real axis of the ξ-plane.

We start from the expression

$$\frac{dz}{d\xi} = A(\xi - a_1)^{\phi_1}(\xi - a_2)^{\phi_2} \ldots (\xi - a_n)^{\phi_n}, \tag{7.16}$$

where A is a complex constant and $a_1, a_2, \ldots, a_n, \phi_1, \phi_2, \ldots, \phi_n$ are real numbers such that $a_n > a_{n-1} > \ldots > a_2 > a_1$. The real numbers $a_1, a_2, \ldots, a_n$ are plotted on the real axis of the ξ-plane in Fig. 7.8a. If ξ_1 is a real number then

$$\arg(\xi - a_r) = \begin{cases} 0, \xi > a_r \\ \pi, \xi < a_r. \end{cases} \tag{7.17}$$

We know that

$$\arg(dz/d\xi) = \arg A + \phi_1 \arg(\xi_1 - a_1) + \ldots + \phi_n(\xi - a_n) \tag{7.18}$$

and so on writing $\theta_r = \arg(dz/d\xi)$ when $a_r < \xi < a_{r+1}$ and using (7.17) and (7.18) we have

$$\theta_{r+1} - \theta_r = -\pi\phi_{r+1}.$$

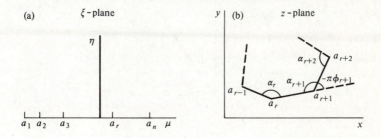

FIG. 7.8. Schwarz–Christoffel transformation

But

$$\arg\frac{dz}{d\xi} = \arg\frac{dx + i\,dy}{d\mu} = \tan^{-1}\frac{dy}{dx},$$

which is the angle that the element dz in the z-plane, into which $d\xi$ is transformed by (7.16), makes with the real axis of the z-plane. Thus as the point ξ traverses the real axis of the ξ-plane, the corresponding point z describes a polygon in the z-plane as shown in Fig. 7.8b. When the point ξ passes from left to right of a_{r+1} in the ξ-plane, the direction of movement of the point z in the z-plane changes by an angle of $-\pi\phi_{r+1}$ in the sense indicated in Fig. 7.8b. If the interior angle of the polygon is α_{r+1} we have $\alpha_{r+1} - \pi\phi_{r+1} = \pi$ and (7.16) becomes after integration

$$z = A\int (\xi - a_1)^{\alpha_1/\pi - 1}(\xi - a_2)^{\alpha_2/\pi - 1} \ldots (\xi - a_n)^{\alpha_n/\pi - 1}\,d\xi + B, \quad (7.19)$$

with B an arbitrary constant. The Schwarz–Christoffel expression (7.19) transforms the real axis of the ξ-plane into a polygon in the z-plane, with interior angles α_r: the modulus and argument of A determine the size and orientation respectively of the polygon and its location is determined by the constant B.

We have been considering only a simply connected polygon, but its boundaries may extend to infinity. It is known that, in mapping a polygon of n sides, three of the values $a_1, a_2, \ldots, a_n$ can be assigned arbitrarily, leaving $n - 3$ to be determined. In the commonly occurring case of a polygon with a vertex at infinity, mapping to $\xi = \infty$, $dz/d\xi$ contains one less term and one less parameter than usual, and only two of the $n - 1$ points $a_1, a_2, \ldots, a_{n-1}$ can be specified arbitrarily.

(ii) *Inverse transformation.* This is frequently of use in the hodograph method. With reference to Fig. 7.9, the point P is mapped on to P' such

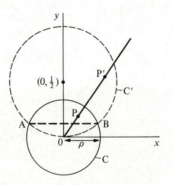

FIG. 7.9. Inverse transformation

that $OP \cdot OP' = \rho^2$ and P' is the image of P in the circle C with centre O and radius ρ. Thus $z' = x' + iy' = \rho^2/(x - iy)$, i.e. with $\bar{z} = x - iy$,

$$z' = \rho^2/\bar{z}. \tag{7.20}$$

Application of (7.20) maps points on the circle C on to themselves, the centre of C maps to a point at infinity, and any circle not centred at O and any straight line map on to a circle or a straight line. A circle C' (Fig. 7.9) passing through the origin O transforms into the straight line passing through the points of intersection AB of the two circles.

(iii) *Zhukovski's function*. This function shares with the hodograph transformation the property that the geometry of the flow region is known in the new plane. Zhukovski (Polubarinova-Kochina 1962; Harr 1962; Bear 1972) defined a function θ_1 given for an isotropic homogeneous medium by

$$\theta_1 = \phi - y$$

and which apart from a multiplying constant is the pressure (see equation (2.5) in §2.2). The function θ_1 is harmonic in the vertical x, y plane and a conjugate function $\theta_2 = \psi + x$ can be defined, so that there is a complex function θ

$$\theta = \theta_1 + i\theta_2 = \zeta + iz, \qquad \zeta = \phi + i\psi, \qquad z = x + iy. \tag{7.21}$$

Since $\phi = y$ along both the free surface and the seepage face and hence $\theta_1 = 0$, these two surfaces map on to the ordinate axis (θ_2) in the θ-plane, with axes θ_1 and θ_2 parallel to x and y axes respectively. The function θ is Zhukovski's function or the complex pressure. We have, by differentiating (7.21),

$$\frac{dz}{d\theta} = -\frac{1}{\tilde{w} - i}, \qquad z = \int \frac{d\theta}{-\tilde{w} + i}, \tag{7.22}$$

so that once the mapping of θ on to $\tilde{w}$ is known, (7.22) yields a relationship between θ and z from which $\phi(x, y)$ and $\psi(x, y)$ follow immediately. Aravin and Numerov (1965) used a Zhukovski function θ defined as $\theta = z - i\zeta$, so that the free and seepage surfaces map on to the real axis of the θ-plane.

(iv) *Hamel's function*. Hamel (1934) introduced a mapping function T defined by

$$T = \tau + i\theta = -\ln(-d^2\zeta/dz^2). \tag{7.23}$$

We have the basic relationships

$$-\frac{d^2\zeta}{dz^2} = -\frac{d}{dz}\left(\frac{d\zeta}{dz}\right) = -\frac{d(d\zeta/dz)/dt}{dz/dt} = \frac{\dot{q}_x - i\dot{q}_y}{n(q_x + iq_y)},$$

where $\dot{q}_x$, $\dot{q}_y$ are time derivatives of q_x, q_y and n is the porosity connecting q_x, q_y with $\dot{x}$, $\dot{y}$ respectively. By defining

$$\dot{q}_x - i\dot{q}_y = nAe^{-i\beta}, \qquad q_x + iq_y = qe^{i\alpha},$$

where $q^2 = q_x^2 + q_y^2$, $(nA)^2 = \dot{q}_x^2 + \dot{q}_y^2$, and α, β are moduli we express $\tau + i\theta$ as $\tau + i\theta = -\ln(A/q) + i(\alpha + \beta)$ and hence

$$\tau = -\ln(A/q), \qquad \theta = \begin{cases} \alpha + \beta, & A/q > 0 \\ \alpha + \beta - \pi, & A/q < 0. \end{cases}$$

The various boundaries of the flow region in the z-plane can now be mapped on to the T-plane. We consider as an example a streamline or an impervious boundary making an angle α with the x-axis. Along such a boundary $dy/dx = q_y/q_x = \tan \alpha$ and hence

$$-\frac{d^2\zeta}{dz^2} = \frac{d(q_x - iq_y)}{d(x + iy)} = \frac{dq_x}{dx}(\cos 2\alpha - i \sin 2\alpha),$$

so that

$$\tau + i\theta = -\ln\{(dq_x/dx)e^{-2i\alpha}\}.$$

For $dq_x/dx > 0$, $\theta = 2\alpha + 2m\pi$; $dq_x/dx < 0$, $\theta = 2\alpha + (m-1)\pi$. In each case

$$\tau = -\ln\{(dq_x/dx)\cos 2\alpha\}.$$

For an equipotential boundary similar arguments yield

$$\theta = \begin{cases} -\tan^{-1}(\cos 2\alpha/\tan 2\alpha) + 2m\pi, & dq_x/dy > 0, \\ -\tan^{-1}(\cos 2\alpha/\tan 2\alpha) + (2m-1)\pi, & dq_x/dy < 0. \end{cases}$$

For a seepage face making an angle α with the x-axis the condition $-q_x \cos \alpha - q_y \sin \alpha = \sin \alpha$ leads to the same expressions for θ as for an equipotential boundary. For a steady free surface, the condition

$$q_x^2 + q_y^2 + q_y = 0, \qquad \text{i.e.} \qquad dq_x/dq_y = -(2q_y + 1)/2q_x,$$

remembering that $dq_x/dq_y = dy/dx$, gives

$$-\frac{d^2\zeta}{dz^2} = \frac{dq_y}{dx}\frac{[-\{(2q_y + 1)/2q_x\} - i]}{1 + i(q_y/q_x)}$$

$$= \frac{dq_y}{dx}\frac{[(2q_xq_y + \frac{1}{2}q_x) + i(q_x^2 - q_y^2 - \frac{1}{2}q_y)]}{q_y}$$

$$= \frac{dq_y}{dx}\frac{e^{-3i\alpha}}{2 \sin \alpha},$$

since along the free surface $q_x = -\sin \alpha \cos \alpha$, $q_y = -\sin^2 \alpha$. For $dq_y/dx > 0$, $\theta = 3\alpha + 2m\pi$; $dq_y/dx < 0$, $\theta = 3\alpha + (2m-1)\pi$. The successive steps in the solution of a flow problem are first to map the z-plane on to the hodograph $\tilde{w}$-plane, and record the values of θ on the various parts of the hodograph boundary by using the expressions just obtained. Next the $\tilde{w}$-plane is mapped on to the ξ-plane ($\xi = \mu + i\nu$), using a modular elliptic function and an intermediate transformation in such a way that all the boundaries of the flow region are mapped on to the real axis of the ξ-plane. Values of $T = \tau + i\theta$ are deduced at all points of the ξ-plane from the values of θ along the real axis by using the Poisson integral

$$iT = \theta - i\tau = -i\tau_0 + \frac{1}{\pi i} \int_{-\infty}^{\infty} \frac{\theta(\mu)(\xi\mu + 1)}{(\mu - \xi)(1 + \mu^2)} \, d\mu$$

where τ_0 is an arbitrary constant. Thus we obtain $T = F_1(\xi)$. From the function which maps $\tilde{w}$ on the ξ-plane we deduce $T = F_2(\tilde{w})$; but $T = -\ln(d\tilde{w}/dz)$ and so we can write

$$z = \int \frac{d\tilde{w}}{\exp\{-F_2(\tilde{w})\}} + C_1 = F_3(\tilde{w}),$$

$$\tilde{w} = F_3^{-1}(z) = -d\zeta/dz, \qquad \zeta = -\int F_3^{-1}(z) \, dz + C_2,$$

from which follow q_x, q_y, $\phi(x, y)$, $\psi(x, y)$ and the equation for the free surface. Muskat (1935) computed solutions for six simple dam problems using Hamel's transformation. The z-plane, $\tilde{w}$-plane, and ξ-plane are shown in Figs. 7.10a,b,c. His solution for z as a function of $\tilde{w}$ is

$$z = C \int \left[\frac{1 - \xi}{\xi(\xi - a)(\xi - b)} \right]^{\frac{1}{2}} \exp\left\{ -\frac{3}{2\pi} \int_0^1 \frac{\beta(\mu) \, d\mu}{\mu - \xi} \right\} d\tilde{w}, \qquad (7.24)$$

where μ is the real axis of the ξ-plane and $\beta(\mu)$ a function evaluated in the process of solution.

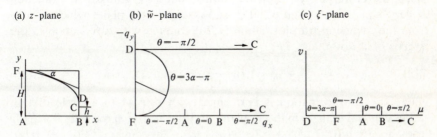

FIG. 7.10. Hamel's transformation for simple dam problem

7.3.3. *Trapezoidal drainage ditch*

An example to illustrate the use of the transformations (i), (ii), (iii) above is afforded by seepage from a trapezoidal ditch in which the water level is

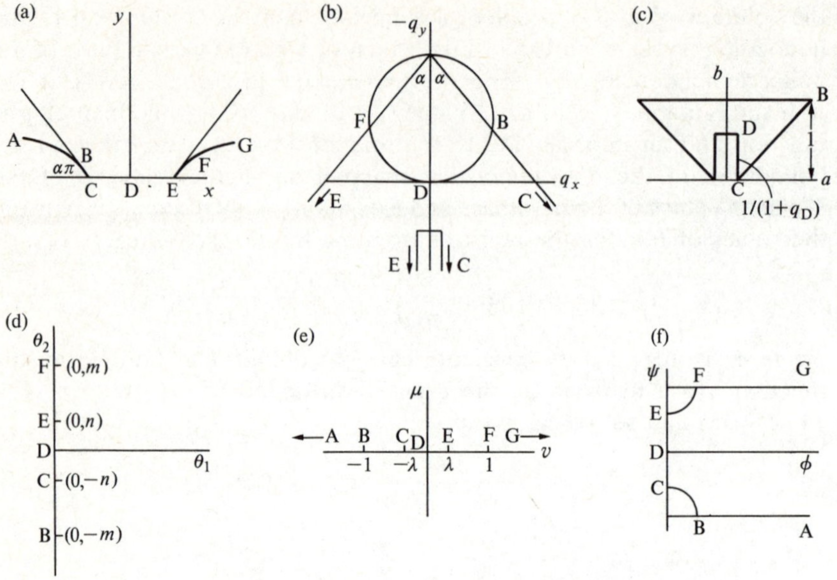

FIG. 7.11. Trapezoidal drainage ditch

negligible. The original solution was given by Vedernikov (1939) and Sokolov (1951). Polubarinova-Kochina (1962) gives a detailed account. The problem in the physical plane is shown in Fig. 7.11a; the hodograph $\tilde{w}$-plane is in Fig. 7.11b. The boundary of the flow region consists of a free surface AB, a seepage surface BC, and a horizontal equipotential base CD (similar statements apply on the right-hand side of the y-axis). Since, therefore, $\phi - y = 0$ on the entire boundary, Zhukovski's function is very convenient. Figure 7.11c shows the $(1/\tilde{w})$ plane which results from the inversion transformation (7.20) coupled with a reflection in the q_x-axis of the $\tilde{w}$-plane, i.e.

$$\eta - \eta_0 = 1/(\widetilde{w - w_0}) = 1/(\tilde{w} - i), \tag{7.25}$$

where the transformation is performed with respect to a circle of unit radius centred at $(0, i)$ in the $\tilde{w}$-plane. Putting $\eta_0 = 0$, $\eta = a + ib$, $\tilde{w} = f + ig$ leads to $f = a/(a^2 + b^2)$, $g = 1 - b/(a^2 + b^2)$. The circle $f^2 + (g - \frac{1}{2})^2 = (\frac{1}{2})^2$ becomes $b = 1$ and the line $g = 1 + f \tan \beta$ maps to $b = a \tan(-\beta)$, as in Fig. 7.11c. From the first of (7.22) it follows that

$$\frac{dz}{d\theta} = -\frac{1}{\tilde{w} - i} = -\eta. \tag{7.26}$$

The appropriate Schwarz–Christoffel transformation (7.19) is

$$\eta = M \int_0^\xi \frac{\xi' \, d\xi'}{(1-\xi'^2)^{\alpha+\frac{1}{2}}(\lambda^2-\xi'^2)^{1-\alpha}} + N,$$

where λ is defined in Fig. 7.11e.

Polubarinova-Kochina (1962) gave the detailed evaluation of the constants M and N. Briefly, the values of η are found using (7.25) at the points $\xi = 0$ and $\xi = 1$, which correspond to points D and B in the hodograph plane. At $\xi = 0$, $\tilde{w} = -iq_D$, where $q_D = q_y$ at D, and so $N = i/(q_D + 1)$. For $\xi = \lambda$, $\tilde{w} = \infty$, i.e. $\eta = 0$, and we have

$$\eta = M \int_\lambda^\xi \frac{\xi' \, d\xi'}{(1-\xi'^2)^{\alpha+\frac{1}{2}}(\lambda^2-\zeta'^2)^{1-\alpha}}.$$

To find real values of the integrand for $\lambda < \xi < 1$, we note that in passing from CD to CB by contouring C in the lower half-plane anti-clockwise through angle π we have

$$(\xi-\lambda)^{1-\alpha}e^{\pi i(1-\alpha)} = -(\xi-\lambda)^{1-\alpha}e^{-\pi i \alpha}$$

and so η becomes

$$\eta = -Me^{\pi i \alpha} \int_\lambda^\xi \frac{\xi' \, d\xi'}{(1-\xi'^2)^{\alpha+\frac{1}{2}}(\xi^2-\lambda^2)^{1-\alpha}}.$$

For $\xi = 1$ and point B in the hodograph plane we find

$$\eta_1 = \frac{ie^{\pi i \alpha}}{\cos \pi \alpha}$$

which leads to

$$\frac{i}{\cos \pi \alpha} = -M \int_\lambda^1 \frac{-\xi \, d\xi}{(1-\xi^2)^{\alpha+\frac{1}{2}}(\xi^2-\lambda^2)^{1-\alpha}} = -MJ, \qquad (7.27)$$

where after the substitution $1-\xi^2 = \lambda'^2 x$, with $\lambda^2 = 1-\lambda'^2$

$$J = \frac{1}{2\lambda'} \int_0^1 x^{-\alpha-\frac{1}{2}}(1-x)^{\alpha-1} \, dx = \frac{1}{2\lambda'} \frac{\pi^{\frac{1}{2}}\Gamma(\alpha)}{\cos \pi \alpha \Gamma(\alpha+\frac{1}{2})},$$

following some manipulation of β and Γ functions. Thus M is determined from (7.27) and

$$\frac{dz}{d\theta} = -\eta = -2Ai\sqrt{(1-\lambda^2)} \int_0^\xi \frac{\xi' \, d\xi'}{(1-\zeta'^2)^{\alpha+\frac{1}{2}}(\lambda^2-\xi'^2)^{1-\alpha}} + \frac{i}{q_D+1},$$

$$\qquad (7.28)$$

where $A = \Gamma(\alpha+\frac{1}{2})/\pi^{\frac{1}{2}}\Gamma(\alpha)$.

Next the θ-plane shown in Fig. 7.11d must be mapped on to the ξ-plane. The complex potential ζ-plane is shown in Fig. 7.11f. The flow region corresponds to the half-plane in the θ-plane and using the facts that for $\xi = 0$, $\phi = 0$, $\psi = \frac{1}{2}Q$, $x = 0$, and for $\xi = 1$, $\phi + y = 0$, $\psi = 0$, $x = x_F$, the linear transformation leads to

$$d\theta = i(\tfrac{1}{2}Q + x_F)\, d\xi, \tag{7.29}$$

where Q is the total flow through CE in Fig. 7.11a. Integration of (7.28) using (7.29) and (7.22) gives

$$z = -A(1-\lambda^2)^{\frac{1}{2}}(x_F + Q) \int_0^\xi d\xi_2 \int_0^{\xi_2} \frac{\xi_1\, d\xi_1}{(1-\xi_1^2)^{\alpha+\frac{1}{2}}(\lambda^2 - \xi_1^2)^{1-\alpha}} + \frac{x_F + Q}{2(q_D + 1)}\xi. \tag{7.30}$$

From this we can relate z to the complex potential ζ via the Zhukovski function θ. Vedernikov (1939) examined the ditch with vertical sides in more detail and Sokolov (1951) extended the analysis of (7.30) by making further transformations.

7.4. Polubarinova-Kochina's solution for the simple dam

Polubarinova-Kochina (1938, 1939) and Risenkampf (1940) applied the theory of linear differential equations to some problems of ground-water flow. A general treatment is given by Polubarinova-Kochina (1962) and applied to the problem of the simple dam and other more difficult problems. The hodograph and complex potential planes are still involved but solutions of the hypergeometric equation are introduced. Here we confine attention to the simple dam problem depicted again for conveni-ence in Fig. 7.12a, together with the hodograph $\bar{w}$-plane, the complex potential ζ-plane, and an auxiliary ξ-plane in Figs. 7.12b,c,d respectively. We first review some relevant properties of the hypergeometric equation

$$z(1-z)\frac{d^2Y}{dz^2} + \{c - (a+b+1)z\}\frac{dY}{dz} - abY = 0, \tag{7.31}$$

and its solutions in the form of hypergeometric series. Equation (7.31) has three regular singularities at the points $z = 0$, 1, and ∞. There are three kinds of solutions in the forms of power series: (i) a solution in a series of powers of z, centre $z = 0$, and valid within a circle of unit radius; (ii) a solution in a series of powers of $1 - z$, centre $z = 1$, and valid within a circle of unit radius; (iii) a solution in a series of powers of $1/z$, centre $z = 0$, and valid outside a circle of unit radius. The two fundamental

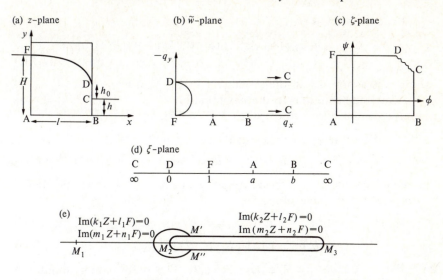

(a) z-plane   (b) $\tilde{w}$-plane   (c) ζ-plane

(d) ξ-plane

C	D	F	A	B	C
∞	0	1	a	b	∞

(e) $\text{Im}(k_1 Z + l_1 F) = 0$
$\text{Im}(m_1 Z + n_1 F) = 0$
M_1
M'
M_2
M''
$\text{Im}(k_2 Z + l_2 F) = 0$
$\text{Im}(m_2 Z + n_2 F) = 0$
M_3

FIG. 7.12. Polubarinova-Kochina's solution for a simple dam

solutions of type (i) developed from a series of the general form

$$y = z^\rho (c_0 + c_1 z + c_2 z^2 + \ldots) \tag{7.32}$$

are found to be

$$Y_1(\rho_1) = c_0 \left\{ 1 + \frac{a \cdot b}{c \cdot 1} z + \frac{a(a+1)b(b+1)}{c(c+1) \cdot 1 \cdot 2} z^2 \right.$$

$$\left. + \frac{a(a+1)(a+2)b(b+1)(b+2)}{c(c+1)(c+2) \cdot 1 \cdot 2 \cdot 3} z^3 + \ldots \right\} \tag{7.33a}$$

and

$$Y_2(\rho_2) = c_0 z^{1-c} \left\{ 1 + \frac{(a-c+1)(b-c+1)}{(2-c) \cdot 1} z \right.$$

$$\left. + \frac{(a-c+1)(a-c+2)(b-c+1)(b-c+2)}{(2-c)(3-c) \cdot 1 \cdot 2} z^2 + \ldots \right\}, \tag{7.33b}$$

where $\rho_1 = 0$, $\rho_2 = 1 - c$ are the exponents for the two solutions and a, b, c are the parameters of the differential equation. The series in (7.33a) is the hypergeometric series represented by the symbol $F(a, b; c; z)$. Thus the series between the brackets in the second solution (7.33b) is represented by $F(a-c+1, b-c+1; 2-c; z)$. The general solution valid in the region

$|z| < 1$ is expressed as

$$Y = C_1 F(a, b; c; z) + C_2 z^{1-c} F(a-c+1, b-c+1; 2-c; z),$$

$$(7.33c)$$

where it is assumed that the parameter c is not an integer.

Fundamental solutions in the neighbourhood of $z = 1$ (case (ii) above) are represented by

$$Y_3 = F(a, b; a+b+1-c; 1-z) \qquad \text{and}$$

$$Y_4 = (1-z)^{c-a-b} F(c-b, c-a; c-a-b+1; 1-z), \qquad (7.34)$$

i.e. the exponents are $\rho = 0$ and $c-a-b$. For case (iii), in the neighbourhood of $z = \infty$ we have solutions

$$Y_5 = z^{-a} F(a, a-c+1; 1+a-b; 1/z),$$

$$Y_6 = z^{-b} F(b, b-c+1; 1+b-a; 1/z), \qquad (7.35)$$

and the exponents for $1/z$ are $\beta = a, b$. This information about the solutions of (7.31) are summarized by the Riemann P-equation

$$Y \equiv P \left\{ \begin{matrix} 0 & \infty & 1 & \\ 0 & a & 0 & z \\ 1-c & b & c-a-b & \end{matrix} \right\}. \qquad (7.36)$$

The singular points of the equation are placed in the first row and the exponents are directly beneath them; the independent variable is indicated in the fourth column. There is a standard transformation that leads to a Riemann function which has one vanishing exponent about each finite singular point, namely

$$P \left\{ \begin{matrix} a_1 & a_2 & \dots & a_n & \infty & \\ \alpha_1 & \alpha_2 & \dots & \alpha_n & \alpha & z \\ \beta_1 & \beta_2 & \dots & \beta_n & \beta & \end{matrix} \right\} = (z-a_1)^{\alpha_1}(z-a_2)^{\alpha_2} \dots (z-a_n)^{\alpha_n}$$

$$\times P \left\{ \begin{matrix} a_2 & a_2 & \dots & a_n & \infty & \\ 0 & 0 & \dots & 0 & \alpha + \alpha_1 + \dots + \alpha_n & z \\ \beta_1 - \alpha_1 & \beta_2 - \alpha_2 & \dots & \beta_n - a_n & \beta + \alpha_1 + \dots + \alpha_n & \end{matrix} \right\}. \qquad (7.37)$$

We now have to relate these solutions of the hypergeometric series to seepage problems. In particular we need to deduce the exponents ρ at the singular points on the real axis in the ξ-plane from the boundary conditions which are prescribed on the different segments of the real axis. Polubarinova-Kochina (1962) observed that the commonly occurring flow regions are confined by straight line boundaries and free surfaces. Furthermore, the boundary conditions are expressed by equating linear functions of the coordinates of the boundaries to zero or linear functions of the coordinates and the velocity potential and stream functions, ϕ and ψ. Along each boundary, therefore, the two conditions to be satisfied can be written

$$k_1 x + \ell_1 y + m_1 \phi + n_1 \psi = p, \qquad k_2 x + \ell_2 y + m_2 \phi + n_2 \psi = q,$$

where the coefficients are constants. More conveniently we write

$$\text{Im}(kz + \ell\zeta) = p, \qquad \text{Im}(mz + n\zeta) = q, \tag{7.38}$$

where k, ℓ, m, n are complex, p, q are real numbers. When the boundaries of the flow region are mapped on to the real axis of the ξ-plane, the corresponding values of ξ are real. On introducing the functions

$$F = d\zeta/d\xi, \qquad Z = dz/d\xi,$$

we obtain by differentiating in (7.38)

$$\text{Im}(kZ + \ell F) = 0, \qquad \text{Im}(mZ + nF) = 0. \tag{7.39}$$

We now seek to find two functions F, Z of the complex variable ξ which satisfy conditions of type (7.39) on segments of the real ξ-axis. Figure 7.12e shows two neighbouring segments M_1M_2 and M_2M_3 with the two conditions displayed for each segment. In order to examine the behaviour of F and Z when a singular point, say M_2, is contoured we use the standard result that for a function $f(z)$ which is analytic in a region of the upper half-plane adjacent to the real axis and continuous up to a segment along which $\text{Im}\{f(z)\} = 0$, the analytic continuation in the lower half-plane is such that in conjugate points the function takes conjugate values. To apply this result to the linear combinations of F and Z associated with M_1M_2 (Fig. 7.12e), a cut is made along segment M_2M_3 and the points on the upper and lower sides of the cut are denoted by M' and M''. The values of F and Z at point M'' after contouring the singular point M_2 are written as F^* and Z^*. Thus we have

$$k_1 Z^* + \ell_1 F^* = \bar{k}_1 \bar{Z} + \bar{\ell}_1 \bar{F}, \qquad m_1 Z^* + n_1 F^* = \bar{m}_1 \bar{Z} + \bar{n}_1 \bar{F}. \tag{7.40}$$

With regard to the segment M_2M_3 the conditions can be rewritten

$$k_2 Z + \ell_2 F - \bar{k}_2 \bar{Z} - \bar{\ell}_2 \bar{F} = 0, \qquad m_2 Z + n_2 F - \bar{m}_2 \bar{Z} - \bar{n}_2 \bar{F} = 0. \tag{7.41}$$

After some manipulation to eliminate $\bar{F}$ and $\bar{Z}$ from (7.40) and (7.41) we can write Z^*, F^* in the form

$$Z^* = \beta_1 Z + \beta_2 F, \qquad F^* = \gamma_1 Z + \gamma_2 F \tag{7.42}$$

where β_1, β_2, γ_1, γ_2 are constants involving the k_1, ℓ_1, etc. in a complicated way, i.e. F and Z undergo a linear transformation when contouring a singular point. We can interpret (7.42) to mean that F and Z

can be expressed as linear combinations of two functions U and V which undergo the transformations

$$U^* = \lambda_1 U, \qquad V^* = \lambda_2 V, \tag{7.43a}$$

where λ_1, λ_2 are the roots of the characteristic equation associated with (7.42), i.e.

$$\begin{vmatrix} \beta_1 - \lambda & \beta_2 \\ \gamma_1 & \gamma_2 - \lambda \end{vmatrix} = 0. \tag{7.44}$$

In the case of a double root of (7.44) the transformations become

$$U^* = \lambda_1 U, \qquad V^* = \beta_1 U + \lambda_2 V. \tag{7.43b}$$

After considerable manipulation the characteristic equation can be written in terms of k_1, $\bar{k}_1$, ℓ_1, $\bar{\ell}_1$, k_2, $\bar{k}_2$, etc. and becomes

$$\begin{vmatrix} k_1\lambda & \ell_1\lambda & \bar{k}_1 & \bar{\ell}_1 \\ m_1\lambda & n_1\lambda & \bar{m}_1 & \bar{n}_1 \\ k_2 & \ell_2 & \bar{k}_2 & \bar{\ell}_2 \\ m_2 & n_2 & \bar{m}_2 & \bar{n}_2 \end{vmatrix} = 0. \tag{7.45}$$

If we now consider the behaviour of the functions $(\xi - \xi_0)^{\alpha_1}$ and $(\xi - \xi_0)^{\alpha_2}$ on contouring the singular point $\xi = \xi_0$, corresponding to M_2 in Fig. 7.12e, through an angle 2π and choose

$$\alpha_1 = \frac{\ln \lambda_1}{2\pi i}, \qquad \alpha_2 = \frac{\ln \lambda_2}{2\pi i} \tag{7.46}$$

we find

$$e^{2\pi i \alpha_1}(\xi - \xi_0)^{\alpha_1} = \lambda_1(\xi - \xi_0)^{\alpha_1}, \qquad e^{2\pi i \alpha_2}(\xi - \xi_0)^{\alpha_2} = \lambda_2(\xi - \xi_0)^{\alpha_2}.$$

Thus the quotients $U/(\xi - \xi_0)^{\alpha_1}$, $V/(\xi - \xi_0)^{\alpha_2}$ remain single-valued functions and, having in mind that α_1, α_2 are only determined within an integer we can develop Laurent series for a regular singularity with no negative powers and write

$$U = (\xi - \xi_0)^{\rho_1} \sum_{k=0}^{\infty} a_k (\xi - \xi_0)^k, \tag{7.47a}$$

$$V = (\xi - \xi_0)^{\rho_2} \sum_{k=0}^{\infty} b_k (\xi - \xi_0)^k + \alpha U \ln (\xi - \xi_0), \tag{7.47b}$$

in which a_0, b_0 are non-zero. When $\alpha_1 - \alpha_2$ is not zero nor an integer, then we take $\alpha = 0$. Polubarinova-Kochina (1962) derived (7.45) directly for the particular case of α_1 and α_2 real.

We now seek to establish the exponents about singular points which may occur at the intersections of neighbouring boundary segments in the simple dam problem, for example. We consider the following combinations.

Case 1: Equipotential dam face and impervious base. Let the dam face make an angle $\pi\alpha$ with a linear impervious base (Fig. 7.13a). If we assume that the point A maps on to the origin of the coordinates of the ξ-plane (Fig. 7.13b), the conformal mapping of z on to the ξ-plane has the form $z = \xi^\alpha(a_0 + a_1\xi + \ldots)$. The function $Z = dz/d\xi = \xi^{\alpha-1}(b_0 + b_1 + \ldots)$, i.e. Z belongs to the class of solutions of (7.31) with exponent $\alpha - 1$. In the complex potential plane, $\zeta = \phi + i\psi$, lines of constant ϕ and ψ meet in point A (Fig. 7.13c) so that $\zeta = \xi^{\frac{1}{2}}(c_0 + c_1\xi + \ldots)$ and the function $F = d\zeta/d\xi = \xi^{-\frac{1}{2}}(\ell_0 + \ell_1\xi + \ldots)$ has an exponent $-\frac{1}{2}$. When $\pi\alpha = \pi/2$ both exponents have the value $-\frac{1}{2}$ and the point A on the velocity hodograph is an ordinary, non-singular point since $\tilde{w} = q_x - iq_y = F/Z$.

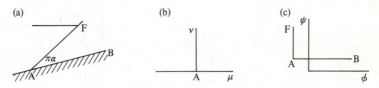

FIG. 7.13. Equipotential face and impervious base

Case 2: Seepage face joins outlet face. In Fig. 7.14a,b the point C is an ordinary point in the z-plane and hence for Z we have

$$Z = a_0 + a_1\xi + \ldots.$$

But in the hodograph $\tilde{w}$-plane C is at infinity (Fig. 7.14b) being the intersection of the two parallel lines DC and BC. Thus about C we have $\tilde{w} = c_0 \ln \xi$ but, following the general rule that for a singular point C at infinity in the ξ-plane the exponents are to be increased by two, so the required exponent of each of Z and F is 2.

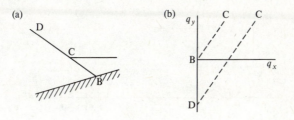

FIG. 7.14. Seepage face joins outlet face

Case 3: Free surface joins a linear seepage surface. For the z- and $\tilde{w}$-planes shown in Figs. 7.15a,b the conditions on the real axis of the ξ-plane are as in Fig. 7.15c. By inserting appropriate values into (7.45) we obtain the characteristic equation

$$\begin{vmatrix} \lambda & \lambda & 1 & -i \\ \lambda e^{-\pi i \alpha} & 0 & e^{\pi i \alpha} & 0 \\ 1 & i & 1 & -i \\ c & 1 & -ci & 1 \end{vmatrix} = 0,$$

which has roots $\lambda_1 = 1$, $\lambda_2 = e^{2\pi i(\alpha + \frac{1}{2})}$ and exponents 0 and $\alpha - \frac{1}{2}$ follow from (7.46) for the point D.

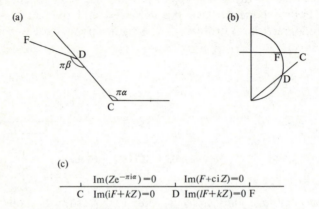

(a)

(b)

(c)

	$\text{Im}(Ze^{-\pi i \alpha}) = 0$		$\text{Im}(F + ciZ) = 0$	
C	$\text{Im}(iF + kZ) = 0$	D	$\text{Im}(lF + kZ) = 0$ F	

FIG. 7.15. Free surface joins linear seepage surface

Case 4: Free surface intersects inlet face. For the z-plane as in Fig. 7.16a, the conditions on the ξ-plane are given in Fig. 7.16b, assuming no

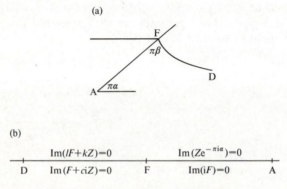

(a)

(b)

	$\text{Im}(lF + kZ) = 0$		$\text{Im}(Ze^{-\pi i \alpha}) = 0$	
D	$\text{Im}(F + ciZ) = 0$	F	$\text{Im}(iF) = 0$	A

FIG. 7.16. Free surface intersects inlet face

infiltration. The characteristic equation is

$$
\begin{vmatrix}
\lambda & i\lambda & 1 & -i \\
0 & \lambda & 0 & 1 \\
e^{-\pi i \alpha} & 0 & e^{\pi i \alpha} & 0 \\
0 & i & 0 & -i
\end{vmatrix} = 0,
$$

with roots $\lambda = e^{\pi i}$, $e^{2\pi i \alpha}$ leading to exponents for the point F, $-\frac{1}{2}$, $-\alpha$ from (7.46).

We are now able to find the exponents for the boundary intersections in the simple dam problem of Fig. 7.12a and hence to find the appropriate functions F and Z. Points $A(\xi = a)$ and $B(\xi = b)$ correspond to Case 1 above with $\alpha = \frac{1}{2}$ so that these are both ordinary points. At point C we have Case 2 and the exponents are 2, 2 for the point at $+\infty$ in the ξ-plane. At the separation point D, Case 3 applies with $\alpha = \frac{1}{2}$, giving exponents 0, 0. Finally, by applying Case 4 to point F with $\alpha = \frac{1}{2}$ we obtain exponents $-\frac{1}{2}$, $-\frac{1}{2}$.

Thus we see that the functions F and Z are different branches of the function $\tilde{Y}$ represented by Riemann's P-equation as

$$
\tilde{Y} = P \begin{Bmatrix}
0 & 1 & a & b & \infty & \\
0 & -\frac{1}{2} & -\frac{1}{2} & -\frac{1}{2} & 2 & \xi \\
0 & -\frac{1}{2} & \frac{1}{2} & \frac{1}{2} & 2 &
\end{Bmatrix}. \tag{7.48}
$$

By applying (7.37) to $\tilde{Y}$ we find

$$
\tilde{Y} = (1-\xi)^{-\frac{1}{2}}(\xi - a)^{-\frac{1}{2}}(\xi - b)^{-\frac{1}{2}} P \begin{Bmatrix}
0 & 1 & a & b & \infty & \\
0 & 0 & 0 & 0 & \frac{1}{2} & \xi \\
0 & 0 & 1 & 1 & \frac{1}{2} &
\end{Bmatrix}
$$

$$
= Y / \sqrt{\{(1-\xi)(\xi - a)(\xi - b)\}}. \tag{7.49}
$$

The new function Y has only three singular points at $\xi = 0, 1, \infty$ and we write

$$
Y = P \begin{Bmatrix}
0 & 1 & \infty & \\
0 & 0 & \frac{1}{2} & \xi \\
0 & 0 & \frac{1}{2} &
\end{Bmatrix}. \tag{7.50}
$$

Thus in (7.36) the values of the parameters corresponding to (7.50) are $a = b = \frac{1}{2}$, $c = 1$ and hence the differential equation (7.31) takes the particular form

$$
\xi(1-\xi)\frac{d^2 Y}{d\xi^2} + (1-2\xi)\frac{dY}{d\xi} - \frac{1}{4}Y = 0. \tag{7.51}
$$

If we denote by U and V the linearly independent solutions of (7.51) we can write

$$F = \frac{AU + BV}{\sqrt{\{(1-\xi)(\xi-a)(\xi-b)\}}}, \qquad Z = \frac{CU + DV}{\sqrt{\{(1-\xi)(\xi-a)(\xi-b)\}}}.$$

(7.52)

About the point $\xi = 0$ the solution U is obtained by putting $a = b = \frac{1}{2}$, $c = 1$ in (7.33a) above and is

$$U = F(\tfrac{1}{2}, \tfrac{1}{2}, 1, \xi) = 1 + (\tfrac{1}{2})^2 \xi + \left(\frac{1.3}{2.4}\right)^2 \xi^2 + \left(\frac{1.3.5}{2.4.6}\right)^2 \xi^3 + \ldots,$$

(7.53)

which is seen to differ from the complete elliptic integral of the first kind only by a multiplying constant. Thus we may write as one of the particular solutions of (7.51)

$$U = K(\xi) = \int_0^{\pi/2} \frac{d\theta}{\sqrt{\{1 - \xi \sin^2 \theta\}}}, \qquad k^2 = \xi,$$

(7.54)

where ξ is equated to the square of the modulus k.

Since (7.51) is not changed by substituting $1 - \xi$ for ξ we can take as a solution

$$V = K(1-\xi) = K', \qquad (k')^2 = 1 - \xi,$$

(7.55)

which is regular about $\xi = 1$ but has a logarithmic singularity about $\xi = 0$. Thus $U = K(\xi)$ and $V = K(1-\xi)$ are linearly independent solutions about $\xi = 0$.

From (7.35) we can find that

$$U_1 = (1/\sqrt{\xi})K(1/\xi), \qquad V_1 = (1/\sqrt{\xi})K(1 - 1/\xi)$$

(7.56)

are linearly independent solutions of (7.51) convergent for $|\xi| > 1$ and by substituting $1 - \xi$ for ξ we obtain

$$U_2 = \frac{1}{\sqrt{(1-\xi)}} K\left(\frac{1}{1-\xi}\right), \qquad V_2 = \frac{1}{\sqrt{(1-\xi)}} K\left(-\frac{\xi}{1-\xi}\right),$$

(7.57)

which converge outside a circle of unit radius, centre $\xi = 1$.

In order to relate these various functions to the hodograph function $\bar{w} = q_x - iq_y$ we note that the hodograph flow region in Fig. 7.12b is mapped on to the upper-half ξ-plane in Fig. 7.12d by the relationship of type

$$\bar{w} = \frac{AK(\xi) + BK(1-\xi)}{CK(\xi) + DK(1-\xi)}.$$

Noting that on the segment $(0, 1)$ for ξ we have for $\xi = 0$, $\bar{w} = ki$ (AB = k), for $\xi = 1$, $\tilde{w} = 0$, and for $\xi = \frac{1}{2}$, $\tilde{w} = \frac{1}{2}k + (i/2)k$, we find

$$\tilde{w} = k \frac{K(1-\xi)}{K(\xi) - iK(1-\xi)}. \tag{7.58}$$

Through the relationships

$$K(\xi) - iK(1-\xi) = (1/\sqrt{\xi})K(1/\xi) = \frac{-i}{\sqrt{(1-\xi)}} K\left(\frac{1}{1-\xi}\right),$$

$$K(1-\xi) = (1/\sqrt{\xi})K\left(\frac{\xi-1}{-\xi}\right) = \frac{1}{\sqrt{(1-\xi)}} \left\{ K\left(\frac{1}{1-\xi}\right) - iK\left(\frac{-\xi}{\xi-1}\right) \right\},$$

we rewrite $\tilde{w}$ for the segment $1 < \xi < \infty$ as

$$\tilde{w} = k \frac{K(1-1/\xi)}{K(1/\xi)} \tag{7.58a}$$

and for $-\infty < \xi < 0$

$$\tilde{w} = ki \left\{ K\left(\frac{1}{1-\xi}\right) - iK\left(\frac{-\xi}{1-\xi}\right) \right\} / K\left(\frac{1}{1-\xi}\right), \tag{7.58b}$$

which satisfy the necessary correspondence between the $\tilde{w}$- and ξ-planes.

Now we return to expressions (7.52) for $F = \partial\phi/\partial\xi + i\, \partial\psi/\partial\xi$ and $Z = \partial x/\partial\xi + i\, \partial y/\partial\xi$ and insert the appropriate expressions U_1, V_1, etc. from (7.54–7) for each segment of the real axis of the ξ-plane (Fig. 7.12d) having in mind the local boundary condition on the corresponding boundary segment in the physical plane. Thus for $1 < \xi < a$, which corresponds to the inlet face AF in Fig. 7.12a, where $x = 0$ and ϕ is constant, we write, using (7.56),

$$Z = i\, \partial y/\partial\xi = \frac{-AiK(1/\xi)}{\sqrt{\{\xi(\xi-1)(\xi-a)(\xi-b)\}}}, \tag{7.59a}$$

$$F = i\partial\psi/\partial\xi = \frac{-kAiK(1-1/\xi)}{\sqrt{\{\xi(\xi-1)(\xi-a)(\xi-b)\}}}. \tag{7.59b}$$

By integrating (7.59a) with respect to ξ over the range $1 < \xi < a$ we obtain

$$H = A \int_1^a \frac{K(1/\xi)\, d\xi}{\sqrt{\{\xi(\xi-1)(\xi-a)(\xi-b)\}}}, \tag{7.60a}$$

and similarly (7.59b) gives the flow Q, where

$$Q = kA \int_1^a \frac{K(1-1/\xi)\, d\xi}{\sqrt{\{\xi(\xi-1)(\xi-a)(\xi-b)\}}}. \tag{7.60b}$$

By proceeding similarly for the segment $b < \xi < \infty$, which is BC in Fig. 7.12a, we find

$$h = A \int_b^\infty \frac{K(1/\xi)\,\mathrm{d}\xi}{\sqrt{\{\xi(\xi-1)(\xi-a)(\xi-b)\}}},$$

and for $a < \xi < b$, i.e. the base AB in Fig. 7.12a,

$$\ell = A \int_a^b \frac{K(1/\xi)\,\mathrm{d}\xi}{\sqrt{\{\xi(\xi-1)(\xi-a)(\xi-b)\}}}.$$

The length of the seepage surface, CD in Fig. 7.12a, follows by considering the segment $-\infty < \xi < 0$, and is

$$h_0 = A \int_{-\infty}^0 \frac{K(1/\{1-\xi\})\,\mathrm{d}\xi}{(1-\xi)\sqrt{\{(a-\xi)(b-\xi)\}}}.$$

The following changes of variable were introduced by Polubarinova-Kochina (1962):

$$\alpha = 1/b, \qquad \beta = 1/a, \qquad 0 \leqslant \alpha \leqslant \beta \leqslant 1, \qquad \tau = 1/\xi, \qquad C = 2A/\sqrt{(ab)},$$

and

$$\tau = \alpha + (\beta - \alpha)\sin^2 \Psi, \qquad a < \xi < b;$$
$$\tau = \beta + (1-\beta)\sin^2 \Psi, \qquad 1 < \xi < a; \qquad \tau = \alpha \sin^2 \Psi, \qquad b < \xi < \infty.$$

The expressions for ℓ, H, and h become

$$\ell = C \int_0^{\pi/2} \frac{K[\alpha + (\beta - \alpha)\sin^2 \Psi]}{\sqrt{\{1 - \alpha - (\beta - \alpha)\sin^2 \Psi\}}}\,\mathrm{d}\Psi, \tag{7.61}$$

$$H = C \int_0^{\pi/2} \frac{K[\beta + (1-\beta)\sin^2 \Psi]}{\sqrt{\{\beta - \alpha + (1-\beta)\sin^2 \Psi\}}}\,\mathrm{d}\Psi, \tag{7.62}$$

$$h = C\alpha^{\frac{1}{2}} \int_0^{\pi/2} \frac{K(\alpha \sin^2 \Psi)\sin \Psi}{\sqrt{\{(1 - \alpha \sin^2 \Psi)(\beta - \alpha \sin^2 \Psi)\}}}\,\mathrm{d}\Psi. \tag{7.63}$$

By putting $1/(1-\xi) = \cos^2 \psi$, $\alpha_1 = 1 - \alpha$, and $\beta_1 = 1 - \beta$ in the integral for h_0 we obtain

$$h_0 = C \int_0^{\pi/2} \frac{K(\cos^2 \Psi)\sin \Psi \cos \Psi}{\sqrt{\{(1 - \alpha_1 \sin^2 \Psi)(1 - \beta_1 \sin^2 \Psi)\}}}\,\mathrm{d}\Psi. \tag{7.64}$$

The height of the separation point above the dam base is $h_s = h + h_0$. Cryer (1976a) pointed to errors in the expressions (7.62) and (7.64) in Polubarinova-Kochina (1962) and his corrections have been included.

To find the equation of the free surface consider the segment $0 < \xi < 1$

and observe that

$$Z = \frac{\partial x}{\partial \xi} + i \frac{\partial y}{\partial \xi} = -A \frac{K(\xi) - iK(1-\xi)}{\sqrt{\{(1-\xi)(a-\xi)(b-\xi)\}}}, \tag{7.65a}$$

$$F = \frac{\partial \phi}{\partial \xi} = \frac{-kAK(1-\xi)}{\sqrt{\{(1-\xi)(a-\xi)(b-\xi)\}}}. \tag{7.65b}$$

By separating real and imaginary parts in (7.65a) and substituting $\xi = \sin^2 \Psi$ we obtain

$$x = \ell - C \int_0^\Psi \frac{K(\sin^2 \Psi)\sin \Psi}{\sqrt{\{(1-\alpha \sin^2 \Psi)(1-\beta \sin^2 \Psi)\}}} \, d\Psi, \tag{7.66a}$$

$$y = h + h_0 + C \int_0^\Psi \frac{K(\cos^2 \Psi)\sin \Psi}{\sqrt{\{(1-\alpha \sin^2 \Psi)(1-\beta \sin^2 \Psi)\}}} \, d\Psi, \tag{7.66b}$$

for $0 \leqslant \Psi \leqslant \pi/2$.

TABLE 7.1

Parameters in Polubarinova-Kochina's solution

H	ℓ	h	h_s	α	β
24	16	4	12.713 2	0.095 126	0.465 367
24	16	0	12.567 4	0	0.416 089
0.322	0.162	0.084	0.204 460	0.098 554	0.199 823

TABLE 7.2

Coordinates (x, y) of points on free boundary from Polubarinova-Kochina's solution:
$H = 1, h = \frac{1}{6}, \ell = \frac{2}{3}$

$24x$	Ordinate y
0	1.000 000
1	0.989 449
2	0.975 507
3	0.959 378
4	0.941 347
5	0.921 523
6	0.899 935
7	0.876 558
8	0.851 318
9	0.824 092
10	0.794 698
11	0.762 864
12	0.728 189
13	0.690 035
14	0.647 290
15	0.597 565
16	0.529 459

Polubarinova-Kochina (1962) gives series expansions of some of the integrals in her solution and presents graphical results for dams of various dimensions. Cryer (1976a) evaluated the values quoted in Table 7.1. Ozis (1981) used Cryer's values of α and β in the first line of Table 7.1 and evaluated the position of the free boundary. His results are given in Table 7.2 normalised to $H = 1$, $\ell = \frac{2}{3}$, $h = \frac{1}{6}$; they were obtained on a CDC 7600 computer using single-precision arithmetic and a Gauss algorithm with 64 points to evaluate the integrals. The separation point is given by $h_s = 0.5295$. These values agree to within one in the fourth decimal place with those quoted by Elliott and Ockendon (1982), who also give solutions for the other two sets of parameters in Table 7.1.

Polubarinova-Kochina (1962) gave details of the cases of a trapezoidal dam with and without evaporation and of seepage in two media of different permeabilities. She also introduced a general theory for more complicated problems in which more than three singular points exist (Polubarinova-Kochina 1939, Nos. 5, 6).

8. Numerical solution of free-boundary problems

8.1. Numerical solutions

A NUMBER of free-boundary problems were formulated in Chapter 2 and frequent reference was made to this present chapter for more details of how numerical solutions can be obtained. Various numerical methods have been devised and they fall into two broad classes. In the first class the techniques are applied to the problem as originally formulated in the physical plane. In the second class the problem is recast in a different form, e.g. by introducing the Baiocchi variable or some suitable change of coordinates, and the transformed problem is solved numerically. The choice of method is influenced by various factors, apart from the problem itself, such as the development of more powerful numerical techniques for solving partial differential equations and the computing facilities available. It is not really possible to make a well-founded statement that one method is in general superior to all other methods, though more restricted claims can be made in relation to certain types of problem especially if some one aspect of the solution is of particular interest.

8.2. Trial free-boundary methods

The simple dam problem illustrated in Fig. 2.1 and formulated by equations (2.7–12) is taken as the model for describing the numerical techniques. All trial free-boundary methods are variations on one basic theme. An initial estimate is made of the free boundary FD in Fig. 2.1, to be denoted for convenience by $\Gamma^{(0)}$. A first approximation $\phi_h^{(0)}(x, y)$ is computed of the solution ϕ^0 of the boundary value problem defined in the seepage region by the appropriate partial differential equation and the conditions on the 'fixed' boundaries together with one of the two conditions on the free boundary. An improved approximation to the free boundary, $\Gamma^{(1)}$, is taken as the line on which the remaining free boundary condition is satisfied according to the first approximate solution, ϕ_h^0. Thus, when an approximation $\Gamma^{(k)}$ has been obtained for the problem of Fig. 2.1 and $\Omega^{(k)}$ is the corresponding seepage region, $\phi_h^{(k)}$ is computed from (2.7), i.e. $\nabla^2 \phi^{(k)} = 0$, in $\Omega^{(k)}$ satisfying the conditions (2.8), (2.10), (2.11), and the second of (2.9) on $\Gamma^{(k)}$. Given $\Gamma^{(k)}$ and $\phi_h^{(k)}$, a new trial free boundary, $\Gamma^{(k+1)}$, is found by requiring that the first of (2.9) should be satisfied on $\Gamma^{(k+1)}$. The iterative process is terminated when successive Γs agree to a specified accuracy.

It is convenient to discuss the two parts of the iterative cycle separately even though, in practice, the solving of the elliptic differential equation on the approximated seepage region $\Omega^{(k)}$ and the movement of the trial free boundary from $\Gamma^{(k)}$ to $\Gamma^{(k+1)}$ are interrelated.

8.2.1. Computation of approximate trial solutions

The numerical solution of fixed boundary-value problems is a specialist topic which has been extensively studied. A useful survey is edited by Gladwell and Wait (1979). A discretized form of a problem can be based on finite differences or finite elements. In each case what is ultimately required is the solution of a system of algebraic equations, not necessarily linear, and again there is a choice between direct and iterative methods of solution. Introductory textbooks by Smith (1978) on finite-difference methods and by Davies (1980) on finite elements include references to more comprehensive accounts and Gladwell and Wait (1979) is a valuable source-book. Cryer (1976b) included many references in a brief review of different methods and of the influence of developments in computing-facilities. More details about finite differences and free-boundary problems are given by Cryer (1970) and Mogel and Street (1974) and about finite element calculations by Neuman and Witherspoon (1970), Taylor and Brown (1967), Finn (1967), Larock and Taylor (1976), and Bathe and Khoshgoftaar (1979).

Shaw and Southwell (1941) first used finite-difference relaxation techniques in a trial free-boundary method. Many similar calculations followed and are described by Southwell (1946), Allen (1954), Bickley (1964). Digital computation was first attempted by Young *et al.* (1955) and Arms and Gates (1957). After a short period of equation solving by computer with adjustment of the free boundary by hand, a computer was virtually always used for the whole operation after 1960. Cryer (1976b) listed a number of the early papers on trial free-boundary solutions based on finite differences.

The matrix which results when the partial derivatives are replaced by finite-difference ratios at every point of a grid covering the seepage region, for example, is usually sparse and well structured. Iterative methods, e.g. Gauss–Seidel or SOR, which take advantage of the special matrix features, are described for example by Varga (1962), Young (1971), and Yanenko (1971), and suitable direct methods by Bunch and Rose (1976), Reid (1971), and Buzbee and Dorr (1974). Both methods are discussed by Reid (1979). The relative efficiencies of these methods when applied to free boundary problems reflect their performance in fixed boundary problems. One or two special features may also influence their usefulness, however. Thus, an approximate solution $\phi_h^{(k-1)}$ provides a better than usual initial approximation to $\phi_h^{(k)}$, since the only change is in Γ, which is expected to be small. If the algebraic equations are non-linear it is likely that an iterative method will be more efficient in a free boundary context. Care in ordering the equations may improve the efficiency of a direct method since many of the computations needed to carry out a Gaussian elimination, for example, for $\phi_h^{(k)}$, will have been performed to find $\phi_h^{(k-1)}$.

As to the choice of grid, it is clear that if the grid lines are equally spaced then the free boundary will in general pass between grid points. Special finite-difference formulae for unequal intervals have to be used in the neighbourhood of the boundary (Smith 1978). Instead, unequally spaced horizontal and vertical grid lines can be chosen such that they intersect on the free boundary. Figure 8.1a shows an example of a grid in which the horizontal step length is constant but the vertical steps are variable. (Varga 1962; Aitchison 1972). When the trial free boundary is moved, the new boundary grid points are the intersections of the new boundary with the same equally spaced vertical grid lines. Near a corner or singularity accuracy can sometimes be maintained by introducing unequal spacings in the vicinity. Alternatively, a graded grid can be used in which equal grid spacings are retained in any region but the grid is refined near a singularity. Cryer (1976*b*) quoted Rippin and Davidson (1967) as having used a graded grid in a computer implementation of a trial free-boundary method. An example of the use of refined and irregular grids near a singularity is provided by Mogel and Street (1974). Cryer (1970) developed a program which generated a grid automatically.

For the simple dam problem there is little to choose between finite differences and finite elements. As usual, however, the preference for

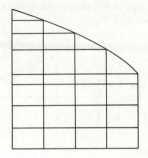

FIG. 8.1(a). Variable vertical grid spacings

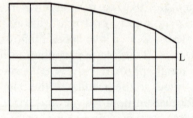

FIG. 8.1(b). Fixed elements below L; variable quadrilaterals above L

finite elements is more marked for irregularly shaped domains and for three-dimensional problems. Special elements are needed near singularities in order to maintain accuracy. We consider first a finite-element approach to the simple dam problem using a fixed mesh in which only the nodes on the free boundary are adjusted in the iterative process. For example, in Fig. 8.1b, rectangular elements below some line L remain fixed and quadrilaterial elements above L fit the free boundary, which is approximated by piecewise linear segments. Any suitable approximating functions can be used in the elements, providing special consideration is given to the singularity at the separation point D. If the free boundary is relatively flat and a good initial approximation is used so that only small adjustments to the successive trial boundaries are needed, this approach is easy to implement. Only the coordinates of the elements near the free boundary are changed but L must be chosen below D. This is all right if the position of D is known reasonably well *a priori*. Otherwise an iterated approximation to D may fall below L and a new start is needed. On the other hand the iteration algorithm must not force the approximations to D to lie above L. A safe choice is for L to pass through C but accuracy may be impaired, as it will be if the elements above L become distended. Neuman and Witherspoon (1970) introduced alternative elements above L in which the mesh lines were not necessarily vertical. The difficulties with disintended elements can also be reduced by inserting fixed horizontal mesh lines which do not meet the surface (Fig. 8.1c). Nodes such as Z

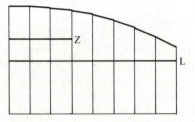

FIG. 8.1(c). Partial horizontal grid lines

need interpolation or other special treatment and complicate the computer program. More details of this approach to free-boundary problems are given by Taylor and Brown (1967) and Finn (1967).

An alternative approach is the use of variable meshes with fixed vertical mesh lines, but with nodes whose positions on the vertical lines are related in some simple automatic way to the positions of the nodes on the free boundary. One useful way is to preserve the number of nodes on any vertical nodal line. As the position of the free boundary changes the nodes always divide the vertical mesh lines into a constant number of equal intervals. This idea was used in the time-dependent problems of Stefan type in §4.3.3. It avoids the need for non-uniform elements near the free boundary but the program can be expensive because the full set of finite-element equations must be set up anew at each iteration.

Again, the finite element method leads to a system of linear equations, where the relevant matrix is sparse and has a band structure but in general has much less structure than the corresponding matrix for the finite-difference method (Zienkiewicz *et al.* 1965, 1966, 1971). Both direct and iterative methods of solution have been tried and key references are to be found in Cryer (1976*b*) and Reid (1979). George (1971) discussed the automatic generation of the finite-element grid and reviewed previous attempts.

In principle, any general method of solving elliptic equations with fixed boundary values can be used in a trial free-boundary method. The investigation of boundary value problems by integral equation techniques is well established. Baker (1979) discussed integral equation techniques with the solution of elliptic equations particularly in mind and gave basic references including Jaswon and Symm (1977) and Baker (1977) for numerical treatments. Cryer's (1976*b*) discussion of integral equations and singularities in the context of trial free-boundary methods started from the pioneer work of Trefftz (1916) and gave several references to the solution of various free-boundary problems by boundary integral methods. Details of classical integral-equation formulations and numerical solutions are given in the context of the degenerate free-boundary problems of electrochemical machining in §8.7.

Here we mention briefly the extension of these methods to what are commonly called 'boundary element methods', because they combine the ideas of classical integral equations and finite elements. By any method we can formulate an equation such as (8.111) in §8.7 but then instead of inserting constant values for ϕ and ϕ' along the sides of the approximating polygonal boundary, we incorporate the more general base functions and element shapes familiar in standard finite-element techniques. Thus ϕ and ϕ' can have linear or higher-order variations within a boundary element which can itself be curved. General introductions to boundary element methods are given by Brebbia (1980 pp. 3–33) and Wrobel and Brebbia (1981), and applications to porous flow free-boundary problems by Brebbia and Chang (1979), Brebbia and Dominguez (1977), Radojkovic (1980), Awater (1980), and Bischoff (1980). The last author includes particularly water movement in non-homogeneous media and refers to Butterfield (1978) and to Liggett (1977) for an iterative procedure to improve the free-boundary position. Lennon (1980) solved a number of free-surface problems in porous media in two and three space dimensions and with axial symmetry by using boundary elements. He includes axisymmetric and three-dimensional solutions in aquifers of infinite and finite depths, ponds, and wells. Some comparisons with the linearized perturbation solutions of Dagan (1967) are made. Approximate solutions obtained by replacing a well, for example, by an axial line sink are found to be adequate for practical purposes away from the wall of the well. A multiple-well problem is solved with four wells at positions with respect to coordinates (x, y, z) given by $(d, 0, z)$, $(0, d, z)$, $(-d, 0, z)$, $(0, -d, z)$, where z is vertical, d is constant, and $-2 < z < -1$. An example takes $d = 1.2$, the well at $(d, 0, z)$ to be a pumping well with a rate $+0.6\pi$, and the other three are re-charge wells with rates -0.15π, -0.4π, and -0.05π respectively. Numerous references include Wu (1976) on the boundary integral method using various approximation techniques, Liggett and Liu (1977, 1979) on unsteady flows, and Liu and Liggett (1977, 1978, 1979), the second paper being on an efficient numerical method for two-dimensional steady ground-water problems.

In finite-element methods the dependent variable u is approximated by

$$u_N = \sum_{i=1}^{N} \alpha_i \phi_i,$$

where the ϕ_i are piecewise polynomial finite-element basis functions. If on the other hand the functions ϕ_i are defined over the whole region we have a global method described by Delves (1979). Thus in a global Galerkin method the functions could be chosen to satisfy all or some of the boundary conditions and the coefficients α_i determined by collocation or least squares, for example. So-called regional methods employ a global expansion over a few sub-regions in more complicated regions, for example, for which the construction of sensible global basis functions is difficult or convergence is slow. Details of implementing global and regional methods are outlined by Delves (1979) with examples. Cryer (1976b) referred to global methods and quoted an application by Kirkham (1964) to the problem of axisymmetric porous flow towards a fully penetrating well. Also Mason and Farkas (1972) considered a freshwater canal in the centre of a long island with an impervious surface.

Cryer (1976*b*) also mentioned several graphical and analogue methods, including the Hele-Shaw analogue discussed in §2.12.1.

8.2.2. *Moving the free boundary*

Once the approximate solution $\phi_h^{(k)}$ of the elliptic equation has been obtained for an estimated position $\Gamma^{(k)}$ of the free boundary there remains the problem of moving the free boundary to an improved approximation position, $\Gamma^{(k+1)}$, such that the remaining free-boundary condition is satisfied. Of course, this means in general that the first condition is no longer satisfied and further iteration is needed. There appears to be no general rule for deciding which boundary condition should be used in solving the elliptic equation and which for moving the free boundary. The choice depends on the problem and sometimes one is clearly more convenient or efficient than the other.

In porous flow problems where the condition $\phi = y$ is to be satisfied on the free boundary, an improved position, $\Gamma^{(k+1)}$, of the boundary can simply be taken as the line on which $\phi = y$, found by interpolation on the approximate solution, $\phi^{(k)}$, of the boundary-value problem. Thus Aitchison (1972), for example, moved the boundary points along the vertical grid lines. To find the new value of y_j, the intercept of the free boundary with the jth vertical grid line, $j = 0, 1, \ldots, n$, the values of $\phi_h^{(k)} - y$ along the jth grid line are examined. The zero of $\phi_h^{(k)} - y$ is found by linear interpolation and taken as the value of y_j.

Cryer (1976*b*) classified methods of moving the boundary as local, integral, or global.

(i) *Local methods.* In a local method adjustments to $\Gamma^{(k)}$ are made at individual points on the basis of the error, $C\phi^{(k)}$, where the second boundary condition is written $C(\phi) = 0$. Thus if $C\phi_h^{(k)}$ is computed at m points $P_j^{(k)}$, $1 \le j \le m$, and if $C\phi_h^{(k)} \ne 0$ at $P_j^{(k)}$, then the point $P_j^{(k)}$ is moved to $P_j^{(k+1)}$ where the boundary condition is better satisfied in some sense. In general, the displacement is programmed by conditions

$$C\phi_h^{(k)} \quad \text{at} \quad P_j^{(k+1)} = 0, \qquad P_j^{(k+1)} - P_j^{(k)} = \alpha_j^{(k)} \mathbf{d}_j^{(k)}, \qquad (8.1)$$

where $\mathbf{d}_j^{(k)}$ is a specified unit vector and $\alpha_j^{(k)}$ is a parameter. Various ways of choosing both $\mathbf{d}$ and α have been tried (Cryer, 1976*b*).

Possible choices for $\mathbf{d}_j^{(k)}$ are (a) the outward normal to $\Gamma^{(k)}$ at $P_j^{(k)}$; (b) the unit vector in one of the coordinate directions; (c) the outward conormal to $\Gamma^{(k)}$ at $P_j^{(k)}$, e.g. if the elliptic equation is

$$\frac{\partial}{\partial x}\left(f_1 \frac{\partial \phi}{\partial x}\right) + \frac{\partial}{\partial y}\left(f_2 \frac{\partial \phi}{\partial y}\right) = 0$$

and the outward normal is $\mathbf{n} = (n_1, n_2)$, the unit outward conormal is the vector

$$\frac{(f_1 n_1, f_2 n_2)}{\{(f_1 n_1)^2 + (f_2 n_2)^2\}^{\frac{1}{2}}}$$

with the possible advantage that the boundary conditions are often expressed in terms of the conormal; or (d) some arbitrarily specified $\mathbf{d}_j^{(k)}$.

A Newton-type interpretation of $\alpha_j^{(k)}$ can be used by defining

$$f_j^{(k)}(\alpha) = C\phi^{(k)}(\mathbf{P}_j^{(k)} + \alpha \mathbf{d}_j^{(k)}) \tag{8.2}$$

and computing or estimating $(df/d\alpha)_{\alpha=0}$ so that

$$\alpha_j^{(k)} = -f^{(k)}(0) \bigg/ \left\{ \frac{d}{d\alpha} f_j^{(k)}(\alpha) \right\}_{\alpha=0}. \tag{8.3}$$

Alternatively, if $C\phi(\mathbf{P})$ is of the form $C\phi(\mathbf{P}) \equiv F\{\phi(\mathbf{P}), \phi_n(\mathbf{P}), \mathbf{P}\} = 0$, $\alpha_j^{(k)}$ has been defined by

$$F\{\phi^{(k)}(\mathbf{P}_j^{(k)}), \phi_n^{(k)}(\mathbf{P}_j^{(k)}), \mathbf{P}_j^{(k)} + \alpha_j^{(k)}\mathbf{d}_j^{(k)}\} = 0, \tag{8.4}$$

i.e. the dependence of $\phi^{(k)}$ on $\mathbf{P}$ is neglected between $\mathbf{P}_j^{(k)}$ and $\mathbf{P}_j^{(k+1)}$. If the free-boundary condition C for porous flow problems is written $C = \phi - y = 0$, where ϕ is velocity potential, (8.4) with $\mathbf{d}_j^{(k)}$ chosen in the direction of the y-axis leads to $\mathbf{P}_j^{(k+1)} = (x_j^{k+1}, y_j^{(k+1)})$, where $x_j^{(k+1)} = x_j^{(k)}$ and $y_j^{(k+1)} = \phi_h^{(k)}(\mathbf{P}_j^{(k)})$. Neumann and Witherspoon (1970) used this idea and also incorporated certain refinements. Similarly for porous problems Taylor (1966) and Kealy and Williams (1971) wrote C as $p = 0$, with p being pressure and then

$$\mathbf{P}_j^{(k+1)} - \mathbf{P}_j^{(k)} = \beta p_h^{(k)}(\mathbf{P}_j^{(k)})\mathbf{d}_j^{(k)}, \tag{8.5}$$

where the factor β and $\mathbf{d}_j^{(k)}$ are arbitrarily chosen. A similar but more general choice is $\alpha_j^{(k)} = G\{C\phi(\mathbf{P}_j^{(k)})\}$, where G is an arbitrary function; (8.5) corresponds to $G(C) = \beta C$.

For $C\phi(\mathbf{P})$ written in the Dirichlet form

$$C\phi(\mathbf{P}) = \phi(\mathbf{P}) + \gamma_1(\mathbf{P}) = 0,$$

(8.3) becomes

$$\alpha_j^{(k)} = -\frac{\phi^{(k)}(\mathbf{P}_j^{(k)}) + \gamma_1(\mathbf{P}_j^{(k)})}{\dfrac{\partial}{\partial \alpha}\{\phi^{(k)}(\mathbf{P}) + \gamma_1(\mathbf{P})\}_{\mathbf{P}=\mathbf{P}_j^{(k)}}}.$$

The quantity $\partial \gamma_1 / \partial \alpha$ is readily computed and if the first free-boundary condition is written $\partial \phi / \partial \alpha + \gamma_2 = 0$, $\partial \phi^{(k)} / \partial \alpha = -\gamma_2$. Otherwise $\partial \phi^{(k)} / \partial \alpha$ must be estimated, probably numerically. Cryer (1976b) quoted several authors who have used this method. He also quoted a number of instances in which the Neuman form, $C\phi(\mathrm{P}) = \phi_n(\mathrm{P}) + \gamma_2(\mathrm{P}) = 0$, was used as the second free-boundary condition, and concluded that for both Dirichlet and Neuman conditions local methods of moving the boundary have been used very successfully.

Doha (1977) compared numerically various finite-difference methods of iterating the position of the free boundary. He considered that the effect of the change in the domain on the solution should be taken into account by using the total derivative of ϕ in computing $C\phi^{(k)}$.

Baiocchi *et al.* (1973*a*) applied a different method of local adjustment of the free boundary in the rectangular dam problem with a sheetpile, formulated in §2.3.2 (Fig. 2.3). They used the physical condition that the discharge across any vertical section of the dam in the seepage region betwen the dam base and the free surface is constant. Thus

$$q(\Omega, x, f(x)) = -\int_0^{f(x)} \frac{\partial}{\partial x} \phi(x, y) \, \mathrm{d}y = -\int_0^{y_s} \frac{\partial}{\partial x} \phi(0, y) \, \mathrm{d}y = q(\Omega, 0, y_s),$$

where $y_s = \mathrm{AG}$. Along GF, $\partial \phi_h^{(k)} / \partial y$ can be computed from the kth iteration of the solution $\phi_h^{(k)}$ and by interpolation or extrapolation the point where $\partial \phi / \partial y = 0$ is determined. This is the first point of $\Gamma^{(k+1)}$.

For the approximation $\Gamma^{(k)}$ in general the flow rate q across different vertical sections will not be constant, i.e. $q(\Omega^{(k)}, x, f(x)^{(k)}) \neq q(\Omega^{(k)}, 0, y_s)$. On any vertical grid line, $x = ih$, we can compute $q(\Omega^{(k)}, x_i, y)$ by integrating the nodal values on this line, and then, by interpolation or extrapolation, determine a point $y_i^{(k+1)}$ such that $q(\Omega^{(k)}, x_i, y_i^{(k+1)}) = q(\Omega^{(k)}, 0, y_s)$. The points $(x_i, y_i^{(k+1)})$ are the vertices of the polygonal $(k+1)$th approximation to the free boundary, $\Gamma^{(k+1)}$. Baiocchi *et al.* (1973*a*) got better results by only partial adjustment of the free boundary in the early stages. Thus if $f_1^{(k+1)}$ was the result of the above process they took $f^{(k+1)} = f^{(k)} + \alpha(f_1^{(k+1)} - f^{(k)})$, $0 < \alpha < 1$, for the first few approximations, if they were not able to start with a good estimate. Their method converged satisfactorily. Special care and treatment was needed in the region of the sheetpile where the streamlines are near to vertical. When the sheetpile covers most of the inlet face, a very fine mesh was needed in that region in order to compute the flow-through, q, sufficiently accurately. In such a situation these authors preferred the variational method described in §2.3.2.

France *et al.* (1971) considered the free-boundary seepage problem as being the final steady-state approached by a certain transient problem. They moved the boundary through successive approximating positions by attaching to it a transient condition which included the normal velocity of the boundary. In many problems, of course, the approach to the steady state is itself of interest but Taylor *et al.* (1973) applied the method to situations in which the transient solution has no physical meaning.

They started from the general equation for seepage flow or irrotational hydrodynamic flow

$$\frac{\partial}{\partial x}\left(k_x \frac{\partial \phi}{\partial x}\right) + \frac{\partial}{\partial y}\left(k_y \frac{\partial \phi}{\partial y}\right) + \frac{\partial}{\partial z}\left(k_z \frac{\partial \phi}{\partial z}\right) + Q + c\frac{\partial \phi}{\partial t} = 0, \qquad (8.6a)$$

where the ks are permeabilities, for example, Q is a specified in-flow and c expresses the specific capacity of the medium for fluid. The usual fixed boundary conditions are

$$\phi = \phi_b(x, y, z, t) \qquad (8.6b)$$

on the part Γ_1 of the whole boundary Γ, and on the part Γ_2

$$\ell_x k_x \frac{\partial \phi}{\partial x} + \ell_y k_y \frac{\partial \phi}{\partial y} + \ell_z k_z \frac{\partial \phi}{\partial z} = q(\phi, x, y, z, t), \qquad (8.6c)$$

where ℓ_x, ℓ_y, ℓ_z are the direction cosines of the outward normal on the boundary of the relevant domain. In free-boundary conditions both types of condition have to be satisfied simultaneously on the 'free' part of the boundary. If a steady state has not been reached an appropriate form of moving boundary condition is

$$\ell_x k_x \frac{\partial \phi}{\partial x} + \ell_y k_y \frac{\partial \phi}{\partial y} + \ell_z k_z \frac{\partial \phi}{\partial z} = q_0 + V_n s_f = q_n \qquad (8.6d)$$

say, where V_n is the normal velocity of the boundary at any instant, q_0 allows for precipitation or other supply of fluid to the boundary, and s_f in the absence of capillarity is the volumetric porosity. If the coordinates of points on the free surface are denoted by x_f, y_f, z_f, all functions of time, then $V_n = \ell_x \dot{x}_f + \ell_y \dot{y}_f + \ell_z \dot{z}_f$, where $\dot{x}_f = dx_f/dt$, etc.

Taylor *et al.* (1973) proceed with the usual finite-element discretization with

$$\phi_a = \sum_{i=1}^{n} N_i \phi_i,$$

where usually the ϕ_i are the nodal values of ϕ and N_i are appropriate shape functions defined piecewise element by element. Conditions (8.6b) on Γ_1 are automatically satisfied. Discretization of (8.6a,b) based on a Galerkin, weighted-residual formulation requires

$$\int_{\Omega} N_j \left[\frac{\partial}{\partial x}\left(k_x \frac{\partial \phi_a}{\partial x}\right) + \frac{\partial}{\partial y}\left(k_y \frac{\partial \phi_a}{\partial y}\right) \right.$$

$$\left. + \frac{\partial}{\partial z}\left(k_z \frac{\partial \phi_a}{\partial z}\right) + Q + c\frac{\partial \phi_a}{\partial t} \right] d\Omega = 0, \qquad j = 1, 2, \ldots, m, \ m < n,$$

which becomes after the usual manipulation

$$\mathbf{H}\boldsymbol{\phi} + \mathbf{c}\frac{d\boldsymbol{\phi}}{dt} + \mathbf{F} = 0, \tag{8.7a}$$

where $H_{ji} = \sum_e h_{ji}^e$, $j = 1, 2, \ldots, m$, with e denoting a particular finite element,

$$h_{ji}^e = -\int_{\Omega^e} \left(\frac{\partial N_j}{\partial x} k_x \frac{\partial N_i}{\partial x} + \frac{\partial N_j}{\partial y} k_y \frac{\partial N_i}{\partial y} + \frac{\partial N_j}{\partial z} k_z \frac{\partial N_i}{\partial z} \right) d\Omega,$$

$$c_{ji} = \sum_e c_{ji}^e, \qquad c_{ji}^e = \int_{\Omega^e} N_j c N_i \, d\Omega, \qquad j = 1, 2, \ldots, m,$$

$$F_j = \sum_e F_j^e, \qquad j = 1, 2, \ldots, m,$$

$$F_j^e = -\int_{\Omega^e} \left(\frac{\partial N_j}{\partial x} k_x \frac{\partial N_i}{\partial x} + \ldots \right) \phi_i \, d\Omega + \int N_j q_n \, d\Gamma + \int_{\Omega^e} N_j Q \, d\Omega,$$

in which the first term of F_j is only included when $\phi_i = \phi_b$. In unconfined seepage problems and irrotational hydrodynamic flow $\mathbf{c} = 0$ and (8.7a) becomes

$$\mathbf{H}\boldsymbol{\phi} + \mathbf{F} = 0. \tag{8.7b}$$

In transient free-surface problems there is a portion Γ_3 of the boundary Γ where at an initial time ϕ is prescribed but the boundary is moving according to the condition (8.6d) which must be additionally satisfied. In such cases Taylor *et al.* (1973) continue by solving the quasi-static problem defined by (8.7b) at an initial time with ϕ prescribed on the part of the boundary Γ_3 assumed to have a known initial position. Then at the nodes on the free boundary, Γ_3, R_i values are calculated, where

$$R_i = \sum h_{ji}\phi_i + \int_{\Omega} N_i Q \, d\Omega,$$

and the condition on Γ_3,

$$F_i = R_i + \int_{\Gamma_3} N_i q_n \, d\Gamma = 0, \tag{8.7c}$$

allows the unknown q_n to be computed and hence V_n follows from (8.6d), i.e. $q_n = q_0 + V_n s_f$ as follows.

Let the free-boundary nodes be constrained to move along prescribed lines, e.g. along the vertical coordinate lines z so that $z_f = z(t)$. If $z_f = \sum N_i z_i$ then the component of $\dot{z}_f$ in the direction of V_n is $\ell_z \dot{z}_f = \ell_z \sum N_i \dot{z}_i$ and using (8.7c) and $q_n = q_0 + V_n s_f$ we have

$$\int_{\Gamma_3} N_i \left(q_0 + s_f \sum N_j \dot{z}_j \right) d\Gamma = -R_i. \tag{8.7d}$$

On writing

$$L_{ij} = \sum_e \ell_{ij}, \qquad \ell_{ij} = \int_{\Gamma_3} N_i s N_j \, d\Gamma,$$

$$R_{0i} = \sum_\ell \int_{\Gamma_3} N_i q_0 \, d\Gamma,$$

(8.7a) gives a system of equations

$$\dot{z}_f = -\mathbf{L}^{-1}(\mathbf{R} - \mathbf{R}_0),$$

and finally an improved free-boundary position follows by moving each boundary node by $\Delta z_f = \dot{z}_f \, \Delta t$. The transient computation proceeds step by step Δt till the steady state is achieved which is the solution of the appropriate elliptic problem with $\partial \phi / \partial t = 0$.

Taylor *et al.* (1973) solve three transient seepage problems: radial flow towards a well following rapid drawdown, porous flow through a dam with a toe drain following rapid lowering of the inlet reservoir, and a waterfall problem. Figure 8.2 illustrates the approach to the steady state in the second problem starting from a supposed known initial steady state for a reservoir height of 150 feet which drops suddenly to 50 feet.

(ii) *Integral methods.* In an integral method of moving the free boundary the condition, $\mathbf{C}\phi = 0$, is expressed in an implicit form such as

$$F\{\phi(x), \phi_x(s), \phi_y(s), \mathbf{x}(s), \dot{\mathbf{x}}(s)\} = 0, \qquad (8.8)$$

where $\mathbf{x}$ denotes (x, y), the free boundary is $\mathbf{x} = \mathbf{x}(s)$ in parametric form, and $\dot{\mathbf{x}}(s)$ denotes the derivatives of $\mathbf{x}(s)$. Thus, given $\Gamma^{(k)}$ and $\phi^{(k)}$, the curve $\Gamma^{(k+1)}$ is obtained by approximate integration of the differential equation for $\mathbf{x}(s)$:

$$F\{\phi_h^{(k)}(s), \phi_{x,h}^{(k)}(s), \phi_{y,h}^{(k)}(s), \mathbf{x}^{(k+1)}(s), \dot{\mathbf{x}}^{(k+1)}(s)\} = 0 \qquad (8.9)$$

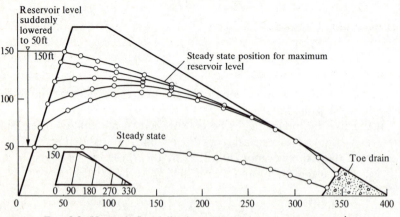

FIG. 8.2. Unsteady flow following rapid drawdown of the reservoir

where $\phi_h^{(k)}(s)$, etc., denote approximations at the point $\mathbf{x}^{(k+1)}(s)$, obtained by interpolation if the point lies inside the seepage region and by extrapolation if it is outside. Applications of the method often take $\phi^{(k)}(s)$ to be the value of $\phi^{(k)}$ at the grid point nearest to $\mathbf{x}^{(k+1)}(s)$. Cryer (1976*b*) reported a number of uses of this method. In the simple dam problem, for example, an approximation to the velocity field, say $\mathbf{v}_h^{(k)}$, can be determined by differentiating the approximation to the velocity potential, $\phi_h^{(k)}$. The free boundary is a streamline for the condition $C\phi = \phi_n = 0$ and so an improved approximation, $\Gamma^{(k+1)}$, can be obtained by integrating the velocity field $\mathbf{v}^{(k)}$.

Cryer (1976*b*) concluded from the practical evidence available that trial free-boundary methods as described so far are stable and converge satisfactorily for porous flow problems. Any local method can be used and smoothing is not required. Where there has been a suspicion of instability (Taylor and Brown 1967; Neuman and Witherspoon 1970; Kealy and Busch 1971) it seems to have occurred in the region where the free boundary joins the seepage surface. The trouble may be associated with the precise nature of the boundary conditions there, which can include the type of singularity discussed by Aitchison (1972) and in §8.3.

Shaw and Southwell (1941) asserted that in porous flow problems, if the initial guess $\Gamma^{(0)}$ for the free boundary is horizontal then successive trials $\Gamma^{(k)}$ converge monotonely downwards. They based a proof on the maximum principle.

Cryer (1976*b*), however, gives examples of instabilities in more general free-boundary problems for which global methods may be preferred.

(iii) *Global methods*. In a global method the solution is computed for a number of independent choices of $\Gamma^{(k)}$ and then, in principle, it is possible to use inverse interpolation or some other technique to determine the new trial boundary $\Gamma^{(k+1)}$. For example, the least-squares error in the boundary condition on $\Gamma^{(k+1)}$ can be minimized on the evidence of the dependence of this error on the parameters defining the set of chosen $\Gamma^{(k)}$. Even so, Fox and Sankar (1973) advise against local improvement of mesh points by solving an inverse interpolation problem to find a new position of just one mesh point on $\Gamma^{(k+1)}$, leaving all other points untreated. In order to avoid the oscillations of the mesh points on $\Gamma^{(k+1)}$ which local improvement can produce, they propose that all mesh points should be improved simultaneously. Their algorithm involves solving the appropriate fixed boundary problem for $n+1$ different choices of Γ initially but then only one new fixed-boundary problem has to be solved per iteration. Both the problem and the algorithm devised by Fox and Sankar (1973) have features of general interest and are now described in more detail.

The problem is about axisymmetric incompressible cavitational flow past a circular disc placed coaxially inside a cylinder. With reference to Fig. 8.3a the formulation is given by

$$\frac{\partial^2 \psi}{\partial r^2} - \frac{1}{r}\frac{\partial \psi}{\partial r} + \frac{\partial^2 \psi}{\partial z^2} = 0, \tag{8.10}$$

$$\left.\begin{aligned}
\psi &= 1 \quad \text{on AB,} \\
\psi &= 0 \quad \text{on FED,} \\
\partial\psi/\partial z &= 0 \quad \text{on AF} \quad \text{and} \quad \text{BC}
\end{aligned}\right\} \tag{8.11}$$

$$\psi = 0, \qquad (1/r)\,\partial\psi/\partial\nu = q \quad \text{on the free boundary DC,} \tag{8.12}$$

where ν denotes the normal and q is a constant whose value has to be determined as part of the problem. Figure 8.3a gives the dimensions used in the numerical solution. In the undisturbed stream, as $z \to -\infty$ the conditions $\psi = r^2$, $(1/r)(\partial\psi/\partial r) = 2$, are assumed.

Quite apart from difficulties associated with the singularity at the separation point D, which Fox and Sankar discuss, the free boundary is not satisfactorily amenable to local adjustments. Because q is not known the first of (8.12) is used in solving the equation (8.10) and the second of (8.12) reserved for the boundary adjustment. In order to avoid uncontrolled oscillations and lack of convergence Fox and Sankar use numerically a generalized *regula falsi* method (Gauss 1809), well known for solving an equation in a single variable. For a set of n simultaneous equations

$$f_1(x_1, x_2, \ldots, x_n) = 0, \qquad f_2(x_1, x_2, \ldots, x_n) = 0, \ldots, f_n(x_1, x_2, \ldots, x_n) = 0,$$
$$\tag{8.13}$$

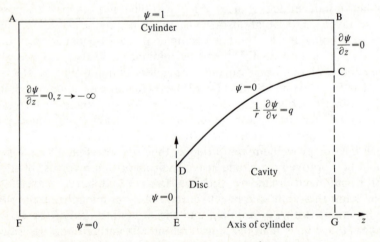

FIG. 8.3(a). $FE = \infty$, $AF = 1$, $EG = 1$, $ED = \frac{1}{8}$ in example

it is assumed that all the f_j, $j = 1, 2, \ldots, n$, can be computed at $n+1$ 'points' $P = (x_1^{(i)}, x_2^{(i)}, \ldots, x_n^{(i)})$, $i = 1, 2, \ldots, n+1$. In the one-dimensional solution of $f(x) = 0$, the value of x is determined at which the straight line

$$ax + b = f \quad \text{(for the required } x, f = 0)\tag{8.14}$$

joining the points $(x_1, f(x_1))$ and $(x_2, f(x_2))$, intersects the x-axis. The relationships

$$ax_1 + b = f(x_1), \qquad ax_2 + b = f(x_2)\tag{8.15}$$

lead to the determinantal equation

$$\begin{vmatrix}(x - x_1) & f(x_1) \\ (x - x_2) & f(x_2)\end{vmatrix} = 0\tag{8.16}$$

for x. In place of (8.14) and (8.15), Fox and Sankar take the n equations

$$\left.\begin{aligned}a_{11}x_1 + a_{12}x_2 + \ldots + b_1 &= 0 \\ \cdots \\ a_{n1}x_1 + a_{n2}x_2 + \ldots + b_n &= 0\end{aligned}\right\}\tag{8.17}$$

and the $n(n+1)$ equations

$$\left.\begin{aligned}a_{11}x_1^{(i)} + a_{12}x_2^{(i)} + \ldots + b_1 &= f_1(P_i) \\ a_{21}x_1^{(i)} + a_{22}x_2^{(i)} + \ldots + b_2 &= f_2(P_i) \\ \cdots \\ a_{n1}x_1^{(i)} + a_{n2}x_2^{(i)} + \ldots + b_n &= f_n(P_i)\end{aligned}\right\}\ i = 1, 2, \ldots, n+1\tag{8.18}$$

respectively and solve (8.18) to give an 'improved' point $P = (x_1, x_2, \ldots, x_n)$. Typical rows of n determinantal equations corresponding to (8.16) become

$$\left.\begin{aligned}|x_1 - x_1^{(i)}, & \quad f_1(P_i), \quad f_2(P_i), \ldots, f_n(P_i)| = 0 \\ |x_2 - x_2^{(i)}, & \quad f_1(P_i), \quad f_2(P_i), \ldots, f_n(P_i)| = 0 \\ & \cdots \\ |x_n - x_n^{(i)}, & \quad f_1(P_i), \quad f_2(P_i), \ldots, f_n(P_i)| = 0\end{aligned}\right\}\ i = 1, 2, \ldots, n+1\tag{8.19}$$

The solution of (8.19) is

$$x_j = \frac{|x_j^{(i)}, f_1(P_i), \ldots, f_n(P_i)|}{|1, f_1(P_i), \ldots, f_n(P_i)|}, \qquad j = 1, 2, \ldots, n,\tag{8.20}$$

where the numerator and denominator are determinants of which the ith row is shown in each case and i goes from 1 to $n+1$. The equations (8.20) give the improved approximate solutions of equations (8.13). One of the previous P_i is replaced by this new P and the process repeated till satisfactory convergence is reached.

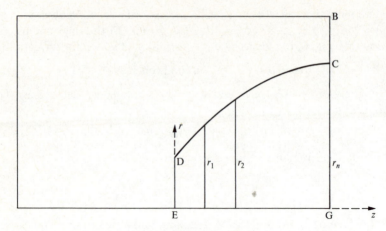

FIG. 8.3(b). Set of radii

Figure 8.3b shows a set of radii $r_1, r_2, \ldots, r_n$ from the z-axis to the free boundary, along vertical mesh lines. These values of r are the true solutions of the algebraic approximations to the n equations

$$f_j = (1/r_j)\, \partial\psi_j/\partial\nu - q = 0, \qquad j = 1, 2, \ldots, n, \qquad (8.21)$$

where each $\partial\psi_j/\partial\nu$ is a function of all the r_j. Because q is unknown in this problem some estimated constant q_e is chosen for each step of the *regula falsi* method. An assumed free boundary in the form of part of an ellipse through D with major axis along EG in Fig. 8.3b and minor axis CG for some position of C gives a set of values $r_j^{(i)}$, $j = 1, 2, \ldots, n$, which specify a particular starting point P_i. The required $n+1$ starting points are obtained by taking $n+1$ different positions for C. The boundary value problem given by equations (8.10), (8.11), and the first of (8.12) is solved for each of the $n+1$ 'points' and hence the f_j in (8.21) are computed for some $q = q_e$. A computed *regula falsi* curve, i.e. a better 'point' calculated from (8.20), follows for this q_e. The simultaneous treatment of the equations eliminates the disturbances which arise with local movement of individual mesh points on the free boundary. The new curve is extremely smooth and yields good approximations to the derivatives in (8.21) when finite-difference methods are applied to each elliptical boundary problem corresponding to the interpolated *regula falsi* boundary for each j from 1 to n. Thus the average value q_a of $(1/r_j)(\partial\psi_j/\partial\nu)$ and the maximum deviation d from the average can be estimated. Results shown in Table 8.1 for each of seven values of q_e indicate that the best result, i.e. the one for which d is smallest, is about $q_a = 2.11$. This value could be improved by repeating the computation with this value for q_e and with one of the ellipses replaced by this particular *regula falsi* curve. Just one extra

TABLE 8.1
*Average values q_a for
estimated constants q_e*

q_e	q_a	d
2.00	2.03	0.13
2.05	2.07	0.07
2.10	2.11	0.01
2.15	2.17	−0.08
2.20	2.22	−0.17
2.25	2.28	−0.26
2.30	2.35	−0.36

boundary-value problem would have to be solved and the new information added to Table 8.1 would yield a better q_a and so on. In fact, Fox
and Sankar did not carry out these additional calculations. They considered the value $q_a = 2.11$ to be as accurate as their mesh size justified,
especially having in mind the untreated singularity at D.

8.2.3. Garabedian's modified boundary conditions

So far, the two conditions on the free boundary have been used in the
form in which they express directly the physical situation. We now
explore the possibility that a modified form of the conditions may have
advantages in a trial free-boundary method. The following is based on
work by Garabedian (1956) and Cryer (1976b, 1970).

Consider a problem in which the elliptic differential equation has the
form

$$L\phi = a_{11}\phi_{xx} + 2a_{12}\phi_{xy} + a_{22}\phi_{yy} + b_1\phi_x + b_2\phi_y = 0 \qquad (8.22)$$

and the two conditions on the free boundary, Γ, are

$$\beta\phi = \beta_1\phi_n + \alpha_1\phi - \gamma_1 = 0, \qquad (8.23)$$
$$C\phi = \beta_2\phi_n + \alpha_2\phi - \gamma_2 = 0, \qquad (8.24)$$

where β_i, α_i, γ_i are smooth functions of x and y. In order that the two
conditions shall be independent we require that

$$\begin{vmatrix} \beta_1 & \alpha_1 \\ \beta_2 & \alpha_2 \end{vmatrix} \neq 0. \qquad (8.25)$$

We shall use (8.23) to compute $\phi_n^{(k)}$ from (8.22) and the conditions on the
fixed boundaries and (8.24) to move $\Gamma^{(k)}$ to $\Gamma^{(k+1)}$. Instead of the
operators β and C defined in (8.23), (8.24) we introduce equivalent

operators $\hat{\beta}$, $\hat{C}$, where

$$\hat{\beta} = e_{11}\beta + e_{12}C \quad \text{on } \Gamma$$
$$= \beta \qquad \text{on the fixed boundary,} \tag{8.26}$$

$$\hat{C} = e_{21}\beta + e_{22}C \quad \text{on } \Gamma, \tag{8.27}$$

and the *e*s are smooth functions satisfying a determinantal condition like (8.25). The aim is to choose $\hat{\beta}$, $\hat{C}$ so as to improve convergence of a trial free-boundary method.

When $\Gamma^{(k)}$ on which (8.23) is satisfied by $\phi_h^{(k)}$ is moved to $\Gamma^{(k+1)}$ so that (8.24) is satisfied, in general (8.23) will no longer be satisfied. Garabedian's idea was to construct $\hat{\beta}$ so that $\hat{\beta}\phi$ is insensitive to movements of Γ, i.e. $\hat{\beta}\phi$ is stationary with respect to normal displacements of Γ, so that

$$(\hat{\beta}\phi)_n = 0 \quad \text{on } \Gamma. \tag{8.28}$$

The coefficients e_{ij} can be chosen such that

$$\hat{\beta}\phi = (\phi_n - \gamma_{11}) + \tau(\phi - \gamma_{12}) = 0, \tag{8.29}$$
$$\hat{C}\phi = \phi - \gamma_{12} = 0, \tag{8.30}$$

where τ is an arbitrary function, and hence

$$(\hat{\beta}\phi)_n = \phi_{nn} - (\gamma_{11})_n + \tau\{\phi_n - (\gamma_{12})_n\}$$

because of (8.30). Thus (8.28) holds if

$$\tau = \frac{-\phi_{nn} + (\gamma_{11})_n}{\gamma_{11} - (\gamma_{12})_n} \tag{8.31}$$

since (8.29), (8.30) imply $\phi_n = \gamma_{11}$. If the unit tangent to Γ is denoted by t, s is the distance along Γ, and K its curvature, we have on Γ

$$\phi_n = \gamma_{11}, \qquad \phi_t = \phi_s = (\gamma_{12})_s \tag{8.32}$$

and as special cases of the Frenet formulae

$$\phi_{tt} - K\phi_n = \phi_{ss} = (\gamma_{12})_{ss}, \qquad \phi_{nt} + K\phi_t = \phi_{ns} = (\gamma_{11})_s. \tag{8.33}$$

The elliptic equation (8.22) can be written as

$$a'_{11}\phi_{nn} + 2a'_{12}\phi_{nt} + a'_{22}\phi_{tt} + b'_1\phi_n + b'_2\phi_t = 0, \tag{8.34}$$

where the coefficients a'_{ij} and b'_i are linear combinations of a_{ij} and b_i. The function τ can now be expressed in terms of the γ_{ij} and their derivatives and a_{ij}, b_i, K.

An important special case occurs when (8.22) is Laplace's equation and γ_{11}, γ_{22} are constants. Then (8.34) becomes $\phi_{nn} + \phi_{tt} = 0$ and $\tau = K$. In practice, the curvature K is not known since Γ is unknown, but τ can be replaced by $\tau^{(k)}$ in moving $\Gamma^{(k)}$ to $\Gamma^{(k+1)}$, and $\phi^{(k)}$ is the solution of

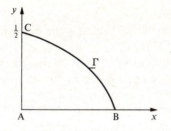

FIG. 8.4. Model free-boundary problem

$L(\phi^{(k)}) = 0$ with the boundary condition

$$\hat{\beta}^{(k)}\phi^{(k)} = (\phi_n^{(k)} - \gamma_{11}) + \tau^{(k)}(\phi^{(k)} - \gamma_{12}) = 0.$$

Cryer (1970) found that convergence was improved by using the Garabedian form of boundary conditions $\hat{\beta}$ and $\hat{C}$ but was not able to verify quadratic convergence in general. He did, however, analyse the following very simple model problem: find ϕ satisfying $\phi_{xx} + \phi_{yy} = 0$ in the domain ABC in Fig. 8.4, where the free boundary Γ is the line BC and is constrained to pass through the fixed point C; the other conditions are

$$\beta\phi_n = \begin{cases} \phi - (1-y) = 0 & \text{on AC,} \\ \phi_n - 1 = 0 & \text{on AB.} \\ \phi_n + \sqrt{10} = 0 & \text{on BC,} \end{cases}$$

$$C\phi = \phi - \tfrac{1}{2} = 0 \quad \text{on BC.}$$

On Γ, y is to be a monotone decreasing function of x. It is easily verified that a solution is given by

$$\phi = 1 - y - 3x, \quad \text{and the equation of the free boundary } \Gamma \text{ is } y = \tfrac{1}{2} - 3x.$$

Since BC turns out to be a straight line with curvature $K = 0$, the boundary conditions are seen to be already in the Garabedian form.

A trial free-boundary approach to this simple problem can assume straight-line approximations to Γ passing through C, i.e. $\Gamma^{(k)}$ is $y = \tfrac{1}{2} + c^{(k)}x$. The problem $\phi_{xx}^{(k)} + \phi_{yy}^{(k)} = 0$ within ABC, $\beta\phi^{(k)} = 0$ on the boundary ABC has the exact solution

$$\phi^{(k)} = 1 - y + \frac{x}{c^{(k)}}[\{10(c^{(k)})^2 + 10\}^{\frac{1}{2}} - 1].$$

The second condition $C\phi^{(k)} = 0$ on $\Gamma^{(k+1)}$ is satisfied exactly if $c^{(k+1)}$ is defined to be

$$c^{(k+1)} = [\{10(c^{(k)})^2 + 10\}^{\frac{1}{2}} - 1]/c^{(k)},$$

so that here the approximate solutions $\phi^{(k)}$ and the approximate $\Gamma^{(k)}$'s are known exactly. Furthermore, by defining $f(c) = \{(1+c^2)\}^{\frac{1}{2}} - \sqrt{10}$, we find that $c^{(k+1)} = c^{(k)} - f(c^{(k)})/f'(c^{(k)})$, which is the Newton iterative formula for the solution of $f(c) = 0$ from the initial guess $c^{(0)}$. Since $f(c)$ is convex for $c \leq 0$ and $f(0) < 0$ it follows that for any initial guess $c^{(0)} < 0$ the sequence of the trials $\Gamma^{(k)}$ converges quadratically to the true free boundary. Thus the simple problem lends support to Garabedian's claim (1956) that his rewriting of the boundary conditions leads to quadratic convergence and points to the connection with Newton's method. Cryer (1968) found that numerical results for a more complicated problem in stellar evolution behaved like those for the model problem which was chosen to resemble it.

8.3. Boundary singularities

Singularities commonly occur on a fixed boundary in boundary-value problems. They may be associated with sudden changes in the direction of the boundary, as at a re-entrant corner, or with mixed boundary conditions. Much attention has been paid to these singularities and methods include mesh refinement for both finite differences and finite elements, and the use of modified finite-difference approximations or singular finite elements in which the local analytical form of the singularity is somehow incorporated. Various other techniques are based on integral equations, power series, dual series, Fourier series, and removal of the singularity. Conformal transformation methods have proved particularly efficient and highly accurate for the solution of elliptic problems. Good summaries together with other basic references are given by Fox (1979) for finite differences, by Wait (1979) for finite elements, Delves (1979) for global and regional methods, and by Scheffler and Whiteman (1979) for conformal-mapping techniques. Many other references to these and other methods are quoted by Furzeland (1977b).

In addition to such generally occurring singularities, free-boundary problems often have a singularity at the separation point where the free boundary meets a fixed boundary, as, for example, at point D in Fig. 2.1 for the simple dam problem. Aitchison (1972) used complex variable methods to determine the shape of the free boundary near the separation point and incorporated the local analytical solution into a finite-difference scheme in order to improve its accuracy.

Starting with the dam problem defined by equations (2.7–11) in Chapter 2, but taking $y_1 = 1$, $y_2 = d$, and $x_1 = L$ in Fig. 2.1, Aitchison (1972) moved the origin of the (x, y) coordinates to D and added a constant to ϕ so that $\phi_D = 0$.

In terms of the stream function ψ, the condition $\partial \phi / \partial n = 0$ on DF becomes $\psi = 0$, and the problem is to determine the complex variable

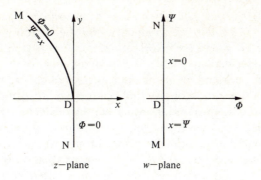

Fig. 8.5(a). Singular separation point D

$w = \phi + i\psi$ as a function of the complex variable $z = x + iy$. To find the equation of the free boundary it is the inverse problem that has to be solved. Introducing $W = w + iz$, the problem is to find z in terms of $W = \Phi + i\Psi$ and then $\Phi = \phi - y$ and $\Psi = \psi + x$ (Fig. 8.5a). The solution is required in the half-plane $\Phi \geqslant 0$, and is

$$z = \frac{W}{\pi} \ln(-iW) + i \sum_{k=1}^{\infty} A_k \exp(\tfrac{1}{2}i\pi k) W^k$$

where the A_k are real constants. Hence

$$x + iy = \frac{\Phi + i\Psi}{\pi} \{\ln|W| + i(\theta - \pi)\} + i \sum_{k=1}^{\infty} A_k \exp\{\tfrac{1}{2}i\pi k\}(\Phi + i\Psi)^k,$$

where $\tan \theta = \Psi/\Phi$. On the free boundary $\theta = -\tfrac{1}{2}\pi$ and so

$$x + iy = \frac{i\Psi}{\pi} \{\ln(-\Psi) - i\pi\} + i \sum_{k=1}^{\infty} A_k(-\Psi)^k$$

$$= \Psi + \frac{i\Psi}{\pi} \ln(-\Psi) + i \sum_{k=1}^{\infty} B_k \Psi^{-k},$$

where $B_k = (-1)^k A_k$, and therefore

$$x = \Psi, \qquad y = \frac{\Psi}{\pi} \ln(-\Psi) + \sum_{k=1}^{\infty} B_k \Psi^k.$$

The equation of the free boundary becomes

$$y = \frac{x}{\pi} \ln(-x) + \sum_{k=1}^{\infty} B_k x^k, \tag{8.35}$$

which has the infinite derivative and infinite curvature required at the separation point D.

Aitchison (1972) described two ways of incorporating the analytical expression (8.35) for the free boundary and the corresponding solution for the potential ϕ into a finite-difference solution. A trial free-boundary method was described in §8.2.2 in which the zero of $\phi - y$ along any vertical grid line is found by linear interpolation along the line and taken to be the new intersection of the free boundary with that ordinate. This method cannot be used to find the value y_n, the ordinate of the separation point D, because $\phi = y$ along the whole of CD in Fig. 2.1. Instead, y_n has to be found by fitting a suitable curve through $y_{n-1}, y_{n-2}, \ldots$. Such a curve is given by equation (8.35). The position of the origin of coordinates for (8.35) is not known and so the curve based on r terms of the infinite sum can be fitted through $r+1$ of the points y_j, conveniently taken as $y_{n-r-1}, y_{n-r}, \ldots, y_{n-1}$ say. Aitchison (1972) quoted numerical results for the free-boundary curve for r having values 0, 1, and 2 and concluded that $r = 1$ gave satisfactory results for most applications. Grid sizes chosen for the finite-difference calculations were 1/24, 1/36, and 1/48.

An alternative procedure included use of values of ϕ from the analytical solution at grid points in the neighbourhood of the singularity. In this case, only one term of the infinite sum was taken, so that

$$z = (W/\pi)\ln(-iW) + AW, \tag{8.36}$$

where $W = w + iz$. The value of A is determined during the calculation of y_n and can be taken as known. In order to find ϕ at a grid point $P(x, y)$ a form of Newtonian iteration was used based on the function $f(W)$, where

$$f(W) = (W/\pi)\ln(-iW) + AW - z_P, \tag{8.37}$$

where z_P is a known constant for a selected point P. The iterative cycle needed to find the root of $f(W) = 0$ is defined by

$$f(W_i) = (W_i/\pi)\ln(-iW_i) + AW_i - z_P,$$

$$f'(W_i) = \frac{\ln(-iW_i) + 1}{\pi} + A,$$

$$W_{i+1} = W_i - f(W_i)/f'(W_i).$$

A good enough starting value is provided by the values calculated for the previous free-boundary position. The value of W yielded by the Newton iteration satisfies (8.37) with $z = z_P$, and $w = W - iz$ and hence ϕ_P follow immediately.

Aitchison (1972) used ϕ values determined in this way at the grid points marked by crosses shown in Fig. 8.5b, and the curve $y = (x/\pi)\ln(-x) + Ax$ was fitted through the points R and Q to give values of y_{n-1}, y_n, and A. For a particular grid size the value of y_n obtained by using the second method, which included the analytical values of ϕ, was

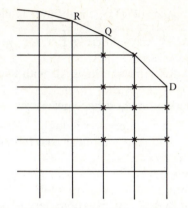

FIG. 8.5(b). Special ϕ-values near separation point

judged to be better than that from the first method. Ordinates of points on the free boundary obtained by this trial free-boundary method are compared in Table 8.7 with later values based on a Baiocchi transformation and method coupled with the analytic treatment of the singular separation point (Aitchison 1977).

Thatcher and Askew (1982) demonstrated that the difficulty of determining the separation point D in Fig. 2.1, for example, can be avoided by basing a trial free-boundary method on the stream function introduced in §2.2.2. They reformulated the problem expressed by equations (2.7–12) in a slightly modified form of equations (2.16). The known value of the velocity potential, ϕ, along $\partial\Omega_1$, which is the part of the boundary BCDFA in Fig. 2.1, is denoted by $f(s)$, where

$$f(s) = \begin{cases} y_1 & \text{on FA,} \\ y & \text{on CDF,} \\ y_2 & \text{on BC.} \end{cases}$$

The equations to be satisfied by the stream function $\psi(x, y)$ are

$$\partial\phi/\partial x = \partial\psi/\partial y, \qquad \partial\phi/\partial y = -\partial\psi/\partial x,$$

$$\nabla^2\psi = 0 \quad \text{in } \Omega_{\Gamma(x)},$$

$$\partial\psi/\partial n = -\partial f/\partial s \quad \text{in } \partial\Omega_1,$$

$$\psi = \text{constant on AB}, \qquad \psi = \text{constant on } \Gamma(x),$$

where $\Omega_{\Gamma(x)}$ is the flow region ABCDF when the free boundary is $\Gamma(x)$. Since ψ is only uniquely determined up to a constant, it becomes convenient to take $\psi = 0$ on AB in the present method. Thatcher and

Askew (1982) seek the pair $\{\psi, \Gamma(x)\}$ such that ψ minimizes the functional

$$I(v) = \int_{\Omega_{\Gamma(x)}} |\nabla v|^2 \, d\Omega + 2 \int_{\partial \Omega_1} (\partial f/\partial s) v \, ds,$$

over trial functions v that are zero along AB with $\Gamma(x)$ chosen so that ψ is constant there.

The trial free-boundary method becomes:

  (i) choose Γ_0 and set the iteration index $p = 0$,
 (ii) triangulate Ω_{Γ_p} and choose a linear trial function v_p which satisfies the condition $v_p = 0$ on AB,
(iii) find ψ_p that minimizes the functional $I(v)$,
 (iv) find Γ_{p+1} by interpolating or extrapolating along the vertical mesh lines, such that on the new free-boundary nodes ψ_p is constant and equal to ψ_p at F.
  (v) set $p = p + 1$ and go to (ii) if the procedure has not converged sufficiently.

The flow region Ω_{Γ_p} is triangulated by constructing vertical lines, parallel to the y-axis, and having an equal number of equally spaced nodes along each vertical line, including the sections of the dam faces, AF and BCD. One of the nodes is at C. The jth nodes along neighbouring vertical lines are connected and each of the resulting quadrilaterals is triangulated by a diagonal as usual.

The constant value of the exact solution ψ along the free boundary is the discharge q per unit length through the dam and the constant value evaluated by the trial free-boundary method was found by Thatcher and Askew to approach the true value as the convergence criterion was made smaller. Their ordinates of the free boundary for the simple dam agreed better with those of Aitchison (1972) than with Aitchison (1977) except at the separation point. Results are also quoted for the same dam but with $y_2 = 0$, for a vertical-stratified dam studied by Oden and Kikuchi (1980), and for a rotationally-symmetric dam of interest in porous-well problems (Cryer and Fetter 1979).

8.4. Methods using variable interchange

Various authors have sought to avoid some of the difficulties associated with trial free-boundary methods by interchanging the dependent variable with one or more of the independent variables, e.g. in porous flow problems the flow potential is interchanged with one of the space variables x, y, which becomes the new, dependent variable to be computed. Finn and Varoğlu (1977) proposed a variational formulation of porous flow through a trapezoidal dam which incorporated a variable domain of integration. In order to generate a finite-element mesh which automatically adjusted to the different trial positions of the free boundary, they specified the nodes by chosen, fixed values of ϕ and y. The values of x both within the seepage region and on the free boundary, computed on this fixed, finite-element mesh are the solutions of variational equations. Crank and Ozis (1980) transformed their problem by the same interchange of variables and solved a transformed equation in a new fixed domain in which the free boundary became a known straight-line boundary. Again one of the space variables was computed in the transformed domain and simultaneously on the free boundary in a single iterative process. A well-established procedure in two-dimensional fluid-flow problems is to interchange both the potential function ϕ and the stream function ψ with the two space variables x and y. The problem is reformulated in the ϕ, ψ plane and x, y values are computed in what is often a fixed, rectangular region. This 'inverse' approach has been explored extensively in relation to free-boundary problems culminating in a three-dimensional application described by Jeppson (1972) in a paper which gives references to earlier work. These different uses of variable interchange are now described in more detail.

8.4.1. Adaptive variational formulation

Finn and Varoğlu (1977) introduced their finite-element method on a variable domain through the example of the trapezoidal dam shown in Fig. 8.6. The depths of the reservoirs on the inlet and outlet sides are denoted by H and h respectively, and y_C is used for the y coordinate of the separation point C. The velocity potential ϕ satisfies Laplace's equation in the seepage region ABCDE and also the boundary conditions

$$\left.\begin{aligned}
\partial\phi/\partial y &= 0 \quad \text{on AE,} \\
\phi &= y, \qquad \partial\phi/\partial n = 0 \quad \text{on BC,} \\
\phi &= y \quad \text{on CD,} \\
\phi &= H \quad \text{on AB,} \\
\phi &= h \quad \text{on DE,}
\end{aligned}\right\} \tag{8.38}$$

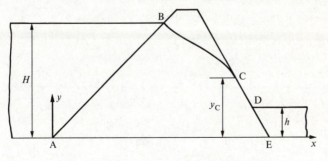

FIG. 8.6. Seepage through a trapezoidal dam

which have the same physical significance as in the problems formulated in Chapter 2.

Finn and Varoğlu (1977) considered the functional I defined by

$$I = \iint_\Omega \{(\partial\phi/\partial x)^2 + (\partial\phi/\partial y)^2\} \, dx \, dy, \tag{8.39}$$

where the domain Ω is the seepage region, and both $\phi(x, y)$ and Ω may vary. The variation of the domain Ω is constrained by the fact that the separation point C can only move along the outlet face CE. Only the free boundary BC and the seepage face CD can be varied; the rest of the boundary of Ω is fixed.

Courant and Hilbert (1953) showed that the first variation of I when the domain of integration can be varied takes the form

$$\delta I = -2 \iint_\Omega \nabla^2\phi \, \bar{\delta}\phi \, dx \, dy + 2 \int_{ABCDE} \left(\frac{\partial\phi}{\partial x}\frac{\partial x}{\partial n} + \frac{\partial\phi}{\partial y}\frac{\partial y}{\partial n} \right) \bar{\delta}\phi \, ds$$

$$+ \int_{ABCDE} \left\{ \left(\frac{\partial\phi}{\partial x}\right)^2 + \left(\frac{\partial\phi}{\partial y}\right)^2 \right\} \left(\frac{\partial x}{\partial n} \delta x + \frac{\partial y}{\partial n} \delta y \right) ds, \quad (8.40)$$

where $\bar{\delta}\phi$ denotes the variation at a fixed point (x, y) and so

$$\bar{\delta}\phi = \delta\phi - \frac{\partial\phi}{\partial x} \delta x - \frac{\partial\phi}{\partial y} \delta y. \tag{8.41}$$

The line integrals in (8.40) are taken round the boundary of Ω, indicated by ABCDE; n is the outward normal and s is measured along the boundary. Admissible functions $\phi(x, y)$ must satisfy the boundary conditions on CD, AB, DE given in (8.38). The admissible variations of the domain Ω are $\delta x = \delta y = 0$ on AB and DE, $\delta y = 0$ on AE, and so (8.40) reduces after a slight rearrangement of terms to

$$\delta I + \int_{BC} \left(\frac{\partial\phi}{\partial s}\right)^2 (\delta x \, dy - \delta y \, dx) = -2 \iint_\Omega \nabla^2\phi \, \bar{\delta}\phi \, dx \, dy$$

$$+ 2 \int_{AE} \frac{\partial\phi}{\partial n} \bar{\delta}\phi \, ds + 2 \int_{CB} \frac{\partial\phi}{\partial n} \bar{\delta}\phi \, ds$$

$$+ \int_{CB} \left(\frac{\partial\phi}{\partial n}\right)^2 (\delta x \, dy - \delta y \, dx). \quad (8.42)$$

If δU is defined by

$$\delta U = \delta I + \int_{BC} \left(\frac{\partial\phi}{\partial s}\right)^2 (\delta x \, dy - \delta y \, dx), \tag{8.43}$$

the function $\phi(x, y)$ which satisfies the boundary conditions on CD, AB, DE in (8.38) and for which δU defined in (8.43) is zero for all admissible variations δx, δy, $\bar{\delta}\phi$ can also be shown to satisfy $\nabla^2 \phi = 0$ in Ω and $\partial \phi / \partial n = 0$ on BC and AE. Thus the variational formulation is equivalent to the boundary-value problem with conditions (8.38).

In order to set up a finite-element mesh for the discretization of the variational problem Finn and Varoğlu specify the nodes by chosen fixed values of ϕ and y. Thus, at a typical node i, ϕ and y take arbitrary fixed values $\phi_i (h \leq \phi_i \leq H)$ and $y_i (0 \leq y_i \leq \phi_i)$, except at the separation point C (Fig. 8.6) at which $\phi_C = y_C$ is an unknown value to be determined. The x-coordinates of the nodes on the boundary segments AB, DE, and CD are known. The x-coordinates on the free surface are unknown except for x_C which is related to y_C through the equation of the exit face of the dam. Thus the problem is to determine the x-coordinates of the nodes inside the flow region Ω and on the impermeable base AE and the free surface BC, excluding the end points; the unknown associated with the node at the separation point C is taken to be y_C, making, say, $n+1$ unknowns in all. The discretized form of (8.43) can now be written

$$\delta U = \delta \sum_{ijk} I_{ijk} + \sum_{ij} \delta \Gamma_{ij} \tag{8.44}$$

where

$$I_{ijk} = \iint_{ijk} \left\{ \left(\frac{\partial \phi}{\partial x}\right)^2 + \left(\frac{\partial \phi}{\partial y}\right)^2 \right\} dx\, dy, \quad \delta\Gamma_{ij} = \int_{ij} \left(\frac{\partial \phi}{\partial s}\right)^2 (\delta x\, dy - \delta y\, dx). \tag{8.45}$$

The first summation in (8.44) is carried out over all the area elements in the seepage region Ω, and the second over all the line elements on the free surface BC; the direction from i to j is the direction from B to C along the free surface.

The unknown function $x(\phi, y)$ is approximated by a linear function of ϕ and y, which is most conveniently expressed in the form

$$\phi(x, y) = [a_i + Y_i x + X_i y, \ a_j + Y_j x + X_j y, \ a_k + Y_k x + X_k y] \begin{bmatrix} \phi_i \\ \phi_j \\ \phi_k \end{bmatrix} \frac{1}{2A_{xy}}$$

$$\tag{8.46}$$

for the typical element ijk where the order of the nodes is taken counterclockwise. The nomenclature used in (8.46) and in what follows is set out in Table 8.2.

TABLE 8.2

Definition of symbols for a typical area element with vertices $i(x_i, y_i)$, $j(x_j, y_j)$, $k(x_k, y_k)$ taken counterclockwise and $\phi_i = \phi(x_i, y_i)$, $\phi_j = \phi(x_j, y_j)$, $\phi_k = \phi(x_k, y_k)$

	Indices		
Symbols	i	j	k
a	$x_i y_k - x_k y_j$	$x_k y_i - x_i y_k$	$x_i y_j - x_j y_i$
X	$x_k - x_j$	$x_i - x_k$	$x_j - x_i$
Y	$y_j - y_k$	$y_k - y_i$	$y_i - y_j$
Φ	$\phi_k - \phi_j$	$\phi_i - \phi_k$	$\phi_j - \phi_i$
$2A_{xy}$		$x_i Y_i + x_j Y_j + x_k Y_k$	
$2A_{\phi x}$		$\phi_i X_i + \phi_j X_j + \phi_k X_k$	
$2A_{\phi y}$		$\phi_i Y_i + \phi_j Y_j + \phi_k Y_k$	

Since $\partial\phi/\partial x = A_{\phi y}/A_{xy}$ and $\partial\phi/\partial y = A_{\phi x}/A_{xy}$ and $\phi = y$ on BC, equations (8.45) become

$$I_{ijk} = (A_{\phi x}^2 + A_{\phi y}^2)/A_{xy},$$

$$\delta\Gamma_{ij} = \frac{(\Delta y)^2}{(\Delta x)^2 + (\Delta y)^2}\{\tfrac{1}{2}\Delta y(\delta x_i + \delta x_j) - \tfrac{1}{2}\Delta x(\delta y_i + \delta y_j)\}, \quad (8.47)$$

by using the trapezoidal integration and inserting $(\partial\phi/\partial s)_m^2 = (\Delta y)^2/(\Delta x^2 + \Delta y^2)$, $m = i, j$, where for the element ij on BC, $\Delta y = y_j - y_i$, $\Delta x = x_j - x_i$. The discretized form of δU can be written in terms of variations of the discrete unknowns y_c and x_m, $1 \le m \le n$, as

$$\delta U = \sum_{m=1}^{n} (f_m \, \delta x_m + f_{n+1} \, \delta y_c), \qquad (8.48)$$

where the coefficients f_m and f_{n+1} are given by

$$f_m = \frac{\partial}{\partial x_m} \sum_{ijk} I_{ijk} + q_m, \qquad m = 1, 2, \ldots, n, \qquad (8.49a)$$

$$f_{n+1} = \frac{\partial}{\partial y_c} \sum_{ijk} I_{ijk} + q_{n+1}, \qquad (8.49b)$$

since

$$\delta I = \sum_{m=1}^{n} \left(\frac{\partial}{\partial x_m} \sum_{ijk} I_{ijk}\right) \delta x_m + \left(\frac{\partial}{\partial y_c} \sum_{ijk} I_{ijk}\right) \delta y_c \qquad (8.50a)$$

and $\sum_{ij} \delta\Gamma_{ij}$ may be written as

$$\sum_{ij} \delta\Gamma_{ij} = \sum_{m=1}^{n} q_m \, \delta x_m + q_{n+1} \, \delta y_c. \qquad (8.50b)$$

The contributions to q_m and q_{n+1} by each line element on BC are determined below. Thus the equation $\delta U = 0$ is discretized to the form

$$f_m(x_1, x_2, \ldots, x_n, y_c) = 0, \qquad m = 1, 2, \ldots, n+1, \qquad (8.51)$$

and it is necessary to evaluate the partial derivatives in (8.49a,b) and to obtain expressions for the contributions to q_m ($m = 1, 2, \ldots, n+1$) from a typical line element. For a typical area element ijk the partial derivatives of δI_{ijk} with respect to x_m can be written

$$\frac{\partial I_{ijk}}{\partial x_m} = \begin{cases} 0, & m \neq i, j, k, \\ -\left[\dfrac{A_{\phi x}}{A_{xy}} \Phi_m + \dfrac{1}{2}\left\{\left(\dfrac{A_{\phi x}}{A_{xy}}\right)^2 + \left(\dfrac{A_{\phi y}}{A_{xy}}\right)^2\right\} Y_m\right], & m = i, j, k, \end{cases}$$
$$(8.52)$$

by using the nomenclature of Table 8.2. The contributions q_m^{ij} to q_m from a typical line element ij can be written as

$$q_m^{ij} = \begin{cases} 0, & m \neq i, j \\ \dfrac{1}{2} \Delta y \left\{\dfrac{(\Delta y)^2}{(\Delta x)^2 + (\Delta y)^2}\right\}, & m = i, j, \end{cases} \qquad (8.53)$$

by using (8.50b) and the second of (8.47), and recalling that

$$\frac{\partial x_c}{\partial y_c} = \alpha, \qquad \frac{\partial Y_c}{\partial y_c} = 1, \qquad (8.54)$$

where $1/\alpha$ is the slope of the outlet face CE, we obtain

$$\frac{\partial I_{ijk}}{\partial y_c} = \begin{cases} 0, & c \neq i, j, k \\ -\dfrac{1}{2}\left\{\left(\dfrac{A_{\phi x}}{A_{xy}}\right)^2 + \left(\dfrac{A_{\phi y}}{A_{xy}}\right)^2\right\}(X_c + Y_c) \\ \quad + \dfrac{A_{\phi y}}{A_{xy}}(\phi_c + Y_c) + \dfrac{A_{\phi x}}{A_{xy}}(X_c - \alpha\phi_c), & c = i, j, k, \quad (8.55a) \end{cases}$$

and

$$q_{n+1}^{ij} = \begin{cases} 0, & c \neq j, \\ -\dfrac{1}{2} \Delta x \left\{\dfrac{(\Delta y)^2}{(\Delta x)^2 + (\Delta y)^2}\right\}, & c = j. \end{cases} \qquad (8.55b)$$

Finn and Varoğlu (1977) solved the first n of the $n+1$ non-linear algebraic equations

$$f_i(x_1, x_2, \ldots, x_n, y_c) = 0, \qquad i = 1, 2, \ldots, n+1, \qquad (8.56)$$

for an initial trial value of $y_c = y_c^0$, by the standard Newton–Raphson iterative method. The rth stage of the iterative solution using y_c^0 is denoted by $\{x\}_0^r$ and the process is halted when $|\{x\}_0^{r+1} - \{x\}_0^r| \leq \varepsilon$, for some pre-scribed accuracy ε. The solutions $\{x\}_0^{r+1}$ and y_c^0 do not satisfy the remain-ing equation

$$f_{n+1}(x_1, x_2, \ldots, x_n, y_c) = 0$$

and depending on the sign of $f_{n+1}(\{x\}_0^{r+1}, y_c^0)$, y_c^0 is increased or decreased to give a new trial value y_c^1 for the next stage of the iteration cycle, which continues until successive iterates differ by less than a prescribed amount. The initial trial value y_c^0 was taken to be $\frac{1}{2}(H + h)$.

In order to facilitate completely automatic mesh generation, Finn and Varoğlu (1977) started from a rectangular mesh of constant $\bar{\phi}$ and constant $\bar{y}$ lines in a rectangular $\bar{\phi}$, $\bar{y}$ domain (Fig. 8.7) where a nodal point $(\bar{\phi}_i, \bar{y}_i)$ is given by

$$\begin{aligned}
\bar{\phi}_i &= (\phi_i - y_c)/(y_c - H), \qquad y_c \leq \phi_i \leq H, \\
&= (\phi_i - y_c)/(h - H), \qquad h \leq \phi_i \leq y_c,
\end{aligned} \tag{8.57}$$

$$\bar{y}_i = y_i/\phi_i. \tag{8.58}$$

To each node $(\bar{\phi}_i, \bar{y}_i)$ in the $\bar{\phi}$, $\bar{y}$ plane there corresponds a node (ϕ_i, y_i) in the x, y plane depending on the unknowns y_c and x_i. An initial set of trial values of x coordinate values associated with the (ϕ_i, y_i) values is also needed to start the iterative solution. These x values are arbitrary except that they must preserve the sequential ordering of equipotential lines, i.e. if $\phi_1 > \phi_2 > \phi_3$, the trial x values cannot be such as to put ϕ_3 between ϕ_1 and ϕ_2. Finn and Varoğlu (1977) described one satisfactory procedure.

They took the free surface initially to be the straight line BC. The equipotential line $\phi = y_c^0$ is assumed to be the line CF which is normal to BC and intersects the base AE in F (Fig. 8.8a). All other equipoten-tial lines were assumed to be straight initially and were located as follows. The equipotential line $\phi = $ constant corresponding to a specific line $\bar{\phi} = $ constant is taken to intersect the free surface BC or the seepage face CD in a point at which $y = \phi$. The other end of the assumed straight line $\phi = $ constant is at G on AF, where $(H - \phi)/(\phi - y_c^0) = $ AG/GF. A similar

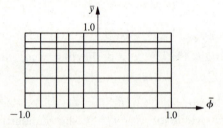

Fig. 8.7. Mesh for numerical examples in the $\bar{\phi}$–$\bar{y}$ plane

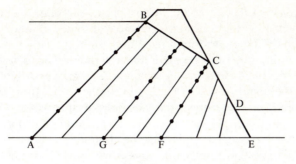

FIG. 8.8(a). Initially assumed equipotential lines and nodes

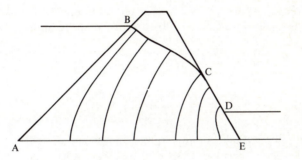

FIG. 8.8(b). Solution for equipotential lines

relationship is used in FE. The initial x values are computed for the chosen ϕ, y values and used to start the Newton–Raphson iterative process. Subsequently, the equipotential lines are modified automatically according to the y_c and x values computed at each stage of the iteration until some convergence criterion is satisfied. The final equipotential lines in a typical solution are shown in Fig. 8.8b.

Finn and Varoğlu (1977) gave a little more computational detail and a graphical solution for a particular dam. Finn and Varoğlu (1980, 1978) applied their method to seepage from an irrigation ditch and to free-surface gravity flow over a spillway, respectively. In the 1980 paper they also solved a Stefan moving-boundary problem by using their variable-domain finite-element method after the method of lines had been used to remove the time dependency (see §4.4).

8.4.2. Transformation of equation and flow region

(i) *Single variable interchange.* The method to be described shares with that of the §8.4.1 the idea of a single variable interchange so that x is to be calculated as a function of ϕ and y in a two-dimensioal problem.

Crank and Ozis (1980), however, transformed the elliptic equation and solved the new equation for $x = x(\phi, y)$ subject to appropriately rewritten conditions on the boundaries of a fixed domain in the ϕ, y plane which is a transformation of the flow region in the x, y plane. The idea is well tried in fluid-flow problems and has been applied to moving boundary problems under the name Isotherm Migration Method (IMM) as described in §5.4. The need to determine the position of the whole of the free boundary in the physical plane is reduced to locating the position of the separation point on a fixed, straight-line boundary in the transformed plane. An iterative algorithm approximates within each single loop both a finite-difference solution of the elliptic partial differential equation and the position of the free boundary.

The simple dam problem of Fig. 2.1 and equations (2.7–12) are used as an example. Crank and Ozis (1980) transformed the appropriate form of (2.7), i.e.

$$\partial^2 \phi / \partial x^2 + \partial^2 \phi / \partial y^2 = 0, \tag{8.59}$$

into

$$\frac{\partial^2 \phi}{\partial y^2} - \frac{\partial^2 x}{\partial \phi^2} \left(\frac{\partial x}{\partial \phi}\right)^{-3} = 0 \tag{8.60}$$

by changing the variables only in the first term of (8.59). They then treated the $\partial^2 \phi / \partial y^2$ term in (8.60) by an interpolation procedure. Later work by Crank and Sabouri, reported by Ozis (1981), showed that both accuracy and convergence were improved by a complete transformation of (8.59) following Boadway's (1976) treatment of fluid-flow problems presented in §5.4.2(ii). With suitable change of nomenclature and by putting $\partial z / \partial t = 0$, the fully transformed form of (8.59) follows immediately from the two-dimensional version of (5.76) and is

$$\frac{\partial^2 x}{\partial y^2} \left(\frac{\partial x}{\partial \phi}\right)^2 + \frac{\partial^2 x}{\partial \phi^2} \left\{1 + \left(\frac{\partial x}{\partial y}\right)^2\right\} - 2 \frac{\partial^2 x}{\partial y \partial \phi} \frac{\partial x}{\partial y} \frac{\partial x}{\partial \phi} = 0. \tag{8.61}$$

Apart from the differences in the finite-difference discretizations of (8.60) and (8.61), Crank and Sabouri used the same algorithm as that of Crank and Ozis (1980). In general, on a boundary $y = g(x)$ we have

$$\phi_n = \frac{g' \phi_x - \phi_y}{[1 + (g')^2]^{\frac{1}{2}}}, \qquad g' = dy/dx, \tag{8.62}$$

where, for example, $\phi_n = \partial \phi / \partial n$ and n is the outward normal to $y = g(x)$. Provided $g' \neq 0$, we have on the free boundary FD, in Fig. 2.1,

$$g' \, \partial \phi / \partial x - \partial \phi / \partial y = 0, \qquad y = \phi. \tag{8.63}$$

But $(\partial \phi / \partial y)_x = -(\partial x / \partial y)/(\partial x / \partial \phi)$ and $(\partial \phi / \partial x)_y = 1/(\partial x / \partial \phi)_y$ so that

$$g' + (\partial x / \partial y)_\phi = 0, \qquad y = \phi. \tag{8.64}$$

It is true that $g' = \infty$ at the separation point D but condition (8.64) is not applied at that point.

On the impervious base AB we have

$$\partial x/\partial y = 0, \qquad y = 0. \tag{8.65}$$

The other boundary conditions are

$$x = 0, \qquad \phi = 1 \quad \text{on AF}, \qquad x = x_1, \qquad \phi = y_2, \qquad 0 \leqslant y \leqslant y_2 \quad \text{on BC}, \tag{8.66}$$

$$x = x_1, \qquad \phi = y \quad \text{on CD}. \tag{8.67}$$

We now wish to solve (8.61) or (8.60), subject to conditions (8.64–67) in the region A'F'D'C'B' in Fig. 8.9 in the ϕ, y plane. All the boundaries are fixed. The original free boundary FD has become the known straight boundary, F'D', $y = \phi$. What we do not know, however, is the position of D' on F'C' corresponding to the separation point D in Fig. 2.1, at which the boundary condition changes from $x = x_1$ to the condition (8.64).

The transformed region in Fig. 8.9 is covered with a mesh of spacings $\delta\phi$, δy and x_{ij} denotes $x(i\,\delta\phi, j\,\delta y)$. For a typical internal node i, j the Jacobi-type iterative process was used to solve (8.61) in the form

$$4[(x_{i+1,j}^n - x_{i-1,j}^n)^2 + 4(\delta y)^2 + (x_{i,j+1}^n - x_{i,j-1}^n)^2]x_{i,j}^{n+1}$$
$$= [2(x_{i,j+1}^n + x_{i,j-1}^n)(x_{i+1,j}^n - x_{i-1,j}^n)^2 + 2(x_{i+1,j}^n + x_{i-1,j}^n)$$
$$\times \{4(\delta y)^2 + (x_{i,j+1}^n - x_{i,j-1}^n)^2\} - \{(x_{i+1,j+1}^n - x_{i-1,j+1}^n)$$
$$- (x_{i+1,j-1}^n - x_{i-1,j-1}^n)\}(x_{i,j+1}^n - x_{i,j-1}^n)(x_{i+1,j}^n - x_{i-1,j}^n)], \tag{8.68}$$

where $x_{i,j}^n$ is the nth iterative value of $x_{i,j}$. Correspondingly, on the boundary F'C', $y = \phi$ we use either $x = x_1$ or

$$x_{j,j}^{n+1} = x_{j,j-1}^n - \frac{2(\delta y)^2}{x_{j+1,j+1}^n - x_{j-1,j-1}^n}$$

from (8.64), where a one-sided difference replacement of $(\partial x/\partial y)_\phi$ is used

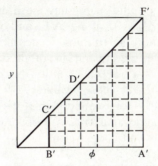

Fig. 8.9. Transformed plane

to avoid fictitious points outside the region. Application of (8.68) to points on the lower boundary $y = 0$, i.e. $j = 0$, introduces ficitious points on the line $j = -1$, one step outside the region; these can be eliminated from (8.68) since (8.65) implies $x_{i,-1} = x_{i,1}$ for all i.

The iteration can be started by assuming the separation point D' to be at one of the mesh points on $F'C'$ in Fig. 8.9. It follows that $x = x_1$ along $B'C'$ and $C'D'$, and we know that $x = 0$ on $A'F'$. For every other mesh point within the region and on the remaining parts of the boundary we have derived an equation. We can carry out one iterative cycle by sweeping along successive j-lines from left to right in Fig. 8.9 in the order $j = 0, 1, 2, \ldots$, where $y = j = 0$ is the lower boundary. The new values of x_{ij} are retained for use in the next cycle subject to the proviso that on the boundary $F'C'$ we take the new value $x_{i,i}^{n+1}$ to be

$$x_{i,i}^{n+1} = \min(x_1, x_{i,i}^{n+1})$$

since we know $x_{i,j} \leqslant x_1$. We proceed with successive iterative loops, improving the solution, which through the $x_{i,i}$ values on $F'C'$ includes the position of the free boundary and the position of the separation point, until some prescribed convergence criterion is satisfied. In the absence of any prior knowledge of the position of D', the iterative process can be started by assuming D' to be at C', its lowest possible position. This ensures that D' is always approached from below its true position in the iterative cycle. The initial values of x along $F'D'$ were taken by Crank and Ozis (1980) to be linear interpolates between $x = 0$ at F' and $x = x_1$ at D'. On each line $y = $ constant they took linear interpolates between the end values on $B'C'D'$ and $A'F'$ for the initial values of x at internal mesh points. Ozis (1981) reported an improved rate of iterative convergence by starting from parabolic interpolated values of x both along $F'D'$ and inside the domain. The parabolic form allows certain physical properties to be taken into account. Thus along the free surface $F'D'$, $\partial\phi/\partial x = 0$ at F' and this can be incorporated in a parabolic expression for the x-values along $F'D'$ which fits $x = 0$ at F', $x = x_1$ at D'. A similar set of parabolic interpolates for initial x-values along any mesh line of constant y can be obtained which has $\partial\phi/\partial x = 0$ along $A'F'$ and which has $x = x_1$ along $B'C'D'$ and x has the interpolated initial values along $F'D'$ just obtained.

For the particular dam used by various other authors in which $x_1 = 2/3$, $y_2 = 1/6$ relative to a water depth of unity on the inlet side, Table 8.3 compares values of $10^4 x$ at chosen values of y on the free boundary, obtained by solving (8.61) for two mesh sizes $h = 1/18$ and $h = 1/30$ with values computed from the analytical solution of Polubarinova-Kochina (1962). Other solutions and comparisons are to be found in Crank and Ozis (1980) and Ozis (1981).

Crank and Ozis (1980) described the use of a 'parabolic tail' in an attempt to improve the position of the estimated separation point by

TABLE 8.3

Comparison of free-boundary position by variable interchange with analytic solution

$10^4 y$	$10^4 x$		
	$h = 1/18$	$h = 1/30$	Analytic
5284	6667	6667	6667
5333	6667	6667	6652
5667	6667	6508	6467
6000	6251	6213	6231
6333	5966	5942	5957
6667	5695	5618	5651
7000	5326	5276	5311
7333	4950	4901	4940
7667	4560	4497	4535
8000	4118	4059	4093
8333	3651	3582	3611
8667	3107	3061	3083
9000	2523	2484	2498
9333	1853	1831	1829
9667	—	1061	1067

ensuring that the gradient of the free boundary, dx/dy, should approach zero at D'. If an accurate solution is required near D', however, Aitchison's singularity solution given in equation (8.35) above needs to be incorporated.

Crank and Ozis (1981) extended their variable-interchange method to a model three-dimensional problem, which was the seepage of fluid through a circular porous dam. Although the problem had circular symmetry Crank and Ozis ignored this and treated it as a general three-dimensional problem in three cartesian space variables, x, y, z. Afterwards, and quite independently, they solved the two-dimensional problem reformulated in the r–z plane in order to compare the two results.

The mathematical formulation of the problem illustrated in Fig. 8.10 is as follows:

$$\nabla^2 \phi = \frac{\partial^2 \phi}{\partial x^2} + \frac{\partial^2 \phi}{\partial y^2} + \frac{\partial^2 \phi}{\partial z^2} = 0 \quad \text{in AEDCLMFG,} \tag{8.69a}$$

$$\phi = H \quad \text{on AEFG,} \tag{8.69b}$$

$$\partial \phi / \partial n = 0 \quad \text{on EDMF,} \tag{8.69c}$$

$$\phi = h \quad \text{on DMPN,} \tag{8.69d}$$

$$\phi = y \quad \text{on CNPL, the seepage surface,} \tag{8.69e}$$

$$\partial \phi / \partial n = 0 \quad \text{on ABDE and GFMK,} \tag{8.69f}$$

$$\phi = y, \qquad \partial \phi / \partial n = 0 \quad \text{on ACLG, the free surface,} \tag{8.69g}$$

where n is the outward normal.

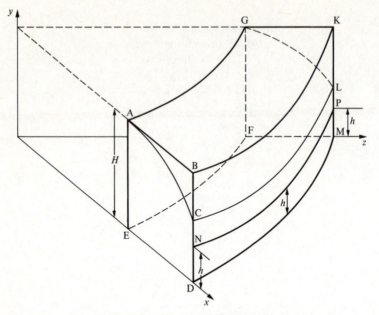

FIG. 8.10. Three-dimensional circular earth dam

We now interchange ϕ and x so that $x = x(\phi, y, z)$ becomes the new dependent variable instead of ϕ. With suitable changes of nomenclature and putting $\partial z/\partial t = 0$, the Boadway (1976) transformation of (8.69a), deduced from (5.76), is

$$\frac{\partial^2 x}{\partial \phi^2}\left\{1+\left(\frac{\partial x}{\partial y}\right)^2+\left(\frac{\partial x}{\partial z}\right)^2\right\}+\left\{\frac{\partial^2 x}{\partial y^2}+\frac{\partial^2 x}{\partial z^2}\right\}\left(\frac{\partial x}{\partial \phi}\right)^2$$

$$-2\frac{\partial^2 x}{\partial \phi\, \partial z}\frac{\partial x}{\partial \phi}\frac{\partial x}{\partial z}-2\frac{\partial^2 x}{\partial \phi\, \partial y}\frac{\partial x}{\partial \phi}\frac{\partial x}{\partial y}=0. \quad (8.70)$$

Equation (8.70) is to be solved in the transformed domain shown in Fig. 8.11 for $x(\phi, y, z)$. The transformed boundary conditions are obtained by the arguments in the two-dimensional problem above. In general, on a surface $y = y(x, z)$ we have

$$\phi_n = \frac{(\partial \phi/\partial x)_{y,z}(\partial y/\partial x)_z-(\partial \phi/\partial y)_{x,z}+(\partial \phi/\partial z)_{x,y}(\partial y/\partial z)_x}{\{(\partial y/\partial x)_z^2+1+(\partial y/\partial z)_x^2\}^{\frac{1}{2}}}, \quad (8.71)$$

where $\phi_n = \partial \phi/\partial n$ and n is the outward normal to $y = y(x, z)$. Provided $\partial y/\partial x \neq 0$ and $\partial y/\partial z \neq \infty$, we have on the free surface A'C'L'G', after slight manipulation,

$$1+\left(\frac{\partial x}{\partial y}\right)_{\phi,z}\left(\frac{\partial x}{\partial y}\right)_z+\left(\frac{\partial x}{\partial z}\right)_{\phi,y}\left(\frac{\partial x}{\partial z}\right)_y=0, \qquad y=\phi. \quad (8.72)$$

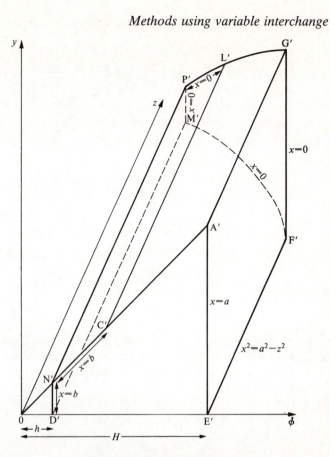

Fig. 8.11. Transformed space

Other boundary conditions are

$$\partial x/\partial y = 0, \qquad y = 0 \quad \text{on the impervious base E'D'M'F',} \qquad (8.73)$$

and

$$\partial x/\partial z = 0, \qquad z = 0 \quad \text{on the side face A'C'N'D'E'.} \qquad (8.74)$$

On the side face GLMF in Fig. 8.10 we know that $x = 0$ everywhere but we do not know the shape of the surface F'G'L'P'M' in Fig. 8.11. In principle, this can be determined by using the second boundary condition on GLMF in Fig. 8.10 which is $\partial \phi/\partial x = 0$, which by a standard change of variables leads immediately to

$$(\partial z/\partial x) = 0, \qquad x = 0 \quad \text{on F'G'L'P'M'.} \qquad (8.75)$$

Condition (8.75) suggests a quadratic relationship for small x on a constant ϕ line in a constant y-plane (Fig. 8.12) of the form

$$z = \alpha x^2 + \beta. \qquad (8.76)$$

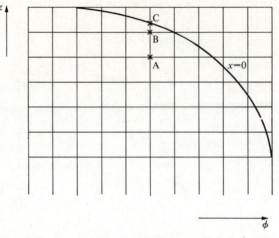

FIG. 8.12. Extrapolation for side of dam $x = 0$ in ϕ–z plane

By extrapolating along AB in Fig. 8.12 using (8.76) where, for example, α, β are determined by fitting the (x, z) values at A, B, the values of z at C can be found as an iterative solution proceeds. The geometric boundary conditions in Fig. 8.11 are

$$x = \surd(a^2 - z^2), \qquad \phi = H, \tag{8.77}$$

$$x = \surd(b^2 - z^2), \qquad \phi = h, \qquad 0 \leqslant y \leqslant h, \tag{8.78}$$

$$x = \surd(b^2 - z^2), \qquad \phi = y \quad \text{on C'N'P'L'.} \tag{8.79}$$

Thus we wish to solve equation (8.70) subject to conditions (8.72–4), (8.75), and (8.77–9) in the (ϕ, y, z) domain of Fig. 8.11. All the boundaries are fixed. The original free surface ACLG in Fig. 8.10 has become the known plane $y = \phi$, A'C'L'G' in Fig. 8.11. We have to find the position of the line C'L' on the plane A'N'P'G' which corresponds to the separation line CL in Fig. 8.10. The shape of the face F'M'P'G' is determined by the extrapolation (8.76). An iterative method of solution, the same in principle as in the two-dimensional problem discussed above for eqn (8.61), can be used, with obvious increase in complication for the three-dimensional problem. The transformed (ϕ, y, z) region is covered with a mesh of spacing $\delta\phi$, δy, δz and $x_{i,j,k}$ denotes $x(i\,\delta\phi, j\,\delta y, k\,\delta z)$. Then if all derivatives in eqn (8.70) are approximated by central finite-differences and the terms in $x_{i,j,k}$ are collected together, a point Jacobi iterative formula corresponding to (8.68) can be written

$$\left[2\left(\frac{1}{\delta y^2}-\frac{1}{\delta z^2}\right)(\delta x_\phi^n)^2 + \frac{2}{\delta\phi^2}\{1+(\delta x_y^n)^2+(\delta x_z^n)^2\}\right]x_{i,j,k}^{n+1}$$

$$= (\delta x_\phi^n)^2\{(\bar{x}_y^n)^2-(\bar{x}_z^n)^2\} + (\bar{x}_\phi^n)^2\{(\delta x_y^n)^2-(\delta x_z^n)^2+1\}$$

$$+ 2\,\delta x_y^n\,\delta x_\phi^n\{x_{i+1,j+1,k}^n - x_{i+1,j-1,k}^n - x_{i-1,j+1,k}^n + x_{i-1,j-1,k}^n\}$$

$$- 2\,\delta x_z^n\,\delta x_\phi^n\{x_{i+1,j,k+1}^n - x_{i+1,j,k-1}^n - x_{i-1,j,k+1}^n + x_{i-1,j,k-1}^n\}. \quad (8.80)$$

For convenience,

$$\delta x_\phi^n = (x_{i+1,j,k}^n - x_{i-1,j,k}^n)/(2\,\delta\phi), \qquad \bar{x}_\phi^n = (x_{i+1,j,k}^n + x_{i-1,j,k}^n)/(\delta\phi)^2,$$

etc. have been introduced and $x_{i,j,k}^n$ is the nth iterative value of $x_{i,j,k}$. On the free surface we have either $x = \sqrt{(b^2-z^2)}$ or

$$x_{i,j,k}^{n+1} = x_{i,j-1,k}^n - \frac{2(\delta y)^2}{x_{i+1,j+1,k}^n - x_{i-1,j-1,k}^n} - \frac{1}{2}\left(\frac{\delta y}{\delta z}\right)^2\frac{(x_{i,j,k+1}^n - x_{i,j,k-1}^n)^2}{x_{i+1,j+1,k}^n - x_{i-1,j-1,k}^n} \quad (8.81)$$

from eqn (8.72); as in the two-dimensional problem, a one-sided difference replacement of $(\partial x/\partial y)_{\phi,z}$ is used to avoid fictitious points. Corresponding modified forms of (8.80) are applied at all mesh points one step below the plane A'N'P'G' and on the lower impervious boundary, $y = 0$. Use of the extrapolation (8.76) leads to unequal mesh sizes in eqn (8.80) for points one step inside the side face F'M'P'G', and in eqn (8.81) for points one step inside the curve P'G' on which $x = 0$. For points in the line A'N', i.e. $z = k = 0$, fictitious points can be avoided when using (8.81) because of (8.74).

The iterative cycle is started by assuming the separation line C'L' to be at N'P', its lowest possible position, unless prior knowledge suggests a better starting position. We thus ensure that the final position of C'L' is approached from below. We then know that $x = \sqrt{(b^2-z^2)}$ on the face D'N'P'M' and on C'N'P'L' and also that $x = \sqrt{(a^2-z^2)}$ on the face E'A'G'F'. For every other mesh point within the region and on the remaining parts of the boundary, an equation has been derived. One iterative cycle involves sweeping along successive j-planes in the order $j = 0, 1, 2, \ldots$, where $j = y = 0$ is the lower boundary. In each consecutive plane each mesh point is covered in the order $i = 0$, $k = 0, 1, 2, \ldots$; $i = 1$, $k = 0, 1, 2, \ldots$, and so on. The new values $x_{i,j,k}^{n+1}$ are retained for use in the next cycle subject to the proviso that on the boundary plane we take the new value $x_{i,j,k}^{n+1}$ to be

$$x_{i,j,k}^{n+1} = \min(\sqrt{(b^2-z^2)}, x_{i,j,k}^{n+1})$$

since we know that $x_{i,j,k} \leq \sqrt{(b^2-z^2)}$. On F'G'P'M' new values of z are retained for use in the next cycle. Values of the solution and of the separation line are simultaneously iterated in each loop until a convergence limit is achieved and then the highest mesh point on each z-line on the plane A'N'P'G' at which $x = \sqrt{(b^2 - z^2)}$ provides the best approximation to the separation that can be obtained directly from the finite-difference mesh.

Numerical experiments show that the iterative process may not converge or may be very slow for arbitrarily chosen starting values for x. As in the two-dimensional problem, the initial values of x in the free surface were taken to be linear interpolates between $x = \sqrt{(b^2 - z^2)}$ on the line C'L' in Fig. 8.11 and $x = \sqrt{(a^2 - z^2)}$ on A'G' along constant z-lines. On each constant y-plane, the initial values of x at internal mesh points were obtained by parabolic interpolation along each constant z-line, between the known values of x on the surface D'N'C'A'G'L'P'M' and the known values at corresponding points on the surfaces A'E'F'G' and G'L'P'M'F' with the additional condition that $\partial\phi/\partial x$ was taken to be zero on the latter two faces. These parabolic starting values satisfy the physical condition that $\partial\phi/\partial x$ must increase negatively as x increases in this problem. An initial estimate of the shape of the surface M'P'G'F' in Fig. 8.11 on which $x = 0$ is also needed, having in mind that in this surface the physical condition at G in the free surface, which is the ϕ–y plane in Fig. 8.11, is $\partial\phi/\partial z = 0$. Therefore, the intersection of the face M'P'G'F' and the free surface is assumed initially to be a parabola $(z - a)^2 = \gamma(H - \phi)$, with γ chosen so that $\phi = \phi_L$, the value of ϕ at the initial position of L' on the separation line, i.e. $(b - a)^2 = \gamma(H - \phi_L)$. The same parabola can be used between the lines M'P' and F'G' on each constant-y plane to estimate initial z-values on the whole face M'P'G'F' by taking γ to be given by $(b - a)^2 = \gamma(H - h)$. Along L'P', $z = b$.

The formulation of this problem in cartesian coordinates x, y, z was done in order to explore the feasibility of the variable-interchange technique in three dimensions. In fact, the problem has cylindrical symmetry and Laplace's equation for ϕ in the form

$$\frac{\partial^2\phi}{\partial r^2} + \frac{1}{r}\frac{\partial\phi}{\partial r} + \frac{\partial^2\phi}{\partial y^2} = 0 \tag{8.82}$$

is appropriate. Boadway (1976) interchanged the variables r and ϕ and obtained the equation for $r = r(\phi, y)$ to be

$$\frac{\partial^2 r}{\partial y^2}\left(\frac{\partial r}{\partial\phi}\right)^2 + \frac{\partial^2 r}{\partial\phi^2}\left\{1 + \left(\frac{\partial r}{\partial y}\right)^2\right\} - 2\frac{\partial^2 r}{\partial\phi\,\partial y}\frac{\partial r}{\partial\phi}\frac{\partial r}{\partial y} - \frac{1}{r}\left(\frac{\partial r}{\partial\phi}\right)^2 = 0. \tag{8.83}$$

Ozis (1981) followed the treatment of eqn (8.61) above with boundary conditions expressed in cylindrical coordinates to obtain numerical solutions of (8.83). For a particular cylindrical dam as shown in Fig. 8.10, in which $a = 1$, $b = 5/3$, $H = 1$, $h = 1/6$, results obtained by the three-dimensional algorithm in x, y, z coordinates with $\delta\phi = \delta y = 1/12$ and $\delta z = 1/15$ compared favorably with those for the cylindrical formulation (8.83).

No formal study of the convergence of the iterative solution of the non-linear finite difference equations proved possible. Instead, Ozis (1981) quoted numerical experiments which demonstrated reasonable convergence of some of the numerical values obtained for the three-dimensional formulation including the position of the free surface, starting from parabolically interpolated x-values. About 100 iterative cycles were necessary to satisfy a convergence limit of $|x_{i,j,k}^{n+1} - x_{i,j,k}^{n}| \leqslant 10^{-3}$.

(ii) *Multi-variable interchange.* Jeppson (1972) gave an inverse formulation of a three-dimensional dam problem in which he interchanged all three cartesian coordinates, x, y, z, with a potential function and two selected streamline functions. In the same paper Jeppson gave several references to corresponding inverse solutions of two-dimensional inviscid and porous flow problems, including some axisymmetric cases. In three dimensions, Jeppson introduced two stream functions ψ and ψ' defined to be orthogonal surfaces normal to equipotential surfaces and tangential to the velocity vector such that their intersections define the streamlines of the flow. The potential function ϕ is defined in the usual way. The inverse formulation is in the ϕ, ψ, ψ' space in which three non-linear first-order partial differential equations are to be solved for the three new dependent variables $x = x(\phi, \psi, \psi')$, $y = y(\phi, \psi, \psi')$, $z = z(\phi, \psi, \psi')$, subject to appropriately rewritten boundary conditions. In order to proceed with a finite-difference solution Jeppson (1972) combined these equations by differentiation, making the assumption that certain relatively small quantities were known and remained constant so that separate equations for x, y, z could be used in different planes in the transformed space. He discussed factors influencing the choice of the approximated equations.

To illustrate his method, Jeppson (1972) obtained solutions for three-dimensional seepage through a dam with a partial toe drain, as in Fig. 8.13a. The transformed ϕ, ψ, ψ' space is shown in Fig. 8.13b, and a graphical flow pattern in Fig. 8.13c. More detailed sectional pictures are given in Jeppson's paper. No attention was paid to the singularities and stagnation regions which occur near the bottom and vertical sides of the dam at the drain end of the flow. In the iterative procedure adopted, the interior values were allowed to settle to one set of boundary conditions before the latter were themselves improved iteratively. A line SOR method was used.

On the face of it, this inverse method calls for more personal judgment in setting up the equations and more computational effort than the

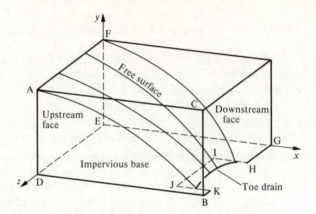

FIG. 8.13(a). Sketch of problem in physical space

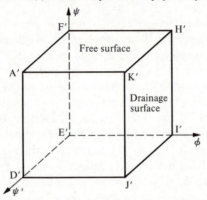

FIG. 8.13(b). Sketch of problem in $\phi\psi\psi'$ space

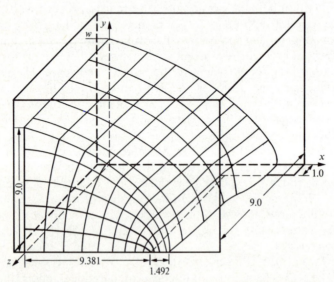

FIG. 8.13(c). Partial isometric plotting of space flow net resulting from solution to problem

single-variable approach of §8.4.2(i). On the other hand, the introduction of the two stream functions ψ and ψ' gives greater flexibility in choosing a suitable ϕ, ψ, ψ' space with all boundaries fixed.

8.5. Variational inequality and complementarity methods

In Chapter 2, a number of free-boundary problems were formulated as elliptic variational inequalities after introduction of the Baiocchi variable. In Chapter 6, moving-boundary problems, many of Stefan type, were expressed as parabolic variational inequalities but discretization of the time derivative led to a succession of elliptic problems. In all cases an equivalent set of differential inequalities can be stated which define an infinite-dimensional linear complementarity problem. When this problem or the variational statement is approximated by finite differences or finite elements, a finite-dimensional linear complementarity problem results which is of the form

$$\mathbf{A}\mathbf{x} \geqslant \mathbf{b}, \qquad \mathbf{x}^{\mathrm{T}}(\mathbf{A}\mathbf{x} - \mathbf{b}) = 0, \qquad \mathbf{x} \geqslant 0, \qquad (8.84)$$

where $\mathbf{x} = (x_1, x_2, \ldots, x_n)^{\mathrm{T}}$ is to be determined, $\mathbf{b}$ is a known n-vector, and $\mathbf{A}$ a known $n \times n$ matrix. The precise form of $\mathbf{A}$ and $\mathbf{b}$ depends on the particular problem and examples are given later in this chapter.

Many methods are available for solving linear complementarity problems (Balinski and Cottle 1978). Cryer (1980), however, pointed out that a linear complementarity statement of a free-boundary problem has special features which prompt a study of special methods of solution. The matrix $\mathbf{A}$ is very large and is sparse with a regular structure, and the solution $\mathbf{x}$ has a large percentage of non-zero terms. Various methods of solving linear equations of the form $\mathbf{A}\mathbf{x} = \mathbf{b}$ associated with fixed boundary-value problems for elliptic equations have been very well studied. Cryer (1980) described various ways of adapting the successive over-relaxation or SOR method to solve the linear complementarity problem (8.84).

The point SOR method for solving the linear equations

$$\mathbf{A}\mathbf{x} = \mathbf{b}$$

starts from an initial guess $\mathbf{x}^{(0)} = (x_i^{(0)})$ and generates a sequence of approximations $\mathbf{x}^{(k)} = (x_i^k)$ in the following steps:

$$\left. \begin{array}{l} \text{(i)} \quad r_i^{(k)} = b_i - \sum_{j<i} a_{ij} x_j^{(k+1)} - \sum_{j \geqslant i} a_{ij} x_j^{(k)}, \\[2mm] \text{(ii)} \quad x_i^{(k+\frac{1}{2})} = x_i^{(k)} + r_i^{(k)}/a_{ii}, \\[2mm] \text{(iii)} \quad x_i^{(k+1)} = x_i^{(k)} + \omega(x_i^{(k+\frac{1}{2})} - x_i^{(k)}). \end{array} \right\} \qquad (8.85)$$

The over-relaxation parameter ω is a fixed constant and $\mathbf{A} = (a_{ij})$, $\mathbf{b} = (b_i)$. There is the well-known theorem that if $\mathbf{A}$ is positive definite the point SOR method converges for all initial guesses $\mathbf{x}^{(0)}$ if and only if $0 < \omega < 2$.

A natural extension with the linear complementarity (8.84) in mind is to extend (8.85) to the successive steps

$$\left. \begin{array}{ll} \text{(i)} & r_i^{(k)} = b_i - \sum_{j<i} a_{ij} x_j^{(k+1)} - \sum_{j \geq i} a_{ij} x_j^{(k)}, \\[2mm] \text{(ii)} & x_i^{(k+\frac{1}{3})} = x_i^{(k)} + r_i^{(k)}/a_{ii}, \\[2mm] \text{(iii)} & x_i^{(k+\frac{2}{3})} = x_i^{(k)} + \omega(x_i^{(k+\frac{1}{3})} - x_i^{(k)}), \\[2mm] \text{(iv)} & x_i^{(k+1)} = \max(0, x_i^{(k+\frac{2}{3})}). \end{array} \right\} \qquad (8.86)$$

This is the point SOR with projection which, provided $\mathbf{A}$ is positive definite, is known to converge for all initial guesses $\mathbf{x}^{(0)}$ if and only if $0 < \omega < 2$. The convergence theorem was proved by Cryer (1971a) after the method had been used heuristically for a long time starting with Christopherson and Southwell (1938). The point SOR with projection is often referred to as the Cryer algorithm. A convergence theorem for semi-definite matrices $\mathbf{A}$ was considered by Glowinski (1971) and Eckhardt (1975). When $\omega = 1$ we have in (8.86) a point Gauss–Seidel method which Cryer (1980) equates to the problem of finding $x_i^{(k+1)}$ which minimizes the quadratic form

$$J_i(x_i) = J(x_1^{(k+1)}, x_2^{(k+1)}, \ldots, x_{i-1}^{(k+1)}, x_i, x_{i+1}^{(k)}, \ldots, x_n^{(k)})$$

over the non-negative real numbers x_i, where

$$J(\mathbf{x}) = \tfrac{1}{2}\mathbf{x}^{\mathrm{T}}\mathbf{A}\mathbf{x} - \mathbf{b}^{\mathrm{T}}\mathbf{x}, \qquad (8.87)$$

and gives several references to studies of this algorithm. Mangasarian (1977) gave a generalized point SOR with projection algorithm for the solution of symmetric linear complementarity problems and examined its convergence properties.

In the point SOR method each equation is temporarily satisfied or over satisfied by changing one component of the solution vector. In a block SOR method the equations for a selected group of points or components are solved simultaneously, with other values kept constant for the time being. The corresponding block SOR with projection can be written

$$\left. \begin{array}{ll} \text{(i)} & \mathbf{R}_i^{(k)} = \mathbf{B}_i - \sum_{j<i} \mathbf{A}_{ij} \mathbf{X}_j^{(k+1)} - \sum_{j \geq i} \mathbf{A}_{ij} \mathbf{X}_j^{(k)}, \\[2mm] \text{(ii)} & \mathbf{A}_{ii} \mathbf{X}_i^{(k+\frac{1}{3})} = \mathbf{A}_{ii} \mathbf{X}_i^{(k)} + \mathbf{R}_i^{(k)}, \\[2mm] \text{(iii)} & \mathbf{X}_i^{(k+\frac{2}{3})} = \mathbf{X}_i^{(k)} + \omega(\mathbf{X}_i^{(k+\frac{1}{3})} - \mathbf{X}_i^{(k)}), \\[2mm] \text{(iv)} & \mathbf{X}_i^{(k+1)} = P_i(\mathbf{X}_i^{(k+\frac{2}{3})}), \end{array} \right\} \qquad (8.88)$$

where the positive-definite matrix $\mathbf{A}$ is partitioned into $\mathbf{A} = (\mathbf{A}_{ij})$, where $\mathbf{A}_{ii}$ is square; $\mathbf{x} = (\mathbf{X}_i)$ and $\mathbf{b} = (\mathbf{B}_i)$ are corresponding partitions of $\mathbf{x}$ and $\mathbf{b}$. In order to define the projection operator P_i a factorization of $R^n = \prod_{i=1}^N R_i^{N_i}$ is used where $\mathbf{A}_{ii}$ is an $N_i \times N_i$ matrix and $\mathbf{B}_i$, $\mathbf{X}_i$ are N_i-vectors, and correspondingly $R_+^n = \prod_i^N R_{+i}^{N_i}$, where R_+^n is the positive orthant. Then P_+ implies R^{N_i} is projected into $R_+^{N_i}$. Cryer (1980) gave different detailed interpretations of P_i and a list of authors who studied them. The standard implementation of the algorithm (8.88) performs steps (i), (ii), and (iii) without imposing non-negativity and finally the projection of step (iv) is performed.

A modified block SOR algorithm due to Cottle (1979), Cottle *et al.* (1978) proceeds in a different order as follows:

(i) Solve the linear complementarity problem

$$\mathbf{A}_{ii}\mathbf{X}_i^{(k+\frac{1}{2})} \geqslant \mathbf{C}_i^{(k)} \equiv \mathbf{B}_i - \sum_{j<i} \mathbf{A}_{ij}\mathbf{X}_j^{(k+1)} - \sum_{j\geqslant i} \mathbf{A}_{ij}\mathbf{X}_j^{(k)},$$

$$\mathbf{X}_i^{(k+\frac{1}{2})} \geqslant 0,$$

$$(\mathbf{X}_i^{(k+\frac{1}{2})})^{\mathrm{T}}(\mathbf{A}_{ii}\mathbf{X}_i^{(k+\frac{1}{2})} - \mathbf{C}_i^{(k)}) = 0, \tag{8.89}$$

(ii) $\mathbf{X}_i^{(k+1)} = \mathbf{X}_i^{(k)} + \omega_i^{(k+1)}(\mathbf{X}_i^{(k+\frac{1}{2})} - \mathbf{X}_i^{(k)}),$

where $\omega_i^{(k+1)} = \max\{\bar{\omega} : 0 < \bar{\omega} < \omega \text{ and } X_i^{(k)} + \bar{\omega}(X^{(k+\frac{1}{2})} - X_i^{(k)}) \geqslant 0\}.$

Cottle *et al.* (1978) proved that for $\mathbf{A}$ positive definite their modified block SOR algorithm converges for $0 < \omega < 2$. Cryer (1980) referred to other authors for earlier work on different ranges of ω. Since the modified algorithm requires the solution of a complementarity problem at every iteration it looks potentially inefficient in general. However, Cottle and Sacher (1977) produced a very efficient algorithm for free-boundary problems in which the matrices $\mathbf{A}_{ii}$ are often tri-diagonal M-matrices, i.e. $a_{ij} \leqslant 0$ for $i \neq j$ and $\mathbf{A}^{-1}$ exists and is non-negative. They demonstrated their algorithm through the following example, which is also reproduced by Cryer (1980).

The linear complementarity problem (8.84) can be conveniently written as

$$\mathbf{w} = \mathbf{A}\mathbf{x} - \mathbf{b}, \qquad \mathbf{w} \geqslant 0,$$

$$\mathbf{x} \geqslant 0, \qquad \mathbf{x}^{\mathrm{T}}\mathbf{w} = 0,$$

and a particular example arising from an elliptic free-boundary problem would have

$$\mathbf{A} = \begin{bmatrix} 2 & -1 & 0 & 0 \\ -1 & 2 & -1 & 0 \\ 0 & -1 & 2 & -1 \\ 0 & 0 & -1 & 2 \end{bmatrix}, \quad \mathbf{b} = \begin{bmatrix} 2 \\ -10 \\ -2 \\ 10 \end{bmatrix}.$$

Start by taking $\mathbf{x}=0$ and $\mathbf{w}=\mathbf{Ax}-\mathbf{b}$ so that the usual linear programming tableau becomes

Slack variables (all zero)

		1	x_1	x_2	x_3	x_4
	$w_1=$	-2	2	-1	0	0
Active	$w_2=$	10	-1	2	-1	0
variable	$w_3=$	2	0	-1	2	-1
	$w_4=$	-10	0	0	-1	2

If w were non-negative already, we should have a solution. But, in fact, $w_1=2x_1-x_2-2=-2<0$ since all slack variables are zero. If x is the solution of the linear complementarity problem then $w_1=2x_1-x_2-2\geqslant0$ so that $x_1\geqslant\frac{1}{2}(x_2+2)>0$. Therefore $x_1>0$ and by the complementarity condition, $w_1=0$ so that the next step is to interchange w_1 and x_1 in the tableau. This produces

	1	w_1	x_2	x_3	x_4
$x_1=$	1	$\frac{1}{2}$	$\frac{1}{2}$	0	0
$w_2=$	9	$-\frac{1}{2}$	$\frac{3}{2}$	-1	0
$w_3=$	2	0	-1	2	-1
$w_4=$	-10	0	0	-1	2

Now w_4 is negative and so we interchange w_4 and x_4 to obtain

	1	w_1	x_2	x_3	w_4
$x_1=$	1	$\frac{1}{2}$	$\frac{1}{2}$	0	0
$w_2=$	9	$-\frac{1}{2}$	$\frac{3}{2}$	-1	0
$w_3=$	-3	0	-1	$\frac{3}{2}$	$-\frac{1}{2}$
$x_4=$	5	0	0	$\frac{1}{2}$	$\frac{1}{2}$

and then finally interchanging w_3 and x_3 leads to

	1	w_1	x_2	w_3	w_4
$x_1=$	1	$\frac{1}{2}$	$\frac{1}{2}$	0	0
$w_2=$	7	$-\frac{1}{2}$	$\frac{5}{6}$	$-\frac{2}{3}$	$-\frac{1}{3}$
$x_3=$	2	0	$\frac{2}{3}$	$\frac{2}{3}$	$\frac{1}{3}$
$x_4=$	6	0	$\frac{1}{3}$	$\frac{1}{3}$	$\frac{2}{3}$

The solution is therefore

$$x_1=1, \qquad x_2=0, \qquad x_3=2, \qquad x_4=6$$
$$w_1=0, \qquad w_2=7, \qquad w_3=w_4=0.$$

Cottle and Sacher (1977) showed that their algorithm terminates in a finite number of steps. Once a component x_i has become active it remains active, so that the number of interchanges cannot exceed n $(i \leq n)$. Also columns corresponding to slack components w_i need not appear because a slack variable never becomes active. Thus only $O(n)$ storage locations and arithmetic operations are needed.

Cryer (1980) remarked that the numerical efficiency of the various algorithms depends strongly on the effort involved in the projection step. This is small in point SOR and the modified block SOR algorithm but substantial in block SOR with projection because the projection is in the A_{ii} norm. He concluded that the latter method is not competitive. On the basis of numerical experiments in problems relating to journal bearings and the dam problem, Cottle (1974) inclined to the view that the modified block SOR algorithm is more efficient than point SOR when a substantial fraction of the components of the solution are zero. Extensive numerical comparisons of iterative methods were reported by Cottle and Goheen (1976). Elliott and Ockendon (1982) substantiate the efficiency of the Cottle fast algorithm in the oxygen consumption problem.

Cryer (1971a) established that the optimum value of the over-relaxation parameter ω is always smaller for the linear complementarity algorithm than for the corresponding linear equation SOR scheme, though the latter value is often used in complementarity problems. In Chapter 6 Elliott's (1976) use of Carre's (1961) technique for estimating an optimum ω is discussed (p. 233).

Cryer (1980) discussed the application of an experimental parallel computer, the pilot DAP (Distributed Array Processor), to the solution of complementarity problems. A red–black chess board ordering of the grid points was found necessary in order that SOR should be suitable for parallel computation.

8.5.1. Finite-difference and finite-element forms of the Cryer algorithm

(i) *Two dimensional.* When finite differences or finite elements are used to discretize the free-boundary problems formulated in Chapter 2, explicit expressions can be substituted for the coefficients a_{ij} and b_{ij} in the Cryer algorithm (8.86). For example, on a mesh with $\Delta x = h_1$ and $\Delta y = h_2$ the standard finite-difference form of $\nabla^2 w = 1$ in (2.35) can be written

$$2w_{ij}\left[\frac{h_1^2 + h_2^2}{h_1^2 h_2^2}\right] = \frac{1}{h_1^2}(w_{i-1,j} + w_{i+1,j}) + \frac{1}{h_2^2}(w_{i,j-1} + w_{i,j+1}) - 1.$$

The corresponding form of the SOR with projection algorithm (8.86) slightly recast becomes

$$w_{i,j}^{(k+\frac{1}{2})} = \frac{h_1^2 h_2^2}{2(h_1^2 + h_2^2)} \left[\frac{1}{h_1^2} (w_{i-1,j}^{(k+1)} + w_{i+1,j}^{(k)}) + \frac{1}{h_2^2} (w_{i,j-1}^{(k+1)} + w_{i,j+1}^{(k)}) \right]$$

$$w_{i,j}^{(k+1)} = \max\{0, w_{i,j}^{(k)} + \omega(w_{i,j}^{(k+\frac{1}{2})} - w_{i,j}^{(k)})\}. \tag{8.90}$$

A finite-element discretization of (2.36), minimized with respect to the $w_{i,j}$ with (2.42) in mind, leads to a set of algebraic equations for the values of $w_{i,j}$ at the nodes. If, for example, linear approximating functions are used with rectangular finite elements the Cryer algorithm becomes (Bruch 1980)

$$w_{i,j}^{(k+\frac{1}{2})} = \frac{3}{4} \frac{h_1 h_2}{h_2^2 + h_1^2} \left[\frac{2h_2^2 - h_1^2}{3h_1 h_2} (w_{i+1,j}^{(k+1)} + w_{i-1,j}^{(k)}) + \frac{2h_1^2 - h_2^2}{3h_1 h_2} (w_{i,j+1}^{(k+1)} + w_{i,j-1}^{(k)}) \right.$$

$$\left. + \frac{h_2^2 + h_1^2}{6h_1 h_2} (w_{i+1,j+1}^{(k+1)} + w_{i-1,j+1}^{(k+1)} + w_{i-1,j-1}^{(k)} + w_{i+1,j-1}^{(k)}) - h_1 h_2 \right],$$

$$w_{i,j}^{(k+1)} = \max\{0, w_{i,j}^{(k)} + \omega(w_{i,j}^{(k+\frac{1}{2})} - w_{i,j}^{(k)})\}. \tag{8.91}$$

Results for one simple dam obtained by Baiocchi *et al.* (1973a) using finite differences and the algorithm (8.90) are extracted in Table 8.4 and compared with analytical values quoted by Ockendon and Elliott (1982) derived from the expressions (7.66a,b) (Polubarinova-Kochina, 1962). The finite-difference computations did not include special treatment of the singular separation point and this may be why they are consistently lower than the analytical solution. Ockendon and Elliott (1982) compared finite-element results based on algorithm (8.91) with the corresponding analytical solution for the dam most commonly adopted as a model for

TABLE 8.4

Ordinates of the free surface for simple dam problem: $H = 0.322$, $h = 0.084$, $\ell = 0.162$; $h_1 = \ell/40$, $h_2 = H/60$ in (8.90)

x	Analytic	Finite diff
0	0.3220	0.3220
0.02025	0.3160	0.3112
0.04050	0.3075	0.3005
0.06075	0.2972	0.2951
0.08100	0.2851	0.2790
0.10125	0.2709	0.2683
0.12150	0.2542	0.2468
0.14175	0.2339	0.2307

TABLE 8.5
Ordinates of the free surface for simple dam problem $H = 1$, $h = 1/6$, $\ell = 2/3$

		Finite element size	
$24x$	Analytic	1/24	1/48
0	1.0000	1.0000	1.0000
1	0.9894	0.9869	0.9885
2	0.9755	0.9735	0.9743
3	0.9593	0.9579	0.9586
4	0.9413	0.9378	0.9406
5	0.9215	0.9202	0.9207
6	0.8999	0.8966	0.8992
7	0.8765	0.8747	0.8757
8	0.8513	0.8496	0.8505
9	0.8240	0.8224	0.8235
10	0.7946	0.7934	0.7936
11	0.7628	0.7605	0.7607
12	0.7281	0.7256	0.7281
13	0.6900	0.6879	0.6900
14	0.6472	0.6467	0.6475
15	0.5975	0.6016	0.5932
16	0.5294 (separation point)		

mesh sizes 1/12, 1/24, and 1/48. Their comparisons are reproduced in Table 8.5 for the two finer meshes.

As a sequel to her 1972 paper discussed in §8.3, Aitchison used finite elements and the algorithm (8.91) to solve the simple dam problem formulated as a minimization problem. If the origin of coordinates is moved to the separation point and one term in the infinite series used, the equation of the free surface is $y = (x/\pi)\ln(-x) + \alpha x + \beta$, where the constants α, β are to be determined.

TABLE 8.6
Effect of fitting through r points

Mesh size 1/60		Mesh size 1/120	
r	y_n	r	y_n
2	0.5426	10	0.5294
3	0.5408	11	0.5302
4	0.5346	12	0.5294
5	0.5306	13	0.5290
6	0.5302	14	0.5289
7	0.5305	15	0.5290
8	0.5308	16	0.5290

TABLE 8.7

Comparison of boundary ordinates

x	(a)	(b)	(c)
0.50000	0.7333	0.7250	0.7282
0.5333	0.7000	0.7000	0.6980
0.5667	0.6667	0.6667	0.6651
0.6000	0.6304	0.6300	0.6286
0.6333	0.5876	0.5868	0.5866
0.6667	0.5302	0.5289	0.5300

(a), (b) minimization method with mesh sizes
1/60, 1/20 respectively (Aitchison 1977).
(c) trial free-boundary iterative method with
mesh size 1/60 (Aitchison 1972).

Aitchison (1977) quoted preliminary numerical values of free-boundary
ordinates before smoothing and illustrated the stepwise nature of the
whole boundary. Then a least-squares straight line was fitted through
points $\bar{y}_i$, where $\bar{y} = y - (x/\pi)\ln(-x)$. Table 8.6 shows the effect of the
number of points included in the fitting on y_n the ordinate of the
separation point. The best straight line is also used to improve internal
free-boundary ordinate values within the range of the fitting.

Table 8.7 compares boundary ordinates for $0.5000 \leqslant x \leqslant 0.6667$ obtained by Aitchison's minimization method (1977) and her iterative
method (1972) coupled with the singular solution in each case. The
earlier computations were revised and extended to include a smaller grid
size.

Tables 8.8–17, 19, 20 contain numerical results extracted from tables

TABLE 8.8

*Ordinates $f_2(x)$ of the sea-
water boundary for a coastal
aquifer (§2.5). Length of
aquifer = a = 1000; height of
aquifer above sea level = y_1 =
2, $\rho = (\rho_f - \rho_s)/\rho_f = -0.03$,
$f_2(0) = y_1/\rho = -66.66$.   Mesh
points 80×60*

x	$-f_2(x)$	x	$-f_2(x)$
0	66.66	700	36.91
100	63.23	750	33.47
200	59.80	800	30.04
300	56.36	850	25.46
400	51.78	900	20.88
500	47.21	950	15.16
600	42.63	987	8.30

TABLE 8.9

Ordinates of free boundary for vertically stratified dam: $y_1 = 1.33$, $y_2 = 0$. *Permeability coefficients* $k' = 0.197$ *in* $0 \leq x < 1$, $k'' = 1.46$ *in* $1 < x \leq 2$ (§2.8.1); *mesh points* 60×100 *in each half*

x	y	x	y
0.00	1.330	1.00	0.611
0.10	1.263	1.10	0.452
0.20	1.230	1.20	0.412
0.30	1.197	1.30	0.385
0.40	1.130	1.40	0.332
0.50	1.097	1.50	0.299
0.60	1.030	1.60	0.268
0.70	0.984	1.70	0.232
0.80	0.904	1.80	0.199
0.90	0.811	1.90	0.133
1.00	0.611	1.95	0.099

TABLE 8.10

Ordinates of free boundary for vertically stratified dam: $y_1 = 1.33$, $y_2 = 0$. *Permeability coefficients* $k' = 1.46$ *in* $0 \leq x < 1$, $k'' = 0.197$ *in* $1 < x \leq 2$ (§2.8.1); *mesh points* 20×40 *in each half*

x	y	x	y
0	1.333	1.0	1.230
0.1	1.296	1.1	1.197
0.2	1.296	1.2	1.163
0.3	1.263	1.3	1.130
0.4	1.263	1.4	1.064
0.5	1.263	1.5	0.997
0.6	1.263	1.6	0.931
0.7	1.263	1.7	0.864
0.8	1.230	1.8	0.764
0.9	1.230	1.9	0.665
1.0	1.230	1.95	0.598

TABLE 8.11

Ordinates of free boundary for horizontally stratified dam: $y_1 = 0.721$, $y_2 = 0.253\,95$, $a = 1.0$. *Permeability coefficients* $k' = 0.0296$ *in* $0 \leq y < 0.496$, $k'' = 0.737$ *in* $0.496 < y \leq 0.721$ (§2.8.1); *Mesh points* 40×20 *in each part*

x	y	x	y
0	0.721	0.55	0.631
0.050	0.709	0.60	0.619
0.10	0.698	0.65	0.608
0.15	0.687	0.70	0.597
0.20	0.687	0.75	0.586
0.25	0.676	0.80	0.574
0.30	0.664	0.85	0.563
0.35	0.664	0.90	0.541
0.40	0.653	0.95	0.529
0.45	0.642	0.975	0.518
0.50	0.631		

TABLE 8.12

Ordinates of free boundary for horizontally stratified dam: $y_1 = 0.814$, $y_2 = 0.128\,88$, $a = 1.0$. *Permeability coefficients* $k' = 0.737$ *in* $0 \leq y < 0.323$, $k'' = 0.0296$ *in* $0.323 < y \leq 0.814$ (§2.8.1); *mesh points* 40×20 *in lower part*; 40×40 *in upper part*

x	y	x	y
0.00	0.814	0.55	0.531
0.05	0.789	0.60	0.494
0.10	0.777	0.65	0.458
0.15	0.752	0.70	0.421
0.20	0.728	0.75	0.372
0.25	0.703	0.80	0.323
0.30	0.678	0.85	0.290
0.35	0.654	0.90	0.274
0.40	0.629	0.95	0.226
0.45	0.593	0.975	0.209
0.50	0.568		

presented by Baiocchi *et al.* (1973*a*) and obtained by using the Cryer algorithm (8.90).

(ii) *Three dimensional.* Caffrey and Bruch (1979) extended the finite-difference version of the Cryer algorithm (8.90) to three dimensions using 7-point formulae for the Laplacian operator. Denoting by $w_{i,j,k}$ the value of $w(x, y, z)$ at the grid point $(i \, \Delta x, \, j \, \Delta y, \, k \, \Delta z)$ the SOR scheme with projection becomes

$$w_{i,j,k}^{(n+\frac{1}{2})} = \frac{1}{2\Delta} \left\{ \frac{1}{\Delta x^2} \left[w_{i-1,j,k}^{(n)} + w_{i+1,j,k}^{(n+1)} \right] + \frac{1}{\Delta y^2} \left[w_{i,j-1,k}^{(n+1)} + w_{i,j+1,k}^{(n)} \right] \right.$$

$$\left. + \frac{1}{\Delta z^2} \left[w_{i,j,k-1}^{(n)} + w_{i,j,k+1}^{(n+1)} \right] - 1 \right\},$$

$$w_{i,j,k}^{(n+1)} = \max[0, \, w_{i,j,k}^{(n)} + \omega(w_{i,j,k}^{(n+\frac{1}{2})} - w_{i,j,k}^{(n)})], \qquad (8.92)$$

where $\Delta = 1/\Delta x^2 + 1/\Delta y^2 + 1/\Delta z^2$.

The finite-element discretization can be applied to the minimum formulation of the variational inequality (2.86) for example. It is required to find w such that $J(w) \leq J(v)$ for all v where

$$J(v) = \frac{1}{2} \iiint_D \left\{ \left(\frac{\partial v}{\partial x} \right)^2 + \left(\frac{\partial v}{\partial y} \right)^2 + \left(\frac{\partial v}{\partial z} \right)^2 + v \right\} \mathrm{d}x \, \mathrm{d}y \, \mathrm{d}z. \qquad (8.93)$$

In an element of the mesh, w is defined as usual by

$$w = \hat{w}(x, y, z) = [N]\{w\}^e = [N_i, N_j, \ldots, N_p]\{w\}^e \qquad (8.94)$$

where $N_i, N_j, \ldots, N_p$ are piecewise linear shape functions, which can be expressed in terms of the nodal values of w. Insertion of the approximation (8.94) for v and minimizing gives typically

$$\iiint_D \left(\frac{\partial N_I}{\partial x} \frac{\partial \hat{w}}{\partial x} + \frac{\partial N_I}{\partial y} \frac{\partial \hat{w}}{\partial y} + \frac{\partial N_I}{\partial z} \frac{\partial \hat{w}}{\partial z} \right) + N_I \right) \mathrm{d}x \, \mathrm{d}y \, \mathrm{d}z = 0. \qquad (8.95)$$

A typical finite-element used by Caffrey and Bruch (1979) in the form of a rectangular prism is shown in Fig. 8.14. It includes 26 neighbouring nodes to approximate w at any node. For convenience the following nomenclature is introduced:

$$a = 2/\Delta x, \quad b = 2/\Delta y, \quad c = 2/\Delta z, \quad 18abc = 1/V,$$

$$E_1 = -(a^2 + b^2 + c^2)V, \quad E_8 = -32E_1;$$

$$E_2 = (-4a^2 - 4b^2 + 2c^2)V, \quad E_3 = (-4a^2 + 2b^2 - 4c^2)V,$$

$$E_4 = (2a^2 - 4b^2 - 4c^2)V; \quad E_5 = (-16a^2 + 8b^2 + 8c^2)V,$$

$$E_6 = (8a^2 - 16b^2 + 8c^2)V, \quad E_7 = (8a^2 + 8b^2 - 16c^2)V.$$

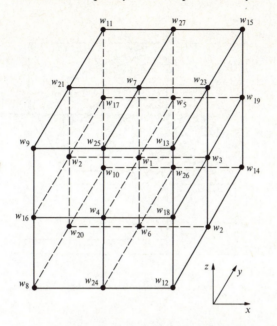

FIG. 8.14. Three-dimensional finite element molecule

In Fig. 8.14 we identify $w_{i,j,k} = w_1$, $w_{i-1,j,k} = w_2$, $w_{i-1,j-1,k-1} = w_8$, etc. so that the SOR with projection can be written

$$w_1^{(n+\frac{1}{2})} = -E_8\{E_1[w_8^{(n)} + w_9^{(n)} + w_{10}^{(n)} + w_{11}^{(n+1)} + w_{12}^{(n)} + w_{13}^{(n+1)} + w_{14}^{(n)}$$
$$+ w_{15}^{(n+1)}]$$
$$+ E_2[w_{16}^{(n)} + w_{17}^{(n+1)} + w_{18}^{(n+1)} + w_{19}^{(n)}]$$
$$+ E_3[w_{20}^{(n)} + w_{21}^{(n+1)} + w_{22}^{(n)} + w_{23}^{(n+1)}]$$
$$+ E_4[w_{24}^{(n)} + w_{25}^{(n+1)} + w_{26}^{(n)} + w_{27}^{(n+1)}] + E_5[w_2^{(n)} - w_3^{(n+1)}]$$
$$+ E_6[w_4^{(n)} + w_5^{(n+1)}] + E_7[w_6^{(n)} + w_7^{(n+1)}] + 8/(abc)\},$$
$$w_{i,j,k}^{(n+1)} = \max[0, w_{i,j,k}^{(n)} + \omega(w_{i,j,k}^{(n+\frac{1}{2})} - w_{i,j,k}^{(n)})]. \tag{8.96}$$

With regard to the boundary conditions, since the w-values on the impervious bottom of the dams considered by Caffrey and Bruch (1979) are known to satisfy Laplace's equation, the bottom boundary is approximated by an appropriate two-dimensional finite-difference or finite-element discretization. The values of w along the intersections of the bottom and the inlet and exit faces are known Dirichlet data, but along the intersections with the impervious sides a zero normal derivative exists. The fictitious points outside the region are eliminated in the usual way, i.e. they are equated to the corresponding node just inside the boundary. The w-values all over the dam bottom are then used as Dirichlet data for

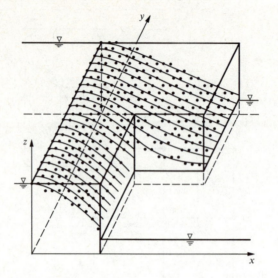

FIG. 8.15. Free surface location for first dam configuration

the internal computations throughout the dam. These start at one impervious side, with similar treatment of all fictitious nodes; and similarly on the opposite side at the end of the computation.

The free surface computed in the first dam studied by Caffrey and Bruch (1979) is shown in Fig. 8.15. The details of the dam are shown in Fig. 8.16a. Calculations by finite differences and finite elements showed only minor differences. A mesh size of $\Delta x = \Delta y = \Delta z = 1$ m gave a total of 2541 nodes and to satisfy a convergence criterion

$$\max_{ijk} |w_{i,j,k}^{(n+1)} - w_{i,j,k}^{(n)}| < 0.001,$$

45 iterations and a computer time of 7.74 seconds were needed with finite differences compared with 46 iterations in 15.62 seconds using finite elements; in both cases compilation and linkage editing time were excluded. In both cases the relaxation factor ω was 1.75. Convergence and accuracy were examined numerically.

The second dam (Fig. 8.16b) was studied only by finite differences with grid spacings $\Delta x = 20$ m, $\Delta y = 3$ m, $\Delta z = 60$ m, and 15 708 nodes. To satisfy a convergence criterion of 1.00 required 131 iterations in 83.99 seconds with a relaxation factor of 1.75 again. Caffrey and Bruch (1979) made comparisons with results for a similar but not identical dam obtained by France (1974) using finite elements.

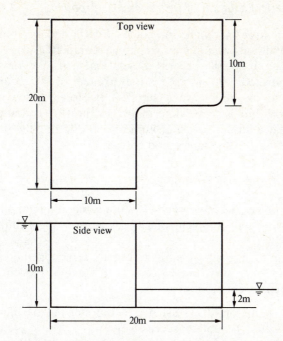

FIG. 8.16(a). Three-dimensional dam, first numerical example

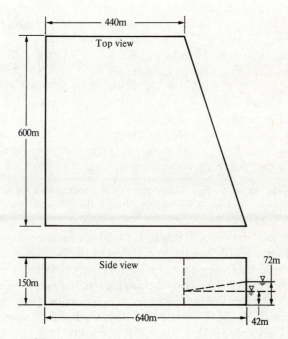

FIG. 8.16(b). Three-dimensional dam, second numerical example

(iii) *Compatability condition.* In some free-boundary problems the boundary conditions contain an unknown parameter which has to be determined as part of the process of solution such that some additional criterion is satisfied. An example is the problem of the rectangular dam with a sheetpile on the inlet face (see §2.3.2). Here the flow rate q is not known *a priori*, but is to be determined so that the condition (2.57) is satisfied by the solution. An outer iteration on q to find the root of (2.57) can be based on the secant method so that

$$q^{(r+1)} = q^{(r)} - \frac{q^{(r)} - q^{(r-1)}}{f_h(q^{(r)}) - f_h(q^{(r-1)})} f_h(q^{(r)}). \qquad (8.97)$$

Baiocchi *et al.* (1973*a*) and Bruch (1980) described a method of combining (8.97) with the complementarity projection algorithms (8.90) or (8.91).

An initial guess for $w_{q^{(0)},i,j}$ is made, usually zero, together with a guess at $q = q^{(0)}$. In fact, two initial guesses, $q^{(0)}$ and $q^{(1)}$, will be needed. Successive steps are as follows:

(i) Solve (8.90) or (8.91) using $q^{(0)}$ and starting the inner iteration with $w_{q^{(0)},i,j}^{(0)}$ to obtain successively $w_{q^{(0)},i,j}^{(1)}$, $w_{q^{(0)},i,j}^{(2)}$, etc. till some prescribed convergence criterion is satisfied. Then $f_h(q^{(0)})$ is evaluated from (2.57) by using $w_{q^{(0)},1,G}$ and $w_{q^{(0)},0,G}$ and its convergence criterion examined. As in the derivation of (2.57), these are values of w at $x = \Delta x$ and $x = 0$ on the horizontal mesh line through G in Fig. 2.3 obtained with estimate $g^{(0)}$.

(ii) Using $w_{q^{(0)},i,j}$ as the initial guess for $w_{q^{(1)},i,j}^{(0)}$ solve (8.90) or (8.91) again but using the second guessed value $q^{(1)}$ which is to be larger than $q^{(0)}$. The solution $w_{q^{(1)},i,j}$ is used now to evaluate $f_h(q^{(1)})$ from (2.57) and insertion of $q^{(0)}$, $q^{(1)}$, $f_h(q^{(0)})$ and $f_h(q^{(1)})$ into (8.97) yields a further estimate, $q^{(2)}$.

(iii) The scheme is repeated starting with $q^{(2)}$ and $w_{q^{(1)},i,j}$ to obtain $w_{q^{(2)},i,j}$ and $f_h(q^{(2)})$ and so on till the convergence criteria for both inner and outer iterations are satisfied.

Baiocchi *et al.* (1973*a*) used this double-iterative scheme with the finite-difference algorithm (8.90) to obtain the boundary ordinates for a dam with a sheetpile having the specifications set out in Table 8.13. The iteration details are given in Table 8.14. Similar information is presented by Baiocchi *et al.* (1973*a*) for two different dams. Tables 8.15 and 8.16 give extracted results for the case of relatively smallest sheetpile. Solutions for both dams were also obtained by a trial free-boundary method which took a longer computing time to achieve the same accuracy.

Bruch and Caffrey (1979) used the compatability condition and an SOR iteration to obtain both finite-difference and finite-element solutions

TABLE 8.13

Ordinates of the free surface for a dam with a sheetpile on the inlet face: $y_1 = 4$, $y_2 = 1.5$, $a = 10$, $c = 0.1$ (§2.3.2); mesh points 60×60

x	y	x	y
0.0	2.666	5.0	2.200
0.5	2.666	5.5	2.133
1.0	2.600	6.0	2.066
1.5	2.600	6.5	2.000
2.0	2.533	7.0	1.933
2.5	2.533	7.5	1.866
3.0	2.466	8.0	1.800
3.5	2.400	8.5	1.733
4.0	2.333	9.0	1.666
4.5	2.266	9.5	1.533
5.0	2.200	9.833	1.466

TABLE 8.14
Iteration details for Table 8.13

Outer iteration	$q^{(r)}$	$f_h(q^{(r)})$	Inner iteration SOR
0	0.025	0.134 43	68
1	0.0384	0.126 77	51
2	0.2597	0.015 01	89
3	0.2894	0.001 24	88
4	0.2921[a]	0.000 009	62

[a] The authors quote 1 per cent agreement with earlier values obtained by Maione and Franzeiti (1969) using experimental and approximate analytic methods.

TABLE 8.15

Ordinates of the free surface for a dam with a sheetpile on the inlet face: $y_1 = 1.5$, $y_2 = 0.3$, $a = 1$, $c = 0.5$ (§2.3.2); mesh points 40×60

x	y	x	y
0.00	1.050	0.50	0.950
0.05	1.050	0.55	0.925
0.10	1.050	0.60	0.900
0.15	1.025	0.65	0.875
0.20	1.025	0.70	0.850
0.25	1.025	0.75	0.825
0.30	1.000	0.80	0.775
0.35	1.000	0.85	0.725
0.40	0.975	0.90	0.700
0.45	0.975	0.95	0.625
0.50	0.950	0.975	0.600

TABLE 8.16
Iteration details for Table 8.15

Outer iteration	$q^{(r)}$	$f_h(q^{(r)})$	Inner iteration SOR
0	0.58	0.030 1	74
1	0.61	0.025 2	61
2	0.7653	0.002 0	98
3	0.7792	0.000 10	77
4	0.7799	0.000 00035	43

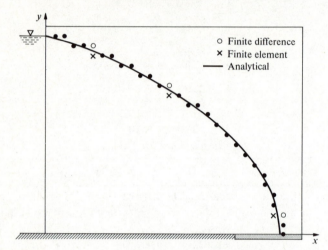

FIG. 8.17(a). Comparison between analytical and numerical results for a dam without an impervious sheetpile

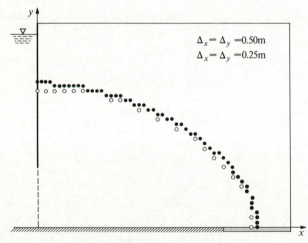

FIG. 8.17(b). Comparison between numerical results for a dam with an impervious sheet-pile along part of the reservoir face

TABLE 8.17

Ordinates of the free surface for a dam with sloping inlet face (§2.3.4): $y_1 = 1$, $y_2 = 0$, $a_1 = 1$, $a = 3$; square mesh size 0.05

x	y^a	y^{*a}	x	y	y^*
0.0	1.00	1.00	1.0	0.65	0.67
0.1	0.95	0.95	1.1	0.60	0.64
0.2	0.90	0.92	1.2	0.60	0.60
0.3	0.85	0.88	1.3	0.55	0.57
0.4	0.85	0.85	1.4	0.50	0.53
0.5	0.80	0.82	1.5	0.45	0.48
0.6	0.80	0.80	1.6	0.40	0.43
0.7	0.75	0.77	1.7	0.35	0.38
0.8	0.70	0.74	1.8	0.30	0.32
0.9	0.70	0.70	1.9	0.25	0.23
1.0	0.65	0.67	1.95	0.20	0.17

[a] y^* computed by a trial free-boundary method; y by complementarity with compatibility iteration.

for a dam with a sheetpile and a toe drain (see §2.3.3). Where there is a toe drain but no sheetpile the compatability condition is not required and there is an analytical solution (Harr, 1962, p. 223). In Fig. 8.17a Bruch and Caffrey compared their three solutions for the toe drain without a sheetpile and Fig. 8.17b shows the effect of introducing a sheetpile.

Problems in dams of some other important geometric shapes involve the compatability condition. The dam with a slanting inlet face (Fig. 2.5) is such a case solved by Baiocchi *et al.* (1973a) using a suitably modified form of the double-iteration cycle and finite differences. An extract of their results is compared in Table 8.17, with values computed by a trial free-boundary method. The latter do not exhibit the stepwise behaviour

TABLE 8.18

Iteration details for dam with sloping inlet face and toe drain (§2.3.5): $a_1/y_1 = 1.0$, $a/y_1 = 3.16$; square mesh size $\Delta x/y_1 = \Delta y/y_1 = 0.05$

Outer iteration	$q^{(r)}$	$f_h(q^{(r)})$	Inner iteration SOR
0	0.230 00	0.016 79	77
1	0.250 00	0.028 98	54
2	0.202 47	0.000 79	63
3	0.201 15	0.000 12	24

TABLE 8.19
Ordinates of the free surface for a dam with a downward sloping base (§2.3.6): $y_1 = 2$, $y_2 = 1.4$, $y_3 = 1$, $a = 2$; *mesh sizes* $\Delta x = 0.05$, $\Delta y = 0.025$. *After 5 iterations of the secant method* $q^{(5)} = 0.323$

x	y	x	y
0.0	2.000	1.0	1.700
0.1	1.950	1.1	1.675
0.2	1.925	1.2	1.650
0.3	1.900	1.3	1.625
0.4	1.875	1.4	1.575
0.5	1.850	1.5	1.550
0.6	1.825	1.6	1.525
0.7	1.800	1.7	1.500
0.8	1.775	1.8	1.450
0.9	1.725	1.9	1.425
1.0	1.700	1.95	1.400

TABLE 8.20
Ordinates of the free surface for a dam with an upwards sloping base (§2.3.6): $y_1 = 2$, $y_2 = 1.3$, $y_3 = 1$, $a = 1$; *square mesh size* 0.025. *After 5 iterations of the secant method* $q^{(5)} = 0.635$

x	y	x	y
0.00	2.000	0.50	1.800
0.05	1.955	0.55	1.775
0.10	1.950	0.60	1.750
0.15	1.950	0.65	1.725
0.20	1.925	0.70	1.700
0.25	1.900	0.75	1.650
0.30	1.875	0.80	1.625
0.35	1.875	0.85	1.600
0.40	1.850	0.90	1.550
0.45	1.825	0.95	1.500
0.50	1.800	0.975	1.475

of the variational solution. Corresponding results were obtained by the same method by Sloss and Bruch (1978) for two different dams each with a toe drain. They compared finite-difference solutions in the one case with an analytic solution by Polubarinova-Kochina (1962) and in the other case with previous results by Jeppson (1968c) and Shaug and Bruch (1976) obtained by a different numerical method. Table 8.18 shows iteration details for the first case.

As final examples of the use of the double-iterative scheme incorporating a compatability condition we give in Tables 8.19 and 8.20 free-boundary positions for two dams with sloping bases (see §2.3.6) extracted from the results of Baiocchi *et al.* (1973a).

8.5.2. Axisymmetric problems

Relatively few solutions seem to have been obtained for cases of two-dimensional axisymmetric porous flow. Reference was made in §2.7 to the work of Cryer and Fetter (1977). Their theoretical analysis of an axisymmetric well referred to a variable $k(x, y) = xK$, where K is the permeability of the soil and $k(x, y) = \exp[f(x) + g(y)]$, but their numerical example took $K = 1$, i.e. $f(x) = \ln x$ and $g(y) = 0$. Their solution for a

Baiocchi variable w required the minimization of

$$J(w) = \int_D x(w_x^2 + w_y^2 + 2w)\, dx\, dy,$$

subject to boundary conditions discussed in §2.7. Linear finite elements on a triangulated domain were used in which, because the solution varies most rapidly near the well, the subdivisions were taken to be uniform in the y-direction and logarithmic in the x-direction. Thus the coordinates of the grid points were

$$y_j = jH/m, \qquad 0 \leq j \leq m; \qquad x_i = r \exp[(i/n)\ln(R/r)], \qquad 0 \leq i \leq n.$$

The integer m was always chosen to be a multiple of 4 so that the corner D was a grid point where w is not smooth. The Cryer algorithm was used to solve the discretized complementarity problem. The numerical example quoted by Cryer and Fetter (1977) referred to a well of radius $r = 4.8$ sunk in a soil of depth $H = 48$ and radius $R = 76.8$; the water level in the well is maintained by pumping at a constant height $h_w = 12$ and the free boundary meets the wall of the well at a height h_s. Table 8.21a, extracted from data given by Cryer and Fetter for a coarse mesh with $m = 4$, $n = 6$, shows values of the Baiocchi variable w. The position of the free boundary is shown by the lowest zero in each column. Because the approximate solution is identically zero on the vertical line $x = r$, the height h_s at which the boundary meets the well cannot be determined directly. Cryer and Fetter took the ordinate of the free boundary on the vertical gridline adjacent to the well as an approximation to h_s, i.e. in Table 8.21a, $h_s = 36$. Values of h_s obtained from finer grids are given in Table 8.21b and results of other authors using different methods are in Table 8.21c.

Elliott (1976) used linear finite elements and a rectangular triangular mesh with equal spacing in the x and y directions to solve the same minimization problem for a different well with parameters: $r = 1/25$, $H = 1$, $R = 21/25$, and $h_w = 1/5$. His free boundary ordinates, smoothed by quadratic extrapolation on the w-values, are reproduced in Table 8.22.

The split field method described in §2.3.8 was pioneered by Remar *et al.* (1982) to handle seepage problems from an arbitrarily shaped axisymmetric pond into a porous medium with a drain at a finite depth. The central part of the bottom of the pond must be horizontal as in Fig. 2.14a. The regions Ω_ψ and Ω_w are each extended so that there is an overlap zone of one mesh length, $A''A'B'B''$. Introducing $\Omega = \Omega_\psi + \Omega_w$ and $\Omega_R = \Omega_w + \Omega_{\text{ext}} + \gamma_{\text{FC}}$, where γ_{FC} is the free boundary, a Baiocchi variable

TABLE 8.21a

Solution of well problem for $m = 4$, $n = 6$ (§2.7): $r = 4.8$, $H = 48$, $R = 76.8$, $h_w = 12$; values of Baiocchi variable w are tabulated

				x			
y	4.80	7.62	12.10	19.20	30.48	48.38	76.80
48	0	0	0	0	0	0	0
36	0	0	0	0	3.0	31.6	72.0
24	0	17.6	43.6	81.3	133.1	204.1	288.0
12	0	88.6	182.0	283.1	394.4	517.1	648.0
0	72.0	252.0	432.0	612.0	792.0	972.0	1152.0

TABLE 8.21b

Values of h_s for decreasing grid lengths (see Table 8.21a)

m	n	h_s	Number of iterations
4	6	36	20
8	12	30	40
16	24	30	70
32	48	30	120
64	96	30	230

TABLE 8.21c

Other computed values for h_s (see Table 8.21a)

Author	Method[a]	h_s
Hall (1955)	TFB: FD	34
Taylor and Luthin (1969)	TFB: FD	34
Neuman and Witherspoon (1970)	TFB: FE	30
Neuman and Witherspoon (1971)	Time-dependent FD	30

[a] TFB = trial free boundary; FD = finite differences; FE = finite elements.

$w(r, z)$ is defined by

$$w(r, z) = \int_{FP} \{-(\tilde{\phi} - z)\, dz - (\tilde{\psi}/r)\, dr\},$$

where $\tilde{\phi} = \phi$ in $\bar{\Omega}$, $\tilde{\phi} = z$ in $\bar{\Omega}_{\text{ext}}$, and $\tilde{\psi} = \psi$ in $\bar{\Omega}$, $\tilde{\psi} = 0$ in Ω_{ext}. Within the Ω_R region w-values are determined on a suitable mesh by a successive

TABLE 8.22

Ordinates of free surface for axisym-
metric well for different mesh sizes Δx;
$r = 1/25$, $H = 1$, $R = 21/25$, $h_w = 1/5$,
$x = 1/25 + i/20$

i	$\Delta x = 1/10$	$\Delta x = 1/20$	$\Delta x = 1/40$
1		8212	8158
2	8456	8394	8406
3		8610	8620
4	8719	8739	8793
5		8931	8943
6	9056	9076	9084
7		9198	9213
8	9292	9290	9329
9		9427	9435
10	9506	9533	9540
11		9619	9629
12	9658	9699	9719
13		9742	9797
14	9855	9849	9865
15		9934	9931

over-relaxation scheme with projection, sweeping to the left along mesh lines and downwards, starting from the upper right-hand corner. In the Ω_ψ region, a successive over-relaxation scheme determines ψ.

To deal with the overlap zone, the Ω_w region is swept to its last internal vertical mesh line, $i = M$ in Fig. 2.14b, and values of ψ are then calculated along this line from the relationship $\psi = -rw_r$, suitably approximated by

$$\psi_{M,j} = \tfrac{1}{2} M(3w_{M,j} - 4w_{M+1,j} + w_{M+2,j}).$$

These $\psi_{M,j}$ values provide the Dirichlet boundary data for the Ω_ψ region in which the SOR solution for ψ can now proceed. Values of w at $i = M - 1$, the boundary of the Ω_w region, are then required for the next stage of the iteration cycle, the updating of w-values by SOR with projection in Ω_w. In the definition of w given above, an integration path is chosen which starts vertically downwards from F to the line $j \Delta z$ being evaluated and continues horizontally to the point $P((M-1) \Delta r, j \Delta z)$, leading to the approximation

$$w_{M-1,j} = w_{M,j} + \frac{1}{2} \left(\frac{\psi_{M-1,j}}{M-1} + \frac{\psi_{M,j}}{M} \right).$$

The discharge rate q is not known *a priori* but has to be determined so that a compatability condition $f_h(q) = 0$ is satisfied where $f_h(q) = (w_q)_{F-h_2} - \frac{1}{2}h_2^2$ and $F - h_2$ denotes the first mesh point vertically below F. The underlying physical assumption is that the velocity of flow is assumed to be zero at the point $F - \frac{1}{2}h_2$. The iterative cycle is performed for two estimated values of q and successive estimates are based on the secant method. The method will not converge if $f_h(q)$ becomes negative. Special formulae for calculating w on the bottom of the pond, the choice of an optimum relaxation parameter and the influence of the singularity on the axis $r = 0$ are discussed by Remar *et al.* (1982). They present numerical results for several examples and favourable graphical comparisons with results obtained by Jeppson (1968*a,c*).

8.6. A general algorithm for partially unsaturated flow

Alt (1979, 1980*a,b*) proposed and used a new algorithm to be applied to saturated–unsaturated flow in a medium of any shape which can be inhomogeneous and anisotropic for any boundary conditions. We give here an outline of his formulation on which the algorithm is based. More details of the mathematical properties, theorems, and conditions are given in Alt's papers.

The boundary $\partial\Omega$ of the porous medium Ω, shown in Fig. 8.18, has three parts: S^0 is the boundary with the atmosphere, S^+ is in contact with water, and $\partial\Omega - S^0 - S^+$ is impervious. The boundary data are given by a function u^0 which is non-negative and together with its first and second derivatives square summable on Ω. The pressure is normalized so that the capillary pressure is zero (§2.10), i.e. the atmospheric pressure may be strictly positive. The permeability of the anisotropic, inhomogeneous medium is denoted by $a > 0$, a matrix function, strictly elliptic. In the particular discretized formulation later, isotropy is assumed. Alt (1979,

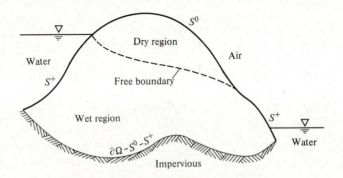

FIG. 8.18. Arbitrary-shaped porous medium Ω

1980a,b) established that there are functions u and γ with $\chi\{(u>0)\}\leqslant$ $\gamma\leqslant1$ which satisfy the inequality

$$\int_{\Omega}\nabla(v-u)a(\nabla u+\gamma\mathbf{e})\geqslant0 \tag{8.98}$$

for all functions v where u and v are square summable together with their first and second derivatives, $v=u^0$ on S^+, $v\leqslant u^0$ on S^0, χ is the characteristic function, i.e. $\chi=1$, $u>0$, $\chi=0$, $u=0$, and $\mathbf{e}$ is the vertical unit vector in the direction of gravity. Alt (1980a) extended (8.98) to include a prescribed flux function on part of the boundary to allow for a leaky region in the base of a dam, for example. The assumption is that unsaturated flow is controlled by an extension of Darcy's law, i.e. $\mathbf{q}=-a(\nabla u+\gamma\mathbf{e})$ so that the inequality (8.98) is equivalent to

$$\nabla\cdot[a(\nabla u+\gamma\mathbf{e})]=0 \quad\text{in }\Omega, \qquad u=u^0 \quad\text{on }S^+,$$

$$(\nabla u+\gamma e)\cdot\mathbf{v}=0 \quad\text{on }\partial\Omega-S^0-S^+,$$

$$u\leqslant u_0, \qquad (\nabla u+\gamma e)\cdot\mathbf{v}\leqslant0 \quad\text{on }S^0,$$

$$(u-u^0)(\nabla u+\gamma e)\cdot\mathbf{v}=0 \quad\text{on }S^0,$$

where $\mathbf{v}$ is the unit outward normal to the relevant surface. The function γ is here a non-linear function of the degree of saturation, which is characteristic of the porous medium (Bear 1972, p. 496). The case of a capillary fringe is included under $u^0>0$ on S^0.

In a homogeneous medium the solution consists only of saturated regions where $u>0$ and $\gamma=1$ and dry regions where $\gamma=0$. Alt (1979) proved that in these cases the inequality $\nabla\cdot(a\mathbf{e})\geqslant0$ is always satisfied. In the saturated regions the function u is the solution of an elliptic partial differential equation. For general inhomogeneous media, however, the condition $\nabla\cdot(a\mathbf{e})<0$ may hold somewhere and then an unsaturated region may exist in which $0<\gamma<1$. In this region $u=0$ and the flow is determined by the gravitational force and not by the pressure gradient. The function γ is a solution of a first-order equation of flow. Alt (1979) analyses a simple example of an unsaturated region in a horizontally stratified inhomogeneous medium with a flow pattern closely resembling that shown in Fig. 8.26.

The physical significance of the condition $\nabla\cdot(a\mathbf{e})<0$ is that it permits the continuity of flow condition to be satisfied between a saturated and an unsaturated region. Thus suppose the saturated and unsaturated regions lie respectively above and below a horizontal surface on which the permeability changes from a_s in the saturated to a_u in the unsaturated region as in Alt's (1979) example and $a_s<a_u$. The rate of flow to the boundary from above is the vertical component of $a_s\,(\nabla u+\gamma_s\mathbf{e})=q_s$ where

$\gamma_s = 1$. The rate of flow away from the boundary into the unsaturated medium is $a_u \gamma_u \mathbf{e} = q_u$ with $\gamma_u < 1$. Continuity of flow requires $q_s = q_u$, which is possible if $a_u > a_s$ but not otherwise.

For the discretization, functions ϕ_h^i, χ_h^i are defined such that

$$\phi_h^i \in H^{1,2}(\Omega) \cap L^\infty(\Omega), \qquad i \in P_h,$$

are linearly independent with

$$\phi_h^i \geq 0 \quad \text{and} \quad \sum_i \phi_h^i(x) = 1 \quad \text{for} \quad x \in \Omega_h, \tag{8.99}$$

where a sequence of positive numbers h converges to zero and P_h is a finite set, and

$$\chi_h^i \in L^\infty(\Omega), \qquad i \in P_h, \sum_i \chi_h^i(x) = 1 \quad \text{for} \quad x \in \Omega_h, \tag{8.100}$$

are characteristic functions. The corresponding finite-dimensional spaces of functions are

$$H_h = \left\{ v = \sum_i v^i \phi_h^i, \, v^i \text{ real numbers} \right\},$$

$$L_h = \left\{ \gamma = \sum_i \gamma^i \chi_h^i, \, \gamma^i \text{ real numbers} \right\}.$$

Boundary data are given by

$$u_h^0 \in H_h, \qquad u_h^0 \geq 0, \qquad P_h^0, P_h^+ \subset P_h \quad \text{disjoint},$$

and the corresponding class of admissible functions is

$$M_h(u_h^0) = \{ v \in H_h, v^i = u_h^{0i} \text{ for } i \in P_h^+, v^i \leq u_h^{0i} \text{ for } i \in P_h^0 \}.$$

It is assumed in order to prove convergence that the data Ω, S^0, S^+ do not depend on h. A discrete version of the inequality (8.98) is introduced by defining

$$a_h^{ij} = \int_{\Omega_h} \nabla \phi_h^i a_h \nabla \phi_h^j, \qquad e_h^{ij} = \int_{\Omega_h} \chi_h^i \nabla \phi_h^i a_h \mathbf{e}, \tag{8.101}$$

and imposing the following restrictions:

$$H_h \text{ contains all linear functions in } \Omega_h, \tag{8.102}$$

$$a_h^{ii} > 0 \quad \text{and} \quad e_h^{ii} \geq 0 \quad \text{for} \quad i \in P_h, \tag{8.103}$$

$$a_h^{ij} \leq 0 \quad \text{and} \quad e_h^{ij} \leq 0 \quad \text{for} \quad i, j \in P_h, i \neq j. \tag{8.104}$$

Then there is a function u_h in $M_h(u_h^0)$ with $u_h^i \geq 0$ for $i \in P_h$ and a function $\gamma_h \in L_h$ with $0 \leq \gamma_h^i \leq 1$ with $\gamma_h^i = 1$ if $u_h^i > 0$, such that

$$\int_{\Omega_h} \nabla(v - u_h) a_h (\nabla u_h + \gamma_h \mathbf{e}) \geq 0 \tag{8.105}$$

for every $v \in M_h(u_h^0)$.

Alt's numerical algorithm is closely related to the existence proof for (8.105) which is as follows. By making use of the restrictions (8.102–104) and inserting (8.101) it is seen that (8.105) is equivalent to

$$\sum_i \sum_j v^i (u_h^i a_h^{ij} + \gamma_h^j e_h^{ij}) \geq 0$$

for every $v \in H_h$ and with $v^i = 0$ for $i \in P_h^+$ and $v^i \leq u_h^{0i} - u_h^i$ for $i \in P_h^0$, which in turn implies

$$\sum_j a_h^{ij} u_h^i + e_h^{ij} \gamma_h^j \begin{cases} = 0 & \text{for} & i \notin P_h^+ \cup P_h^0 \quad \text{or} \quad i \in P_h^0 \quad \text{and} \quad u_h^i < u_h^{0i}, \\ \leq 0 & \text{for} & i \in P_h^0 \quad \text{and} \quad u_h^i = u_h^{0i}. \end{cases}$$

$$(8.106)$$

Now a new variable $w_h(i) = u_h^i + \gamma_h^i$ is introduced and $w_h(i) \geq 0$ because of the definitions of u and γ. The variables u, γ are regained through the expressions

$$u_h^i = \max(w_h(i) - 1, 0), \qquad j_h^i = \min(w_h(i), 1). \qquad (8.107)$$

In order to write (8.106) in terms of w_h, the functions $C_h^i(w)$, $B_h^i(w)$ are defined by

$$C_h^i(w) = \sum_{j \neq 1} \{(-a_h^{ij})\max(w(j) - 1, 0) + (-e_h^{ij})\min(w(j), 1)\}$$

$$(8.108a)$$

$$B_h^i(w) = \begin{cases} (1/a_h^{ii})(C_h^i(w) - e_h^{ii}) + 1, & \text{if} & C_h^i(w) \geq e_h^{ii}, \\ (1/e_h^{ii})C_h^i(w), & \text{if} & 0 \leq C_h^i(w) < e_h^{ii}, \end{cases}$$

where $i, j \in P_h$. Then remembering that $\gamma_h^i = 1$ if $i \in P_h^+$ or $e_h^{ii} = 0$, the existence of u_h and γ_h is equivalent to the existence of a vector $\mathbf{w}_h$ satisfying

$$w_h(i) = u_h^{0i} + 1, \qquad i \in P_h^+,$$
$$w_h(i) = \min(u_h^{0i} + 1, B_h^i(w_h)), \qquad i \in P_h^0 \qquad (8.109)$$
$$w_h(i) = B_h^i(w_h), \qquad i \notin P_h^+ \cup P_h^0,$$

where $w_h(i)$ are components of the vector $\mathbf{w}_h$ at the points i.

If this is written $\mathbf{w}_h = A_h^i(\mathbf{w})$, the mapping A will be continuous and monotone by (8.104) so that $w_1(i) \geq w_2(i)$ implies $A_h^i(w_1) \geq A_h^i(w_2)$ for every $i \in P_h$ and fixed points of A_h can be found by an iterative procedure if at least one solution of A_h can be found below and one above the fixed point. Because $A_h^i(0) \geq 0$, the vector $\mathbf{w} = 0$ is a lower solution and Alt (1980*b*) also established an upper solution of A. This completes the proof and suggests an iterative numerical algorithm based on (8.109). Alt (1979, 1980*a*) showed that at least a subsequence of solutions of the discrete problem converges to a solution of the continuous problem.

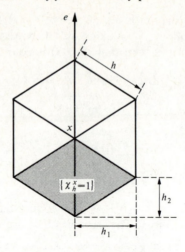

FIG. 8.19. Finite element

Alt (1980a,b) referred to a general program for two-dimensional problems using a finite-element discretization based on a uniform triangulation as in Fig. 8.19. For mesh points x he took $\chi_h^x = 1$ to be the shaded region in Fig. 8.19, and ϕ_h^x to be piecewise linear with $\phi_h^x(x) = 1$, $\phi_h^x(y) = 0$ for mesh points $y \neq x$. Then if $\bar{\Omega}$ comprises a number of closed triangles, P_h is defined as the set of mesh points in $\bar{\Omega}$. If S^+ and S^0 consist of some sides of these triangles then P_h^+ and P_h^0 denote mesh points on the corresponding sides. The boundary data function u_h^0 is taken to be piecewise linear. Conditions (8.99–103) are satisfied but (8.104) hold if and only if $h_2/h_1 \leq 1$ (Fig. 8.19). Alt's conditions for convergence are fulfilled provided u_h^0 does not depend on h and $C \leq h_2/h_1$ since this implies $\{x + (\pm h_1, -h_2)\} \subset Q_h(\{x\})$, where $Q_h(I) = \{j \in P_h,\ a_h^{ii} \leq C(-a_h^{ij})$ for some $i \in I\}$.

The program can be used for polyhedral domains Ω and piecewise-linear functions u^0 for all positions of S^+ and S^0 in both homogeneous and inhomogeneous media. For given h_1, h_2 the program produces an approximating domain Ω_h, though this situation was not included in the convergence proof referred to above. An iterative procedure based on the equations (8.109) was used, starting with the lower and upper solutions $w_-^0 = 0$ and w_+^0 introduced in the proof of the existence theorem for a solution of (8.105) and computing

$$w_-^{(k+1)} = A_h(w_-^{(k)}), \qquad w_+^{(k+1)} = A_h(w_+^{(k)}).$$

Because A_h does not satisfy the strong maximum principle it is permissible that at some stage in the iteration $w_-^{(k)} = w_+^{(k)} = 0$ in certain regions which are the dry regions of the porous medium. The mesh points in such

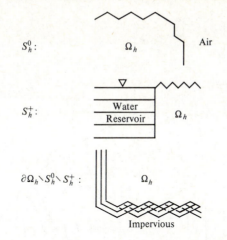

$S_h^0:$ Ω_h Air

$S_h^+:$ Water Reservoir Ω_h

$\partial\Omega_h \setminus S_h^0 \setminus S_h^+:$ Ω_h

Impervious

FIG. 8.20. Key to different boundary segments

regions need not be considered in further iterations. Furthermore, unless u_h is positive everywhere, the discrete solution w_h has regions or narrow strips where $0 < \gamma_h < 1$. This must be so if the solution of the continuous problem includes unsaturated flow, but this only happens if $\nabla \cdot (a\mathbf{e}) < 0$ somewhere. If the solution of the continuous problem has a smooth free boundary, the corresponding discrete solution has a small strip of size h, where $0 < \gamma < 1$, instead of a sharp free boundary. Alt presented a number of graphical solutions including some with a capillary fringe and others for inhomogeneous media. The following selected examples illustrate that the method can cope with flow regions of any shape and with flow in inhomogeneous porous media in which unsaturated regions appear. Figure 8.20 shows how the boundary conditions are indicated in the diagrams to follow. Where γ values in the range $0 < \gamma < 1$ appear they are indicated by small vertical lines as shown in Fig. 8.21. The continuous lines represent the piezometric head in the saturated regions, i.e. the function $u(\mathbf{x}) + \mathbf{x} \cdot \mathbf{e}$. The values of w_h do not define the free boundaries precisely but their approximate shapes are drawn as a guide.

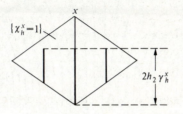

$\{\chi_h^x = 1\}$

$2h_2 \gamma_h^x$

FIG. 8.21. Depicting γ values, $0 < \gamma < 1$

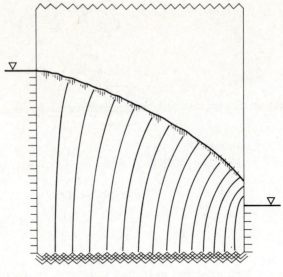

FIG. 8.22. Simple dam

Figure 8.22 shows Alt's numerical solution for the simple dam problem for which, of course, a Baiocchi-transformation solution is also available. Figure 8.23 illustrates the effect of introducing a sheetpile on the inlet face. Figure 8.24 demonstrates the power of the method in a flow region of a different shape. In Figs 8.25 and 8.26 the permeability is given by $a = 1$ in the upper half, i.e. above the level at half the inlet reservoir height, and by $a = 3$, $a = 9$ respectively in the lower half. Figure 8.26 shows two saturated regions connected by an unsaturated zone and a

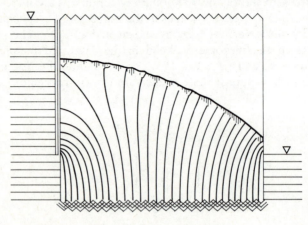

FIG. 8.23. Effect of sheetpile

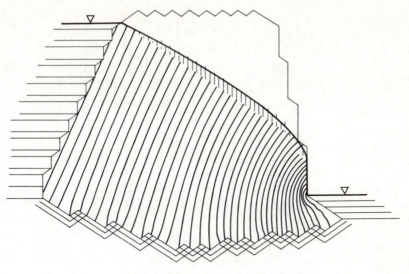

FIG. 8.24. More general dam shape

dry region exists. This is an example of the existence of a region where $0 < \gamma < 1$ when the condition $\nabla \cdot (a\mathbf{e}) < 0$ holds as discussed in relation to solutions of (8.98). With regard to the second region of saturated flow at the bottom of the dam, the function γ can be interpreted physically as a precipitation on its upper surface. Alt (1980a,b) shows other graphical solutions together with computational details such as numbers of mesh points, convergence limits and running times for his two basic programs

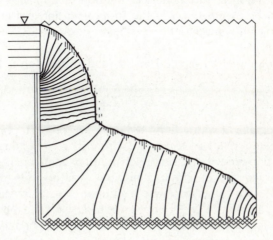

FIG. 8.25. Horizontally stratified dam: $a = 1$ in upper half; $a = 3$ in lower half

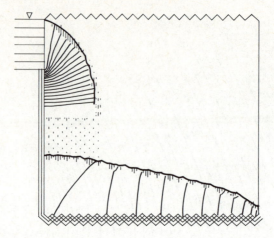

FIG. 8.26. Horizontally stratified dam: $a = 1$ in upper half; $a = 9$ in lower half

one for inhomogeneous media and a simpler version for homogeneous media.

8.7. Integral equation methods

Reference was made in §8.2.1 to integral equation methods as one of several possible ways of computing trial solutions for successive estimates of the position of the free boundary in trial and error methods. Several references to integral-equation formulations of electrochemical machining problems were given in §2.12.3. The present section contains a brief general description of the numerical solution of Laplace's equation in a fixed domain by expressing it in the form of an integral equation. Then some developments in relation to problems of electrochemical machining are described.

The integral equation method in two dimensions solves the equation

$$\nabla^2 \phi = \frac{\partial^2 \phi}{\partial x^2} + \frac{\partial^2 \phi}{\partial y^2} = 0, \tag{8.110}$$

in a plane domain D with a finite boundary, L, on each point of which ϕ or its normal derivative $\partial \phi / \partial n$ are prescribed. The domain may be bounded or unbounded and simply or multiply connected, provided only that L must consist of closed contours with no cuts or cusps. The boundary data are to be piecewise continuous except possibly for jumps in the normal derivative $\partial \phi / \partial n$ at corners of L. No value need be given at such points nor at other discrete points where the boundary condition changes.

The method to be described starts from Green's formula (the third identity) in the form

$$\int_L \{\phi'(\mathbf{q})\ln|\mathbf{q}-\mathbf{p}| - \phi(\mathbf{q})\ln'|\mathbf{q}-\mathbf{p}|\}\,dq = \eta(\mathbf{p})\phi(\mathbf{p}), \qquad (8.111)$$

where $\mathbf{p}, \mathbf{q}$ are vectors specifying points of the plane and on L respectively; the prime denotes differentiation at the point $\mathbf{q}$ along the inward normal to the domain and dq is the differential increment of L at $\mathbf{q}$. If $\mathbf{p}$ is a point in D then $\eta(\mathbf{p}) = 2\pi$, but if $\mathbf{p}$ is a boundary point on L, $\eta(\mathbf{p}) = \Omega(\mathbf{p})$, where $\Omega(\mathbf{p})$ is the internal angle at the point $\mathbf{p}$, i.e. the angle between the tangents to L on either side of $\mathbf{p}$. In the latter case (8.111) becomes (Jaswon and Symm, 1977)

$$\int_L \{\phi'(\mathbf{q})\ln|\mathbf{q}-\mathbf{p}| - \phi(\mathbf{q})\ln'|\mathbf{q}-\mathbf{p}|\}\,dq - \Omega(\mathbf{p})\phi(\mathbf{p}) = 0, \qquad \mathbf{p} \in L. \quad (8.112)$$

This provides a linear relationship between the boundary values of a solution ϕ of (8.110) and those of its normal derivative. Hence, given values of either ϕ or $\partial\phi/\partial n$ at each point of L we have a linear equation for the other corresponding boundary values, or a pair of equations when ϕ is given on a part of L and ϕ' on the remainder. We note that provided L has no cusps, $\eta(\mathbf{p})$ is never zero and so by substituting the solution of (8.112), together with the original boundary data, into (8.111) the value of ϕ at any point $\mathbf{p}$ in D and on L can be obtained. Exceptions to the statement that the integral equation arising from (8.112) has a unique solution are discussed by Jaswon and Symm (1977) and Baker (1979). The general basis of a numerical procedure for solving the integral equations is to divide the boundary L into smooth intervals such that any corners or changes in form of the boundary occur at the points of subdivision. The functions ϕ and ϕ' are approximated by constant values within each interval. The boundary L itself is approximated by a polygon by replacing each interval by two chords which join its end points to the nodal point within it. A system of simultaneous equations results, in which all the coefficients of the discretized ϕ and ϕ' can be computed analytically.

8.7.1. *Two-dimensional annular electrochemical machining problem*

We now discuss in more detail the integral-equation method of solving a quasi-steady-state model of the electrochemical machining process outlined in §2.12.3 used by Christiansen and Rasmussen (1976). Their system and terminology are illustrated in Fig. 8.27, where Γ_0 is the cathode and Γ_1 the anode and the problem is to find a function ϕ satisfying (8.110) in the annular space D subject to boundary conditions

$$\phi = \phi_0 \quad \text{on } \Gamma_0, \qquad \phi = \phi_1 \quad \text{on } \Gamma_1. \qquad (8.113a)$$

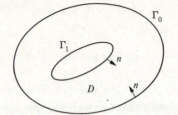

FIG. 8.27. Annular integration domain

The position vector $\mathbf{R}$ for the anode surface Γ_1 is determined by the condition in non-dimensional terms

$$d\mathbf{R}/dt = \nabla \phi_a, \qquad (8.113b)$$

where $\nabla \phi_a$ denotes the value of $\nabla \phi$ at the anode surface. Since therefore the normal gradient of ϕ on the anode surface is the prime objective of solving (8.110) the integral-equation method is especially suitable.

Christiansen and Rasmussen (1976) confined their attention to electrodes which are represented by simple closed piecewise-smooth curves which do not touch each other. Since their problem is wholly of Dirichlet type, by using the relationship $\nabla^2 \ln \rho(\mathbf{p}, \mathbf{q}) = 2\pi \, \delta(\mathbf{p} - \mathbf{q})$, where $\rho(\mathbf{p}, \mathbf{q}) = |\mathbf{p} - \mathbf{q}|$, they simplify (8.111) to

$$\phi(\mathbf{p}) = \phi_0 + \frac{1}{4\pi} \int_{\Gamma_0 \cup \Gamma_1} \ln\{\rho(\mathbf{p}, \mathbf{q})\}^2 \frac{\partial \phi}{\partial n} (\mathbf{q}) \, ds, \quad \text{for} \quad \mathbf{p} \text{ in } D. \tag{8.114}$$

Now $\mathbf{p}$ is moved to either Γ_0 or Γ_1, i.e.

$$\mathbf{p} \rightarrow \mathbf{p}_0 \quad \text{on } \Gamma_1, \qquad \phi(p) \rightarrow \phi(p_0) = \phi_1,$$
$$\mathbf{p} \rightarrow \mathbf{p}_0 \quad \text{on } \Gamma_0, \qquad \phi(p) \rightarrow \phi(p_0) = \phi_0,$$

and (8.114) becomes

$$\frac{1}{4\pi} \int_{\Gamma_0 \cup \Gamma_1} \ln\{\rho(\mathbf{p}_0, \mathbf{q})\}^2 \psi(\mathbf{q}) \, ds = \begin{cases} \phi_1 - \phi_0, & \mathbf{p}_0 \quad \text{on } \Gamma_1, \\ 0, & \mathbf{p}_0 \quad \text{on } \Gamma_0, \end{cases} \tag{8.115}$$

where $\psi(\mathbf{q})$ denotes $\partial \phi(\mathbf{q})/\partial n$ and is the unknown function. Christiansen and Rasmussen (1976) discuss the restriction on the size of the external boundary Γ_0 necessary to ensure a unique solution of (8.115).

To effect a numerical solution the curves Γ_0 and Γ_1 are approximated by polygons with an even number of vertices, $2n_k$, specified on each curve, where the index k takes the value 0 on Γ_0 and 1 on Γ_1. The distances between consecutive points need not be equal to allow for changes in the anode surface with time. The $2n_k$ points are numbered on Γ_0 as $i = \frac{1}{2}, 1, 1\frac{1}{2}, \ldots, n_0 - \frac{1}{2}, n_0$, and on Γ_1 as $i = \frac{1}{2}, 1, 1\frac{1}{2}, \ldots, n_1 - \frac{1}{2}, n_1$.

The coordinates of these points are denoted by $x_{k,i}$ and $y_{k,i}$. Over the interval between two boundary points $j-\frac{1}{2}$ and $j+\frac{1}{2}$ the unknown function $\psi(\mathbf{q})$ is approximated by a constant value, $\psi_{\ell,j}$, where j is an integer varying between 1 and n_ℓ, where $\ell=0$ on Γ_0, 1 on Γ_1. The integrals in (8.115) can now be approximated in the form

$$\sum_{\ell=0}^{1}\int_{\Gamma_\ell}\psi(\mathbf{q})\ln\{\rho(\mathbf{p}_0,\mathbf{q})\}^2\,\mathrm{d}s=\sum_{\ell=0}^{1}\sum_{j=1}^{n_\ell}\psi_{\ell,j}\int_{\Gamma_\ell:j-\frac{1}{2}}^{\Gamma_\ell:j+\frac{1}{2}}\ln\{\rho(\mathbf{p}_0,\mathbf{q})\}^2\,\mathrm{d}s,\quad(8.116)$$

where $\mathbf{p}_0$ is a given parameter point and the integrations are carried out along the two straight line segments joining the points $j-\frac{1}{2}$, $j+\frac{1}{2}$ on curve Γ_ℓ. The integrals in (8.116) are denoted by

$$A_{k,i,\ell,j}=\int_{\Gamma_\ell:j-\frac{1}{2}}^{\Gamma_\ell:j+\frac{1}{2}}\ln\{\rho(\mathbf{p}_{k,i},\mathbf{q})\}^2\,\mathrm{d}s\qquad(8.117)$$

where p_0 has been identified with the n_0+n_1 points in turn, i.e. with points $1, 2, \ldots, n_0$ on Γ_0 and points $1, 2, \ldots, n_1$ on Γ_1. These integrals can all be evaluated analytically. With reference to a typical configuration shown in Fig. 8.28a we write

$$A_{k,i,\ell,j}=\int_{\Gamma_\ell:j-\frac{1}{2}}^{\Gamma_\ell:j}\ln\{\rho(p_{k,i},q)\}^2\,\mathrm{d}s+\int_{\Gamma_\ell,j}^{\Gamma_\ell:j+\frac{1}{2}}\ln\{\rho(p_{k,i},q)\}^2\,\mathrm{d}s$$

and then consider the general situation in Fig. 8.28(b) where C is the point on the curve Γ_k and A and B correspond either to points $j-\frac{1}{2}$ and j or $j+\frac{1}{2}$ on Γ_ℓ. Taking a, b, c as the sides of the triangle ABC and a point D a distance s from A, we see that CD is equal to the quantity ρ in the integrals, i.e. in the integral

$$\int_A^B\ln\{\rho(C,\mathbf{q})\}^2\,\mathrm{d}s.$$

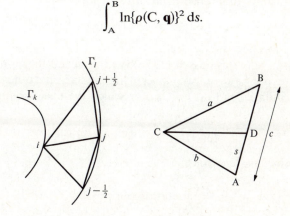

FIG. 8.28(a). Typical configuration
FIG. 8.28(b). Integration triangle

The cosine rule gives $\rho^2 = s^2 + b^2 + s(a^2 - b^2 - c^2)/c$, from which we get

$$\int \rho^2 \, ds = \left(s + \frac{a^2 - b^2 - c^2}{2c}\right) \ln \rho^2 + \frac{d}{c} \tan^{-1} \frac{2cs + a^2 - b^2 - c^2}{d} - 2s,$$

where $d = \{[(a+b)^2 - c^2][c^2 - (a-b)^2]\}^{\frac{1}{2}}$ and eventually

$$\int_0^c \ln \rho^2 \, ds = \frac{a^2 - b^2 + c^2}{2c} \ln a^2 + \frac{-a^2 + b^2 + c^2}{2c} \ln b^2$$

$$+ \frac{d}{c} \tan^{-1} \frac{d}{a^2 + b^2 - c^2} - 2c.$$

As j assumes successively the values between 1 and n_ℓ, the different distances appear in cyclic order as the quantities a, b, c, and so lead to a fast numerical procedure as Christiansen and Rasmussen observe.

Finally, an algebraic approximation to the integral equation (8.115) can be written as

$$\sum_{\ell=0}^{1} \sum_{j=1}^{n_\ell} \psi_{\ell,j} A_{k,i;\ell,j} = \begin{cases} 4\pi(\phi_1 - \phi_0), & k = 1, \\ 0 & , \quad k = 0 \end{cases} \tag{8.118}$$

$$i = 1, 2, \ldots, n_0, \qquad i = 1, 2, \ldots, n_1,$$

where

$$A_{k,i;\ell,j} = \int_{\Gamma_\ell : j - \frac{1}{2}}^{\Gamma_\ell : j + \frac{1}{2}} \ln \rho^2(q_{k,i}, q) \, ds. \bullet$$

The system of linear equations was solved by a standard routine. From values of $\partial \phi / \partial n$ at points on the anode surface $\nabla \phi_a$ can be determined and used in a finite-difference form of (8.113b), i.e.

$$R_{n+1} = R_n + \Delta t \, \nabla \phi_a, \tag{8.119}$$

to determine the new anode shape at time $(n+1) \Delta t$. Details were given by Christiansen and Rasmussen (1976), who checked their procedure by applying it to the case of two concentric circular electrodes with non-dimensional radii c_0 and c_1, $c_1 < c_0$, for which the exact solution is

$$\phi = \phi_0 + \frac{\ln(r/c_0)}{\ln(c_1/c_0)} (\phi_1 - \phi_0), \qquad c_1 \leqslant r \leqslant c_0. \tag{8.120}$$

On the interior anode surface

$$\left. \frac{\partial \phi}{\partial r} \right)_{r=c_1} = \frac{\phi_1 - \phi_0}{c_1 \ln(c_1/c_0)}. \tag{8.121}$$

When $N = n_0 = n_1 = 24$ and the exterior curve is far from the critical one mentioned above ($c_0 = 1$), the values of $\partial \phi / \partial n$ for $\phi_0 = 0$, $\phi_1 = 1$ are in

TABLE 8.23

Percentage errors in $\partial\phi/\partial n$ for two concentric circular electrodes

N	Interior circle $c_1 = 0.05$	Exterior circle $c_0 = 0.10$	Interior circle $c_1 = 5.00$	Exterior circle $c_0 = 10.00$
6	1.35	0.88	1.37	1.84
12	0.31	0.19	0.31	0.43
24	0.07	0.04	0.07	0.10

excellent agreement with analytical values. Comparisons are displayed in Table 8.23 taken from Christiansen and Rasmussen (1976). Satisfactory comparisons were also found with the analytical solution for two confocal elliptic electrodes with semi-axes $(a_0, b_0) = c(\cosh \mu_0, \sinh \mu_0)$ and $(a_1, b_1) = c(\cosh \mu_1, \sinh \mu_1)$, where $0 < \mu_1 < \mu_0$ and c is a constant. The exact solution gives

$$\phi = \phi_0 + \frac{\mu_0 - \mu}{\mu_0 - \mu_1} (\phi_1 - \phi_0), \qquad \mu_1 \leqslant \mu \leqslant \mu_0.$$

By combining (8.121) with (8.113b) and taking $\phi_1 - \phi_0 = 1$ we have, on reverting to the two circular electrodes,

$$\frac{dR(t)}{dt} = \frac{c_0}{R(t)\ln R(t)}$$

and hence

$$t = \tfrac{1}{4}[R_0^2 - R^2 + R^2 \ln(R/c_0)^2 - R_0^2 \ln(R_0/c_0)^2],$$

where t is a non-dimensional time and $R_0 = R(t)$ at $t = 0$. For the example $c_0 = 10.0$ and $R_0 = 9.0$, Christiansen and Rasmussen found $R(t) = 7.437\,400$ for $t = 2.5$. Table 8.24 shows the relative errors of several approximations obtained by their numerical procedure.

TABLE 8.24

Errors in percentages of the interior radius at $t = 2.5$ for two concentric circular electrodes with initial radii 9.0 and 10.0

Δt	$N = 12$	$N = 18$	$N = 24$
0.1	3.24	2.89	2.78
0.05	1.89	1.54	1.44
0.025	1.23	0.89	0.78
0.0125	0.91	0.57	0.46

Further satisfactory comparisons were made with a perturbation solution obtained by Rasmussen and Christiansen (1976) for an elliptical anode placed symmetrically inside a circular cathode. In this problem the non-dimensional radius of the cathode is 10.0 and the semi-axes of the anode are 9.5 and 9.0. Rasmussen and Christiansen (1976) considered $0.25 \sin 2\theta$ to be a small perturbation on the circle with radius 9.25. The same problem was studied by Crowley (1979) (see §6.2.7(i)) and by Elliott (1980) (see §2.12.3) and all four sets of results are collected together in Table 8.25 taken from Elliott (1980). Both Crowley and Elliott make the point that this is not an ideal example for fixed-domain methods as the amplitude of oscillation is small compared with reasonable mesh sizes δr. They both started from a sinusoidally perturbed anode. Nevertheless the comparisons are good, especially for the average radius. The original papers give results for different mesh sizes and for other machining problems.

Hansen (1980, 1983) extended and modified the method of Christiansen and Rasmussen (1976) just described in axial symmetry. The electrode surfaces are assumed to be axially symmetric and concentric; the cathode is finite and the anode infinite in extent. Each electrode surface

TABLE 8.25

Comparison of results for elliptic anode inside circular cathode

Time	C^a $\Delta r = 0.0625$ $\Delta\theta = \pi/20, \Delta t = 0.01$	CR^a $2N = 48$ $\delta t = 0.025$	RC^a $2N = 48$	E^a $\Delta r = 0.125$ $\Delta\theta = \pi/64$
Average radius				
0.0	9.25	9.25	9.25	9.25
0.5	8.719	8.71	8.727	8.722
1.0	8.320	8.34	8.356	8.351
1.5	8.031	8.03	8.048	8.045
2.0	7.781	7.76	7.778	7.776
2.5	7.531	7.52	7.533	7.532
3.0	—	7.29	7.307	7.304
3.5	—	7.08	7.095	7.093
Amplitude of oscillation				
0.0	0.25	0.25	0.25	0.025
0.5	0.156	0.15	0.151	0.147
1.0	0.117	0.11	0.118	0.121
1.5	0.094	0.10	0.100	0.095
2.0	0.094	0.09	0.089	0.087
2.5	0.094	0.08	0.080	0.079
3.0	—	0.07	0.074	0.074
3.5	—	0.07	0.069	0.066

[a] C = Crowley (1979), enthalpy; CR = Christiansen and Rasmussen (1976), integral equation, numerical; RC = Rasmussen and Christiansen (1976), perturbation; E = Elliott (1980), variational inequality; $2N$ = number of points on each electrode.

may consist of ring-shaped regions in which either Dirichlet or Neuman type boundary conditions apply. There may be a finite number of circular common edges where the tangent planes to the surfaces meet at an exterior angle β, $0 \leqslant \beta \leqslant 2\pi$. The unions of the generators of the regions on the cathode surface where the Dirichlet condition is $\phi = \phi_C$ are denoted by C_{CD}, and those with the Neuman condition, $\partial\phi/\partial N = 0$, by C_{CN}. The corresponding notation for the regions of the anode surface is C_{AD} and C_{AN}.

For the cylindrical symmetry the appropriate fundamental solution can be written

$$G(\mathbf{r}, \mathbf{r}') = (\pi R)^{-1} K(4rr'/R^2) \tag{8.122}$$

where $\mathbf{r}$ has coordinates r, z and those of $\mathbf{r}'$ are r', z', $R = \{(r+r')^2 + (z-z')^2\}^{\frac{1}{2}}$, and K is the elliptic integral of the first kind. By using (8.122) and proceeding from Green's second identity in the usual way we obtain

$$\phi(\mathbf{r}') - 1 = \int_{C_{AD}+C_{CD}} G\frac{\partial\phi}{\partial N}\, ds - \int_{C_{AN}} \frac{\partial G}{\partial N}(\phi-1)\, ds - \int_{C_{CN}} \frac{\partial G}{\partial N}\phi\, ds.$$

We now derive an integral equation by taking $\mathbf{r}'$ to be a point on the boundary and using the boundary conditions. We express it as

$$\int_C B(s', s) y(s)\, ds + v(s') y(s') = u(s'), \tag{8.123}$$

where C is the union of all the boundaries, i.e. $C = C_{CD} \cup C_{CN} \cup C_{AD} \cup C_{AN}$ and s the arc length along C. The functions in (8.123) are defined by

$$B(s', s) = \begin{cases} G(\mathbf{r}', \mathbf{r})r, & \mathbf{r} \in \bar{C}_{CD} \cup \bar{C}_{AD} \\ -\dfrac{\partial G}{\partial N}(\mathbf{r}', \mathbf{r})r, & \mathbf{r} \in C_{CN} \cup C_{AN} \end{cases}$$

where $\mathbf{r} = \mathbf{r}(s)$, $\mathbf{r}' = \mathbf{r}(s')$, $\bar{C}_{CD}$ and $\bar{C}_{AD}$ are the closures of C_{CD} and C_{AD} respectively, and $\partial/\partial N$ denotes differentiation in the direction of the inward normal to the surface;

$$y(s) = \begin{cases} \partial\phi/\partial N, & \mathbf{r} \in \bar{C}_{CD} \cup \bar{C}_{AD}, \\ \phi - 1, & \mathbf{r} \in C_{AN}, \\ \phi, & \mathbf{r} \in C_{CN}; \end{cases}$$

$$v(s') = \begin{cases} 0, & \mathbf{r}' \in \bar{C}_{CD} \cup \bar{C}_{AD}, \\ \tfrac{1}{2}(\beta/\pi) - 1, & \mathbf{r}' \in C_{CN} \cup C_{AN} \end{cases};$$

$$u(s') = \begin{cases} 0, & \mathbf{r} \in C_{AD} \cup C_{AN}, \\ -1, & \mathbf{r} \in C_{CD} \cup C_{CN}. \end{cases}$$

On a smooth part of $C_{CN} \cup C_{AN}$ the exterior angle β between the tangents is π so $v = -\frac{1}{2}$.

Hansen (1980, 1983) introduced local and analytic solutions at points on the boundary C where intervals with different boundary conditions meet, in order to handle the singularities there. At such points and at points corresponding to edges, y is singular. Local expansions of the type

$$\phi = \sum_{n=1}^{\infty} a_n \rho^{(n-\frac{1}{2})(\pi/\beta)} \sin(n-\tfrac{1}{2})\pi\theta/\beta, \tag{8.124}$$

are commonly incorporated into the solution at an edge point on C for example, at which $\phi = 0$ on one side where $\theta = 0$ and $\partial\phi/\partial n = 0$ on the other side where $\theta = \beta$; $a_1, a_2, \ldots$ are constants, and ρ is the distance from the edge (Motz 1946; Fox 1979).

The integral equation is to be satisfied only at a finite number of collocation points which in general are chosen to include all the singularities. If s_- and s_+ denote two neighbouring singularities on C, and s_1, $s_2, \ldots, s_M$ are the collocation points in the interval between, so that $s_1 = s_-$ and $s_M = s_+$, the function y in the closed interval s_- to s_+ is written as

$$y(s) = (s_+ - s)^{\alpha_+}(s - s_-)^{\alpha_-}f(s),$$

where the constants α_- and α_+ are such that f is finite and non-zero. The function $f(s)$ is approximated by a function which is continuous between s_- and s_+ and is a first degree polynomial in each sub-interval between neighbouring nodal points s_j and s_{j+1}, for $2 \le j \le M-2$. In the intervals s_- to s_2 and s_{M-1} to s_+, the approximating function is determined from the first two terms of the appropriate local expansion of type (8.124). If all the angles β in a particular problem are rational fractions of π, y(s) can be replaced in the integral of (8.123) by an analytical function of σ where $\sigma^m = s$. When s' is not equal to s_j or s_{j+1} the integral is evaluated by Simpson's rule. For $s' = s$, $B(s', s)$ has a logarithmic singularity, but the integral of the logarithmic term can be evaluated by a quadrature formula which gives the integral of a function $P(s)\ln|s|$ exactly if P is a second degree polynomial. In the neighbourhood of such a point s_n, therefore, C is approximated by a parabola passing through s_{n-1}, s_n, and s_{n+1}. The anode surface extends to infinity and so for the infinite interval beyond the last collocation point f is derived from an asymptotic solution of Laplace's equation. Finally, at an edge point P with an angle β between the tangents associated with $C_{AN} \cup C_{CN}$, v changes discontinuously from $v = \frac{1}{2}(\beta/\pi) - 1$ to $v = -\frac{1}{2}$ where the boundary is smooth in the neighbourhood of P. Hansen (1983) avoided the numerical difficulty by omitting such a point P from the selected collocation points. From the solution of the system of linear algebraic equations obtained by Gaussian elimination, the values of $\partial\phi/\partial N$ were evaluated at the collocation points and the changes $(\Delta r, \Delta z)$ in the anode coordinates obtained from the equivalent of (8.119). Hansen evaluated results for a geometry resembling an experimental one available to him and made the comparisons mentioned in §2.12.3.

References

Aguirre-Puente, J. and Frémond, M. (1976). In *Lecture notes in mathematics Vol. 503* (ed. P. Germain and B. Nayroles) pp. 137–47. Springer-Verlag, Berlin.

Aitchison, J. (1972). *Proc. R. Soc.* **A330,** 573–80.

—— (1975). Package for integration of large sets of ordinary differential equations. Internal Report Edinburgh University.

—— (1977). *J. Inst. Math. Appl.* **20,** 33–44.

Albasiny, E. L. (1956). *Proc. Instn. elect. Engrs.* **103B,** 158–62.

Albrecht, J., Collatz, L., and Hoffman, K.-H. (eds.) (1980). *Proceedings of the Oberwolfach Conference on Free Boundary Problems.* (To appear.)

Allen, D. N. de G. (1954). *Relaxation methods in engineering science.* McGraw-Hill, New York.

—— Severn, R. T. (1962). *Q. Jl. Mech. appl. Math.* **15,** 53–62.

Alt, H. W. (1977). *Archs. ration. Mech. Analysis* **64,** 111–26.

—— (1979). *J. reine angew. Math.* **305,** 89–115.

—— (1980a). *Num. Math.* **36,** 73–98.

—— (1980b). In Magenes (1980) Vol. I, pp. 89–108.

Andrews, J. G. and Atthey, D. R. (1975a). *J. Inst. Math. Appl.* **15,** 59–72.

—— —— (1975b). In Ockendon and Hodgkins (1975). pp. 38–53.

Aravin, V. I. and Numerov, S. N. (1965). *Theory of motion of liquids and gases in undeformable porous media.* (English translation by A. Moscona.) Israel Programme for Scientific Translation, Jersalem.

Arms, R. J. and Gates, L. D. (1957). Report No. 1533, US Naval Proving Ground, Dahlgren, Virginia.

Atthey, D. R. (1972). D. Phil. Thesis, University of Oxford.

—— (1974). *J. Inst. Math. Appl.* **13,** 353–66.

—— (1975). In Ockendon and Hodgkins (1975). pp. 182–91.

Awater, R. (1980). In Brebbia (1980). pp. 191–207.

Baiocchi, C. (1971). *C.r. hebd. Séanc. Acad. Sci., Paris* **273,** 1215–17.

—— (1972). *Annali. Mat. pura appl.* **92,** 107–127.

—— (1975a). *Proc. Int. Cong. Math. Vancouver* **2,** 237–43.

—— (1975b). *Cuad. Inst. Mat. Beppo Levi* **8**.

—— (1975c). *Boll. U. mat. ital.* **11,** 589–613.

—— (1976). *C.r. hebd. Séanc. Acad. Sci., Paris* **283,** 29–32.

—— (1980a). In Magenes (1980). Vol. I, pp. 175–91.

—— (1980b). In *Variational inequalities and complementarity problems.* (eds. R. W. Cottle, F. Gianessi and J. L. Lions). Chapter 2. Wiley, New York.

—— Brezzi, F., and Comincioli, V. (1976). *Proc. 2nd Int. Symp. Finite element methods in flow problems, Santa Margherita Ligure,* pp. 409–20.

—— Comincioli, V., Guerri, L., and Volpi, G. (1973a). *Calcolo* **10,** 1–85.

—— —— Magenes, E., and Pozzi, G. A. (1973b). *Annali Mat. pura appl.* **97** (4), 1–82.

—— —— —— —— (1976). *Proc. Int. Autumn course, Math. Numerical methods in fluid dynamics, Int. Centre for Theoretical Physics, Trieste.*

—— —— Maione, U. (1975). *Meccanica* **10,** 151–5.

—— —— (1977) In *Proc. Symp. on Hydrodynamic diffusion and dispersion in porous media, Pavia.* pp. 315–28.

—— Friedman, A. (1977). *Annali Mat. pura appl.* **114,** 337–93.

—— Magenes, E. (1974). *Russ. math. Survs* **29,** 50–9.

—— —— (1975). *Problemi di Frontiera Libera in Idraulica. Ac. Naz. Lincei* **217,** 395–421.

—— Pozzi, G. A. (1976). *Appl. Math. Optim.* **2** (4), 304–14.

Baker, C. T. H. (1977). *The numerical treatment of integral equations.* Clarendon Press, Oxford.

—— (1979). In Gladwell and Wait (1979). Chapter 8.

Balinski, M. L. and Cottle, P. W. (1978). *Complementarity and fixed-point problems.* North Holland, Amsterdam.

Bankoff, S. G. (1964). Adv. chem. Engng **5,** 75–150.

Bathe, K.-J. and Khoshgoftaar, M. R. (1979). *Int. J. numer. anal. Meth. Geomech.* **3,** 13–22.

Baumeister, J., Hoffman, K.-H., and Jochum, P. (1980). *J. Inst. Math. Appl.* **25,** 99–109.

Bear, J. (1972). *Dynamics of fluids in porous media.* Elsevier, New York.

—— (1980). In Magenes (1980). Vol. I, pp. 33–77.

—— Zaslavsky, D., and Irmay, S. (1968). *Physical principles of water percolation and seepage.* UNESCO, Paris.

Bell, G. E. (1978). *Int. J. Heat Mass Transfer* **21,** 1357–61.

—— (1979). *Int. J. Heat Mass Transfer* **22,** 1681–6.

—— (1982). *Int. J. Heat Mass Transfer* **25,** 587–9.

Benci, V. (1973). *Rend. Acad. Naz. Lincei* **54**(8), 10–15.

—— (1974). *Annali Mat. para appl.* **100**(4), 191–209.

Bensoussan, A. and Lions, J. L. (1973). *C.r. hebd. Séanc. Acad. Sci., Paris,* **A–B, 276,** 1189–92.

—— —— (1978). *Applications des inequations variationelles en controles stochastique.* Dunod, Paris.

—— —— Papanicolau, G. (1975). *C.r. hebd. Séanc. Acad. Sci., Paris* **218A,** 317–22.

Berger, A. E. (1976). *Revue Fr. Automat. Informat. Rech. Operation.: Anal Numer.* **10,** 29–42.

—— Ciment, M., and Rogers, J. C. W. (1975). *SIAM J. numer. Anal.* **12,** 646–72.

—— —— —— (1979). *SIAM J. numer. Anal.* **16,** 563–87.

—— Falk, R. S. (1977). *Maths Comput.* **31,** 619–29.

Bickley, W. G. (1964). *Non-linear problems of engineering.* (ed. W. F. Ames). Academic Press, New York.

Bischoff, H. (1980). In Brebbia (1980). pp. 208–26.

Boadway, J. D. (1976). *Int. J. numer. Meth. Engng.* **10,** 527–33.

Boieri, P. and Gastaldi, F. (1980). In Magenes (1980). Vol. I, pp. 193–6.

Boley, B. A. (1961). *J. Math. Phys.* **40,** 300–313.

—— (1963). *Q. appl. Math.* **21**(1), 1–11.

—— (1964). *Q. Jl. Mech. appl. Math.* **17**(3), 253–69.

—— (1968). *Int. Jl. Engng. Sci.* **6,** 89–111.

—— (1972). *Nucl. Engng. Des.* **18,** 377–99.

—— (1973). *Int. J. Heat Mass Transfer* **16,** 2035–41.

—— (1975). In Ockendon and Hodgkins (1975). pp. 150–72.

—— (1978). In Wilson *et al.* (1978). pp. 205–31.

—— Estenssoro, L. (1977). *Mech. Res. Comm.* **4,** 271–279.

—— Yagoda, H. P. (1969). *Q. appl. Math.* **27,** 223–46.

—— —— (1971). *Proc. R. Soc.* **A323,** 89–110.

Bonacina, C., Comini, G., Fasano, A., and Primicerio, M. (1973). *Int. J. Heat Mass Transfer* **16,** 1825–32.

Bonnerot, R. and Jamet, P. (1974). *Int. J. numer. Meth. Engng.* **8,** 811–20.

—— Jamet, P. (1975). *J. comput. Phys.* **18,** 21–45.

—— —— (1977). *J. comput. Phys.* **25,** 163–81.

—— —— (1979). *J. comput. Phys.* **32,** 145–67.

—— —— (1981). *J. comput. Phys.* **41,** 357–88.

Boreli, M. (1955). *Trans. Am. Geophys. Un.* **36,** 664–72.

Brauner, C., Frémond, M., and Nicolaenko, B. (1983). In Fasano and Primicerio (1983). pp. 365–79.

Brebbia, C. A. (1980). *New developments in boundary element methods.* CMLP Publications Ltd, Southampton.

—— Chang, O. (1979). *Adv Engng Software J.* **I,** No. 3.

—— Dominguez, J. (1977). *Appl. math. model.* **I.**

Brezis, H. (1970). In *Symposia in Pure Math.* **18,** 28–38. American Mathematical Society, Providence R.I.

—— (1972*a*). *J. Math. pures appl.* **51,** 1–168.

—— (1972*b*). *C.r. hebd. Séanc. Acad. Sci., Paris* **274A,** 310–312.

—— Kinderlehrer, D., and Stampacchia, G. (1978). *C.r. hebd. Séanc. Acad. Sci., Paris* **287,** 711–14.

Bruch, J. C. (1973). *J. Hydraul. Div. ASCE* **99,** HY3, 395–403.

—— (1979*a*). *Int. J. num. anal. Meth. Geomech.* **3,** 23–36.

—— (1979*b*). *J. Hydrol.* **41,** 31–41.

—— (1980). *Adv. Water Res.* **3,** 65–80; 115–24.

—— Caffrey, J. (1979). In *The mathematics of finite elements and applications III.* (ed. J. R. Whiteman) pp. 123–33. Academic Press, New York.

—— Mirnateghi, A. (1982). In *Proc. 4th. Int. Symp. on The finite element method in flow problems, Chuo University, Tokyo, Japan* (Ed. T. Kawai) pp. 665–72.

—— Sayle, F. C., and Sloss, J. M. (1978). *J. Hydrol.* **36,** 247–59.

—— Sloss, J. M. (1978). *Water Res. Res.* **14**(1), 119–24.

—— —— (1981). *In Numerical methods for coupled problems. Proceedings of the International Conf. Swansea, Wales* (eds. E. Hinton, P. Bettess, R. W. Lewis). pp. 567–77.

—— —— Remar, J. (1982). In *The mathematics of finite elements and applications IV.* (ed. J. Whiteman) pp. 251–8. Academic Press, New York.

Buckmaster, J. (1979). *Combust. Sci. Technol.* **20,** 33–40.

Budak, B. M. and Gapenko, Yu. L. (1971). *Trudy Vycisl. Centra Moskov. Gos. Univ.* 235–313. (In Russian).

—— Moskal, M. S. (1970). *Dokl. Akad. Nauk. SSSR* **191,** 751–4. (In Russian).

—— —— (1971). Solutions of Stefan problems. *Proc. of Comp. Centre of Moscow State University.* 87–113. (In Russian).

—— Soboleva, E. N., and Uspenskii, A. B. (1965). *USSR comput. Math. math. Phys.* **5**(5), 59–76.

Budhia, H. and Kreith, F. (1973). *Int. J. Heat Mass Transfer* **16,** 195–211.

Bunch, J. R. and Rose, D. J. (1976). *Sparse matrix computations.* Academic Press, New York.

Butterfield, R. (1978). *An application of the BEM to potential flow problems in generally inhomogeneous bodies. Recent advances in boundary element methods.* Pentech Press, London.

Buzbee, B. L. and Dorr, F. W. (1974). *SIAM J. numer. Anal.* **11,** 753–63.

Caffarelli, L. A. (1976). *Archs ration. Mech. Analysis* **63,** 77–86.

—— Friedman, A. (1978*a*). *Archs. ration. Mech. Analysis* **68**(2), 125–54.

—— —— (1978*b*). *Ind. Univ. Math. J.* **27,** 551–80.

Caffrey, J. and Bruch, J. C. (1979). *Adv. Water Res.* **2,** 167–76.

—— —— (1980). *Proc. 3rd. Int. Conf. Finite elements in flow problems Vol. I, Banff, Canada.* pp. 280–9.

Cannon, J. R. (1978). In Wilson *et al.* (1978). pp. 3–24.

—— Douglas, J., and Hill, C. D. (1967). *J. math. Mech.* **17,** 21–33.

—— Hill, C. D. (1967). *J. math. Mech.* **17,** 1–19; 433–441.

—— —— (1970). *Ind. Univ. Math. J.* **20,** 429–54.

—— Primicerio, M. (1973). *SIAM J. math. Anal.* **4,** 141–8.

Capriz, G. and Cimatti, G. (1980). In Magenes (1980) Vol. II, pp. 117–31.

Carbone, L. and Valli, A. (1976). *Rend. Ac. Naz. Lincei* **61**(8), 161–4.

—— —— (1977). *Appl. Math. Opt.* **4,** 1–4.

Carré, B. A. (1961). *Comp. J.* **4,** 73–78.

Carslaw, H. S. and Jaeger, J. C. (1959). *Conduction of heat in solids.* Clarendon Press, Oxford.

Chadam, J. and Ortoleva, P. (1982). Private communication.

Chalmers, B. (1964). *Principles of solidification.* Wiley, New York.

Chambré, P. L. (1956). *Q. Jl. Mech. appl. Math.* **9,** 224–33.

Charni, I. A. (1951). *Dokl. Akad. Nauk SSSR* **79,** 937–40.

Chernousko, F. L. (1970). *Int. chem. Engng.* **10,** No. 1, 42–8.

Christiansen, S. and Rasmussen, H. (1976). *J. Inst. Math. Appl.* **18,** 295–307.

Christopherson, D. G. (1941). *Proc. Inst. mech. Engrs.* **146,** 126–35.

—— Southwell, R. V. (1938). *Proc. R. Soc.* **A168,** 317–50.

Chu, W. H. (1971). *J. comput. Phys.* **8,** 392–408.

Chuang, Y. K. and Ehrich, O. (1974). *Int. J. Heat Mass Transfer* **17,** 945–53.

—— Reinisch, D., and Schwerdfeger, K. (1975). *Met. Trans.* **6A,** 235–8.

—— Szekely, J. (1971). *Int. J. Heat Mass Transfer* **14,** 1285–94.

—— —— (1972). *Int. J. Heat Mass Transfer* **15,** 1171–4.

Churchill, S. W. and Evans, L. B. (1971). *J. Heat Transfer* **93C,** 234–6.

—— Gupta, J. P. (1977). *Int. J. Heat Mass Transfer* **20,** 1251–3.

Ciavaldini, J. F. (1975). *SIAM J. numer. Anal.* **12,** 464–87.

Ciment, M. and Guenther, R. B. (1974). *Appl. Anal.* **4,** 39–62.

—— Sweet, R. A. (1973). *J. Comput. Phys.* **12,** 513–25.

Citron, S. J. (1962). *Proc. Fourth U.S. Nat. Congress of Appl. Mech. ASME,* 1221–7.

Collatz, L. (1978). In *Dundee Conference on Numerical Analysis* (ed. G. A. Watson) *Lecture Notes in Mathematics Vol. 630.* Springer-Verlag, Berlin.

—— (1980). In *Magenes* (1980). Vol. II, pp. 147–59.

Comincioli, V. (1974). *Pubbl. Lab. Anal. numer. CNR* **79,** Pavia.

—— (1975). *Appl. Math. Opt.* **1,** 313–36.

—— (1980). In Magenes (1980). Vol. I, pp. 197–208.

—— Guerri, L. (1976). *Comput. Methods appl. Mech. Eng.* **7,** 153–78.

—— Guerri, L., and Volpi, G. (1971). *Pubbl. Lab. Anal. numer., CNR* **17,** Pavia.

—— Torelli, A. (1979). *Calcolo* **16,** 93–124.

Comini, G., del Guidice, S., Lewis, R. W., and Zienkiewiez, O. C. (1974). *Int. J. numer. Meth. Engng.* **8,** 613–24.

Constable, T. B. and Evans, N. T. S. (1975). *Respir. Physiol.* **25,** 175–90.

—— —— (1976). *Adv. Exp. Med. Biol.* **75,** 611–18.

Cooley, R. L. (1971). *Water Resources Res.* **7,** 1607–25.

Cottle, R. W. (1974). Tech. Report No. SOL 74-13, Department of Operational Research, Stanford University.

—— (1979). *Proc. Comput. Meth. Appl. Sci. Engng. Springer Lec. Notes No.* **704,** 37–52.

—— Goheen, M. S. (1976). Tech. Rep. No. SOL 76–7, Department of Operational Research, Standord University.

—— Golub, G. H., and Sacher, P. S. (1978). *Appl. Math. Optim.* **4,** 347–363.

—— Sacher, P. S. (1977). *Appl. Math. Optim.* **3,** 321–40.

Couch, E. J., Keller, H. H., and Watts, J. W. (1970). *J. Can. Petrol. Technol.* **2**(2), 107–11.

Courant, R. and Hilbert, D. (1953). *Methods in mathematical physics Vol. I.* pp. 260–64. Interscience Publishers, New York.

Crank, J. (1957a). *Q. Jl. Mech. appl. Math.* **10,** 220–231.

—— (1957b). *Trans. Faraday Soc.* **53,** 1083–1091.

—— (1975a). *The mathematics of diffusion.* Clarendon Press, Osford.

—— (1975b). In Ockendon and Hodgkins (1975) pp. 192–207.

—— (1981). In *Numerical methods in heat transfer.* (ed. R. W. Lewis, K. Morgan, and O. C. Zienkiewicz) pp. 177–200. Wiley, New York.

—— Crowley, A. B. (1978). *Int. J. Heat Mass Transfer* **21,** 393–8.

—— —— (1979). *Int. J. Heat Mass Transfer* **22,** 1331–7.

—— Gupta, R. S. (1972a). *J. Inst. Math. Appl.* **10,** 19–33.

—— —— (1972b). *J. Inst. Math. Appl.* **10,** 296–304.

—— —— (1975). *Int. J. Heat Mass Transfer* **18,** 1101–7.

—— Ozis, T. (1980). *J. Inst. Math. Appl.* **26,** 77–85.

—— —— (1981). Brunel University Math. Report TR/03/81.

—— Phahle, R. D. (1973). *Bull. Inst. Math. Appl.* **9,** 12–14.

Crowley, A. B. (1978). *Int. J. Heat Mass Transfer* **21,** 215–17.

—— (1979). *J. Inst. Math. Appl.* **24,** 43–57.

—— (1980). In Albrecht *et al.* (1980).

—— (1981). Numerical solution of flame front free-boundary problems. The Royal Military College of Science: College Report 81004. Also in Fasano and Primicerio (1983) pp. 493–504.

—— Ockendon, J. R. (1977). *J. Inst. Math. Appl.* **20,** 269–81.

—— —— (1979). *Int. J. Heat Mass Transfer* **22,** 941–6.

Cryer, C. W. (1968). MRC Techn. Summary Rep. 894, University of Wisconsin, Madison.

—— (1970). *J. Ass. comput. Mach.* **17,** 397–411.

—— (1971a). *SIAM J. Control* **9,** 385–92.

—— (1971b). *Math. Comput.* **25,** 435–43.

—— (1976a). MRC Techn. Summary Rep. 1657, University of Wisconsin, Madison.

—— (1976b). MRC Techn. Summary Rep. 1693, University of Wisconsin, Madison.

—— (1977). MRC Techn. Summary Rep. 1793, University of Wisconsin, Madison.

—— (1980). In Magenes (1980). Vol. I, pp. 109–31.

—— Fetter, H. (1977). MCR Techn. Summary Rep. 1761, University of Wisconsin, Madison.

—— —— (1979). In *Constructive methods for non-linear boundary value problems* (ed. J. Albrecht, L. Collatz, and K. Kirchgassner). Birkhauser, Basle.

Dagan, G. (1967). *J. Geophys. Res.* **72**(4), 1183–93.

Dahmardah, H. O. and Mayers, D. F. (1983). *IMA J. numer. Analysis* **3**, 81–5.

Damlamian, A. (1977). *Commun. partial diff. Eqns.* **2**, 1017–44.

—— (1979). *Recent methods in non-linear analysis* (ed. E. De Giorgi, *et al.*) pp. 25–8. Pitagora Editrice, Bologna.

—— (1980). In Magenes, E. (1980) Vol. I, pp. 267–275.

Danckwerts, P.V. (1950). *Trans. Faraday Soc.* **46**, 701–12.

Darcy, H. (1856). *Les fontaines publiques de la ville de Dijon.* Paris.

Datzev, A. (1970). Sur le problème linéaire de Stefan. *Mémoires de Sciences Physique, Fascicule* **69**. Gautier-Villars, Paris.

Davies, A. J. (1980). *The finite element method: a first approach.* Clarendon Press, Oxford.

Davis, G. B. and Hill, J. M. (1982). *J. Inst. Math. Appl.* **29**, 99–111.

Delves, L. M. (1979). In Gladwell and Wait (1979) Chapter 6.

Dix, R. C. and Cizek, J. (1970). In *Heat Transfer Vol I, 4th Int. Heat Transfer Conf., Paris, Versailles.* Elsevier, Amsterdam.

Doha, E. H. (1977). Ph.D. Thesis, University of Lancaster.

Douglas, J. and Gallie, J. M. (1955). *Duke math. J.* **22**, 557–71.

Duda, J. L., Malone, M. F., Notter, R. H., and Ventras, J. S. (1975). *Int. J. Heat Mass Transfer* **18**, 901–10.

Dupuit, J. (1863). *Études théoriques et pratiques sur le movement des eaux dans les canaux découverts et à travers les terrains perméables.* 2nd edn. Dunod, Paris.

Durack, D. and Wendroff, B. (1977). *Nucl. Sci. Engng.* **64**, 187–91.

Duvaut, G. (1973). *C.r. hebd. Séanc. Acad. Sci., Paris* **276A**, 1461–3.

—— (1975). In Ockendon and Hodgkins (1975). pp. 173–81.

—— Lions, J. L. (1976). Inequalities in mechanics. *Grund der math. Wiss* **219**. Springer, Berlin.

Eckhardt, U. (1975). *Z. angew. Math. Mech.* **55**, 237–8.

Ehrlich, L. W. (1958). *J. Ass. comput. Mach.* **5**, 161–76.

Elliott, C. (1976). D.Phil. thesis, Oxford University.

—— (1980). *J. Inst. Math. Appl.* **25**, 121–31.

—— (1981). *IMA J. numer. Analysis* **1**, 115–25.

—— Janovsky, V. (1979). In *The mathematics of finite elements and applications III.* (ed. J. R. Whiteman) pp. 97–104. Academic Press.

—— —— (1981). *Proc. R. Soc. Edin.* **88A**, 93–107.

—— —— (1983). *IMA J. numer. Analysis* **3**, 1–9.

—— Ockendon, J. R. (1982). *Weak and variational methods for moving boundary problems.* Research Notes in Mathematics 59. Pitman, London.

Evans, N. T. S. and Gourlay, A. R. (1977). *J. Inst. Math. Appl.* **19**, 239–51.

Evans, G. W., Isaacson, E. and Macdonald, J. K. L. (1950). *Q. Jl. appl. Math.* **8**, 312–20.

Eyres, N. R., Hartree, D. R., Ingham, J., Jackson, R., Sarjant, R. J., and Wagstaff, S. M. (1946). *Phil. Trans. R. Soc.* **A240**, 1–57.

Fasano, A. (1980). In Magenes (1980). Vol. II, pp. 237–47.

—— Primicerio, M. (1973). *Z. angew. Math. Mech.* **53**, 341–8.

—— —— (1977). *J. math. Anal. Appl.* **57**, 694–723; **58**, 202–31; **59**, 1–14.

—— —— (1979a). *J. math. Anal. Appl.* **72**, 247–273.

—— —— (1979b). *Riv. mat. Univ. Parma* **5/2**, 615–34.

—— —— (1979c). *Riv. mat. Univ. Parma* **5/2**, 635–45.

—— —— (1979d). *J. Inst. Math. Appl.* **23**, 503–17.

—— —— (1981). *Q. appl. Maths.* **38,** 439–60.

—— —— (eds.) (1983). *Free boundary problems—theory and applications.* Vols I and II. Research Notes in Mathematics 78 and 79. Pitman, London.

—— —— Rubinstein, L. (1980). *J. Inst. Math. Appl.* **26,** 327–47.

Federenko, R. P. (1975). *USSR Comput. Math. math. Phys.* **15**(5), 246–51.

Ferriss, D. H. (1975). In Ockendon and Hodgkins (1975). pp. 251–5.

—— Hill, S. (1974). Report NAC45, National Physical Laboratory, Teddington.

Finn, W. D. L. (1967). *J. Soil Mech. Fdns. Div., Am. Soc. civ. Engrs.* **93,** SM6, 41–8.

Finn, W. D. L. and Varoğlu, E. (1977). In *Finite elements in water resources.* (eds. W. G. Grey, G. F. Pinder and C. A. Brebbia) pp. 3.115–31. Pentech Press, London.

—— —— (1978). *J. Comp. Fluids* **6,** 103–14.

—— —— (1980). In *Proc. First Int. Conf. on Numerical methods for non-linear problems, Swansea,* pp. 623–48.

Fisher, I. and Medland, I. C. (1974). In *Finite Meth. Engng.* pp. 767–83. Clarendon Press, Australia.

Fitz-Gerald, J. M. and McGeough, J. A. (1969). *J. Inst. Math. Appl.* **5,** 387–408; **5,** 409–21.

Fitz-Gerald, J. M. and McGeough, J. A. (1970). *J. Inst. Math. Appl.* **6,** 102–10.

Fix, G. J. (1978). In Wilson *et al.* (1978). pp. 109–28.

Fleishman, B. A., Gingrich, R., and Mahar, T. J. (1978). In Wilson *et al.* (1978). pp. 41–55.

—— Mahar, T. J. (1977). *Nonlin. Analysis* **1,** 561–9.

Forsyth, P. and Rasmussen, H. (1979). *J. Inst. Math. Appl.* **24,** 411–24.

Fox, L. (1975). In Ockendon and Hodgkins (1975). pp. 210–41.

—— (1979). In Gladwell and Wait (1979). Chapters 3, 20.

—— Sankar, R. (1973). *J. Inst. Math. Appl.* **12,** 49–54.

France, P. W. (1974). *J. Hydrol.* **21,** 381–98.

France. P. W., Parekh, C. J., Peters, J. C., and Taylor, C. (1971). *Proc. Am. Soc. civ. Engrs.* **97,** 165–79.

Frank, F. C. (1950). *Proc. R. Soc.* **A201,** 586–99.

Frémond, M. (1974). Frost propagation in porous media. In *Computational methods in non-linear mechanics.* University of Texas at Austin.

—— (1980). Méthodes variationnelles en calcul des structures. Cours de D. E. A. École Nationale des Ponts et Chaussées, Paris.

Friedman, A. (1964). *Partial differential equations of parabolic type.* Prentice Hall, New Jersey.

—— (1968). *Trans. Am. math. Soc.* **133,** 51–87; 89–114.

—— (1975). *J. funct. Anal.* **18,** 151–76.

—— (1976). *Ind. Univ. Math. J.* **25,** 577–92.

—— (1979). *SIAM Rev.* **21,** 213–21.

—— Jensen, R. (1977). *Archs. ration. Mech. Analysis* **67,** 1–24.

—— Kindlehrer, D. (1975). *Ind. Univ. Math. J.* **24,** 1005–35.

—— Torelli, A. (1977). *Nonlinear Analysis* **1**(5), 503–45; (1978) **2,** 513–18.

Furzeland, R. M. (1974). M.Sc. dissertation, Brunel University.

—— (1977*a*). Brunel University Math. Report TR/76.

—— (1977*b*). Ph.D. Thesis, Brunel University; also Brunel University Math. Report TR/77.

—— (1979*a*). *Bull. Inst. Math. Appl.* **15,** 172–6.

—— (1979*b*). Analysis and computer packages for Stefan problems. Internal Report, Oxford University Computing Laboratory.

—— (1980). *J. Inst. Math. Appl.* **26,** 411–29.

Galib, T. A., Bruch, J. C., and Sloss, J. M. (1981). *Int. J. bio-med. Comput.* **12,** 157–80.

Garabedian, P. R. (1956). *Bull. Am. math. Soc.* **62,** 219–35.

Gastaldi, F. (1979). *Boll. U. mat. ital.* **16A**(5), 148–56.

Gauss, C. F. (1809). Werke VII, 150–2.

Geiman, H. and Rubinstein, L. (1974). *Bull. math. Biophys.* **36,** 379–401.

Gelder, D. and Guy, A. G. (1975). In Ockendon and Hodgkins (1975). pp. 71–90.

Gelinas, R. J., Doss, S. K., and Miller, K. (1981). *J. comput. Phys.* **40,** 202–49.

George, J. A. (1971). Tech. Report No. 208, Computer Science Department, Stanford University.

—— Damle, P. S. (1975). *Int. J. numer. Meth. Engng.* **9,** 239–45.

Gilardi, G. (1976). *Boll. U. mat. ital.* **13B,** 138–159.

—— (1977). *Annali. Mat. pura appl.* **113**(14), 1–17.

—— (1979). *Commun. on partial diff. Eqns.* **4,** 1099–122.

—— (1980). In Magenes (1980) Vol. I, pp. 209–17.

Gladwell, I. and Wait, R. (eds.) (1979). *A survey of numerical methods for partial differential equations.* Clarendon Press, Oxford.

Glasser, D. and Kern, J. (1978). *A.I.Ch.E.Jl.* **24**(2), 161–70.

Glowinski, R. (1971). La methode de relaxation. Pendiconte di Mathematica 14, Universita Roma.

—— (1976). Report No. 76006, Laboratoire Analyse Numerique, Universite Paris VI.

Gnanadoss, A. A. and Osborne, M. R. (1964). *Q. Jl. Mech. appl. Math.* **17,** 241–6.

Goodling, J. S. and Khader, M. S. (1974). *J. Heat Transfer* **96,** 114–15.

Goodman, T. R. (1958). *Trans. ASME* **80,** 335–42.

—— (1961). *J. Heat Transfer, Trans. ASME*(*C*) **83,** 83–6.

—— (1964). In *Advances in heat transfer.* (ed. T. F. Irvine and J. P. Hartnett). Academic Press, New York.

—— Shea, J. J. (1960). *J. appl. Mech.* **27,** 16–24.

Goodrich, L. E. (1978). *Int. J. Heat Mass Transfer* **21,** 615–21.

Gordon, W. J. and Hall, C. A. (1973). *Int. J. numer. Meth. Engng.* **7,** 461–77.

Gourlay, A. R. (1970). *J. Inst. Math. Appl.* **6,** 375–90.

Gupta, R. S. (1973). Ph.D. Thesis, Brunel University.

—— (1974). *Comput. Meth. appl. Mech. Engng.* **4,** 143–52.

—— Kumar, D. (1980). *Comput. Meth. appl. Mech. Engng.* **23,** 101–9.

—— —— (1981*a*). *Int. J. Heat Mass Transfer* **24,** 251–9.

—— —— (1981*b*). *Comput. Meth. appl. Mech. Engng.* **29,** 233–9.

Hager, W. W. (1980). In Magenes (1980). Vol. II, pp. 333–45.

Hall, H. P. (1955). *Houille Blanche* **10,** 8–35.

Hamel, G. (1934). *Z. angew. Math. Mech.* **14,** 129–57.

Hansen, E. B. (1980). Methoden und Verfahren der mathematischen Physik (eds. B. Brosowski and E. Martenson). Vol. 21. Peter D. Lang, Frankfurt am Main, Germany.

—— (1983). In Fasano and Primicerio (1983). pp. 516–25.

—— Holm, A. M. (1980). *Z. angew. Math. Mech.* **60,** 249–51.

—— Hougaard, P. (1974). *J. Inst. Math. Appl.* **13,** 385–98.

Hantush, M. S. (1964). *Advances of hydroscience Vol. I.* Academic Press, New York.

Harr, M. E. (1962). *Groundwater and seepage.* McGraw-Hill, New York.

Heitz, W. L. and Westwater, J. W. (1970). *Int. J. Heat Mass Transfer* **13,** 1371–5.

Hodgkins, W. R. and Waddington, J. F. (1975). In Ockendon and Hodgkins (1975). pp. 26–37.

Hoffman, K.-H. (ed.) (1977). Frei Universitat (Berlin) Fachbereich Mathematik Volumes I (preprint No. 22), II (No. 28), III (No. 34).

—— Sprekels, J. (1982). Math. Inst. Universitat Ausburg Preprint No. 1.

—— Niezgodka, M. (1983). In Fasano and Primicerio (1983). pp. 431–62.

Höhn, W. (1978). *Int. Ser. Num. Math.* **39,** 191–213 eds. J. Albrecht, L. Collatz, and G. Hammerlin). Birkhauser Verlag.

Horie, Y. and Chehl, S. (1975). *J. Crystal Growth* **29,** 248–56.

Hornung, U. (1978). In *ISNM volume on Oberwolfach Conference on Numerical solution of differential equations, May 1977.* (eds. J. Albrecht, L. Collatz, and G. Hammerlin). Birkhauser Verlag.

Horvay, G. (1962). *Proc. 4th U.S. Nat. Congress Appl. Mech. Vol. 2.*

—— (1965). *Int. J. Heat Mass Transfer* **8,** 195–243.

—— Cahn, J. W. (1961). *Acta Metall.* **9,** 695–705.

Hougaard, P. (1977). Report S.9. Danish Centre for Appl. Math. and Mech.

Howarth, J. A. and Poots, G. (1976). *Mech. Res. Commun.* **3,** 509–14.

Hsu, C. F., Sparrow, E. M., and Patankar, S. V. (1981). *Int. J. Heat Mass Transfer* **24,** 1335–45.

Huber, A. (1939). *Z. angew. Math. Mech.* **19,** 1–21.

Ichikawa, Y. (1978). M.S. Thesis, University of Texas at Austin.

—— Kikuchi, N. (1979). *Int. J. numer. Meth. Eng.* **14,** 1197–1220.

Imber, M. and Huang, P. N. S. (1973). *Int. J. Heat Mass Transfer* **16,** 1951–4.

Ivantsov, G. P. (1947). *Dokl. Akad. Nauk. SSSR* **58,** 567. Mathematical Physics. Translated by G. Horvay. Report No. 60-RL-(2511M), G. E. Research Lab. Schenectedy, New York, 1960.

Jamet, P. (1978). *SIAM J. Numer. Anal.* **15,** 912–28.

—— (1979). MRC Technical Summary Report 1985, University of Wisconsin, Madison.

Jaswon, M. A. and Symm, G. T. (1977). *Integral equation methods in potential theory and electrostatics.* Academic Press, London.

Jensen, R. (1977). *Ind. Univ. Math. J.* **26,** 1121–35.

—— (1980). In Magenes (1980). Vol. I, pp. 133–49.

Jeppson, R. W. (1968*a*). *Water Resources Res.* **4,** 1277–88.

—— (1968*b*). *Water Resources Res.* **4,** 435–45.

—— (1968*c*). *J. Irrig. Drain. Div. ASCE* **94,** No. IR1, Proc. Paper 5835, 23–37.

—— (1969). *J. Hydraul. Div., Proc. Am. Soc. Civil Engrs* **95,** 364–81.

—— (1972). *J. Engng. Mech. Div. ASCE* **98,** 789–812.

Jerome, J. W. (1976). In *Dundee Conference on Numerical Analysis.* (ed. G. A. Watson). Lecture Notes in Mathematics Vol. 506. Springer-Verlag, Berlin.

—— (1977). *J. Diff. Eqn.* **26,** 240–61.

Jiji, L. M. (1970). *J. Franklin Inst.* **289,** 281–91.

Johnson, C. (1976). *SIAM J. Numer. Anal.* **13,** 599–606.

Jones, I. P. and Thompson, C. P. (1980). *AERE Harwell Report 9765.* HMSO, London.

Kamenomostskaja, S. L. (1961). Translation for U.S. Army Cold Regions Research and Engng. Lab. from *Mat. Sb.* **53,** 489–514.

Kamin (Kamenomostskaja), S. L. (1976). *Arch. ration. Mech. Analysis* **60,** 171–83.

Kealy, C. D. and Busch, R. A. (1971). *Rep. Invest. RI7477.* U. S. Bureau of Mines, Washington, D.C.

—— Williams, R. E. (1971). *Water Resources Res.* **7,** 143–54.

Kikuchi, N. (1977). *Int. J. numer. Anal. Meth. Geomech.* **1**, 283–97; and *Q. appl. Math.* **35**, 149–63.
—— Ichikawa, Y. (1979). *Int. J. numer. Meth. Engng.* **14**, 1221–39.
Kirkham, D. (1964). *J. Geophys. Res.* **69**, 2537–49.
Koh, J. C. Y., Price, J. F., and Colony, R. (1969). *Prog. Heat Mass Transfer* **2**, 225–47.
Kolodner, I. I. (1956). *Comm. pure appl. Maths* **9**, 1–31.
—— (1957). *Comm. pure appl. Maths* **10**, 220–31.
Kozeny, J. (1931). Wasserkraft und Wasserwirtshaft No. 3.
Kreith, F. and Romie, F. E. (1955). *Proc. Phys. Soc.* **68**, 277–91.
Kumar, D. (1982). Ph.D. Thesis, Roorkee University, U. P. India.
Lamb, H. (1932). *Hydrodynamics.* Cambridge University Press.
Lamé, M. M. and Clapeyron, B. P. E. (1831). *Ann. Chimie Phys.* **47**, 250–6.
Landau, H. G. (1950). *Q. appl. Math.* **8**, 81–94.
Langer, J. S. and Turski, L. A. (1977). *Acta Metall.* **25**, 1113–19; **25**, 1121–37.
Langford, D. (1973). *Int. J. Heat Mass Transfer* **16**, 2424–8.
Lardner, T. J. and Pohle, F. V. (1961). *J. appl. Mech.* **28**, 310–12.
Larock, B. E. and Taylor, C. (1976). *Int. J. numer. Meth. Engng.* **10**, 1143–52.
Lauwerier, H. A. (1955). *Appl. Scient. Res. S. A.* **5**(2, 3).
Lax, P. D. (1954). *Comm. pure appl. Maths* **7**, 159–93.
Lazaridis, A. (1969). Doctoral dissertation, Columbia University.
—— (1970). *Int. J. Heat Mass Transfer* **13**, 1459–77.
Lembke, K. E. (1887). *The Engineer. J.* Ministry Communications Nos. 17–19. (in Russian).
Lennon, G. P. (1980). M.Sc. Thesis, University of Cornell.
Lewy, H. and Stampaccia, G. (1969). *Comm. pure Appl. Maths* **22**, 153–88.
Li, C. H. (1983). *IMA J. numer. Analysis* **3**, 87–107.
Liggett, J. A. (1977). *J. Hydrol. Div. ASCE* **HY4**, 353–65.
—— Liu, P. L.-F. (1977). *Proc. Symp. Appl. Comp. Methods, University of California*, pp. 357–65.
—— —— (1979). *Water Resources Res.* **5**(4), 861–6.
Lightfoot, N. M. H. (1929). *Proc. Lond. Math. Soc.* (2)**31**, 97–116.
Lions, J. L. (1969). *Quelques méthodes de resolution des problèmes aux limites non linéares.* Dunod, Paris.
—— (1976). *Sur quelques questions d'analyse, de mécanique et de contrôle optimal.* Les Presses de l'Université de Montréal, Montréal.
—— Stampacchia, G. (1967). *Comm. pure appl. Maths* **20**, 493–519.
Liu, P. L.-F. and Liggett, J. A. (1977). *Proc. First Int. Conf. Appl. Num. Modelling, University of Southampton.* pp. 559–69.
—— —— (1978). *Water Resources Res.* **14**(3), 385–90.
—— —— (1979). *J. Hydrol. Div. ASCE* **105**, HY3, 171–83.
Longworth, D. (1975). In Ockendon and Hodgkins (1975). pp. 54–61.
Lotkin, M. (1960). *Q. Appl. Math.* **18**, 79–85.
Magenes, E. (ed.) (1980). *Free boundary problems Vols. I and II. Proceedings of a seminar in Pavia (1979).* Instituto Nazionale di Alta Matematica Francesco Severi, Rome.
Maione, U. and Franzeiti, S. (1969). *13th Congress Int. Ass. Hyd. Res. Kyoto.* pp. 191–204.
Mangassarian, O. L. (1977). *J. Optimiz Theory Appl.* **22**, 465–85.
Mason, J. C. and Farkas, L. (1972). In *Inf. Process.* **71**, 1305–1310.
Mastanaiah, K. (1976). *Int. J. numer. Meth. Engng.* **10**, 833–44.
Mauersberger, P. (1965). Pure appl. Geophys. **60**, 101–6.

McGeough, J. A. (1974). *Principles of electrochemical machining*. Chapman and Hall, London.
—— Rasmussen, H. (1974). *J. Inst. Math. Appl.* **13,** 13–21.
Megerlin, F. (1968). *Forsch. Ing.-Wes.* **9,** 40–6.
Meirmanov, A. M. (1981*a*). *Dokl. Akad. Nauk. SSSR,* **258,** 547–9. (In Russian).
—— (1981*b*). *Matem. Sbornik* **115,** 532–43.
Meyer, G. H. (1970). *Numer. Math.* **16,** 248–67.
—— (1973). *SIAM J. numer. Anal.* **10,** 522–38.
—— (1975). *Int. J. numer. Meth. Engng.* **9,** 669–78.
—— (1976). *Brunel University Math. Report* TR/62.
—— (1977*a*). *Int. J. numer. Meth. Engng.* **11,** 741–52.
—— (1977*b*). *J. Inst. Math. Appl.* **20,** 317–29.
—— (1977*c*). *SIAM Rev.* **19,** 17–34.
—— (1978*a*). Inst. Comp. Math., Brunel University, Report 78–13.
—— (1978*b*). In Wilson *et al.* (1978). pp. 73–89.
—— (1978*c*). *J. appl. Math. Comp.* **4,** 283–306.
—— (1978*d*). Inst. Comp. Math., Brunel University, Report 78–12.
—— (1978*e*). *Numer. Math.* **29,** 329–44.
—— (1980). In Magenes (1980). Vol. I, pp. 151–73.
—— (1981). *Int. J. Heat Mass Transfer* **24,** 778–81.
Miellou, J. C. (1971). *C.r. hebd. Séanc. Acad. Sci. Paris* **273,** 1257–9.
Mikhailov, M. D. (1975). *Int. J. Heat Mass Transfer* **18,** 797–804.
—— (1976). *Int. J. Heat Mass Transfer* **19,** 651–5.
Miller, J. V., Morton, K. W., and Baines, M. J. (1978). *J. Inst. Math. Appl.* **22,** 467–77.
Mogel, T. R. and Street, R. L. (1974). *J. Ship Res.* **18,** 22–31.
Moiseynko, B. D. and Samarskii, A. A. (1965). *USSR Comput. Math. math. Phys.* **5,** 816–27.
Morgan, K., Lewis, R. W., and Zienkiewicz, O. C. (1978). *Int. J. numer. Meth. Engng.* **12,** 1191–5.
Mosco, U. (1973). *Constructive aspects of functional analysis.* (ed. G. Geymonat). Part 22, pp. 499–685. Edizioni Cremonese, Rome.
Motz, H. (1946). *Q. appl. Math.* **4,** 371–7.
Muehlbauer, J. C. and Sunderland, J. E. (1965). *Appl. Mech. Rev.* **8,** pp. 951–9.
Mullins, W. W. and Sekerka, J. (1963). *J. appl. Phys.* **34,** 323–9; (1964). *J. appl. Phys.* **35,** 444–51.
Murray, W. D. and Landis, F. (1959). *J. Heat Transfer, Trans. ASME* (c)**81,** 106–12.
Muskat, M. (1934). *Physics* **5,** 250–64.
—— (1935). *Physics* **6,** 402–15.
—— (1937). *The flow of homogeneous fluids through porous media*. McGraw-Hill, New York; reprinted by J. W. Edwards, Publisher, Inc., Ann Arbor, 1946.
Neuman, S. P. (1975). In *Finite elements in fluids Vol. I* (eds. R. H. Gallagher, J. T. Oden, C. Taylor, and O. C. Zienkiewicz) pp. 201–217. Academic Press, London.
—— Witherspoon, P. A. (1970). *Water Resources Res.* **6,** 889–97; 1376–87.
—— —— (1971). *Water Resources Res.* **7,** 611–23.
Niezgodka, M. (1979). *Control Cybernet.* **8,** 23–42.
—— (1980). In Albrecht *et al.* (1980).
—— (1983). In Fasano and Primicerio (1983). pp. 321–48.
—— Pawlow, I. (1983). *Appl. Maths Optimiz.* **9,** 193–224.
Nitsche, J. A. (1980). In Magenes (1980). Vol. I, pp. 277–318.

Noble, B. (1975). In Ockendon and Hodgkins (1975). pp. 208–9.
Nogi, T. (1974). *Publ. Res. Inst. Math. Sci. Kyoto Univ.* **9,** 543–75.
Numerov, S. N. (1942). *Prikl. Mat i Mekh.* **6.**
Oberkampf, W. L. (1976). *Int. J. numer. Meth. Engng.* **10,** 211–23.
Ockendon, J. R. (1975). In Ockendon and Hodgkins (1975). pp. 138–49.
—— (1978). In Wilson *et al.* (1978). pp. 129–45.
—— (1980). In Magenes (1980). Vol. II, pp. 443–78.
—— Hodgkins, W. R. (eds.) (1975). *Moving boundary problems in heat flow and diffusion.* Clarendon Press, Oxford.
Oden, J. T. and Kikuchi, N. (1978). In Wilson *et al.* (1978). pp. 147–64.
—— —— (1980). *Int. J. Energy Sci.* **18,** 1173–284.
O'Leary, D. P. (1977). Stanford University Report CS-77-638.
Oleinik, O. A. (1960). *Sov. Math. Dokl.* **1**(2), 1350–4.
Outmans, H. D. (1964). *J. geophys. Res.* **69,** 3383–6.
Ozis, T. (1981). Ph.D. Thesis, Brunel University.
Patel, P. D. (1968). *Am. Inst. Aeronautics Astro. J.* **6,** 2454.
Pawlow, I. (1981). In *Proceedings of 7th Internat. Summer School in Berlin (1979).* Ab. der Akad. der Wissen. Der DDR Ab. Math.-Natur.-Tech. No. 2N, 221–231.
Pedroso, R. I. and Domoto, G. A. (1973). *Int. J. Heat Mass Transfer* **16,** 1037–43.
Pekeris, C. L. and Slichter, L. B. (1939). *J. appl. Phys.* **10,** 135–7.
Peletier, L. A. and Gilding, B. H. (1976). *Arch. ration. Mech. Analysis* **61,** 127–40.
Pinkus, O. and Sternlicht, B. (1961). *Theory of hydrodynamic lubrication.* McGraw-Hill, New York.
Polubarinova-Kochina, P. Ya. (1938). *IZv. An (CCCP) Math. Series* No. 3.
—— (1939). *IZv. An (CCCP) Math. Series* Nos. 3, 5, 6.
—— (1962). *Theory of groundwater movement.* Princeton University Press.
Poots, G. (1962*a*). *Int. J. Heat Mass Transfer* **5,** 339–48.
—— (1962*b*). *Int. J. Heat Mass Transfer* **5,** 525–31.
Potter, D. (1973). *Computational physics.* Wiley, New York.
Pozzi, G. A. (1974*a*). *Boll. U. mat. ital.* **9**(4), 416–44.
—— (1974*b*). *Pub. Lab. Analisi Numerica CNR.* **81,** Pavia.
Price, R. H. and Slack, M. R. (1954). *Br. J. appl. Phys.* **5,** 285–7.
Primicerio, M. (1980). In Magenes (1980). Vol. I, pp. 451–60.
—— (1981*a*). Boll. U. mat. ital. **A-18,** 11–68.
—— (1981*b*). *Mushy regions in phase-change problems.* Talk presented at the Workshop on Applied Non-linear Functional Analysis, Berlin, Sept. 1981.
Pukhnachev, V. V. (1980). *Differ. Urav.* **16,** 492–500. (In Russian)
Quateroni, A. and Vinsintin, A. (1980). *Boll. U. mat. ital.* (5) **17-B,** 204–31.
Radojkovic, M. M. (1980). In Brebbia (1980). pp. 155–9.
Rahmer, H., Hoper, J., Heitland, W., Durst, J., and Kessler, M. (1977). In *Advances in experimental medicine and biology* **94,** pp. 643–8. Plenum Press, New York.
Rasmussen, H. and Christiansen, S. (1976). *J. Inst. Math. Appl.* **18,** 149–53.
—— McGeough, J. A. (1981). *J. Inst. Math. Appl.* **27,** 211–20.
—— Salhani, D. (1981). *J. Inst. Math. Appl.* **27,** 307–18.
Rathjen, K. A. and Jiji, L. M. (1971). *J. Heat Transfer, Trans. ASME(C)* **93,** 101–9.
Reid, J. D. (1979). In Gladwell and Wait (1979). Chapters 14, 15.
Reid, J. K. (1971). *Large sparse sets of linear equations.* Academic Press, London.

Remar, J., Bruch, J. C., and Sloss, J. M. (1982). *Numer. Math.* **40,** 143–68.

Richardson, S. (1972). *J. Fluid Mech.* **56,** 609–18.

Riemann-Weber (1912). *Die partiellen Differentialgleichungen der mathematischen Physik* edn. 5, Vol. 2.

Riley, D. S. and Duck, P. W. (1971). *Int. J. Heat Mass Transfer* **20,** 294–6.

—— Smith, F. T., and Poots, G. (1974). *Int. J. Heat Mass Transfer* **17,** 1507–76.

Rippin, D. W. T. and Davidson, J. F. (1967). *Chem. Engr. Sci.* **22,** 217–28.

Risenkampf, B. K. (1940). *Proceedings, State University of Saratovsky* Vol. XV, No. 5, Part 3, Chap. IV.

Rogers, J. C. W. (1977). *J. Inst. Math. Appl.* **20,** 261–8.

—— (1980). In Magenes (1980). Vol. I, pp. 333–82.

Rose, M. E. (1960). *Math. Comput.* **14,** 249–56.

Rubinstein, L. I. (1971). *Trans. Math. Monographs* **27**. Am. Math. Soc., Providence, R.I.

—— (1974). *Bull. Math. Biophys.* **36,** 365–77.

—— (1979). *J. Inst. Math. Appl.* **24,** 259–77.

—— (1980*a*). In Magenes (1980). Vol. I, pp. 383–450.

—— (1980*b*). In Magenes (1980). Vol. II, pp. 507–37.

—— (1982*a*). Private communication.

—— (1982*b*). *IMA J. appl. Math.* **28,** 287–99.

—— Geiman, H., and Shachaf, M. (1980). Rep. No. 104 of the Hebrew University of Jerusalem.

Sackett, G. G. (1971). *SIAM J. numer. Anal.* **8,** 80–96.

Saffman, P. G. and Taylor, G. I. (1958). *Proc. R. Soc.* **A245,** 312–29.

Saitoh, T. (1972). *Mem. Sagami Inst. of Technol.* **6,** 1–14.

—— (1976) *Tech. Rep. Tohoku Univ.* **41,** 61–72.

—— (1978). *J. Heat Transfer* **100,** 294–9.

Schäfer, E. (1977). In Hoffman (1977). Vol. III, pp. 145–60.

Schatz, A. (1969). *J. math. Anal. Appl.* **28,** 569–80.

Scheffler, S. and Whiteman, J. R. (1979). In Gladwell and Wait (1979). Chapter 4.

Shamsundar, N. (1978). In Wilson *et al.* (1978). pp. 165–85.

—— Sparrow, E. W. (1975). *J. Heat Transfer, Trans. ASME* **97,** 333–40.

—— —— (1976). *J. Heat Transfer, Trans. ASME* **98,** 550–7.

Shaug, J. C. and Bruch, J. C. (1976). *Proc. Int. Symp. Finite elements in flow problems, Santa Margherita Ligure.* pp. 435–45.

Shaw, F. S. and Southwell, R. V. (1941). *Proc. R. Soc.* **A178,** 1–17.

Sherman, B. (1970). *Q. appl. Math.* **27,** 427–39.

—— (1971). *J. math. Anal. Appl.* **33,** 449–66.

Shishkin, G. I. (1975). *Dokl. Akad. Nauk. SSSR* **224,** 1276–8.

Sloss, J. M. and Bruch, J. C. (1978). *J. Eng. Mech. Div. Proc. ASCE* **104,** EM5, 1099–1111.

Smith, G. D. (1978). *Numerical solution of partial differential equations: finite difference methods.* Clarendon Press, Oxford.

Sokolov, Ju. D. (1951). *PMM* **15,** No. 6.

Solomon, A. D. (1978). In Wilson *et al.* (1978). pp. 187–202.

—— Wilson, D. G., and Alexiades, V. (1981). Oak Ridge National Laboratory Rep. CSD-84.

—— —— —— (1982). Oak Ridge National Laboratory Rep. CSD-91.

Southwell, R. V. (1946). *Relaxation methods in theoretical physics.* Clarendon Press, Oxford.

Soward, A. M. (1980). *Proc. R. Soc.* **373A,** 131–47.

Sparrow, E. M. and Hsu, C. F. (1981). *Int. J. Heat Mass Transfer* **24,** 1345–57.

—— Ramadhyani, S., and Patankar, S. V. (1978). *J Heat Transfer, Trans. ASME* **100,** 395–402.

Stampacchia, G. (1964). *C.r. hebd. Acad. Séanc. Acad. Sci. Paris* **258,** 4413–16.

—— (1974). *Russ. Math. Surveys* **29**(4), 89–102.

Stefan, J. (1889*a*). *Sber. Akad. Wiss. Wien.* **98,** 473–84.

—— (1889*b*). *Sber. Akad. Wiss. Wien.* **98,** 965–83.

—— (1891). *Ann. Phys. u. Chem.* **42,** 269–86.

Stewartson, K. and Waechter, R. T. (1976). *Proc. R. Soc.* **348A,** 415–26.

Tao, L. N. (1979). *Q. Jl. appl. Math.* **22,** 175–85.

Tayler, A. B. (1975). In Ockendon and Hodgkins (1975). pp. 120–37.

Taylor, C. J., France, P. W., and Zienkiewicz, O. C. (1973). In *The mathematics of finite elements and applications I* (ed. J. R. Whiteman) pp. 313–25. Academic Press, New York.

Taylor, G. I. (1961). In *Problems of continuum mechanics, SIAM. Muskhelishvili Anniversary Vol.* pp. 546–55.

—— Saffman, P. G. (1959). *Q. Jl. Mech. appl. Math.* **12,** 265–79.

Taylor, G. S. and Luthin, J. N. (1969). *Water Resources Res.* **5,** 144–52.

Taylor, R. L. (1966). Axisymmetric and plane flow in porous media. Tech. Report, University of California at Berkeley.

—— Brown, C. B. (1967). *J. Hydraul. Div., Proc. ASCE* **93,** HY2, 25–33.

Thatcher, R. W. and Askew, S. L. (1982). *IMA J. numer. Analysis* **2,** 229–39.

Thomas, L. and Westwater, J. (1963). *Heat Transfer.* Houston Chemical Eng. Progress Symp. Series **41,** 155–64.

Thompson, J. F., Thames, F. C., and Mastin, C. W. (1974). *J. comput. Phys.* **15,** 299–319.

Tichonov, A. N. (1950). *Mat. Sb.* **26,** 35–56. English translation *Am. math. Soc. Transl.* **1,** 440–66 (1962) *Math. Rev.* **11,** 440; **14,** 337.

Tien, L. C. and Churchill, S. W. (1965). *A.I.Ch.E. Jl.* **11,** 790–3.

Tien, R. H. and Geiger, G. E. (1967). *J. Heat Transfer Trans. ASME* **89,** 230–4.

Todsen, M. (1971). *J. Hydrol.* **12,** 177–210.

Torelli, A. (1974). *Rend. Sem. Mat. Padova* **52,** 25–58.

—— (1975). *C.r. hebd. Séanc. Acad. Sci., Paris* **A280,** 353–6.

—— (1977*a*). *Ann. Scuola Norm. Sup. Pisa* **4**(1), 33–59.

—— (1977*b*). *Pubbl. Lab. Analisi Numerica C.N.R.* **148,** Pavia.

Trefftz, E. (1916). *Z. Math. Phys.* **64,** 35–61.

Turland, B. D. (1979). UKAEA (Culham) Report CLM-P582. Also in *Proc. 4th Int. Conf. on Structural Mechanics in Reactor Technology, Berlin Vol. B* (eds. T. A. Jaeger and B. A. Boley). North Holland Publishing for the Commission of the European Community.

—— (1981). Personal communication.

—— (1982). UKAEA (Culham) Report CLM-P662. Also in Fasano and Primicerio (1983). pp. 293–305.

—— Peckover, R. S. (1980) *J. Inst. Math. Appl.* **25,** 1–15.

—— Wilson, G. (1977). UKAEA (Culham) Report RS(77)N42.

Varga, R. (1962). *Matrix iterative analysis.* Prentice-Hall, Englewood Cliffs, NJ.

Vasilev, F. P. (1964). *Sov. Math. Dokl.* **5,** 1109–13.

Vedernikov, V. V. (1939). *Seepage theory and its applications in the field of irrigation and drainage.* State Press.

Verma, R. D. and Brutsaert, W. (1971). *J. Hydraul. Div., ASCE* **97,** NY8, 1213–29.

Visintin, A. (1979). *Boll. U. mat. ital.* (5) **16 B,** 212–37.

—— (1980*a*). In Magenes (1980). Vol. I, pp. 219–27.

—— (1980*b*). *Annali Mat. pura appl.* **124,** 293–320.

—— (1983). In Fasano and Primicerio (1983). pp. 419–30.

Voller, V. R. and Cross, M. (1981*a*). *Int. J. Heat Mass Transfer* **24,** 545–56.

—— —— (1981*b*). *Int. J. Heat Mass Transfer* **24,** 1457–62.

—— —— (1983). *Int. J. Heat Mass Transfer* **26,** 147–50.

Wait, R. (1979). In Gladwell and Wait (1979). Chapters 2, 5.

Weber, H. (1928). *Die Reichweite von Grundwasserabsenkungen mittels Rohrbrunnen.* Springer, Berlin.

Wellford, L. C. and Ayer, R. M. (1977). *Int. J. numer. Meth. Engng.* **11,** 933–43.

Wiener, J. H. (1955). *Br. J. appl. Phys.* **6,** 361–3.

Williams, J. (1979). In Gladwell and Wait (1979). Chapter 17.

Williams, W. E. (1980). *Partial differential equations.* Clarendon Press, Oxford.

Wilson, D. G. (1978). *SIAM J. appl. Math.* **35,** 135–47.

Wilson, D. G. (1982). Oak Ridge National Laboratory Rep. CSD-93.

—— Solomon, A. D., and Alexiades, V. (1982). Oak Ridge National Laboratory Rep. CSD-97.

—— —— Boggs, P. T. (eds.) (1978). *Moving boundary problems.* Academic Press, New York.

—— —— Trent, J. S. (1979). A bibliograph on moving boundary problems with keyword index. Oak Ridge National Laboratory.

Winslow, A. M. (1967). *J. comput Phys.* **2,** 149–72.

Wollkind, D. J. and Segel, L. A. (1970). *Phil. Trans. R. Soc.* **268A,** 351–80.

Wood, A. S., Ritchie, S. I. M. and Bell, G. E. (1981). *Int. J. numer. Meth. Engng.* **17,** 301–5.

Wrobel, L. C. and Brebbia, C. A. (1981). In *Numerical methods in heat transfer.* (ed. R. W. Lewis, K. Morgan, and O. C. Zienkiewicz). Chapter 5. Wiley, New York.

Wu, T. S. and Boley, B. A. (1966). *Siam J. appl. Math.* **14,** 306–23.

Wu, Y. S. (1976). M.Sc. Thesis, University of Kentucky, Lexington, KY.

Yagoda, H. P. and Boley, B. A. (1970). *Q. Jl Mech. appl. Math.* **23,** 225–46.

Yanenko, N. N. (1971). *The method of fractional steps.* Springer, Berlin.

Young, D. M. (1971). *Iterative solution of large linear systems.* Academic Press, New York.

—— Gates, L. D., Arms, R. J., and Eliezer, D. F. (1955). Report No. 1413, U.S. Naval Proving Ground, Dahlgren, Virginia.

Youngs, E. G. (1971). *Water Resources Res.* **7,** 1366–8.

Yuen, W. W. (1980). *Int. J. Heat Mass Transfer* **23,** 1157–60.

—— Kleinman, A. M. (1980). *A.I.Ch.E. Jl.* **26,** 828–32.

Zienkiewicz, O. C. (1971). *The finite element method in engineering science.* 2nd ed. McGraw-Hill, New York.

—— Cheung, Y. K. (1965). *Finite elements in the solution of field problems.* The Engineer, London.

—— Meyer, P., and Cheung, Y. K. (1966). *Proc. Am. Soc. civ. Engrs* **12,** 111–20.

—— Phillips, D. V. (1971). *Int. J. numer. Meth. Engng.* **3,** 519–28.

Zygmund, A. (1959). *Trigonometric series Vol. I.* Cambridge University Press.

Author index

Note: authors named in main text are indexed; no entries are included for the References section (pp. 394–408)

Aguirre-Puente, J. 259
Aitchison, J. 191, 315, 319, 325, 332, 334, 335, 336, 347, 361, 362
Albasiny, E. L. 217, 219, 226
Albrecht, J. 2
Alexiades, V. 141
Allen, D. N. de G. 203, 228, 314
Alt, H. W. 54, 79, 92, 376–7, 379, 380, 383
Andrews, J. G. 11, 151, 152, 155
Aravin, V. I. 287, 295
Arms, R. J. 314
Askew, S. L. 335–6
Atthey, D. R. 5, 11, 151, 152, 155, 221, 223, 224, 228, 237, 240, 241
Awater, R. 318
Ayer, R. M. 173

Baines, M. J. 172, 178
Baiocchi, C. 31, 34, 35, 40, 45, 46, 48, 49, 50, 51, 52, 53, 54, 60–1, 62, 63, 71, 74, 96, 98, 99, 258, 259, 260, 270, 321, 360, 364, 368, 371, 372
Baker, C. T. H. 317, 385
Balinski, M. L. 355
Bankoff, S. G. 2
Bathe, K.-J. 314
Baumeister, J. 188
Bear, J. 31, 63, 76, 283, 285, 286, 287, 288, 289, 291, 295, 377
Bell, G. E. 132, 133, 136, 137, 142, 228, 238, 239
Benci, V. 67, 71
Bensoussan, A. 52, 259, 266, 270
Berger, A. E. 23, 253, 255, 256, 257, 267, 268, 269
Bickley, W. G. 314
Bischoff, H. 318
Boadway, J. D. 204, 344, 348, 352
Boggs, P. T. 2, 163
Boieri, P. 33
Boley, B. A. 2, 122, 123, 124, 125, 126, 127, 132

Bonacina, C. 218, 227, 229, 233
Bonnerot, R. 12, 14, 167, 173, 175, 176, 177, 178, 188, 195, 197, 215, 216, 228–9
Boreli, M. 69
Brauner, C. 228
Brebbia, C. A. 318
Brezis, H. 54, 79, 80, 92, 98, 220, 258, 259, 266, 282
Brezzi, F. 54
Brown, C. B. 69, 314, 317, 325
Bruch, J. C. 22–3, 31, 35, 39, 43, 46, 48, 49, 54, 55, 56, 57, 58, 60, 63, 64, 65, 66, 258, 360, 364, 365, 366, 368, 371, 372
Brutsaert, W. 90
Buckmaster, J. 16, 249
Budak, B. M. 218, 220, 223, 228, 229, 282
Budhia, H. 113
Bunch, J. R. 314
Busch, R. A. 325
Butterfield, R. 318
Buzbee, B. L. 314

Caffarelli, L. A. 57, 71
Caffrey, J. 43, 46, 57, 58, 60, 364, 365, 366, 368, 371
Cahn, J. W. 139
Cannon, J. R. 12, 16, 150, 178, 228
Capriz, G. 96
Carbone, L. 57
Carré, B. A. 359
Carslaw, H. S. 19, 101, 102, 103, 108, 111, 112, 117, 123, 124, 130, 136, 153, 239
Chadam, J. 150
Chalmers, B. 247
Chambré, P. L. 10, 14
Chang, O. 318
Charni, I. A. 34, 284, 287
Chehl, S. 142
Chernousko, F. L. 199
Christiansen, S. 85, 243, 385, 386, 388, 389, 390
Christopherson, D. G. 93, 356
Chu, W. H. 193, 194
Chuang, Y. K. 117, 118
Churchill, S. W. 102, 147, 148, 171
Ciavaldini, J. F. 226, 228

Cimatti, G.   96
Ciment, M.   167–8, 253, 255
Citron, S. J.   187
Cizek, J.   199, 200, 201
Clapeyron, B. P. E.   101
Collatz, L.   2, 33–4, 117
Comincioli, V.   31, 35, 52, 54, 62–3, 76–7, 78, 91
Comini, G.   229
Constable, T. B.   19
Cooley, R. L.   69, 90
Cottle, R. W.   261, 262, 355, 357, 359
Couch, E. J.   229
Courant, R.   338
Crank, J.   2, 10, 19, 21, 110, 118, 120–1, 128, 140, 143, 147, 163, 164, 166, 171, 172, 187, 192, 199, 201, 202, 203, 211, 212, 213, 214, 215, 216, 228, 233, 261, 274, 275, 337, 344, 346–7
Cross, M.   233, 235, 236, 237, 239, 247
Crowley, A. B.   16, 85, 109, 211, 212, 213, 214, 215, 216, 228, 236, 242, 243, 244, 245, 246, 248, 249, 250, 251, 274, 275, 390
Cryer, C. W.   31, 66, 67–8, 69–70, 71, 94, 96, 261, 310, 312, 314, 315, 317, 318, 319, 320, 325, 329, 331, 332, 336, 355, 356, 357, 359, 372, 373

Dagan, G.   318
Dahmardah, H. O.   170, 187, 263
Damlamian, A.   8, 282
Damle, P. S.   186
Danckwerts, P. V.   10, 110
Darcy, H.   32
Datzev, A.   223
Davidson, J. F.   315
Davies, A. J.   314
Davis, G. B.   162
Delves, L. M.   318, 332
Dix, R. C.   199, 200, 201
Doha, E. H.   321
Dominguez, J.   318
Domoto, G. A.   156
Dorr, F. W.   314
Doss, S. K.   180
Douglas, J.   150, 155, 168, 169, 170, 236
Duck, P. W.   139
Duda, J. L.   196, 197
Dupuit, J.   283
Durack, D.   204
Duvaut, G.   38, 84, 258, 259, 270, 275, 276, 280, 281, 282

Eckhardt, U.   356
Ehrich, O.   117
Ehrlich, L. W.   167

Elliott, C.   6, 82, 84, 85, 86, 87, 95, 100, 150, 214, 226, 227, 229, 232, 233, 234, 243, 252, 258, 259, 261, 262, 266, 267, 268, 269, 274, 275, 280, 312, 359, 360, 373, 390
Estenssoro, L.   132
Evans, G. W.   114, 150
Evans, L. B.   102, 148
Evans, N. T. S.   19, 21, 257
Eyres, N. R.   217, 219

Falk, R. S.   257
Farkas, L.   318
Fasano, A.   8, 23–4, 25, 26, 168, 188, 244, 252, 282
Federenko, R. P.   228
Ferriss, D. H.   127, 128, 187, 188, 191, 234
Fetter, H.   31, 66, 67–8, 69, 71, 336, 372, 373
Finn, W. D. L.   314, 317, 336, 337, 338, 339, 341, 342, 343
Fisher, I.   163, 229
Fitz-Gerald, J. M.   109
Fix, G. J.   15
Fleishman, B. A.   150
Forsyth, P.   88, 89
Fox, L.   2, 130, 150, 163, 325, 326, 327, 329, 332, 392
France, P. W.   91, 321, 366
Frank, F. C.   110, 111
Franzeiti, S.   369
Frémond, M.   259, 270, 275, 282
Friedman, A.   33, 71, 74, 75, 90, 150, 220, 222, 227, 270, 272, 281, 282
Furzeland, R. M.   2, 163, 167, 186, 189, 190, 191, 192, 193, 194, 195, 197, 216, 226, 227, 228, 230, 231, 232, 233, 235, 244, 332

Galib, T. A.   22–3
Gallie, J. M.   155, 168, 169, 170, 236
Gapenko, Yu. L.   282
Garabedian, P. R.   329, 332
Gastaldi, F.   33, 99
Gates, L. D.   314
Gauss, C. F.   326
Geiger, G. E.   239
Geiman, H.   8, 26, 29
Gelder, D.   108
Gelinas, R. J.   180
George, J. A.   186, 317
Gilardi, G.   54, 57, 92, 93
Gilding, B. H.   110
Gingrich, R.   150
Gladwell, I.   314
Glasser, D.   127
Glowinski, R.   40, 356

Gnanadoss, A. A. 94
Goheen, M. S. 359
Goodling, J. S. 170
Goodman, T. R. 128, 129, 130, 131, 132, 136, 170
Goodrich, L. E. 238
Gordon, W. J. 195
Gourlay, A. R. 21, 228, 257
Guenther, R. B. 167
Guerri, L. 35, 76–7, 78
Gupta, J. P. 147, 148
Gupta, R. S. 19, 21, 118, 120–1, 128, 143, 166, 169–70, 171–2, 192, 201, 202, 203, 228, 261
Guy, A. G. 108

Hager, W. W. 257
Hall, C. A. 195
Hall, H. P. 374
Hamel, G. 295
Hansen, E. B. 85, 89, 116, 118, 120, 121, 170, 172, 180, 187, 234, 261, 262, 263, 390, 392, 393
Hantush, M. S. 66
Harr, M. E. 31, 42, 43, 66, 71, 283, 291, 295, 371
Heitz, W. L. 171
Hilbert, D. 338
Hill, C. D. 150, 228
Hill, J. M. 162
Hill, S. 127, 128, 187, 188, 191, 234
Hodgkins, W. R. 2, 163, 226, 228
Hoffman, K.-H. 2, 12, 188, 197
Höhn, W. 188
Holm, A. M. 85
Horie, Y. 142
Hornung, U. 243
Horvay, G. 10, 139
Hougaard, P. 85, 116, 118, 120, 121, 170, 172, 180, 187, 234, 261, 262, 263
Howarth, J. A. 150
Hsu, C. F. 196
Huang, P. N. S. 132
Huber, A. 110, 168

Ichikawa, Y. 261, 271, 272, 273, 274, 275, 277, 278, 279
Imber, M. 132
Irmay, S. 289
Ivantsov, G. P. 139

Jaeger, J. C. 19, 101, 102, 103, 108, 111, 112, 117, 123, 124, 130, 136, 153, 239
Jamet, P. 12, 14, 167, 173, 174, 175, 176, 177, 178, 188, 195, 197, 215, 216, 228–9
Janovsky, V. 82, 84
Jaswon, M. A. 317, 385

Jensen, R. 33, 74, 75, 282
Jeppson, R. W. 49, 66, 71, 337, 353, 372, 376
Jerome, J. W. 226, 282
Jiji, L. M. 113, 150, 228
Johnson, C. 259
Jones, I. P. 181

Kamenomostskaja, S. L. 217, 220, 222, 224, 281
Kamin (Kamenomostskaja), S. L. 110
Kealy, C. D. 320, 325
Kern, J. 127
Khader, M. S. 170
Khoshgoftaar, M. R. 314
Kikuchi, N. 71, 259, 261, 271, 272, 273, 274, 275, 277, 278, 279, 336
Kinderlehrer, D. 79, 270, 272
Kirkham, D. 318
Kleinman, A. M. 170
Koh, J. C. Y. 12, 167
Kolodner, I. I. 117
Kozeny, J. 43
Kreith, F. 113, 144, 145, 146
Kumar, D. 169–70, 172

Lamb, H. 81
Lamé, M. M. 101
Landau, H. G. 187
Landis, F. 167, 170, 172
Langer, J. S. 15
Langford, D. 132
Lardner, T. J. 132
Larock, B. E. 314
Lauwerier, H. A. 24
Lax, P. D. 217
Lazaridis, A. 168, 203, 228
Lembke, K. E. 287
Lennon, G. P. 318
Lewy, H. 98
Li, C. H. 228
Liggett, J. A. 318
Lightfoot, N. M. H. 111–12, 113, 117
Lions, J. L. 38, 40, 52, 220, 257, 258, 259, 266, 270, 273, 275, 282
Liu, P. L.-F. 318
Longworth, D. 226, 228, 229, 237
Lotkin, M. 187
Luthin, J. N. 90, 374

McGeough, J. A. 85, 86, 88, 109, 151, 242
Magenes, E. 2, 35, 54
Mahar, T. J. 150
Maione, U. 54, 63, 369
Mangassarian, O. L. 356
Mason, J. C. 318
Mastaniah, K. 187, 191

Mastin, C. W.   193
Mauersberger, P.   71
Mayers, D. F.   170, 187, 263
Medland, I. C.   163, 229
Megerlin, F.   140
Meirmanov, A. M.   223, 282
Meyer, G. H.   8, 85, 91, 96, 167, 181, 182,
    183, 184, 186, 218, 226, 228, 229, 233
Miellou, J. C.   49
Mikhailov, M. D.   16, 110
Miller, J. V.   172, 178, 179, 180
Miller, K.   180
Mirnateghi, A.   66
Mogel, T. R.   314, 315
Moiseynko, B. D.   228, 229
Morgan, K.   229
Morton, K. W.   172, 178
Mosco, U.   38
Moskal, M. S.   223
Motz, H.   392
Muehlbauer, J. C.   2
Mullins, W. W.   15
Murray, W. D.   167, 170, 172
Muskat, M.   16, 31, 297

Neuman, S. P.   66, 71, 90, 314, 316, 320,
    325, 374
Niezgodka, M.   12, 220, 222, 223, 244, 281,
    282
Nitsche, J. A.   188
Noble, B.   133
Nogi, T.   169
Numerov, S. N.   43, 287, 295

Oberkampf, W. L.   193, 194
Ockendon, J. R.   2, 6, 15, 21, 26, 100, 109,
    116–17, 150, 163, 242, 245, 246, 248,
    252, 258, 259, 261, 262, 266, 267, 269,
    274, 275, 312, 359, 360
Oden, J. T.   259, 336
O'Leary, D. P.   261
Oleinik, O. A.   217, 220, 222, 223, 224,
    243, 246, 281
Ortoleva, P.   150
Osborne, M. R.   94
Outmans, H. D.   71, 285
Ozis, T.   312, 337, 344, 346–7, 352, 353

Papanicolau, G.   259
Patankar, S. V.   196
Patel, P. D.   17, 19
Pawlow, I.   281, 282
Peckover, R. S.   151
Pedroso, R. I.   156
Pekeris, C. L.   141, 142
Peletier, L. A.   110
Phahle, R. D.   201

Phillips, D. V.   195
Pinkus, O.   93
Pohle, F. V.   132
Polubarinova-Kochina, P. Ya.   31, 42, 49,
    66, 283, 285, 287, 288, 291, 295, 298,
    299, 300, 302, 304, 310, 312, 346, 360,
    372
Poots, G.   132, 136, 137, 139, 150, 156,
    202, 203
Potter, D.   207
Pozzi, G. A.   35, 74, 75, 260
Price, R. H.   217
Primicerio, M.   7, 23–4, 25, 26, 168, 178,
    188, 244, 252, 281, 282
Pukhnachev, V. V.   282

Quarteroni, A.   81

Radojkovic, M. M.   318
Rahmer, H.   22
Rasmussen, H.   85, 86, 88, 89, 93, 151,
    242, 243, 385, 386, 388, 389, 390
Rathjen, K. A.   113, 228
Reid, J. D.   315, 317
Reid, J. K.   314
Remer, J.   54–5, 56, 66, 70, 373, 376
Richardson, S.   81, 82, 109. 243
Riemann, G. F. B.–Weber, H.   101
Riley, D. S.   139, 156, 159
Rippin, D. W. T.   315
Risenkampf, B. K.   300
Ritchie, S. I. M.   228
Rogers, J. C. W.   54, 100, 253, 255
Romie, F. E.   144, 145, 146
Rose, D. J.   314
Rose, M. E.   141, 217, 220
Rubinstein, L. I.   2, 6, 7, 8, 23–4, 25, 26,
    28–9, 104, 105, 117, 150, 168, 222, 223,
    246, 248, 249, 252

Sacher, P. S.   357, 359
Sackett, G. G.   19, 118
Saffman, P. G.   81, 82
Saitoh, T.   167, 197, 198–9, 228
Salhani, D.   93
Samarskii, A. A.   228, 229
Sankar, R.   325, 326, 327, 329
Schäfer, E.   226
Schatz, A.   21, 192
Scheffler, S.   332
Segel, L. A.   151
Sekerka, J.   15
Severn, R. T.   203, 228
Shachaf, M.   26, 29
Shamsundar, N.   219, 227, 228, 230, 238
Shaug, J. C.   49, 372
Shaw, F. S.   314, 325

Shea, J. J.   130, 131, 132
Sherman, B.   109, 178, 222
Shishkin, G. I.   282
Slack, M. R.   217
Slichter, L. B.   141, 142
Sloss, J. M.   22–3, 43, 49, 56, 63, 64, 66, 372
Smith, F. T.   156
Smith, G. D.   314, 315
Sokolov, Ju. D.   298, 300
Solomon, A. D.   2, 6, 7, 140, 141, 142, 143, 144, 163, 281
Southwell, R. V.   314, 325, 356
Soward, A. M.   161–2
Sparrow, E. M.   196
Sparrow, E. W.   219, 228, 230
Sprekels, J.   12
Stampacchia, G.   40, 57, 60, 79, 98, 100, 257
Stefan, J.   1, 2, 101, 139, 146
Sternlicht, B.   93
Stewartson, K.   156, 157, 158, 159–60, 161, 162
Street, R. L.   314, 315
Sunderland, J. E.   2
Sweet, R. A.   167–8
Symm, G. T.   317, 385
Szekely, J.   117

Tao, L. N.   10
Tayler, A. B.   14, 15, 21, 110, 222
Taylor, C.   314
Taylor, C. J.   91, 321, 322, 323, 324
Taylor, G. I.   81, 82
Taylor, G. S.   90, 374
Taylor, R. L.   69, 314, 317, 320, 325
Thames, F. C.   193
Thatcher, R. W.   335–6
Thomas, L.   6
Thompson, C. P.   181
Thompson, J. F.   193
Tichonov, A. N.   23
Tien, L. C.   171
Tien, R. H.   239
Todsen, M.   90
Torelli, A.   63, 90, 91, 93, 99
Trefftz, F.   317
Trent, J. S.   281
Turland, B. D.   150–1, 199, 206, 207, 209–

10, 211
Turski, L. A.   15

Valli, A.   57
Varga, R.   314, 315
Varoğlu, E.   336, 337, 338, 339, 341, 342, 343
Vasilev, F. P.   169
Vedernikov, V. V.   298, 300
Verma, R. D.   90
Visintin, A.   80–1, 93, 244
Voller, V. R.   233, 235, 236, 237, 239, 247
Volpi, G.   35

Waddington, J. F.   226, 228
Waechter, R. T.   156, 157, 158, 159–60, 161
Wait, R.   314, 332
Weber, H.   287
Wellford, L. C.   173
Wendroff, B.   204
Westwater, J.   6
Westwater, J. W.   171
Whiteman, J. R.   332
Wiener, J. H.   106, 107
Williams, J.   183
Williams, R. E.   320
Williams, W. E.   220
Wilson, D. G.   2, 10, 105, 106, 107, 108, 127, 141, 163, 249, 281, 282
Wilson, G.   199
Winslow, A. M.   193, 197
Witherspoon, P. A.   66, 71, 90, 314, 316, 320, 325, 374
Wollkind, D. J.   151
Wood, A. S.   228
Wrobel, L. C.   318
Wu, T. S.   126
Wu, Y. S.   318

Yagoda, H. P.   126
Yanenko, N. N.   314
Young, D. M.   314
Youngs, E. G.   70
Yuen, W. W.   130, 170

Zaslavsky, D.   289
Zienkiewicz, O. C.   195, 317
Zygmund, A.   161

Subject index

ablåtion
  defined 11
  laser-cutting 151–5
  space vehicle alloys 12
ablation problems
  enthalpy formulation of 252
  integral-equation formulations for 126, 127
  method of lines with SOR algorithm 186
absorption functions, oxygen diffusion 23
accretion 75
  allowance for 286
adaptive meshes 173
adaptive variational formulation, numerical solution 337–43
air–liquid interfaces, enthalpy functions for 251
algorithms
  Alt 92, 376–84
  APT 254
  Cottle 357–9
  Cryer 46, 75, 77, 359–72
  DO2AJF 191
  Fox–Sankar 325–9
  Gear 191
  outer iteration built into 46, 52, 368
  SOR see successive overrelaxation
  also see numerical algorithms
alloy solidification problems 6, 14–16
  enthalpy method 245–9
  Green's functions used in 117, 118
  similarity solutions for 110
  stability analysis of 150–1
alloys
  ablation of 12
  diffusion/heat flow in 14–16
Alt algorithm 92, 376–84
alternating direction method 183–4
alternating method, split-domain problem 56
alternating-phase truncation (APT) method 253–7
analytical solutions 101–62
  seepage problems 283–312
annular electrochemical machining prob-

lems 85–6, 385–93
anode profile, electrochemical machining 89, 385–93
appearing/disappearing phases, Stefan problem 178
approximate methods, analytical, solution 283–8
approximate solutions 139–62
  fictitious diffusivity approximation 147–50
  inverse Stefan problem 144–7
  perturbation method 150–62
  spatial subdivision 142–4
  steady-state approximation 139–42
approximate trial solutions, computation of 314–19
aquifers 60–3, 362
arbitrary-shaped dams 52–4, 376–9
  with toe drain 54–7
asymptotic formulations
  concentrated capacity 26
  perturbation method 150
axisymmetric flow problems 28–9, 31, 66–70, 318, 372–6
  disc-in-cylinder 326–9
  wells 318, 372–6

Baiocchi transformation
  arbitrary-shaped dam 53
    with toe drain 55–6
  axisymmetric well flow 67–8, 373
  capillary-fringed rectangular dams 77
  coastal aquifer 61
  electrochemical machining 86
  evaporation 74–5
  general statement of 96–100
  pond seepage 70, 373–4
  rectangular dam
    with sheetpile 44
    with toe drain 41–2
  simple rectangular dams 34–8
  sloping-base dam 51
  stratified dam 72, 73
  stream functions incorporated into 40–1
  three-dimensional dams 57, 58

Baiocchi transformation—*continued*
  transient dam 90, 92, 93
Baiocchi variables
  form of 98–100
  three-dimensional boundary conditions
    for 59
biological problems, numerical solution
    of 8
biquadratic finite elements 175–6
blob model, mushy region 6
Boadway transformation 204–6, 344, 348
body heating 239–42
body-fitted curvilinear coordinates 89,
    192–5
  example of use of 195–9
boundary conditions
  dam seepage 32–3
  moving interfaces 3–4
  representation of 381
boundary element methods 317
boundary position
  comparison of various methods for deter-
    mination of 233–4
  melting problem 136–7
  oxygen diffusion 20
  solution of 217
boundary singularities 332–6
boundary-fixing coordinate transformation,
    boundary-position determined by 233–4
boundary-fixing methods 187–216
boundary-tracking methods 163–86
boundary-value problems, defined 1
Boussinesq equations 285–7
Brezis formulation, generalized dam
    problem 79–80, 92

canals, seepage from 63–6
capillary fringe 76–9
  degree of saturation for 377
capillary rise 76
capillary tissue, oxygen diffusion in 22
cathodes, electrochemical machining 85,
    88, 390
Cauchy–Riemann equation 34, 41
cavitation
  disc-in-cylinder 325–9
  journal bearing 93–6
chemical activity 246
circular electrodes, numerical solution
    for 388–9
circular porous dam, seepage through 347–
    53
classical Stefan problem *see* Stefan problem
coastal aquifers 60–3
  Cryer algorithm for 362
collapse, solid-phase 12–13

comparison theorem, embedding 127
compatibility conditions 368–72
  arbitrary-shaped dam, with toe drain 56
  dam with inlet face slanted 48, 371–2
  open channel 64, 65
  rectangular dam
    with capillary fringe 78
    with sheetpile 45, 368–71
  sloping-base dam 50
complementarity formulation, two-
    dimensional dam 49
complementarity methods, numerical
    solutions 355–76
computer programs, boundary-fixing itera-
    tive algorithm 191
computers, parallel 359
concentrated thermal capacities 23–9
concentration discontinuity 245
concentric circular electrodes, numerical
    solution for 388–93
conformal transformations 293–7, 298,
    299, 300
conjugate gradient method 261
connections, free-boundary problem 100
conservation of energy, Stefan
    problem 10
conservative difference schemes 207, 210–
    11
constant surface cooling, series solutions
    for 126
constant-rate solidification, temperature
    expression 2, 144–7
convection, Stefan problem 10
convection terms, method of lines 181
convective boundary conditions
  variable time step finite-difference
    method applied to 170
convergence theorem proofs
  Cryer algorithm 356
  enthalpy method 226
convolution integrals 132
convolution theorem, inversion by 155
coordinate transformation 93, 189
cosine integral transforms 127
Cottle fast algorithm 357–9
Crank–Nicolson scheme
  boundary-fixing in 189–90, 191
  complementarity methods used
    with 268–9
  generalization of 173
  truncation methods used with 268–9
Cryer algorithm
  coastal aquifer 362
  definition of 356
  evaporation 75
  finite-difference and finite-element forms
    of 359–72

Cryer algorithm—*continued*
    lubrication cavitation   94
    rectangular dam
        with capillary fringe   77
        with sheetpile   46
        simple dam problem   360–2
        stratified dams   363
        also see successive overrelaxation
crystal growth   142, 251
crystal packing, mushy region   6–7
crystallization, supercooled melt   104
cubic splines method   143, 172
curvilinear coordinates   89, 192–9
    temperatures interchanged with   207
cylinder, freezing inside and outside
    of   170
cylinder problem
    constant-rate solidification   144–5
    enthalpy method for   237–8
cylindrical coordinates, similarity solutions
    in   111
cylindrical pipes, solidification of liquid
    around   137

dams
    arbitrary-shaped   52–4, 376–9
        with toe drain   54–7
    evolution   33, 90–7
    generalized expression   79–81
    impervious base   75, 365
    porous base   75
    rectangular *see* rectangular dams
    simple *see* simple dam
    slanted exit face   54
    slanted inlet face   47–9, 54, 371
        with toe drain   49, 371, 372
    sloping base   50–2
    stratified   71–4, 336, 363, 383–4
    three-dimensional   57–60, 353–5, 366–7
    transient   90–1
Darcy's law   32, 81, 283
decision theory, examples   1
degenerate free-boundary problems   81–93,
    282
degeneration, phase   12–14
degree of saturation
    capillary fringe   377
    representation of   381
dendrite formation   6, 15
density changes, Stefan problem   10
detachment point, defined   32
determinantal equations, trial free-boundary
    method   327
dimensionless heat parameters, Stefan
    problem   9, 156–7
dimensionless latent heat   9, 157
    values of   102, 149

dimensionless solidification time   237
Dirac delta functions   118, 218, 227
Dirichlet boundary data   37–8, 320, 365,
    375
Dirichlet conditions   45, 47, 321, 391
Dirichlet-type problem   386
disc-in-cylinder flow problems   326–9
discharge
    axisymmetric well   69
    coastal aquifer   62, 63
    dam seepage   34
discontinuous enthalpy functions   228–9,
    280
distributed array processor (DAP)   359
ditches, seepage from   63–6, 297–300, 343
divergence theorem proof, enthalpy
    method   219
DO2AJF algorithm   191
double-iterative scheme
    capillary-fringed dam   78
    dam with sheetpile   368
    dam with slanting inlet face   371
    dam with toe drain   371, 372
    sloping-base dams   372
drains, seepage into   63–5
drawdown, transient seepage problems
    following   92, 324
drying of porous region, similarity solutions
    for   110
Duhamel integrals   132
Duhamel's theorem   123
Dupuit approximation   283–5, 287
Duvaut transformation   270, 271, 276

electrochemical machining (ECM)   85–9,
    109, 384–93
electrochemical machining problems
    formulation of   85–9
    integral-equation methods for   384–93
    numerical solutions for   385–93
electron beams, hole-cutting by   151–5
elliptic electrodes, numerical solution
    for   389–90
elliptic equations   1, 30, 81, 150
elliptic variational inequalities
    defined   258
    electrochemical machining   87
    Hele-Shaw injection   84
elliptic variational inequality, solution
    of   265
embedding technique,   122–8
enthalpy functions
    alternative forms of   226–8
    discontinuous   228–9
    form of   217–19
enthalpy methods   217–53
    alloy solidification   15, 245–9

enthalpy methods—*continued*
  conditions not amenable to 252–3
  electrochemical machining 85
  explicit finite-difference schemes 223–6
  moving-boundary position 233–5
  other methods compared with 233–5
  weak solutions of 217, 220–3
equilibrium diagrams, alloy solidification 14–15, 245
equipotential boundary, hodograph plane 289
equipotential lines, variable-interchange 342–3
equipotential transformations 197
error-function solutions 2, 102–3, 112–13, 128
Euler's constant 159
eutectic diagrams 14–15, 245
evaporation
  ablation problems 151–5
  porous flow problems 74–5
evolution dam problems 33, 90–3, 99
exact solutions, defined 101
examples, free and moving boundary problems 1
explicit finite-difference methods
  boundary-fixing transformation 188
  enthalpy 223–6
  fixed grid 168
  isotherm migration 212–13, 216
explicit moving-boundary conditions 190

fictitious boundary conditions 122–3
fictitious diffusivity approximations 147–50
fictitious heat parameters 148
fictitious temperatures, fixed finite-difference grid 167
fictitious-point approximations, boundary-fixing 197
filtration problem, similarity solutions for 110
finite-difference methods
  dam problems 43
  embedding techniques compared with 128
  finite-element methods compared with 315–16, 366
  fixed grid 163–8
  laser hole-cutting 155
  modified grid 168–81
  rectangular dam, with capillary fringe 78
  unequal space interval 164
  variable space-grid 170–3
  variable time step 168–70
finite-element methods
  adaptive mesh 173–81
  dam problems 43
  finite-difference methods compared

with 315–16, 366
transient dam 90, 91, 92
finite elements, triangulation 380
five-sixths rule 180
fixed finite-difference grids 163–8
fixed-domain methods 35–8, 217–82
flame-front problems 249–51
  diffusion/heat flow in 16
flow analogy, Hele-Shaw 16, 81–2, 109, 243
fluid injection 82
fluid transfer, semi-permeable membrane 8
Forchheimer equations 286
Fourier-series solutions
  oxygen diffusion problem 187, 263
  single-phase Stefan problem 117
Fox–Sankar algorithm 325–9
Frechet derivative 188
free-boundary moving methods 319–29
free-boundary problems
  defined 1, 30
  formulated 31–100
  numerical solution of 313–93
freezing, axial line source 111
freezing fronts, coupling 274–5
freezing index 259, 270, 272, 280
Frémond freezing index 259, 270, 272, 280
Frenet formulae 330
front-fixing methods 187–216
front-tracking methods 163–86
fully penetrating wells 67–9, 318
  Dupuit formulae applied to 285

Galerkin formulations 173, 178, 180, 318, 322
Garabedian modified boundary conditions 329–32
gas bubbles
  dissolution of 170
  electrochemical machining 89
gas dynamics, examples 1
Gauss–Seidel iteration 184, 230, 356
Gear algorithm 191
general solutions, Stefan problem 2
generalized enthalpy functions 242, 243
glass manufacture 108
global basis functions 318
global methods, free-boundary moving 325–9
Goodman integral method 128–33, 134
  variable time step finite-difference method compared with 170
graded grids, trial free-boundary method 315
graphical solutions
  concentrated capacity 27–8

graphical solutions—*continued*
  two-dimensional problem 279
  variable-interchange method 343
Green's functions 113, 117–21
  laser hole-cutting 154
Green's theorem
  dam with inlet face slanted 48
  integral-equation method starting
    from 385, 391
  multi-dimensional problem 138, 139
  simple rectangular dam 36, 37
  Stefan phase-change boundary
    conditions 175
grid spacing, trial free-boundary
    method 315–17
ground-water flow 81, 283, 300
ground-water pollution 63

Hamel transformation 295–7
heat balance, ice–water interface 5
heat flow problems
  Green's functions used in 117–21
  integral methods used in 128
heat flux jumps 173, 217
heat-balance integral method 128–33, 134
  with spatial subdivision 133–7
heat-flow equation
  IMM form of 212
  non-dimensional form of 211
Heaviside function, generalized dam
    problem 80
Hele-Shaw flow analogy 16, 81–2, 109
  enthalpy formulation of 243
  Neumann's solution for 109
Hele-Shaw injection 82–5
heterogeneous porous media 70–4
hodograph method 288–93
  conformal transformations 293–7, 298,
    299, 300
  simple dam problems 291, 297
  trapezoidal drainage ditch 297–300
homogenization, Stefan problem 8
hopscotch scheme
  enthalpy formulation 228
  truncation method combined with 257
horizontally stratified dams 71–2, 73–4,
    383–4
hot liquid injection, oil-saturated
    media 23, 28
hydraulic conductivity 32
hyperbolic equations, weak solution of 217
hyperbolic representation of
  heat conduction 117
  moving boundary 113

ice *see* melting ice
ice cubes, heat flow in water of 12

ice formation, steady-state approximation
    of 141–2
images, method of *see* method of images
imbedding methods, invariant 133–4, 184–
    6, 233
immiscible fluids, moving boundary problem
    in 16
impermeable medium
  oscillating temperatures in 27–8
  porous strata in 23–4
impervious boundary, hodograph
    plane 289
impervious-base dams 75, 365
  Boussinesq equations for 286
implicit boundary conditions, Stefan
    problem 19–23, 118
implicit methods
  discontinuous enthalpy functions 228–9
  isotherm-tracking along orthogonal flow
    lines 215–16
  solidification boundary 167–8
implicit moving-boundary conditions 190–
    1
impurities, alloy 14
infiltration 75
inhomogeneities
  axisymmetric flow 67
  stratified media 71
injection, Hele-Shaw 82–5
injection moulding 82
instabilities, moving boundary 150
instability region, alloy 15
insulated cathode sections, electrochemical
    machining 88–9
integral methods, free-boundary mov-
    ing 324–5
integral transforms 114–17
integral-equation formulations, numerical
    solution of 317, 385–93
integral-equation methods 114–39, 384–93
  electrochemical machining 85, 89, 385–
    93
  embedding technique 122–8
  Goodman integral method 128–33, 134
  Green's functions 113, 117–21
  heat-balance method 128–33, 134
    with spatial subdivision 133–7
  integral transforms 114–17
  multi-dimensional problem 137–9
interfaces
  examples of 109–10
  melting 2–4
invariant imbedding methods 133–4, 184–6
  two-phase Stefan problem 233
inverse formulation, coordinate inter-
    change 353–5
inverse Stefan problem 11–12, 144–7

inverse transformations
    hodograph method 294–5
    single-phase Stefan problem 115
iron, recrystallization of 114, 143
iron–carbon alloys, solidification of 118
irregularly shaped domains, finite elements for 316
isoparametric curvilinear coordinates 195
isoparametric finite elements 174
isotherm migration method (IMM) 199–216, 344
    linear complementarity method compared with 274–5
    multi-dimensional 201–16
        complete transformation of variables 204–11
        orthogonal flow lines 211–16
        partial transformation of variables 201–4
    one-dimensional 199–201

Jacobi-type iterative process 345, 350–1
joule heating 239–40
journal bearings, lubrication cavitation in 93–6, 98
jump conditions, flame-front 249–50
jump discontinuities
    enthalpy method 217, 219, 226
    multi-phase crystallization 106
    sloping-base dam with sloping inlet face 51

Kantorovitch expansion 93
Kirchhoff transformation 219, 281
Kozeny transformation 43
Kuhn–Tucker conditions 260, 267

Lagrange multiplier method 261, 274
Lagrangian interpolation formulae 164, 166
lake freezing problem, integral equations for 117
Langer model of alloy solidification 15
Laplace equation
    axisymmetric wells 67
    boundary-fixing 193, 194
    circular porous dam 352
    electrochemical machining 86, 392
    impervious-base dams 365
    integral-equation expression of 384
    simple rectangular dam 32
    trapezoidal dam seepage region 337
    trial free-boundary method 330
Laplace transforms
    approximate solutions 153, 154, 158, 159
    laser hole-cutting 153, 154
    single-phase Stefan problem 114, 115

Laplacian operator, replacement in Baiocchi transformation 97
laser beams, hole formation by 11, 151–5
latent-heat approximations 2, 140, 229
Laurent series 304
linear complementarity methods
    isotherm migration methods compared with 274–5
    truncation methods compared with 267–9
linear complementarity problems
    equations for 357
    Hele-Shar injection 85
    lubrication cavitation 94
    numerical solution of 355–76
linearized forms, generalized Stefan problem 8
lines, method of see method of lines
lubrication, journal bearing 93–6
lumped-parameter interpretation, fixed grid temperatures 167

mapping, hodograph 291–5
Megerlin method 140, 142, 143
melt depths 127
melt-pool growth, reactor substrate 209
melting interfaces 2–4
    position of 136–7
melting-ice problem 2–4, 99
    allowable temperature range for 143
    embedding techniques used for 122
    heat-balance integral method 129
    Neumann's solution 101
method of images 132, 161–2
method of lines
    boundary-position determined by 233–4
    electrochemical machining 85
    journal bearing lubrication 96
    with Ricatti transformation 186
    with SOR algorithm 184–6
    Stefan problem time-dependency removed by 343
    transient dams 91
minimization formulations
    axisymmetric well 69
    capillary-fringed rectangular dam 77
    coastal aquifer 62
    electrochemical machining 87
    simple dam problem 39–40, 361–2
    stratified dam 73, 74
    variational inequality 265–6, 356, 364
modified boundary conditions, Garabedian 329–32
modified grids 168–81
modified heat parameters 248
modified velocity potentials 32, 41, 57

moving, free-boundary   319–29
  global methods of   325–9
  integral methods of   324–5
  local methods of   319–24
moving finite-element method   180–1
moving grid systems   171–3
moving heat sources, similarity solutions
    and   111–14
moving-boundary problems
  defined   1
  formulated   2–29
multi-dimensional problems
  enthalpy functions in   228–35
  enthalpy method for   218–19
  heat-balance integral method applied
    to   137–9
multiple penetration depths   132
multiple-well problems   70, 318
Murray–Landis variable grid   170–1
mushy regions   5–6, 239–41
  alloy solidification   249
  spot welding   5, 237, 239–41
Muskat problem   16

NAG algorithms   191
negative latent heats   109
Neumann conditions   45, 47, 271, 321, 391
Neumann's solution   101–10
Newton linearization   233, 234
Newton–Raphson iteration   341, 343
Newton-type iterative process   136, 320
Newton's three-eighths rule   120
nodal enthalpy determination   236–7
node positioning, finite-element   179–80
non-dimensional variables, Stefan prob-
    lem   9, 156–7
non-linear heat parameters   5–8
non-linear instabilities, alloy solidifica-
    tion   15
non-linear Stefan problems   148
non-orthogonal transformations   193–5
normals to isotherms, geometrical proce-
    dure to determine   213–14
numerical algorithms
  arbitrary-shaped dams   54
  coastal aquifer   62
  dam with inlet face slanted   48
  generalized dam problem   79, 376
  rectangular dam
    with sheetpile   45, 46
    with toe drain and sheetpile   46
  simple dam   34–5, 38
  transient dam   91
numerical methods
  general description of   163
  rectangular dam with capillary fringe   78

simple rectangular dams   35
numerical solutions   313–93
  concentrated capacity   26
  rectangular dam with sheetpile   45–6
  well problem   69

Oberkampf transformations   194, 196, 197
oil production methods   23, 28, 29
one-dimensional problems
  boundary-fixing transformations for
    187–92
  density changes in   10
  finite-element methods applied to 173,
    175–8
  integral-equation formulations in   117
  journal bearing lubrication   96
  melting   129, 187
  method of lines   184–6
  solidification   111
  variable time step finite-difference
    methods for   169
  weak solution for   220–1
one-dimensional Stefan problem
  discontinuous finite-element method
    for   175–8
  truncation method   253–4
  weak solution for   222
one-dimensional two-phase Stefan problem
  finite-element method   173
  method of lines   181–6
  smoothed enthalpy functions in   227
one-parameter integral method   137–9,
    202–3
one-phase *see* single-phase
open channels, seepage from   63–6
orthogonal flow lines, isotherm movement
    along   211–16
orthogonal transformations   192
oscillating temperatures, concentrated
    capacity   27–8
oxygen diffusion problem
  boundary-fixing transformations for
    187–8
  capillary tissue   22, 98–9
  Cottle fast algorithm used in   359
  embedding technique used in   127
  enthalpy method for   244
  finite-element method   178
  Green's functions used in   118
  implicit boundary conditions in   19–23,
    118
  Lagrangian formulae in   166
  moving grid systems   171–3
  truncation method   255–7
  two-dimensional   21

oxygen diffusion problem—*continued*
  variable time step finite-difference method for 170
  variational approach 260, 263–4

parabolic equations 1, 81
parabolic inequalities 258–9, 266
paraffin sediments, melting of 23, 28
parallel computer, complementarity problems solved using 359
partially penetrating wells 69
partially unsaturated flow, algorithm for 376–84
Patel condition 17, 91, 168
  IMM form of 202
  three-dimensional extension of 206
penetration depths 132–3, 142
penultimate nodes, modified grid 179
permeabilities
  stratified porous media 71
  well-soil 67
perturbation methods 150–62, 287, 288
phase boundaries
  conditions on 10, 25, 105, 106
  heat jumps at 173, 217, 221
  isotherm expression of 199, 214
location of 164, 175
  position determined by enthalpy method 235–9
  shape of 126–7, 139
  Stefan problem 2
phase change problem, weak solution of 225
phase changes, concentrated thermal capacities 23–9
phase degeneration 12–14
phase diagrams 14–15
piecewise-constant functions 224–5
point of detachment, defined 32
Polubarinova-Kochina solution 300–12
  change of variables in 310
  equipotential dam face with impervious base 305
  free surface intersecting inlet face 306–7
  free surface joining linear seepage face 306
  free-boundary coordinates 311
  parameters in 311
  seepage face joining outlet face 305
polygonal approximation, integral-equation method 386–7
pond, seepage from 66, 70, 373
porous base dams, infiltration through 75
porous media
  concentrated thermal capacity of 23–4
  examples of flow through 30
  stratification of 71

power series solution, single-phase Stefan problem 116
pre-heating times, laser hole-cutting 153
pre-melting problem, heat-balance method applied to 130
precipitation problem, enthalpy method in 228
pressure functions, generalized dam problem 80
prism, finite-element 364–5
prism solidification
  integral-equation method, 137–9
  isotherm migration method 201–2, 214

quadrilateral finite elements 173
quasi-steady-state models
  dam seepage 287
  electrochemical machining 85, 87–8, 89, 242, 385
  solidification 139
quasi-variational inequality, arbitrary-shaped dam 52, 54

radial coordinates, boundary fixing in 197–8
radiation conditions 123–4
Rayleigh–Ritz expansion 93
reactor substrate, melt-pool growth in 209
recrystallization of iron 114, 143
rectangular dams
  capillary-fringed 76–7
  with sheetpile 44–6, 321, 382
    and toe drain 46
  simple *see* simple dam
  with toe drain 41–4, 324
recurrence relation, heat-balance method 135
regional numerical methods 318
regula-falsi methods 188, 326, 328
reservoir heights, rapid change of 92, 324
Reynolds' equation, journal bearing lubrication 95
Ricatti transformations 182, 186
Riemann *P*-equation 302, 307
Riemann zeta-function 160
Runge–Kutta method 131, 137, 143

saturated flow 30, 31
saturation, degree of 377, 381
Schatz transformation 21
Schwarz–Christoffel transformations 293–4, 299
secant method 368
  arbitrary-shaped dam 56
  open channel seepage 65

seepage faces, hodograph mapping of 290
seepage phenomena, examples of 30, 31
seepage problems, analytical solution
    of 283–312
seepage surfaces, defined 32
semi-inverse methods 22–3
semi-permeable membranes, fluid transfer
    across 8
separation point, disc-in-cylinder 326, 332–
    5
series solutions
    melting ice 125–6
    slab-melting problem 131
shallow-flow approximation method 288
sheetpile, rectangular dam 44–6, 321, 382
shocks, hyperbolic equations involving 217
shooting techniques
    method of lines 182
    rectangular dam with capillary fringe 78
short-time solutions
    melting/solidifying slabs 125–6, 128
    oxygen diffusion problem 120–1
silty soil problem 279
similarity solutions
    cylindrical coordinate 111
    defined 101
    density change 10
    moving heat source 111–14
    multi-phase 106, 107, 108
    Neumann's 101–10
    spherical coordinate 110–11
    supercooled liquid 104, 105, 108
similarity variables
    cylinder solidification 162
    sphere solidification 162
simple dam problem
    Alt's numerical solution 382
    Baiocchi transformation 34–8, 99
    boundary singularities 332–6
    Cryer algorithm 360–2
    Dupuit formulae 284–5
    fixed domain method 35–8
    hodograph mapping of 291, 297
    minimization formulation of 361–2
    numerical solution 313–16
    physical derivation 31–4
    Polubarinova-Kochina solution 300–12
    stream functions 34
        extended to fixed domain 40-1
    variable-interchange method 344–7
    variational inequality formulation 38–40
Simpson formula 120, 176, 392
simultaneous diffusion and heat flow, 14–
    16
single-dimension *see* one-dimension
single-phase problems
    enthalpy method for 219

fixed grid finite-difference method
    for 165–6
heat-balance integral method used
    in 129, 132, 133
    steady-state approximation for 140
single-phase Stefan problem 2–4, 11, 99–
    100, 114
    boundary-fixing transformation for 187
    finite-element method for 173–81
    isotherm migration method for 199–201
    spatial subdivision in 142–3
    variational inequalities for 269–75
single-phase two-dimensional problem
    body-fitted coordinates used in 195–7
singularities
    boundary 332–6
    finite-difference grid spacings near 315
    modified grid finite-element methods
        dealing with 177–8
    separation point 326, 332–5
    special finite elements near 316
sloping-base dams 50–2
    Boussinesq equations for 286
    slanting inlet face 51–2
smoothed enthalpy functions 227, 229–30
smoothed enthalpy method, boundary-
    position determined by 233–4
solid mechanics, examples 1
solid slab problem, Green's functions used
    in 117
solid–liquid zone models 6
solidification
    constant-rate 2, 144–7
    volume change on 103
solidification front, moving heat source
    treatment 111–14
space vehicle, ablation of alloy walls of
    12
spatial subdivision
    heat-balance method 133–7
    steady-state approximation 142–4
sphere problem
    constant-rate solidification 145–6
    solutions of 156–62
spherical coordinates, similarity solutions
    in 110–11
split-domain method
    arbitrary-shaped dam 54–7, 373–6
    open channel 66
    pond seepage 70
spot welding examples 5, 240
square cylinder, melting of 275
square duct, melting around 228
square region, solidification of 199, 279
square-channel freezing 236
stability analysis, melting/solidifica-
    tion 150–1

starting solutions
  isotherm migration method 200
  melting slab 126
steady state
  approximations using 139–42
  axisymmetric wells 66–7
  oxygen diffusion 19, 20
Stefan condition 4–5
  IMM form of 214
  supercooled liquid 104, 108
Stefan jump condition 221, 231
Stefan number 7, 9, 198
Stefan problem
  asymptotic solution 156
  concentrated thermal capacities in 23–9
  convective boundary condition 141
  defined 1–2
  finite-element methods applied to 173–8
  general solution 2
  generalizations 5–29, 281
  homogenization of 8
  implicit boundary conditions 19–23
  inverse 11–12, 144–7
  laser hole-cutting 152
  moving heat-source description 229
  multi-phase 12–14
  non-linear 148
  one-dimensional 117, 173–8
  simple example 2–5
  single-phase *see* single-phase Stefan
  solidification of sphere 156–62
  three-dimensional 16–19
  three-phase 178
  two-dimensional 16–19, 274–5
  two-phase 4, 11, 222, 227, 236–7, 275–81
stellar evolution problem, modified boundary conditions 332
stepwise time history, enthalpy method 238
stiff-system algorithm 191–2
stratified dams 71–4
  Alt's numerical solution for 383–4
  boundary singularities in 336
  Cryer algorithm for 363
stratified media, Dupuit formulae applied to 285
stream functions
  complex notation of 291
  rectangular dam with toe drain 41–2
  simple dam 34, 40–1, 291, 335–6
streamlines 34, 61, 90
successive overrelaxation (SOR) algorithm
  capillary-fringed dam 78
  dam with inlet face slanted 49
  enthalpy method 232–3
  Hele-Shaw injection 85
  method of lines 184–6

modified block 356–9
open-channel seepage 65
oxygen diffusion problem 261–3
point 355–7, 359
sloping-base dam 52
also see Cryer algorithms
supercooled liquids, solidification of 104–9, 111
superheated solids, melting of 109
surface tension effects, Hele-Shaw flow 82

T-plane, hodograph mapping 295–7
Taylor series
  oxygen diffusion problem 166–7
  single-phase Stefan problem 116
  Stefan problem 2, 116
  variable space grid finite-difference method 172
temperature distribution, successive freezing 238–9
temperature profiles
  sphere solidification 156
  steady-state 140, 141–2
  supercooled liquid 108
temperature–time histories, constant-rate solidification 145–6
temperature-dependence
  alloy solidification 15
  enthalpy function 218
  generalized Stefan problem 8
temperature-dependent thermal properties 132, 187
temperatures
  concentrated capacity 27–8
  constant-rate solidification 2, 147
terminal temperature distributions
  solidification of molten sphere 160–1
test functions
  finite-element method 178, 180
  generalized dam problem 80
thermal capacities, concentrated 23–9
three-dimensional dam problems 57–60
  coordinate interchange used in 353–5
three-dimensional problems
  explicit fixed finite-difference grid methods for 168
  method of lines 183
  variable-interchange method for 347–53
three-dimensional Stefan problems 16–19
three-eights rule 120
three-phase problems 12–13
  finite-element method used to solve 178
time steps
  adjustment of 168, 201
  variable 168–70
time-dependence
  electrochemical machining 88–9

time-dependence—*continued*
  general-shaped dams   90, 92–3
  Hele–Shaw injection   82–3
  oxygen diffusion problem   120–1
  Stefan problem   173–8, 317
toe drains   41–4, 46, 49, 324
transformations
  Baiocchi *see* Baiocchi transformation
  Boadway   204–6, 344, 348
  boundary-fixing   187
  conformal   293–7, 298, 299, 300
  coordinate   93, 187, 189
  Duvaut   270, 271, 276
  equipotential   197
  Fourier   117, 187, 263
  Hamel   295–7
  IMM variables
    complete   204–11
    partial   201–4
  integral   114–17
  inverse   115, 294–5
  Kirchhoff   219, 281
  Kozeny   43
  Laplace   114, 115, 153, 154, 158, 159
  multi-variable interchange   353–5
  Oberkampf   194, 196, 197
  Ricatti   182, 186
  Schwarz–Christoffel   293–5, 299
  single variable interchange   343–53
  Winslow   197
  Zhukovski   295, 298, 300
transient problems   90–1, 321
trapezoidal dams, seepage through   312
trapezoidal   drainage   ditch,   hodograph
    method   297–300
trapezoidal rule   183
trial free-boundary methods   313–32
trial-and-error   process,   enthalpy
    method   230
triangular channels, seepage from   63
triangular dam, hodograph planes for   291–2
triangular finite elements   56, 174, 380
truncation methods   253–7
  linear   complementarity   methods   com-
    pared with   267–9
  oxygen diffusion   23, 255–7
two-dimensional   annular   electrochemical
    machining problem   385–93
two-dimensional dams *see* dams
two-dimensional   heat   flow   problem,
    isotherm representation   212, 214
two-dimensional   model,   oxygen
    diffusion   21, 257
two-dimensional problems
  explicit   fixed   finite-difference   grid
    methods for   168

method of lines for   183
two-dimensional   solidification   front,   heat-
    balance   integral   method   applied
    to   137–9
two-dimensional Stefan problems   16–19
  variational   inequality   formulation
    of   274–5
two-dimensional two-phase problems
  linear finite element solution   279
  variational formulation   275
two-phase,   multi-dimensional   problem,   en-
    thalpy method for   218–19
two-phase   melting-slab   problem,   heat-
    balance method applied to   130
two-phase moving boundary problem   16–17
two-phase Neumann problem   101–2, 147
two-phase   solidification   problems,   explicit
    fixed   finite-difference   grid   methods
    168
two-phase Stefan problem   4, 11
  nodal enthalpy   236–7
  variational inequalities   275–81
  weak solution for   222, 227
two-phase   systems,   dimensionless   express-
    ion of   9

unsaturated flow   75–9
unsteady porous flow
  with accretion   286
  dam   90

variable space grids   170–3
variable time step methods   168–70
  enthalpy method   236–7
  Goodman   integral   method   compared
    with   170
variable-interchange methods   336–55
  adaptive variational formulation   337–43
  multi-variable interchange   353–5
  single variable interchange   343–53
variational inequalities   257–82
  arbitrary-shaped dam, with toe drain   56
  axisymmetric flow   67, 68
  capillary-fringed rectangular dam   77
  coastal aquifer   61
  dam with inlet face slanted   47–9
  electrochemical machining   85, 86, 87
  evaporation   75
  general derivation of   100
  Hele-Shaw injection   84
  journal bearing lubrication   96
  mathematical properties of   281–2
  minimization formulation of   265–6
  rectangular dam
    with sheetpile   45, 46
    with toe drain   42–3, 46

variational inequalities—*continued*
  simple rectangular dam problems 38–40
  single-phase Stefan problem 269–75
  sloping-base dam 50
  stratified dam 73
  theoretical properties of 257
  three-dimensional dam 57
  transient dam 92–3
  two-phase Stefan problem 275–81
  well problem 31, 67, 68
variational inequality methods 355
velocity potentials
  boundary singularity 334–5
  complex notation of 291
  Hele-Shaw flow 82
  simple rectangular dam 32
vertically stratified dams 71–3, 336
volume change, solidification 103

water table, seepage into 63, 65
water/air interfaces, saturated flow 30

weak solutions
  capillary-fringed rectangular dam 77
  enthalpy method 217, 220–3
  stratified dam 72
  well problem 67
wedge-shaped space freezing 113–14
welding problems, 5–6, 239–41
well problems 28–9, 31, 66–70, 372–6
  Dupuit formulae applied to 285
  numerical solution of 372–6
  variational inequality solution of 31,
    372–6
wells
  coastal 60–3
  hot-liquid injection into 28
Winslow transformation 197

zero latent heat 147, 148, 150
zero specific heat 109, 242–5
zero-order perturbation solutions,
  laser hole-cutting 153, 154
Zhukovski function 295, 298, 300